Organic Superconductors (Including Fullerenes)
Synthesis, Structure, Properties, and Theory

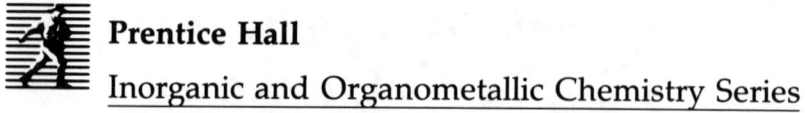
Prentice Hall
Inorganic and Organometallic Chemistry Series

Russell N. Grimes, *Series Editor*

Mingos and Wales, *Introduction to Cluster Chemistry*
Martell and Hancock, *Metal Complexes in Aqueous Solutions*
Williams et al., *Organic Superconductors (Including Fullerenes):
Synthesis, Structure, Properties, and Theory*

Organic Superconductors (Including Fullerenes)
Synthesis, Structure, Properties, and Theory

Jack M. Williams, John R. Ferraro,
Robert J. Thorn, K. Douglas Carlson,
Urs Geiser, Hau H. Wang,
Aravinda M. Kini, and *Myung-Hwan Whangbo

*Chemistry and Materials Science Divisions
Argonne National Laboratory
Argonne, Illinois*

**Department of Chemistry
North Carolina State University
Raleigh, North Carolina*

Prentice Hall, Englewood Cliffs, New Jersey 07632

Library of Congress Cataloging-in-Publication Data

Organic superconductors (Including Fullerenes): synthesis, structure, properties, and theory / Jack M. Williams . . . [et al.].
    p. cm.
  Includes bibliographical references and index.
  ISBN 0-13-640566-5
  1. Organic superconductors. I. Williams, Jack M. (Jack Marvin).
QC611.98.074074 1992
537.6′233—dc20   90-23031
                                                                                                    CIP

Editorial/production supervision and interior design: *Jean Lapidus*
Cover design: *Wanda Lubelska Design*
Copy editor: *Maria Caruso*
Pre-press buyer: *Mary Elizabeth McCartney*
Manufacturing buyer: *Susan Brunke*
Acquisition editor: *Betty Sun*
Editorial assistant: *Maureen Diana*

© 1992 by Prentice-Hall, Inc.
A Simon & Schuster Company
Englewood Cliffs, New Jersey 07632

All rights reserved. No part of this book may be
reproduced, in any form or by any means,
without permission in writing from the publisher.

Printed in the United States of America
10 9 8 7 6 5 4 3 2 1

ISBN 0-13-640566-5

PRENTICE-HALL INTERNATIONAL (UK) LIMITED, *London*
PRENTICE-HALL OF AUSTRALIA PTY. LIMITED, *Sydney*
PRENTICE-HALL CANADA INC., *Toronto*
PRENTICE-HALL HISPANOAMERICANA, S.A., *Mexico*
PRENTICE-HALL OF INDIA PRIVATE LIMITED, *New Delhi*
PRENTICE-HALL OF JAPAN, INC., *Tokyo*
SIMON & SCHUSTER ASIA PTE. LTD., *Singapore*
EDITORA PRENTICE-HALL DO BRASIL, LTDA., *Rio de Janeiro*

*This book is dedicated to our families,
who so generously relinquished family
time so that we might prepare this volume.
JMW especially wishes to thank
Joan, Kelly, Hillary, and Kim for their
continuing interest and support.*

Contents

Foreword   xi

Preface   xiii

Acknowledgments   xv

1 Introduction   1

    List of Symbols and Abbreviations   10

2 Synthesis of Organic-Donor Molecules and Conducting Salts   11

    Introduction   11

    Tetrachalcogenafulvalenes   11

        Symmetrical Tetrachalcogenafulvalenes, 12 / Derivatization of Tetrathiafulvalene and Tetraselenafulvalene, 26 / Unsymmetrical Tetrachalcogenafulvalenes, 30

    Tetrachalcogenoacenes   33

    Miscellaneous Donor Molecules   35

        TTF-Type, 35 / 1, 3-dithiole-2-ylidene Compounds, 38 Polychalcogenaarenes, 39 / Tetraoxafulvalene and Tetraphosphafulvalene, 40

Design and Synthesis of Organic Conducting Salts   41

Experimental Procedures   43

Survey of ET-Based Conductors   45

> ET Salts with Octahedral, Tetrahedral, and Simple Anions, 46
> ET Salts with Linear Anions, 48 / ET Salts with Planar
> Anions, 51 / ET Salts with Polymeric Anions, 52

3 Structural Aspects   65

Introduction   65

Molecular Structure of the Donor Molecules   66

> Bis(ethylenedithio)tetrathiafulvalene, ET, 66 / Other
> Bis(alkylenedithio)tetrathiafulvalenes, 67 / Chalcogen
> Substitutions, 68 / Unsymmetrical Donor Molecules, 70
> Further Modifications of the Outer Rings, 71
> Alkylthio(seleno-, telluro-,)-substituted Tetrathiafulvalenes, 71

Overview of Crystal Structures   73

> BEDT-TTF Salts, 73 / Salts Derived from Other Donor
> Molecules, 73

Crystal Structures of Superconducting ET Salts   80

> β-$(ET)_2X$ Family, 80 / β-$(ET)_2I_3$ at Low Temperature or
> High Pressure, 81 / κ-$(ET)_2Cu(NCS)_2$, 90 / κ-$(ET)_2I_3$, 92
> θ-$(ET)_2(I_3)_{1-x}(AuI_2)_x$, 92 / γ-$(ET)_3(I_3)_{2.5}$, 94
> κ-$(ET)_4Hg_{3-\delta}X_8$ Family (X = Cl, Br), 94 / $(ET)_3Cl_2 \cdot 2H_2O$,
> 95 / α-$(ET)_2(NH_4)Hg(SCN)_4$, 97

Crystal Structures of Superconductors Based on Other Donor
Molecules   98

> κ-$(MDT\text{-}TTF)_2AuI_2$, 98 / $(DMET)_2X$ Family, 99
> β_m-$(BEDO\text{-}TTF)_3Cu_2(NCS)_3$, 100

Latest Developments   101

> κ-$(ET)_2Cu[N(CN)_2]X$, X = Br, Cl, 101

4 Electrical Conductivities and Superconducting
Properties   115

Introduction   115

Normal-State Conductivities   118

> The Nature of Electrical Transport in Organic Salts, 118
> ET Salts with Polyhedral Anions, 123 / ET Salts with Linear
> Anions, 127 / The κ-Phase Salts and Other Organic Metals,
> 134

Superconducting Properties   140

> *The Superconducting Transition Temperature T_c, 140 / The β-Phase Superconductors at Ambient Pressure, 141 / The β-Phase Superconductors under Applied Pressures, 149 The κ-Phase Superconductors at Ambient Pressures, 153 The κ-Phase Superconductors under Applied Pressures, 160 BCS or Non-BCS Superconductivity, 161 / Superconductivity in Other Salts of the ET/I System, 166 / The $\alpha\text{-}(ET)_2(NH_4)Hg(SCN)_4$ and $\beta_m\text{-}(BEDO\text{-}TTF)_3Cu_2(NCS)_3$ Salts, 169*

5 Single Crystal ESR Studies   180

Introduction   180

Room Temperature ESR Studies   181

> *Orientational Studies, 181 / Phase Identification, 185*

Low-Temperature ESR Studies   193

> *Organic Superconductors, 194 / Organic Metals, 199 Organic Semiconductors, 201*

High-Temperature ESR Studies   203

6 Vibrational Spectroscopy   210

Introduction   210

Spectroscopy of Organic CT Superconductors   211

> *Vibrational Assignments for the Neat Donor ET, 211 Vibrational Studies of Charge-Transfer Salts of ET, 213 Polarized Reflectance Spectra of the CT Salts, 214 Micro-FT-IR Reflectance Techniques in the Study of ET Superconductors, 217*

Experimental   224

7 Polarized Reflectance in the Infrared   229

Introduction   229

Models for Optical Conductivities   230

Optical Conductivities   235

> *General Descriptions and Features, 235 / Individual Spectra, 239*

Summary   246

8 Electronic and Structural Properties of Organic Superconductors — 250

Electronic Properties   250

Band Electronic Structure   251

One Orbital per Site, 251 / Many Orbitals per Site, 253

Fermi Surface Nesting and Electronic Instability   255

Superconducting State and Orbital Mixing   257

Structural Properties   259

Structural Factors Affecting Superconductivity   259

β-(BEDT-TTF)$_2$X (X$^-$ = I$_3^-$, AuI$_2^-$, IBr$_2^-$), 259
κ-Phase Superconductors, 265

C-H···Donor and C-H···Anion Interactions and Crystal Packing Patterns   270

Structural Characteristics of Donor-molecule Packing, 270 / Energetics of Conformational Change and Intermolecular Contact Interactions, 274 / Structural Differences in TXF·TCNQ (X = S, Se Te), 275 / Thermal Phase Transitions, 277

Appendix A Bibliography of Charge-Transfer Salts, Organic Superconductors, and Conductors — 284

Appendix B List of Researchers in the Field of Organic Superconductivity — 368

Appendix C Superconducting Fullerenes (C$_{60}$) — 383

Index   393

Foreword

This new series was launched in 1990 with the publication of the volume *Introduction to Cluster Chemistry* by D. M. P. Mingos and D.J. Wales, a beautifully conceived and executed treatment of the whole broad subject of molecular clusters in which the descriptive facts and theoretical foundation of these systems are expertly interwoven. In the present book, this approach is applied to the burgeoning area of organic superconducting materials. It is no accident that the scope of each of these first two volumes crosses many of the traditional boundaries that once delineated inorganic vs. organic chemistry, and chemistry vs. physics. The obliteration of these once-sharp demarcations between research fields is not only the wave of the future; it is the wave of the present. The discovery and development of novel materials inconceivable a few years ago (for example, the new allotrope of carbon based on molecular soccer-ball clusters!) requires interdisciplinary collaborations that once would have seemed improbable: physicists and organic chemists, radiologists and boron hydride chemists, inorganic polymer chemists and medical researchers.

In such a stimulating scientific era, it is not only excusable for a Series Editor to take a very broad view of the definition of "inorganic and organometallic chemistry", it is inescapable. In this spirit, the present volume on new materials constructed from nonmetals in combination with organic ligands, is a most welcome and appropriate addition to the series. Without doubt, there are no better qualified individuals to deal with this topic than the authors of this book, who are in the forefront of this area and, as of this writing, hold the current world record for high-temperature superconductivity in molecular solids based on organic electron-donor ligands.

Russell N. Grimes

Series Editor
Prentice Hall Inorganic and Organometallic Chemistry Series
University of Virginia
Charlottesville, Virginia

Preface

In this book the authors have attempted to include in one volume an up-to-date and detailed discussion of the chemical, structural, electrical, superconducting and electronic band-structure (theoretical) aspects of **organic superconductors**. This field has fascinated scientists in a variety of disciplines since the discovery of the first organic superconductor, in 1979, and new discoveries are reported at an increasingly rapid rate. The superconducting transition temperatures, T_c, are also increasing at a very rapid rate. In addition, being novel materials due to the phenomenon of superconductivity that they exhibit, there is also future promise for practical applications.

This book is intended to serve as a detailed discussion of the main types of known superconducting synthetic organic metals (synmetals). It is written with appropriate introductory material so that the novice, or the research specialist, should find it of great utility. The book is referenced in considerable detail (books, individual articles, reviews, etc.) at the end of each chapter and an extensive bibliography of references, as they relate to specific scientific areas of study, is also presented. Only superconductive organic materials that can be synthesized in the laboratory are discussed. Finally, the book is written by chemists and the main focus is on all important chemical aspects of organic superconductors although important physical measurements, and their interpretation, are also discussed thoroughly.

Chapter 1 is introductory in nature and presents important background information as well as historical information concerning organic superconductors. Chapter 2 discusses the synthesis of organic electron-donor molecules and the corresponding conducting and superconducting charge-transfer salts that can be derived from the donor-molecule species. Chapter 3 explores the structural aspects of organic charge-transfer salt superconductors and includes detailed discussions of various crystal types and how they are correlated with the electrical and superconductive properties. In Chapter 4, the most "physics oriented" chapter, the detailed normal state electrical and superconducting properties are discussed. Chapter 5 is concerned with the use of electron spin resonance (ESR) spectroscopy for the identification and characterization of these novel synmetals while Chapter 6 covers the same material by use of vibrational spectroscopic techniques. Chapter 7 describes the characterization of these salts by use of the infrared polarized specular reflectance spectroscopy technique. Chapter 8 is a summary chapter that discusses the electronic (band-structure theory) aspects of organic superconductors and we also correlate these findings with the associated crystal structures exhibited by these materials. Appendix A includes a very recent bibliography, in which we have attempted to keep the book current up to publication time, and which is indexed according to areas of scientific investigation of these charge-transfer salt superconductors. Appendix B compiles a list of researchers in the field of organic superconductivity. Appendix C, added as we completed the galley proof revisions, contains the latest findings on the fullerene (C_{60}) superconductors which have now demonstrated superconducting transition temperatures (T_c) as high as 45K!

We point out that the focus, thrust and depth given to various topics in the chapters reflect the interests and prejudices of the authors. In every way, we have attempted to make this book an exhaustive study of organic superconductors, one that we hope will be helpful to all those interested in this rapidly growing field, and we apologize for any omission of topics or references.

Jack M. Williams

Acknowledgments

The authors wish to express their thanks and appreciation to their collaborators, the visiting scientists, and the students who have contributed so much to the organic superconductor program. The authors also thank Mrs. V. R. Bowman and Mrs. J. J. Bormet of Argonne National Laboratory for their patience, meticulous care and effort in the preparation of this manuscript. The authors also wish to acknowledge the continuing support of the U.S. Department of Energy, Office of Basic Energy Sciences, Division of Materials Sciences, under Contract W-31-109-ENG-38 (Argonne) and Grant DE-FG05-86-ER45259 (North Carolina State).

1

Introduction

In the past 20 to 25 years, the syntheses and physical characterization of a large number of materials with unusual electrical and magnetic properties has occurred.[1] Among these substances are the organic, charge-transfer (CT) salts, often known as synthetic metals (synmetals). The term synmetals derives from the fact that these materials exhibit the electrical properties of metals even though they most often contain *no metal atoms* in the electrically conducting framework. In 1979, superconductivity was discovered in these compounds when the selenium-based Bechgaard[2a] salts $(TMTSF)_2X$ (X = monovalent anions) were shown to become superconductors at $1-2$ K under an applied pressure of ~ 5 to 12 kbar. Conductivity studies at 1 K and under applied pressure are very difficult. However, in 1981 another derivative, $(TMTSF)_2ClO_4$, was found to be the first *ambient pressure organic superconductor*.[2b]

$$H_3C\underset{H_3C}{\overset{}{\diagup}}\!\!\!\!\!\!\diagdown \underset{Se}{\overset{Se}{=}}\!\!\!\!\!\diagdown\!\!\!\!\!\diagup \underset{Se}{\overset{Se}{=}}\!\!\!\!\!\diagdown \underset{CH_3}{\overset{CH_3}{\diagup}}$$

TMTSF

At this point, it should be noted that the electrical conductivity (σ in Ω^{-1} cm^{-1}) of solids has a range of values greater than 20 orders of magnitude with insulators having $\sigma \cong 10^{-22} - 10^{-15}$ Ω^{-1} cm^{-1}, semiconductors having thermally activated conductivity (conductivity rises with increased temperature, $\sigma \cong 10^{-8}-10^{-2}\ \Omega^{-1}\ cm^{-1}$), and metals having conductivity ($\sigma \cong 10-10^4\ \Omega^{-1}\ cm^{-1}$) that increases with decreased temperature. The remarkable phenomenon of superconductivity is usually characterized by the following three physical properties: (a) sample electrical resistance (ρ) drops to zero or conversely, ($\sigma \rightarrow \infty$), (b) magnetic fields are expelled by the sample (the Meissner effect), and (c) with decreasing temperature, the sample specific heat increases discontinuously upon the onset of superconductivity at T_c, indicating a second-order transition.

During the past ten years, there has seen a very significant international research effort as a result of the search for higher T_c organic superconductors, much stimulated of late by the discovery[3] of the high-T_c copper oxide ceramic superconductors. Thus, now that "high-T_c" ($T_c > 90$ K) materials are known to exist, there is every reason to expect "high-T_c" in organic materials.

Much of the motivation for these increased research activities on superconducting materials has been based on the potential technological applications that may eventually be realized. In this regard, the organic materials are especially light-weight (density $\approx 1.5-2.0$ g/cm^3) compared to copper metal (density $\cong 9$ g/cm^3) and the ceramic superconductors (density $\cong 7$ g/cm^3). Interestingly, copper has high-ambient temperature electrical conductivity ($\approx 10^6\ \Omega^{-1}cm^{-1}$) but it never becomes superconductive even at the lowest temperatures (milli-Kelvin) yet achieved. For the organic superconductors, it is possible that their low density may be useful in electronic uses in automobiles and space vehicles, for example, where weight is often kept to a minimum. They have also been proposed as components for compact high-efficiency electrical motors, and in computers where their very low-heat generation could allow construction of high-density circuits. For example, in 1985 Graff[4] found them useful as photoresists.

From scientific records, it appears that as early as 1911 it was suggested that organic solids might exhibit metallic electrical conductivities even though they might not contain metal atoms (McCoy and Moore[5] 1911, Kraus[6] 1913). Interestingly, McCoy and Moore[5] were ahead of their times when (in 1911, the same year that superconductivity was discovered[7]) they stated:

> "[W]e think, in concluding that the organic radicals in our amalgams are in the metallic state and, therefore, that it is possible to prepare composit metallic substances from non-metallic constituent elements."

Apparently these publications were ahead of their times and much later, in two classic publications, Little (1964, 1965)[8,9] suggested that *organic* solids could not only be conductive, but when properly constructed could become superconductive at room or higher temperatures. Little's suggestions prompted a great deal of research into synmetal conductors. An important step in this direction occurred in 1962, when Melby[10] synthesized the electron acceptor TCNQ (7,7,8,8-tetracyano-p-quinodimethane) and in 1970 TTF (tetrahiafulvalene) was synthesized and became an important electron-donor molecule for use in preparing charge-transfer conductors.[11] (Structures of various donor molecules and acceptor species discussed in this chapter appear in Fig. 1.1a,b.) Then Ferraris et al.,[12] reacted TTF and

TCNQ (an electron-acceptor molecule) and obtained a charge-transfer salt, TTF-TCNQ, which has become known as the *first true organic metal* because it was metallic down to 54 K. The results were significant because it was realized that charge-transfer salts formed by reacting an electron donor and an acceptor molecule could form metallic materials with resulting high electrical conductivities. These findings increased the interest in research directed toward the search for new electron donor/acceptor molecules. In 1974, Bechgaard et al.,[13] synthesized TMTSF (tetramethyltetraselenafulvalene), an electron-donor molecule containing four Se atoms compared to TTF, which contains four S atoms in the organic backbone. TMTSF was subsequently reacted with TCNQ and DMTCNQ (2,4-dimethyl-TCNQ) and even better metals than TTF-TCNQ were formed, and under slight pressure (~ 1.0 kbar) TMTSF-DMTCNQ remained metallic all the way down to liquid He temperatures (Andrieux et al. 1979).[14] In 1979, Bechgaard et al.[2] made

TTF

TCNQ

a giant step in synmetals research when he began[2] to use inorganic monovalent anions in place of TCNQ and prepared a series of salts of the type (TMTSF)$_2$X, where X = monovalent anion, such as PF_6^-, AsF_6^-, TaF_6^-, NbF_6^-, SbF_6^-, ClO_4^-, ReO_4^-, BF_4^-, BrO_4^-, IO_4^-, NO_3^-, FSO_3^-, $CF_3SO_3^-$ and TeF_5^-. These salts quickly became known as the Bechgaard salts, and the materials with the anions, PF_6^-, AsF_6^-, SbF_6^-, TaF_6^-, NbF_6^- and ReO_4^- were found to be superconductors ($T_c = 1-2$ K) but only under applied pressures ($\sim 5-12$ kbar). In 1981, (TMTF)$_2$ClO$_4$ was found to be the first *ambient pressure* organic superconductor with a $T_c = 1.4$ K.[2] Thus, the most speculative suggestions of McCoy and Moore,[5] and Kraus[6] were realized after 60 years!

Research aimed at the search for new electron-donor molecules continued and Mizuno et al.[15] prepared another modification of the TTF molecule, that is, BEDT-TTF, or "ET," [bis(ethylenedithio)tetrathiafulvalene]. ET contains eight sulfur atoms, twice the number of chalcogenide atoms than Se atoms in TMTSF. At this writing, the major number of organic superconductors found to date contain ET as the electron-donor molecule. Following Bechgaard's discovery, inorganic anions were commonly used in CT reactions. It was found that linear, inorganic, symmetrical, monovalent anions (X = I_3^-, IBr_2^-, AuI_2^-) served best to provide the first superconductors at ambient pressure based on ET.[1] At least 20 superconductors of ET salts have now been synthesized, with κ-(ET)$_2$Cu(NCS)$_2$ possessing the third highest superconducting T_c of 10.4 K[16-18], κ-(ET)$_2$Cu[N(CN)$_2$]Br having the second highest T_c of 11.6 K.[33-34], and κ-(ET)$_2$Cu[N(CN)$_2$]Cl having the highest T_c

Donors

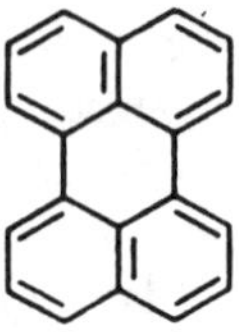

Per
Perylene

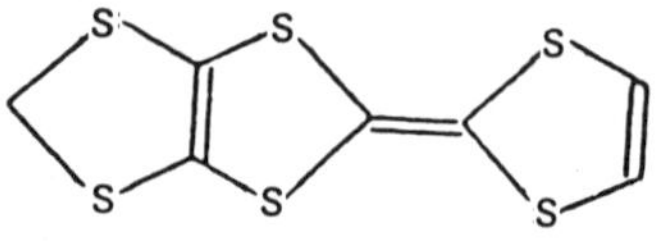

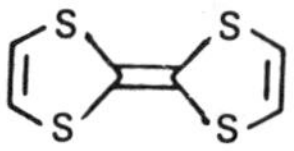

TTF
Tetrathiafulvalene

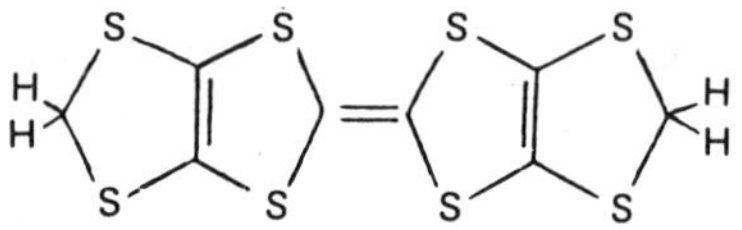

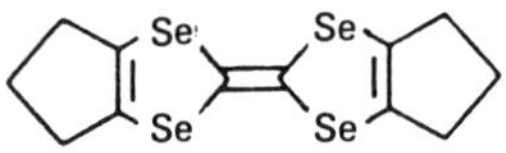

HMTSF
Hexamethylenetetraselenafulvalene

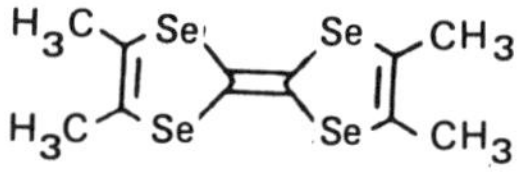

TMTSF
Tetramethyltetraselenafulvalene

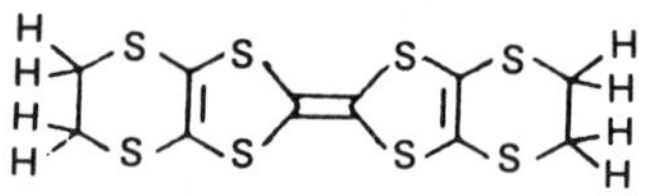

BEDT-TTF (ET)
Bis(ethylenedithio)tetrathiafulvalene

MDT-TTF
Methylenedithiotetrathiafulvalene

BMDT-TTF
Bis(methlenedithio)tetrathiafulvalene

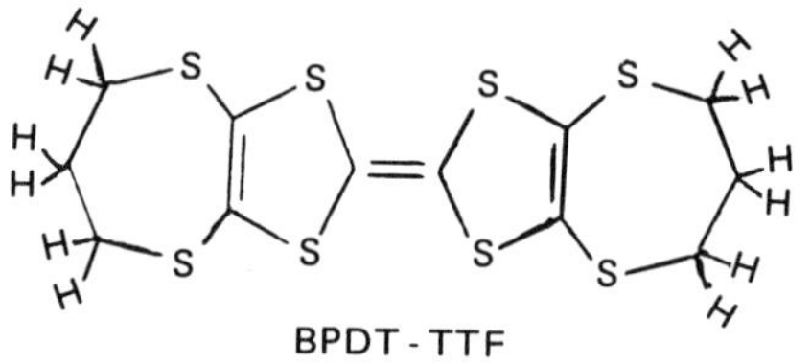

BPDT-TTF
Bis(propylenedithio)tetrathiafulvalene

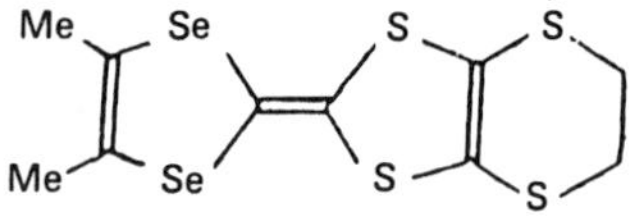

DMET
Dimethyl(ethylenedithio)diselenadithiafulvalene

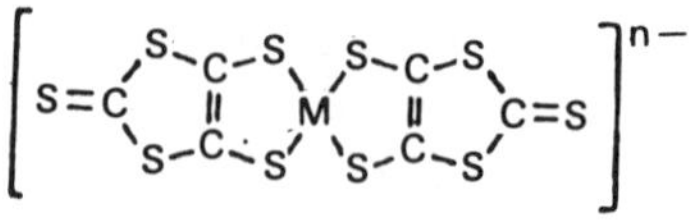

M=Ni, Pd, Pt

$M(dmit)_2^{2-}$

Ligand is

4,5-dimercapto-1,3-dithiole-2-thione

Figure 1.1 (a)

Acceptors

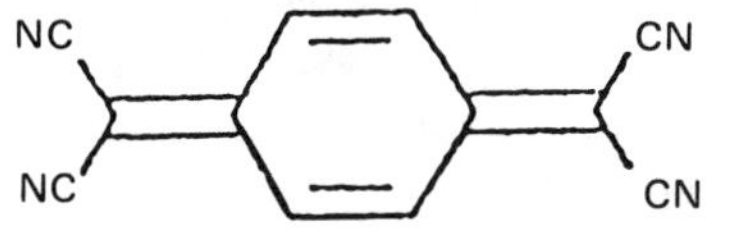

TCNQ

7,7,8,8 tetracyano-p-quinodimethane

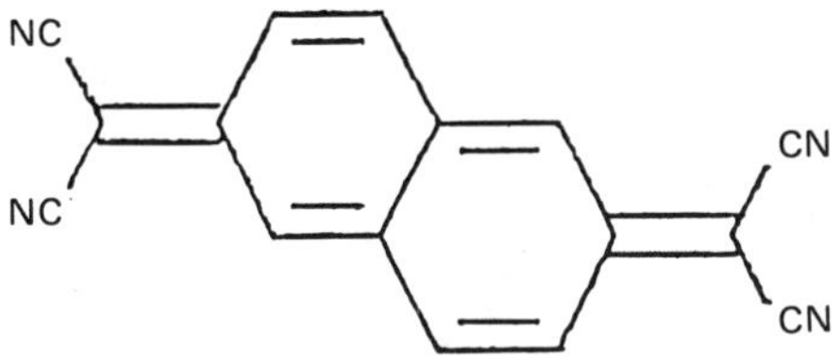

TNAP

11,11,12,12 - tetracyanonaptho-2,6-quinodimethane

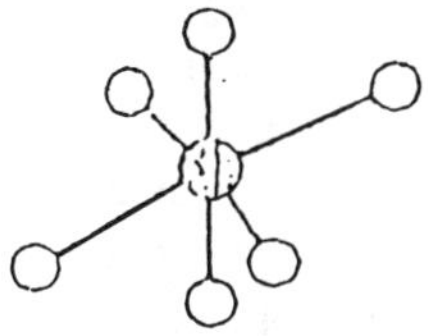

Hexafluorophosphate (PF_6^-)

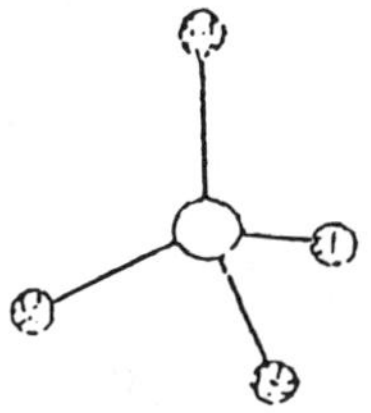

Perchlorate (ClO_4^-)

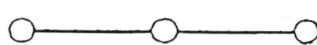

Triiodide (I_3^-)

Figure 1.1 (b) Structures of several donors and acceptor species, and anions found in organic CT salt superconductors.

at 12.8 K.[35] Recently, Wudl et al.,[19] synthesized a new donor containing four O atoms substituted for four of the eight S atoms of ET, and designated as BEDO-TTF. Shortly thereafter, Beno et al., 1990[20, 21] synthesized the first BEDO-TTF superconductor ($T_c \sim 1$ K at ambient pressure) (BEDO-TTF)$_3$Cu$_2$(NCS)$_3$. It should be noted at this point that all of the electron-donor molecules previously mentioned are *symmetrical*.

BEDT-TTF or ET

BEDO-TTF

Research into the synthesis of *un*symmetrical electron-donor molecules has also progressed, and donors such as MDT-TTF (methylenedithiotetrathiafulvalene)[22] and DMET [dimethyl(ethylenedithio)diselenadithiafulvalene][23–26] have been prepared (see Fig. 1.1). The donor MDT-TTF has provided an ambient pressure superconductor (MDT-TTF)$_2$AuI$_2$ (T_c = 4.5 K), and DMET has given seven superconductors, four of which are ambient pressure superconductors, but with T_c's of < 2 K. Four superconductors were also synthesized of the type M(dmit)$_2^{n-}$, where dmit is 4,5-dimercapto-1,3-dithiole-2-thione (see Figure 1.1(a)) and M = Ni or Pd.[27, 28] The latter superconductors are interesting in that they combine a transition metal in a metal-dmit anion complex as the acceptor, and TTF as the counter-ion donor. Figure 1.1a,b shows the structures of several donor molecules and acceptor species, and anions that are found in organic CT salt superconductors. Table 1.1 tabulates the list of organic superconductors known to date and Table 1.2 provides a chronology of synthetic organic conductors and superconductors.

The research and development of new organic CT superconductors continues today in the direction of synthesizing both new electron-donor molecules as well as new acceptor species and anions. In a recent (1989) National Committee report by a panel of experts in the field of superconductors, it was pointed out that in Japan some 100 researchers are studying organic superconductors, and that smaller efforts on these materials are found in other countries such as the USA, Europe, and the Soviet Union to name a few.[29] Thus, the organic superconductor research effort is international in nature. Ginzburg[30, 31] has suggested that the best candidates for new superconductors are *organic materials,* because organic species can be systematically modified, and inorganic layered compounds, particularly intercalated materials. Thus, there is much justification for the continued interest in these novel materials.

Of course, the hope is that superconductors with even higher T_c's (room

TABLE 1.1 Organic Superconductors

Compounds	$T_c(K)$
$(TMTSF)_2ClO_4$	1.4
$(TMTSF)_2X$ $X = PF_6, AsF_6, SbF_6, TaF_6, ReO_4, FSO_3$	~ 1 (5–12 kbar)
$(ET)_2ReO_4$	2 (4.5 kbar)
$\alpha_t-(ET)_2I_3$	7–8
$\alpha\beta-(ET)_2I_3$	2.5–6.9
$\beta-(ET)_2I_3$	1.4
$\beta-(ET)_2IBr_2$	2.8
$\beta-(ET)_2AuI_2$	5.0
$\beta^*-(ET)_2I_3$	8 (0.5 kbar)
$(ET)_3Cl_2\cdot 2H_2O$	2 (16 kbar)
$(ET)_4Hg_3Cl_8$	1.8 (12 kbar)
$(ET)_4Hg_3Cl_8$	5.3 (29 kbar)
$(ET)_4Hg_{2.89}Br_8$	4.3, 6.7 (3.5 kbar)
$(ET)_2Hg_{1.41}Br_4$	2.0
$(ET)_2NH_4Hg(SCN)_4$	1.15
$\gamma-(ET)_3(I_3)_{2.5}$	2.5
$\theta-(ET)_2(I_3)_{1-x}(AuI_2)_x (x<0.02)$	3.6
$\kappa-(ET)_2(I_3)_{1-x}(AuI_2)_x (x<0.006)$	3.6
$\beta-(ET)_{1.96}(MET)_{0.04}I_3$	4.6
$\kappa-(ET)_2Cu(NCS)_2$	10.4
$\kappa-(ET)_2Cu[N(CN)_2]Br$	11.6
$\kappa-(ET)_2Cu[N(CN)_2]Cl$	12.8 (0.3 kbar)
$\kappa-(ET)_2Ag(CN)_2\cdot H_2O$	5.0
$(MDT-TTF)_2AuI_2$	4.5
$(DMET)_2Au(CN)_2$	0.8 (5 kbar)
$(DMET)_2AuCl_2$	0.83
$(DMET)_2AuBr_2$	1 (1.5 kbar), 1.9 (κ-phase)
$(DMET)_2AuI_2$	0.55 (5 kbar)
$(DMET)_2I_3$	0.47
$(DMET)_2IBr_2$	0.59
$(TTF)[Ni(dmit)_2]_2$	1.62 (7 kbar)
$(NMe_4)[Ni(dmit)_2]_2$	5.0 (7 kbar)
$\alpha-(TTF)[Pd(dmit)_2]_2$	onset 1.7 at 22 kbar
$\alpha'-(TTF)[Pd(dmit)_2]_2$	onset 6.42 at 20.7 kbar
$(BEDO-TTF)_3Cu_2(NCS)_3$	1.1

temperature) will eventually be found. Aside from the promise that the low-density organic superconductors offer for device application, synmetals research has united many different disciplines of science (e.g., physics, chemistry, solid state, engineering, theory), and as a result, many large team research efforts have resulted. The study of these materials also necessitates the combination of the expertise of scientists in different disciplines. Finally, no single laboratory possesses all the expertise and instrumentation necessary to characterize completely these materials, and much cooperative research between laboratories has resulted.

It is the purpose of this text to discuss organic superconductors and especially the ET salts and related materials. The book focuses on the *chemical* aspects of organic superconductors and has been authored by scientists conducting research in this area in the Chemical and Electronic Structure group at Argonne National Laboratory and Prof. M.-H. Whangbo at North Carolina State University. An excellent com-

TABLE 1.2 Chronology of Synthetic Organic Conductors and Superconductors

1962	Synthesis of TCNQ
1963	Synthesis of TCNQ Salts [$Cs_2(TCNQ)_3$]
1970	Synthesis of TTF
1973	TTF-TCNQ synthesized: first organic metal
1976	TMTSF synthesized
1978	ET synthesized
1979	TMTSF-DMTCNQ synthesized: $\sigma = 10^5$ ($\Omega^{-1} cm^{-1}$) at 10 kbar and 1 K
1979	Selenium-based Beckgaard salts synthesized: $(TMTSF)_2X$, where X = monovalent anion Superconductors under pressure at T~1 K
1981	$(TMTSF)_2ClO_4$ = first organic SC at ambient pressure at 1.4 K
1983	$(BEDT-TTF)_2ReO_4^-$ SC under pressure and $T_c \cong 1K$—first S-based organic SC
1984–1986	β-$(BEDT-TTF)_2X$, where $X = I_3^-$, IBr_2^-, AuI_2^- Superconductors synthesized, where $T_c = 1.4$, 2.8 and 5.0 K, respectively
1986–1989	$Ni(dmit)_2^{n-}$ salts synthesized with T_c ranging from 1.62–5.9 K at pressures from 7–24 kbar
1987–1988	κ-$(ET)_2Cu(SCN)_2$ synthesized: highest T_c at 10.4 K $(MDT-TTF)_2AuI_2$ synthesized with T_c at 4.5 K $(DMET)_2X$ salts synthesized, where $X = AuCl_2^-$, $AuBr_2^-$, AuI_2^-, $Au(CN)_2^-$, I_3^- and IBr_2^- with T_c's ranging from 0.47–1.9 K. Only the I_3^-, IBr_2^-, $AuCl_2^-$ and the κ-phase of $AuBr_2^-$ salts are ambient pressure superconductors.
1989	BEDO-TTF synthesized and numerous salts prepared New superconductors synthesized, $(ET)_2NH_4Hg(SCN)_4$, T_c at 1.15 K, $(BEDO-TTF)_3Cu_2(NCS)_3$, $T_c = 1.1K$
1990	κ-$(ET)_2Cu[N(CN)_2]Br$, $T_c = 11.6$ K
1990	κ-$(ET)_2Cu[N(CN)_2]Cl$, $T_c = 12.8$ K at 0.3 kbar.

plementary text on the *physical* aspects of organic superconductors has also just appeared.[32] Chapter 2 deals with the syntheses of organic-donor molecules and donor-molecule salts. Chapter 3 discusses the structural aspects of the organic CT conductors and superconductors. Chapter 4 presents a detailed discussion of the conductive and magnetic properties of these materials. Chapter 5 summarizes ESR measurements made for these synmetals. Chapter 6 presents microreflectance data used to characterize these compounds, while Chapter 7 gives a theoretical discussion of reflectance measurements. Chapter 8 contains theoretical discussions of various aspects of these materials. In addition, an appendix appears at the end, consisting of an extensive indexed bibliography on the subject matter covered in the text.

We hope this monograph will serve as a starting point for the interested reader, into the area of synthetic metals, emphasizing especially organic superconductors.

REFERENCES

1. Ferraro, J. R.; Williams, J. M., *Introduction to Synthetic Electrical Conductors*, Academic Press, Orlando, FL (**1987**), pp. 1–337.
2. (a) Bechgaard, K.; Jacobsen, C. S.; Mortensen, K.; Pedersen, M. J.; Thorup, N., *Solid State Commun.* **1980**, *33*, 1119.

(b) Bechgaard, K.; Carneiro, K.; Rasmussen, F. G.; Olsen, K.; Rindorf, G.; Jacobsen, C. S.; Pederson, H. J.; Scott, J. E., *J. Am. Chem Soc.* **1981**, *103*, 2440.
3. Bednorz, J. G.; Müller, K. A. *Z. Phys. B. Conden. Matter* **1986**, *64*, 189.
4. Graff, C., *Ind. Technol.* **1985**, 64.
5. McCoy, H. N.; Moore, W. C., *J. Am. Chem. Soc.* **1911**, *33*, 273.
6. Kraus, H. J., *J. Am. Chem. Soc.*, **1913**, *34*, 1732.
7. Onnes, H. K. *Akad. van Wetenschappen (Amsterdam)* **1911**, *14*, 113, 818.
8. Little, W. A., *Phys. Rev.* **1964**, *134A*, 1416.
9. Little, W. A., *Sci. Am.* **1965**, *212*, 21.
10. Melby, L. R.; Harder, R. J.; Hertler, W. R.; Mahler, W.; Benson, R. E.; Mochel, W. E., *J. Am. Chem. Soc.* **1962**, *84*, 3374.
11. (a) The first well-characterized TTF *derivative* was prepared in 1965: Pninzbach, H.; Berger, H.; Lüttringhaus, A., *Angew. Chem. Int. Ed. Engl.* **1965**, *4*, 436.
    TTF itself was synthesized essentially simultaneously by various research groups of Wudl, Coffen, and Hünig and was documented as follows:
    (b) Dissertation, G. Kiesslich, Würzburg **1968**.
    (c) Coffen, D. L.; Garrett, P. E., *Tetrahedron Lett.* **1969**, 2043.
    (d) Wudl, F.; Smith, G. M.; Hufnagel, E. J., *J. Chem. Soc. Chem. Commun.* **1970**, 1453.
    (e) Zahradník, R.; Cársky, P.; Hünig, S.; Kiesslich, G.; Scheutzow, D., *Int. J. Sulfur Chem.*, *C*, **1971**, *6*, 109.
    (f) Coffen, D. L.; Chambers, J. Q.; Williams, D. R.; Garrett, P. E.; Canfield, N. D., *J. Am. Chem. Soc.* **1971**, *93*, 2258.
    (g) Copper, W. F.; Kennedy, N. C.; Edmonds, J. W.; Nagel, A.; Wudl, F.; Coppens, P., *J. Chem. Soc. Chem. Commun.* **1971**, 889.
    (h) Wudl, F.; Wobschall, D.; Hufnagel, E. J., *J. Am. Chem. Soc.* **1972**, *94*, 671.
    (i) Hünig, S.; Kiesslich, G.; Quast, H.; Scheutzov, D., *Liebigs Ann. Chem.* **1973**, 310.
12. Ferraris, J.; Cowan, D. O.; Walatka, V. J.; Perlstein, J. H., *J. Am. Chem. Soc.* **1973**, *95*, 948.
13. Bechgaard, K.; Cowan, D. O.; Bloch, A. N., *J. Chem. Soc. Chem. Commun.* **1974**, 937.
14. Andrieux, A.; Duromre, C.; Jérome, D.; Bechgaard, K., *J. Phys. Lett.* **1979**, *40*, 381.
15. Mizuno, M.; Garito, A. F.; Cava, M. P., *J. Chem. Soc. Chem. Commun.* **1978**, 18.
16. Urayama, H.; Yamochi, H.; Saito, G.; Nozawa, K.; Sugano, T.; Kinoshita, M.; Sato, S.; Oshima, K.; Kawamoto, A.; Tanaka, J., *Chem. Lett.* **1988**, 55.
17. Gärtner, S.; Gogu, E.; Heinen, I.; Keller, H. J.; Klutz, T.; Schweitzer, D., *Solid State Commun.* **1988**, *65*, 1531.
18. Carlson, K. D.; Geiser, U.; Kini, A. M.; Wang, H. H.; Montgomery, L. K.; Kwok, W. K.; Beno, M. A.; Williams, J. M.; Cariss, C. S.; Crabtree, G. W.; Whangbo M.-H.; Evain, M., *Inorg. Chem.* **1988**, *27*, 965.
19. Suzuki, T.; Yamochi, H.; Srdanov, G.; Hinkelmann, K.; Wudl, F., *J. Am. Chem. Soc.* **1989**, *111*, 3108.
20. Beno, M. A.; Wang, H. H.; Carlson, K. D.; Kini, A. M.; Frankenbach, G. M.; Ferraro, J. R.; Larson, N.; McCabe, G. D.; Thompson, J.; Purnama, C.; Vashon, M.; Williams, J. M.; Jung, D.; Whangbo, M.-H., *Mol. Cryst. Liq. Cryst.* **1990**, *181*, 145.
21. Beno, M. A.; Wang, H. H.; Kini, A. M.; Carlson, K. D.; Geiser, U.; Kwok, W. K.; Thompson, J. E.; Williams, J. M.; Ren, J.; Whangbo, M.-H. *Inorg. Chem.* **1990**, *29*, 1599.

22. Papavassiliou, G. C.; Mousdis, G. A.; Zambounis, J. S.; Terzis A.; Hountas, A.; Hilti, B.; Mayer, C. W.; Pfeiffer, J. *Synth. Met.* **1988**, *27,* B379.
23. Kikuchi, K.; Saito, K; Ikemoto, I.; Murata, K.; Ishiguro, T.; Kobayashi, K. *Synth. Met.* **1988**, *27,* B269.
24. Kikuchi, K.; Kikuchi, M.; Namiki, T.; Saito, K.; Ikemoto, I.; Murata, K.; Ishiguro, T.; Kobayashi, K., *Chem. Lett.* **1987,** 931.
25. Kikuchi, K.; Ishikawa, Y.; Saito, K.; Ikemoto, I.; Kobayashi, K., *Synth. Met.* **1988,** *27,* B391.
26. Kikuchi, K.; Ikemoto, I.; Kobayashi, K., *Synth. Met.* **1987,** *19,* 551.
27. Brossard, L.; Ribault, M.; Valade, L.; Cassoux, P., *Physica* **1986,** *143 B,* 378.
28. Brossard, L.; Ribault, M.; Valade, L.; Cassoux, P. *J. Phys. France* **1989,** *50,* 1521.
29. *Science,* **1989,** *245,* 594 (August).
30. Ginzburg, V. L., *Phys. Today,* **1989,** *42,* 9 (March).
31. Ginzburg, V. L., Kirzhnits, D. A. eds. "High Temperature Superconductors," Consultants Bureau (Plenum), New York, **1982.**
32. Ishiguro, T.; Yamaji, K., "Organic Superconductors," Springer-Verlag, Berlin (1990), pp. 1–288.
33. Williams, J. M.; Kini, A. M., Geiser, U.; Wang, H. H.; Carlson, K. D.; Kwok,W. K.; Vandervoort, K. G.; Thompson, J. E.; Stupka, D. L.; Jung, D.; Whangbo, M.-H. In *Proc. Intl. Conf. Organic Superconductors*; South Lake Tahoe, CA, May 1990, Kresin, V., Little, W., Eds. *Organic Superconductivity*; Plenum Press: New York, **1990;** pp. 33–44.
34. Kini, A. M.; Geiser, U.; Wang, H. H.; Carlson, K. D.; Williams, J. M.; Kwok, W. K.; Vandervoort, K. G.; Thompson, J. E.; Stupka, D. L.; Jung, D.; Whangbo, M.-H.; *Inorg. Chem.,* **1990,** 29, 2555.
35. Williams, J. M.; Kini, A. M.; Wang, H. H.; Carlson, K. D.; Geiser, U.; Montgomery, L. K.; Pyrka, G. J.; Watkins, D. M.; Kommers, J. M.; Boryschuk, S. J.; Strieby Crouch, A. V.; Kwok, W. K.; Schirber, J. E.; Overmyer, D. L.; Jung, D.; Whangbo, M.-H., *Inorg. Chem.,* **1990,** 29, 3262.

LIST OF SYMBOLS AND ABBREVIATIONS

TNAP	11,11,12,12-tetracyanonaptho-2,6-quinodimethane
TCNQ	7,7,8,8-tetracyano-p-quinodimethane
TTF	tetrathiafulvalene
TMTSF	tetramethyltetraselenafulvalene
DMTCNQ	(2,4-dimethyl-TCNQ)
BEDT-TTF or "ET"	Bis(ethylenedithio)tetrathiafulvalene
BEDO-TTF	Bis(ethylenedioxy)tetrathiafulvalene
MDT-TTF	methylenedithiotetrathiafulvalene
DMET	[dimethyl(ethylenedithio)diselenadithiafulvalene]
dmit	4,5-dimercapto-1,3-dithiole-2-thione
T_c	= transition temperature to superconducting state
1 bar	= 0.9869 atm = 10^5 Pa
kbar	= 10^3 bar = 10^8 Pa

2

Synthesis of Organic-Donor Molecules and Conducting Salts

INTRODUCTION

In this chapter, we discuss the organic synthetic aspects of electron-donor molecules. This subject has been excellently reviewed in recent years by several authors.[1-5] The most recent of these reviews, by Schukat et al.,[5] is very comprehensive up to early 1986, as far as the tetrachalcogenafulvalene literature is concerned. Therefore, our emphasis in this section will be on various general synthetic methods which are used in the synthesis of the compounds, rather than to provide a complete bibliographic compilation of tetrachalcogenafulvalene syntheses. Additionally, we will highlight the most recent advances in the tetrachalcogenafulvalene synthesis area, as well as discuss synthetic methods employed in the preparation of other types, i.e., nontetrachalcogenafulvalene types of organic donors.

TETRACHALCOGENAFULVALENES

Tetrachalcogenafulvalenes are the most prolifically studied compounds in the preparation of electrically conducting and superconducting solids. In fact, nearly all the organic superconductors known to date are composed of tetrachalcogenafulvalene derivatives—TMTSF, BEDT-TTF, DMET, MDT-TTF and BEDO-TTF. Therefore, the synthetic methodologies used in the preparation of tetrachalcogenafulvalenes are also very diverse and intensively studied. With the discoveries of superconductivity in cation-radical salts of *un*symmetrical tetrachalcogenafulvalenes, for example, DMET and MDT-TTF, new synthetic methodologies for their exclusive preparation, rather than as mixtures followed by separation, have also been the major focus in recent years. (See Fig. 2.1.)

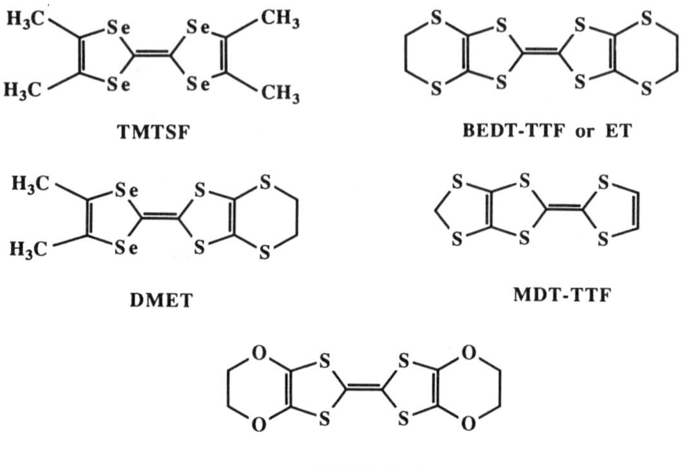

Figure 2.1 Structures of TMTSF, ET, DMET, MDT-TTF, and BEDO-TTF.

Tetrachalcogenafulvalenes are synthesized, to a large extent, by coupling the two halves of the molecule, thus involving the formation of the central C=C bond as the last step. There are other methods which give tetrachalcogenafulvalenes directly, that is, without a coupling reaction of the two halves. Each of these strategies have virtues, as well as limitations, depending on what the substituents on the tetrachalcogenafulvalene core are. We will discuss the general preparative methods of tetrachalcogenafulvalenes, both symmetrical and unsymmetrical.

Symmetrical Tetrachalcogenafulvalenes

Coupling Methods

1. Using Trivalent Phosphorus Reagents

    The most successful and predominantly employed method for the synthesis of tetrachalcogenafulvalene derivatives involves coupling of 1,3-dichalcogenole-2-chalcogenones utilizing trivalent phosphorus reagents (trialkylphosphines, triarylphosphines, trialkylphosphites, and triarylphosphites). (See Fig. 2.2.)

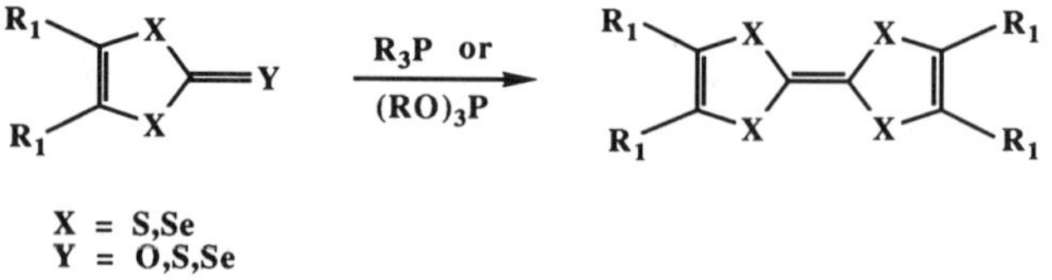

X = S,Se
Y = O,S,Se

Figure 2.2

The reaction, which was originally discovered by Corey, Carey, and Winter in connection with the syntheses of strained *trans*-cycloalkenes,[6] has emerged as a workhorse reaction for the preparation of a variety of tetrachalcogenafulvalenes. Although mechanistic details of the reaction are far from being fully understood, a general reaction pathway can be written as in Fig. 2.3. Some of the intermediates have either been isolated[7,8,9] or trapped[10] in certain cases. The yields are generally very sensitive to the exact nature of substituents on the 1,3-dichalcogenole-2-chalcogenone, chalcogens present, solvent, temperature, and even solubility characteristics of the final product. Therefore, some trial experiments are usually required in order to find the right combination of solvent, temperature, and the phosphorus reagent in order to optimize the yields. (See Fig. 2.3.)

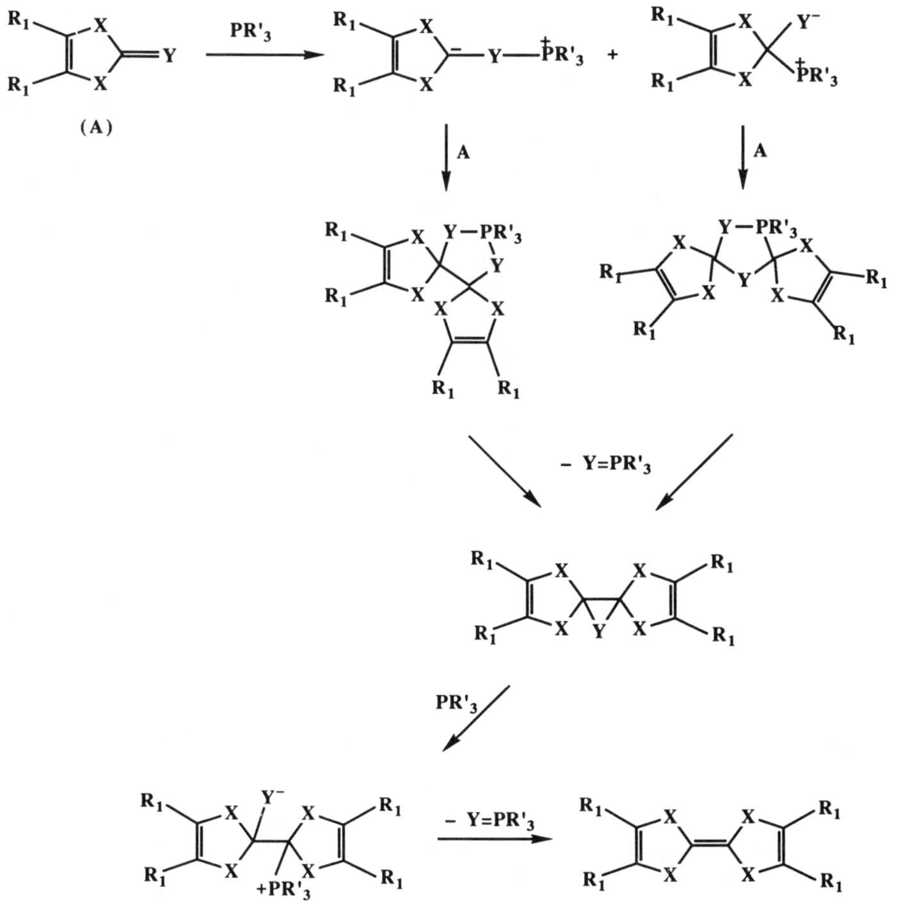

Figure 2.3

A slight variant of this method has been described by Fanghanel and coworkers, wherein 2-alkylthio-, 2-arylthio-, and 2-alkylseleno-1,3-dithioles are converted to tetrathiafulvalenes by reaction with trivalent phosphorus reagents (trialkyl- and triarylphosphines, trialkylphosphites and even phosphoramides) in acetonitrile.[11] Intermediacy of a phosphorane in the reaction, which further reacts with the 1,3-dithiolium salt to yield the product, was postulated. If the reaction was carried out in the presence of methanol, 2-methoxy-1,3-dithioles were isolated as a byproduct, suggesting the possibility of carbene intermediates. (See Fig. 2.4.)

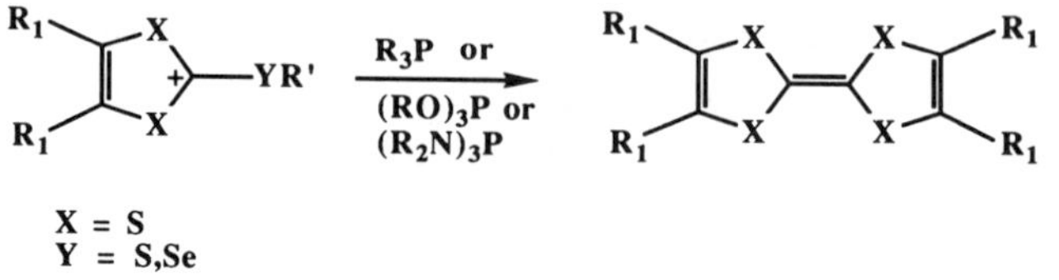

Figure 2.4

2. Photochemical Coupling Method

A variety of 1,3-dithiole-2-thiones with different substituents (electron donating as well as electron withdrawing) can be smoothly converted to the corresponding tetrathiafulvalenes by photochemical irradiation in the presence of hexabutylditin,[12, 13] triethylamine,[14] and trialkylphosphites[14, 15] at room temperature. The postulated mechanism for coupling invokes complexation of these reagents to the excited state of the 1,3-dithiloe-2-thione, followed by sulfur extrusion yielding a carbene intermediate, which then self-couples to give the tetrathiafulvalene. There are no reports yet in the literature on the usefulness of this methodology for tetraselena- and tetratellurafulvalene derivatives. (See Fig. 2.5.)

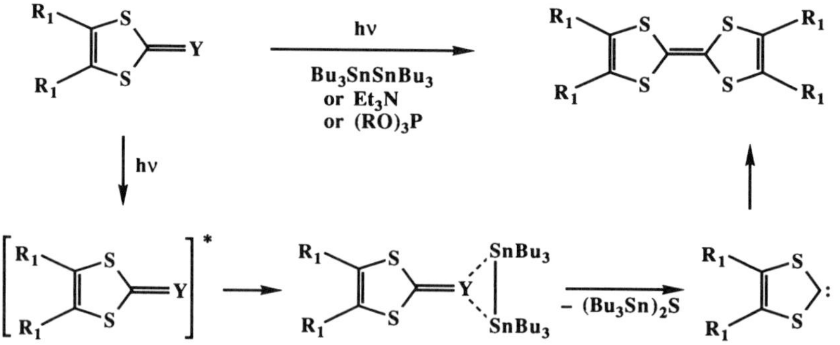

Figure 2.5

3. Coupling by transition metal carbonyls

Coupling of 1,3-diothiole-2-thione and 1,3-diselenole-2-selenone derivatives can also be brought about by transition metal carbonyls in benzene or toluene. The metal carbonyls are either dicobalt octacarbonyl[16–18] or triiron dodecacarbonyl.[9, 19] The reaction proceeds in reasonably good to poor yields with a variety of substituents of both electron-donating and electron-withdrawing character. (See Fig. 2.6.)

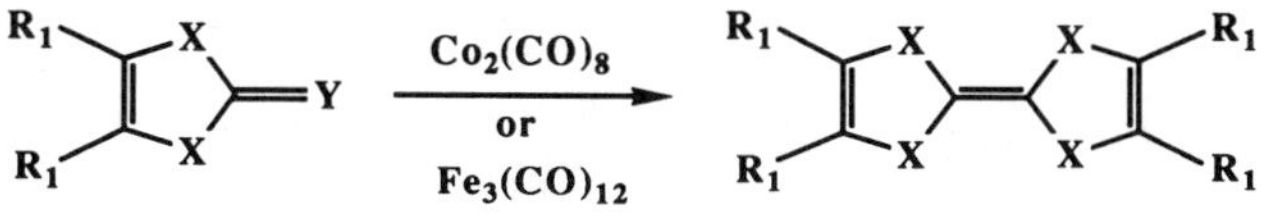

Figure 2.6

4. Electrochemical coupling

A novel synthesis of tetrathiafulvalene derivatives by electrochemical oxidation of phenyl substituted 1,3-dithiole derivatives in the presence of a base was described by Saeva and coworkers.[20] The yields of the reaction are low to moderate and the compatibility of various substituents in dithioles to the electrochemical oxidative coupling has not been explored. The proposed mechanism invokes oxidation of the dithiole to yield a cation radical, proton transfer to the base to yield a 1,3-dithiole radical, which on dimerization produces dihydro-tetrathiafulvalene. An identical reaction sequence (electron and proton transfer) then yields the final product. (See Fig. 2.7.)

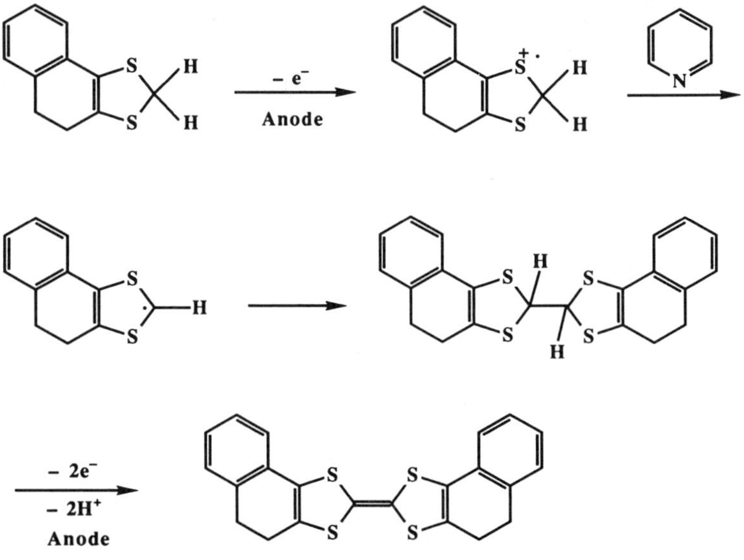

Figure 2.7

Another electrochemical coupling, useful in the preparation of tetrathiafulvalenes, is the reduction of 2-thioalkyl-1,3-dithiolium ions.[21] The electrochemical reduction presumably yields the 2-thioalkyl-1,3-dithiolium radical, which dimerizes to give the hexathioorthooxalates. Thermal decomposition of the hexathioorthooxalates, which can also be synthesized by an alternative method (vide infra), in halogenated solvents such as carbon tetrachloride[21] or 1,1,2-trichloroethane[22] furnishes tetrathiafulvalenes in good yields. (See Fig. 2.8.)

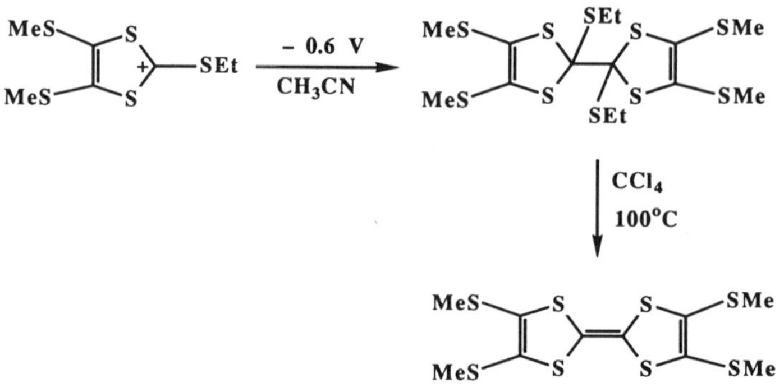

Figure 2.8

5. Deprotonation of 1,3-dithiolium salts

Next to the trivalent phosphorus-mediated coupling method, this is also one of the most often employed preparative procedures for the synthesis of tetrathiafulvalenes. The 1,3-dithiolium salts are stable compounds by virtue of resonance stabilization,[23] and they are accessible by a variety of methods.[5] Deprotonation of 1,3-dithiolium cations with tertiary amines produces 1,3-dithiolium carbenes, which give rise to tetrathiafulvalenes either by dimerization or by reaction with a second 1,3-dithiolium ion followed by a proton loss.[24] This method works well with unsubstituted[25–28] and substituted (with electron-releasing substituents) 1,3-dithiolium ions,[1,5] but not with those with electron-withdrawing substituents. (See Fig. 2.9.)

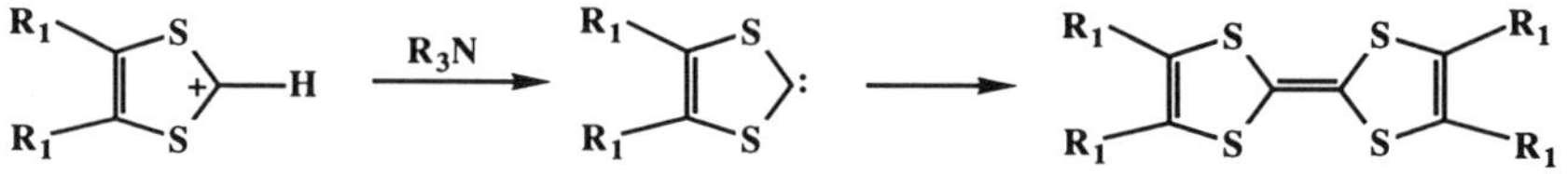

Figure 2.9

Although 1,3-diselenolium salts are known and can be readily prepared, analogous deprotonation fails to yield tetraselenafulvalenes.[29] (See Fig. 2.10.)

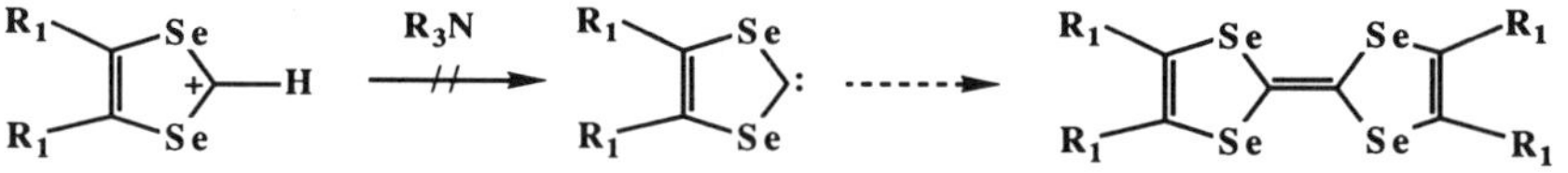

Figure 2.10

6. Thermolysis of 2-tosylhydrazono-1,3-dithioles

Closely related to the previously mentioned method, 1,3-dithiolium carbenes and hence tetrathiafulvalenes, can also be synthesized by thermolysis of 2-tosylhydrazono-1,3-dithioles in the presence of one equivalent of sodium ethoxide.[30] Fairly high temperature (180°C), which is required for this reaction, severely limits its synthetic utility. However, a facile synthesis of a variety of 2-tosylhydrazono-1,3-dithioles has recently been reported.[31] (See Fig. 2.11.)

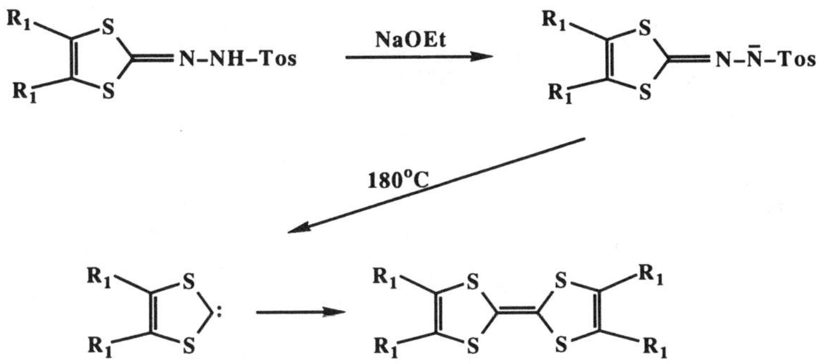

Figure 2.11

7. Thermolysis of 2-alkoxy-1,3-dithioles

Readily accessible 2-alkoxy-1,3-dithioles, either by reaction of benzyne with carbon disulfide in the presence of alcohols[32] or by reaction of acetylenes bearing electron withdrawing groups with carbon disulfide and alcohols,[33] undergo thermal 1,1-elimination of the alcohol to produce tetrathiafulvalenes.[34] The thermolysis, which is carried out either without solvent or in high-boiling solvents such as triglyme, proceeds in moderate yields and presumably via the intermediacy of carbenes. The reaction was found to proceed at a much lower temperature (refluxing benzene) in the presence of strong acids (trichloro- and triflouroacetic acids), and the yields were significantly better.[35, 36] (See Fig. 2.12.)

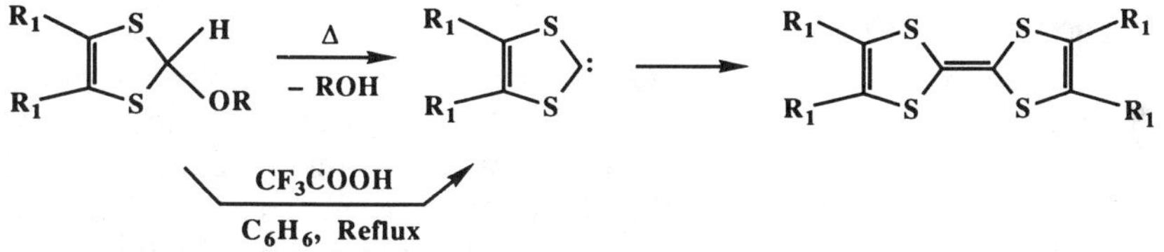

Figure 2.12

8. By reaction of 2-alkylthio-1,3-dithiolium salts with Zinc and Bromine

From the readily available 2-alkylthio-1,3-dithiolium salts (by alkylation of 1,3-dithiole-2-thiones), tetrathiafulvalenes can be obtained usually in good yields by reaction with zinc dust in ethanol and an oxidizing agent (bromine, iodine, peracids, lead dioxide or hydrogen peroxide). The reaction is thought to proceed through hexathioorthooxalate intermediates (obtained by the reduction of dithiolium ion with zinc) rather than carbenes, which are known to form 2-alkoxy-1,3-dithioles under the reaction conditions employed.[37] (See Fig. 2.13.)

9. TSeF by oxidative coupling of 2-methylene-1,3-diselenole

A novel method for the preparation of the unsubstituted tetraselenafulvalene has recently been described by Cava and his associates.[38] This is by far the most convenient method for the preparation of TSeF in reasonably good

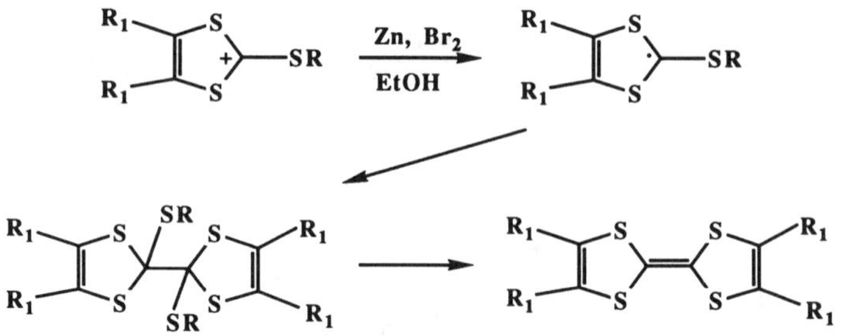

Figure 2.13

quantities, and it does not involve highly toxic and malodorous carbon diselenide. The starting compound 2-methylene-1,3-diselenole is readily accessible in good yields from 1,2,3-selenadiazole by treatment with potassium tert-butoxide in tert-butanol/N,N-dimethylformamide (DMF) mixture. Conversion of 2-methylene-1,3-dielenole to TSeF was achieved by the action of iodine/morpholine in DMF in 30–33% yield. The mechanism of this novel reaction, an overall oxidative dimerization with the loss of ethylene, is unknown and is reported to be under investigation.[38] (See Fig. 2.14.)

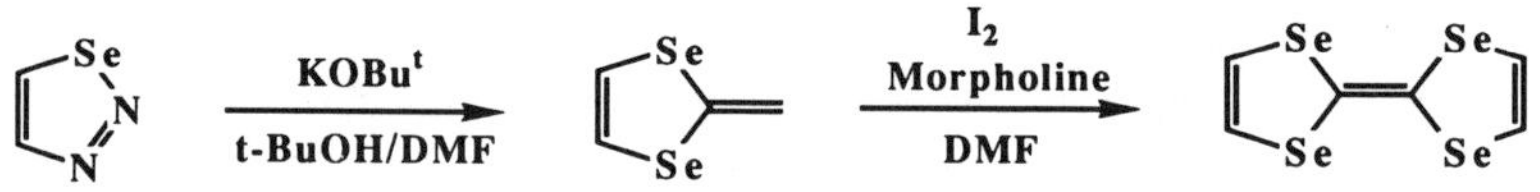

Figure 2.14

10. Thermolysis of hexathiaorthooxalates

Reaction of organolithium reagents (e.g., methyllithium) with 1,3-dithiole-2-thiones results in a thiophilic addition to yield a trithioorthoformate anion intermediate.[39] However, if the ratio of methyllithium:1,3-dithiole-2-thione is 1:2, the intermediate trithioorthoformate anion reacts with the second equivalent of 1,3-dithiole-2-thione in a carbophilic manner yielding a thiolate which can be alkylated to obtain hexathioorthooxalates.[39] Thermolysis of hexathiaorthooxalates in high-boiling halogenated solvents (carbon tetrachloride, 1,2-dichloroethane, 1,1,2-trichloroethane) yields tetrathiafulvalenes in good yields.[40] The conversion, hexathioorthooxalate to tetrathiafulvalenes, was reported to be catalyzed by p-toluenesulfonic acid,[40] although in some other tetrathiafulvalene systems, the acid was found to be detrimental.[22] (See Fig. 2.15.)

The great synthetic potential of this method for the preparation of unsymmetrical tetrathiafulvalenes has also been exploited.[22, 40]

Noncoupling Methods. There are several methods which can produce tetrachalcogenafulvalenes directly and without the need to couple two molecular halves. In fact, all the presently known methods for synthesizing tetratelluraful-valenes involve the noncoupling methodology. Therefore, we will survey some of these methods.

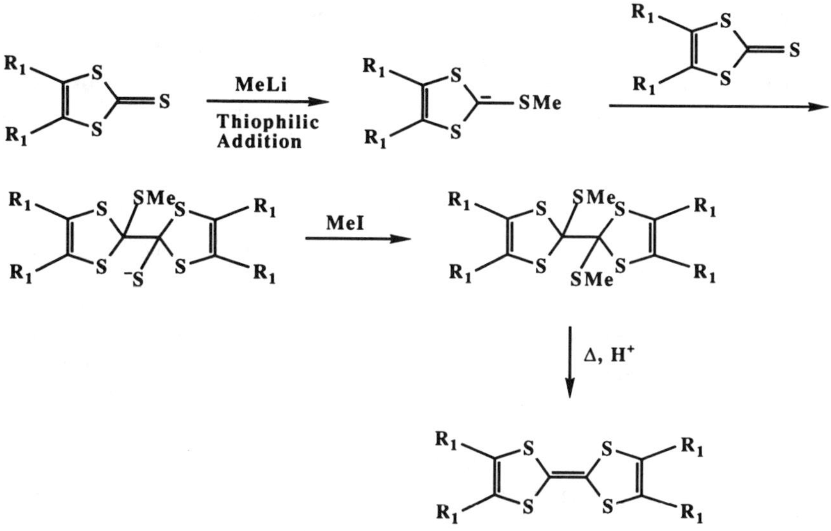

Figure 2.15

1. Reaction of 1,2-chalcogenolates with tetrachloroethylene

The first tetrathiafulvalene derivative reported in the literature dates back to 1926, when Hartley and Smiles reported the preparation of dibenzotetrathiafulvalene or DBTTF, by heating disodium 1,2-benzenedithiolate with tetrachloroethylene.[41] The reaction was subsequently reinvestigated by Bajwa et. al.,[42] and Mizuno et al.,[43] for the synthesis of a variety of substituted dibenzotetrathiafulvalenes. (See Fig. 2.16.)

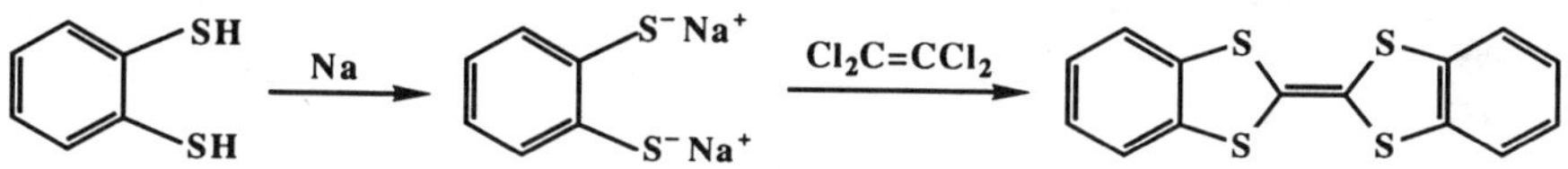

Figure 2.16

More recently, this methodology has been extended successfully for the preparation of tetratellurafulvalenes by the groups of Wudl[44] and Cowan.[45-50] Even a tetraselenafulvalene derivative, hexamethylenetetraselenafulvalene HMTSF, has been synthesized by this method.[51] (See Fig. 2.17.)

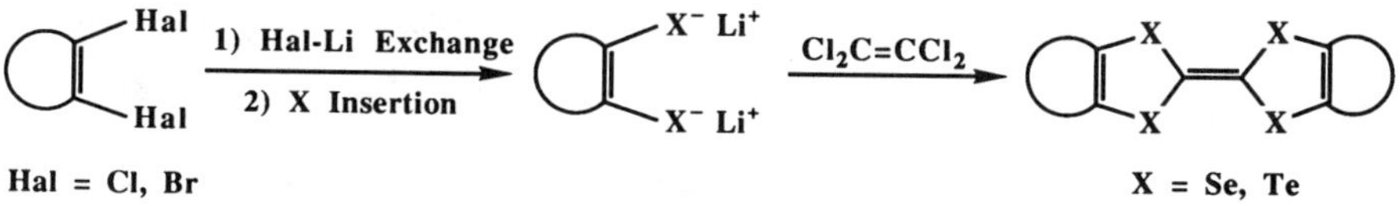

Figure 2.17

In the reaction of 1,2-dichalcogenolates with tetrachloroethylene, the formation of the isomeric compound containing a six-membered ring (e.g., tetrathiatetralin) is, in principle, possible. Such products have not been observed in the case of 1,2-dithiolates (i.e., in tetrathiafulvalene synthesis).[42, 43]

Mizuno et al. have proposed that it is consistent within the framework of Baldwin's rules, although their validity may be doubtful for anions of second and higher row elements.[43] However, in the case of 1,2-ditellurolates, the isomeric six-membered ring compounds have been isolated and characterized.[49, 52] (See Fig. 2.18.)

Figure 2.18 Structures of tetrathiatetralin and tetratelluratetralin.

The use of other synthons in place of tetrachloroethylene has also been explored. For example, the use of tetrabromoethylene in place of tetrachloroethylene is reported to improve the yield of tetratellurafulvalene.[50] Lambert and Christiaens have developed a synthesis of dibenzotetraselenafulvalene by the reaction of disodium 1,2-benzenediselenolate with vinylidene dichloride.[53]

It has also been pointed out that the reaction may consist of several addition/elimination steps and hence may involve acetylenic intermediates,[44, 53] although the overall reaction has the appearance of a simple nucleophilic displacement of halogens with chalcogenolates.

2. Reaction of electron-deficient acetylenes with carbon disulfide and carbon diselenide

Formation of tetrathiafulvalenes, albeit as a minor product (2%), directly by the reaction at 100°C of carbon disulfide with acetylenes bearing at least one electron-withdrawing group (triflouromethyl, methoxycarbonyl) was first reported by Hartzler.[54] The major products in the reaction were compounds A and B. (See Fig. 2.19.) However, the reaction can be driven to the exclusive and

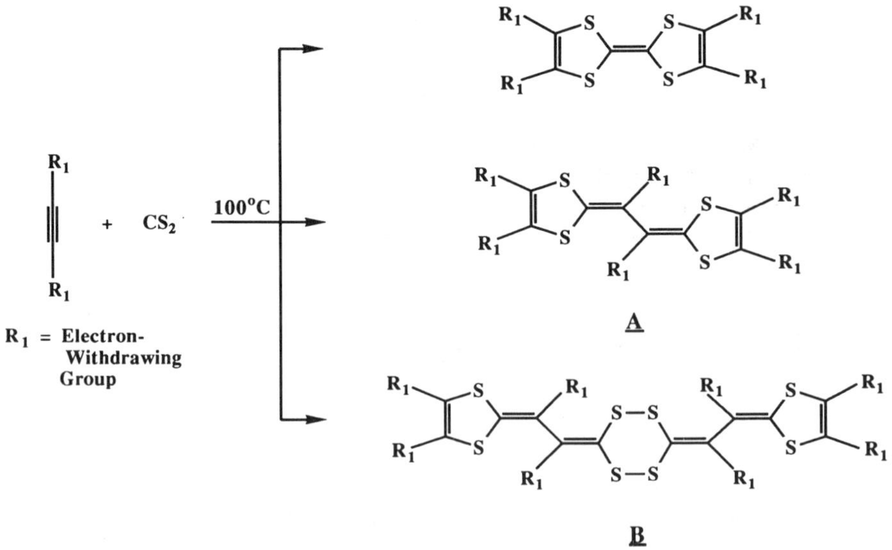

Figure 2.19

quantitative formation of tetrathiafulvalene in the presence of triflouroacetic acid under the same conditions. The addition of carbon disulfide to electron-deficient acetylenes results in a dipolar intermediate, which further cyclizes to give a 1,3-dithiolium carbene. The intermediacy of the carbene has been demonstrated by carbene-capturing reagents (alcohols, phenols, alkenes).[33] In the presence of strong acids, the 1,3-dithiolium carbene is converted, via protonation, to the corresponding 1,3-dithiolium cation, which subsequently adds to a second equivalent of the 1,3-dithiolium carbene giving rise to the protonated tetrathiafulvalene and then the tetrathiafulvalene itself. (See also Fig. 2.20.)

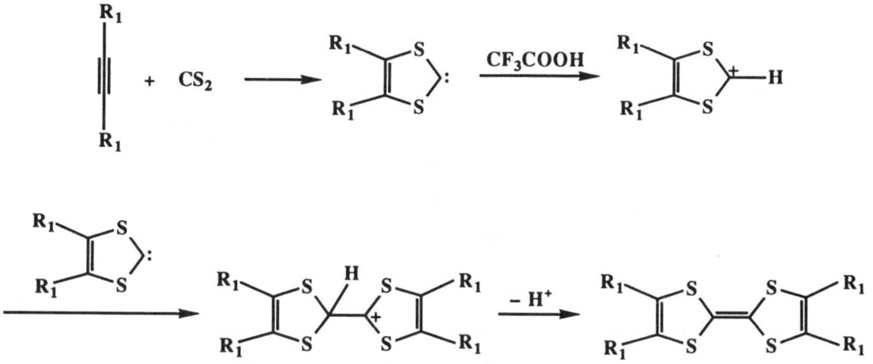

Figure 2.20

Okamoto and coworkers have shown that similar syntheses can be carried out without triflouroacetic acid but under high pressures (5–6 kbar).[55-57] The presence of at least one electron-withdrawing group on the acetylene is essential, so that the dipolar intermediate is resonance stabilized. The dipolar intermediate is then converted to the 1,3-dithiolium carbene and the tetrathiafulvalene by dimerization of the carbene. (See Fig. 2.21.)

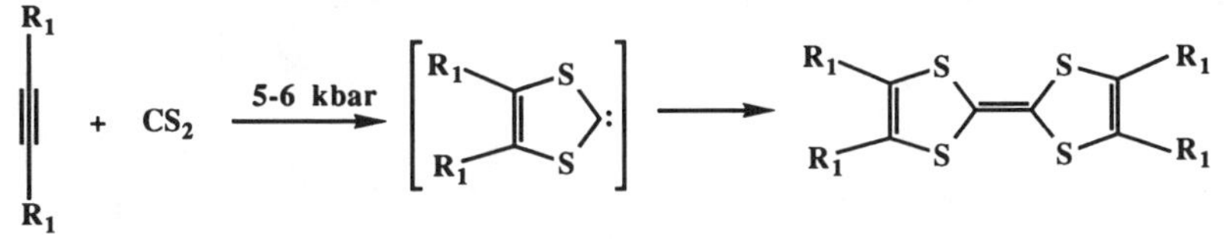

R_1 = COOR, CF_3, $SiMe_3$

Figure 2.21

The high-pressure conditions, in contrast to the triflouroacetic acid-mediated reaction, are also amenable to the reaction of carbon disulfide or carbon diselenide with acetylenes with trimethylsilyl substituents to yield the corresponding trimethylsilyl-substituted tetrathiafulvalenes and tetraselenafulvalenes.[58] These trimethylsilyl derivatives can be further converted into deuterated tetrathiafulvalene and tetraselenafulvalene by reaction with potassium carbonate/potassium bicarbonate buffer in D_2O. The synthetic utility of the trimethylsilyl substituted-TTF and -TSF is also evident from their conversion to the corresponding alkylthio-, alkylseleno- and arylseleno-tetrathiafulvalenes.[58] (See Fig. 2.22.)

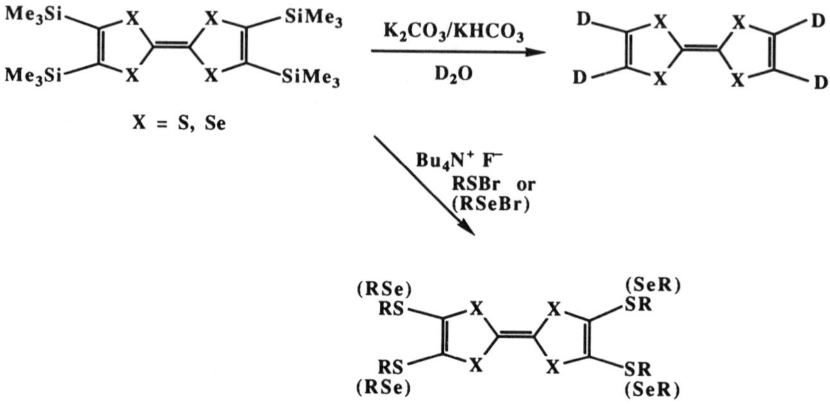

Figure 2.22

Reaction of electron-deficient acetylenes (e.g., methyl propiolate) with carbon disulfide can also be made extremely facile by tributylphosphine. In fact, tetrathiafulvalenes can be synthesized in ca. 21% yield by this method in tetrahydrofuran (THF) at $-30°C$.[59] The reaction is believed to involve a phosphine-carbon disulfide complex, which adds to the acetylene giving an extremely reactive phosphorane intermediate.[60] The phosphorane attacks another molecule of carbon disulfide and a second equivalent of the acetylene, thus yielding the tetrathiafulvalene derivative.[59] (See Fig. 2.23.)

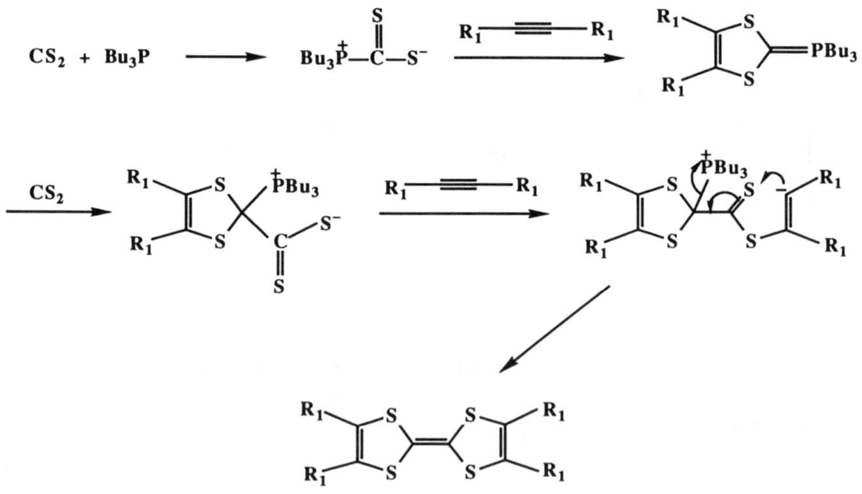

Figure 2.23

Another variant of the method involves the addition reaction of iron and nickel η^2-CS_2 complexes with activated acetylenes to form 1,3-dithiolium carbene complexes of iron[61-63] and nickel.[64] (See Fig. 2.24.)

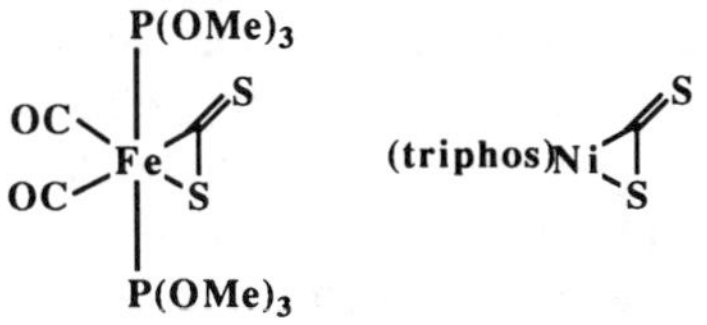

Figure 2.24 η^2-CS_2 complexes of Fe and Ni.

The iron-carbene complexes can be converted into the corresponding tetrathiafulvalene derivatives by several methods (air oxidation,[61] iodine oxidation,[63] thermolysis by refluxing in toluene and decalin,[63] or by electrochemical oxidation[65]). The yields are generally very good and the method involves mild reaction conditions. The only limitation is that it is only applicable to the synthesis of tetrathiafulvalenes with electron-withdrawing substituents. (See Fig. 2.25.)

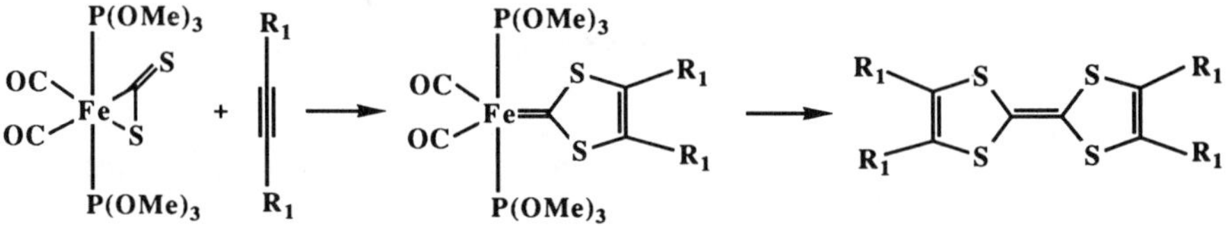

Figure 2.25

On the other hand, the nickel-1,3-dithiolium carbene complex can be converted into the tetrathiafulvalene in 62% yield by treatment with carbon monoxide.[64] (See Fig. 2.26.)

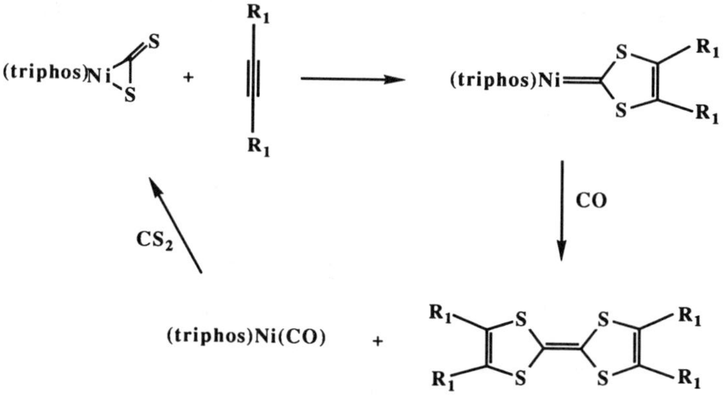

Figure 2.26

3. Noncoupling synthesis utilizing thiapendione

Thiapendione, now a commercially available compound, was used as a starting point for the synthesis of tetrathiafulvalene derivatives by a novel noncoupling method by Schumaker et al.[66] In this method, one of the carbonyl centers is selectively opened under phase-transfer catalysis to yield a dithiolate, which is alkylated with propargyl halides to obtain dipropargyl-1,3-dithiole-2-one. Ring opening of the second carbonyl center with a stronger base (sodium methoxide) generates the propargyl dithiolate intermediate, which spontaneously cyclizes to give di(*exo*-methylene)dihydrotetrathiafulvalene. Under acid catalysis, the *exo*-cyclic double bond can be shifted into the ring giving the tetrathiafulvalene derivatives. The central tetrathioethylene moiety in the product tetrathiafulvalene is derived from the starting compound thiapendione and the outer part of the tetrathiafulvalene is constructed around it, thus making this a unique noncoupling route. In other words, the tetrathiafulvalene molecule is constructed starting from the interior part of the molecule. The method is limited to the syntheses of TTF derivatives containing at least two methyl substituents. (See Fig. 2.27.)

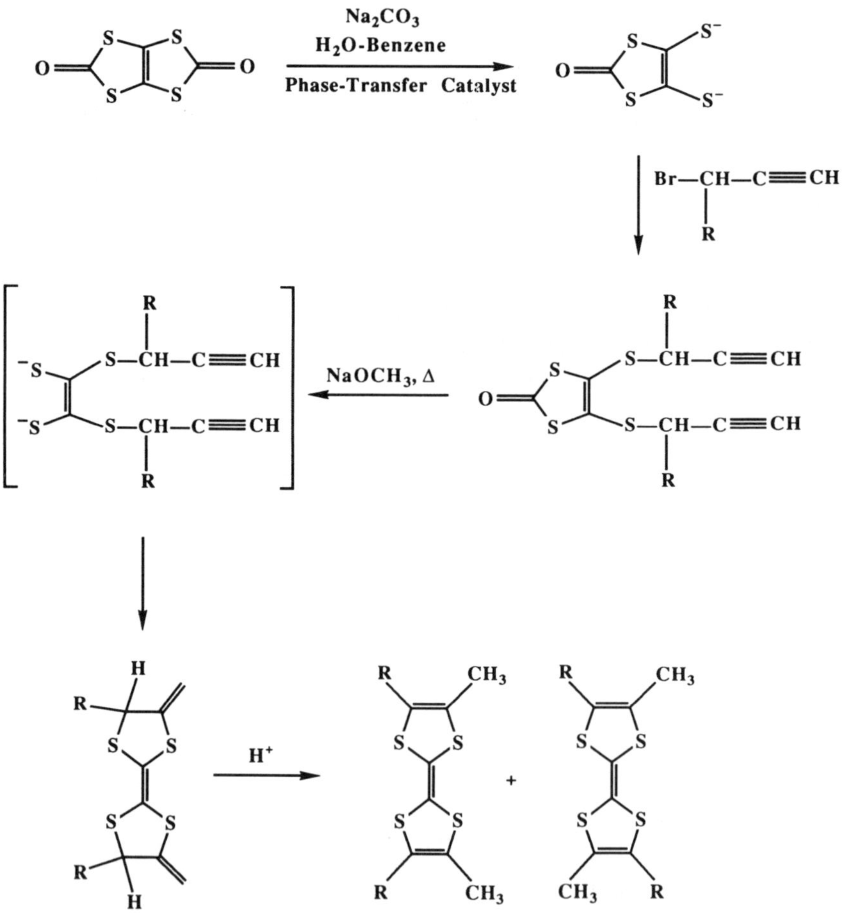

Figure 2.27

Its synthetic potential was also exploited to synthesize fully deuterated tetramethyltetrathiafulvalene, a useful compound necessary for neutron-diffraction studies of its charge-transfer and cation-radical salts.[66] (See Fig. 2.28.)

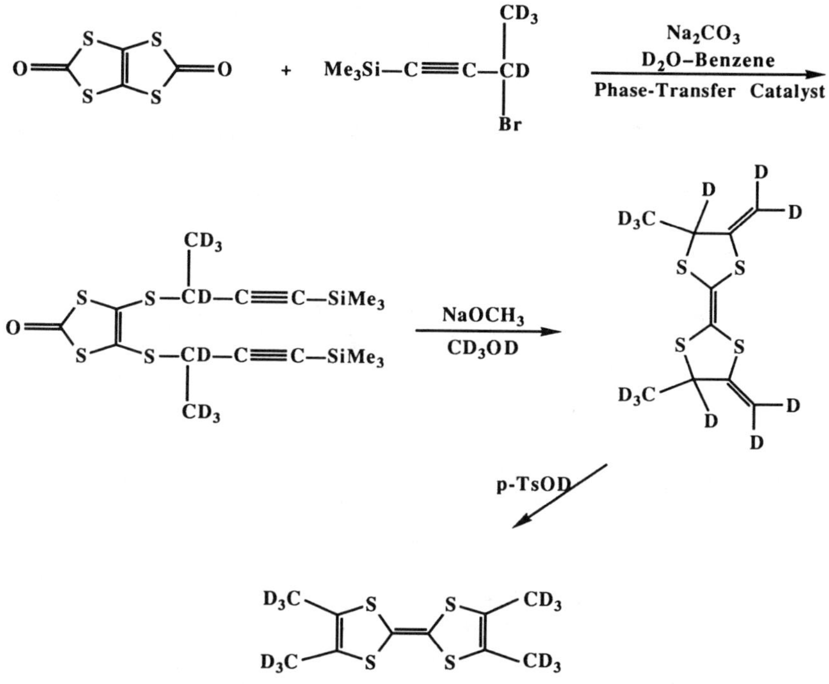

Figure 2.28

4. Reaction of cycloalkynes with carbon disulfide

Tetramethylcycloheptyne, a highly reactive yet stable cycloalkyne, reacts with carbon disulfide to give the corresponding tetrathiafulvalene directly.[67] The addition of carbon disulfide is believed to produce 1,3-dithiolium carbene as an intermediate, which readily dimerizes to the product tetrathiafulvalene. (See Fig. 2.29.)

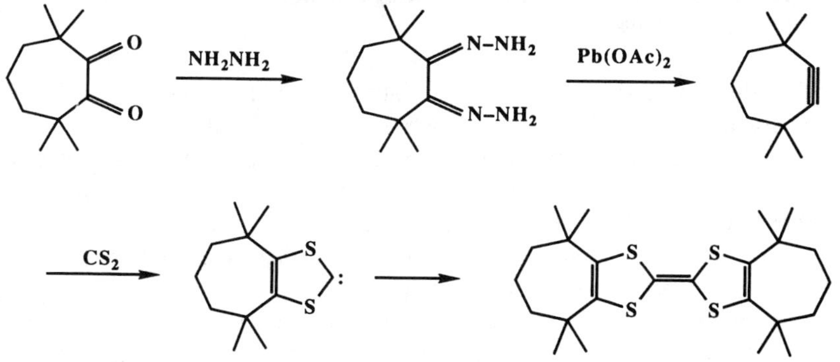

Figure 2.29

Gas-phase reaction of benzyne, generated by the thermolysis of phthalic anhydride at 700°C, with carbon disulfide, however, gives a more complex mixture of products, among which dibenzotetrathiafulvalene has been detected in small quantities.[68] (See Fig. 2.30)

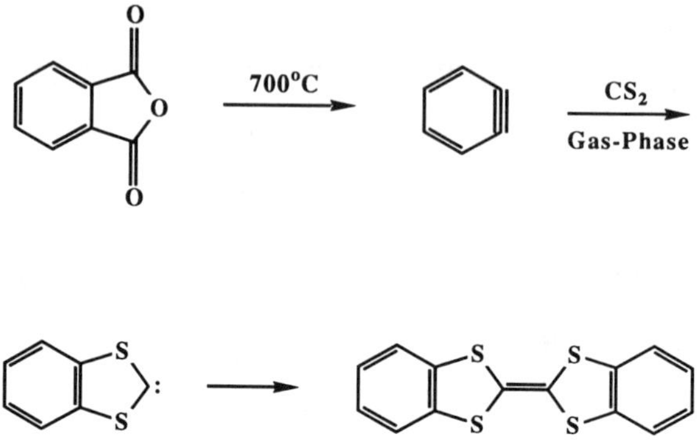

Figure 2.30

The solution phase reaction of benzyne with carbon disulfide yields the 1,3-dithiolium carbene intermediate which, however, reacts with the solvent or the reagent used for the generation of the benzyne. Thus, tetrathiafulvalenes are not produced; instead, 2-acetoxy-1,3-dithiole[69] or 2-alkoxy-1,3-dithioles are obtained.[32] (See Fig. 2.31.)

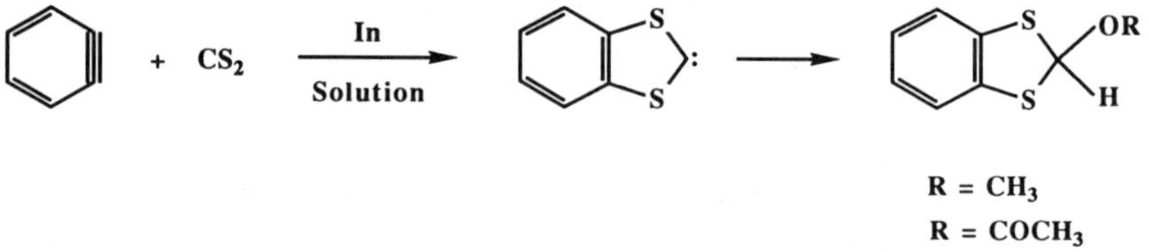

R = CH_3
R = $COCH_3$

Figure 2.31

Derivatization of Tetrathiafulvalene and Tetraselenafulvalene

Tetrathiafulvalene is now commercially available,[70] and many convenient methods for its large-scale preparation are also in the literature.[59, 71, 72] Tetraselenafulvalene can also be prepared conveniently, and in reasonably good quantities, by a method recently described by Cava's group.[38] There have been many reports of synthesizing derivatives of tetrathiafulvalene and tetraselenafulvalene by use of these readily available parent compounds. Some of these methods are particularly attractive for the preparation of functionalized (e.g., with long-chain hydrocarbon substituents with hydrophobic and/or hydrophilic character) tetrathiafulvalenes useful in the fabrication of Langmuir-Blodgett films, for which there is a strong industrial interest.

Pioneering work in chemical studies of tetrathiafulvalene and tetraselenafulvalene with regard to functionalizing the molecules was carried out by Green.[73–75] Tetrathiafulvalene can be cleanly lithiated to monolithioTTF via proton-lithium exchange reaction with either butyllithium or lithium diisopropylamide (LDA) in ether or tetrahydrofuran at −78°C. Once generated, the lithioTTF can be reacted with a wide variety of reagents to obtain functionalized TTF's (Fig. 2.32).[75, 76]

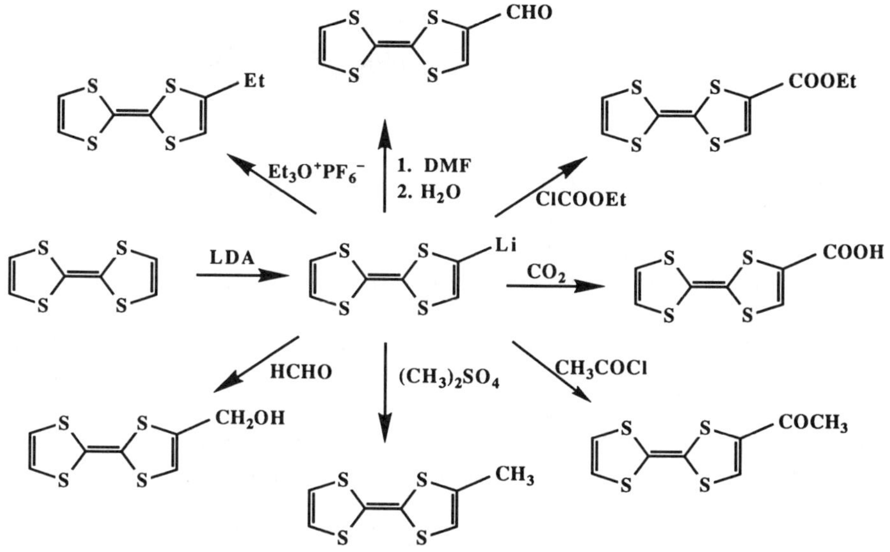

Figure 2.32

MonolithioTTF is stable only at −78°C, and on warming it undergoes redistribution to give multilithiated TTF's, TTF, and an unidentified product. (See Fig. 2.33.)

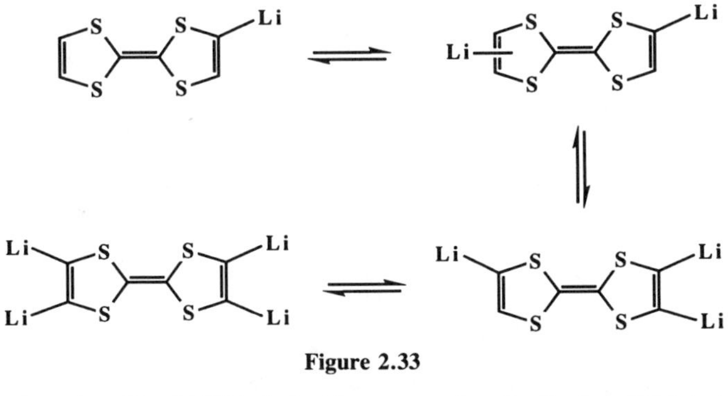

Figure 2.33

Monofunctionalized TTF derivatives can also be further lithiated under the same conditions, and the lithiation can be directed on either of the two heterocyclic rings depending on the functionality. Electron-withdrawing substituents (e.g., −COOEt) directed the lithiation to take place in the same ring, while electron-releasing substituents (e.g., −CH$_3$) directed the lithiation on the other ring.[75] (See Fig. 2.34.)

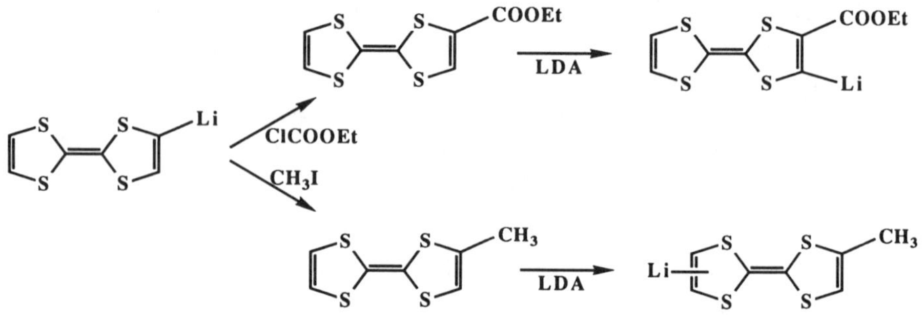

Figure 2.34

By using four equivalents of LDA, Aharon-Shalom et al. showed that proton-lithium exchange of all four protons of TTF occurs giving tetralithioTTF.[77] The synthetic versatility of tetralithioTTF is attested to by its use by several research groups for the preparation of a vast variety of TTF derivatives by subsequent synthetic manipulations.

For example, chalcogen (Se, Te) insertion into the carbon-lithium bond can give rise to tetrachalcogenolato-TTF, which can be alkylated with alkyl halides to obtain tetrakis(alkylchalcogeno)TTF's.[77-81] Alternatively, tetralithioTTF can be treated with dialkyl disulfides and diselenides to obtain tetrakis(alkylthio)- and tetrakis(alkylseleno)TTF's.[79, 81] This latter strategy is particularly useful to obtain tetrakis(arylthio)- and tetrakis(arylseleno)TTF's,[79, 81] which are otherwise not accessible from tetrachalcogenolato-TTFs. (See. Fig. 2.35.)

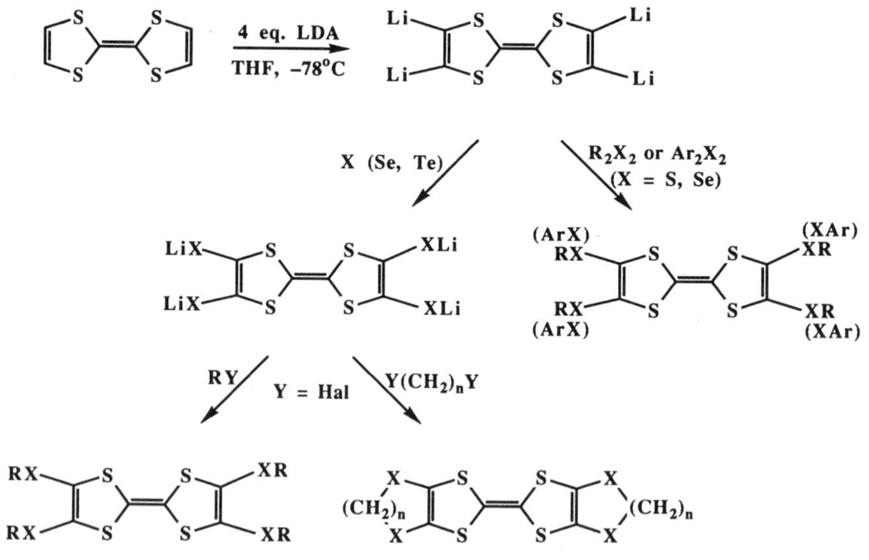

Figure 2.35

Alkylation of tetrakis(chalcogeno)TTF's (chalcogen = Se, Te) with alkylenedihalides to obtain mixed chalcogen analogs of BEDT-TTF was found not to be possible in the earlier studies.[77, 80] However, such alkylative ring-closure can be effected cleanly in highly solvating sovents such as hexamethylphosphoramide (HMPA) to obtain BEDSe-TTF in excellent yield.[82] (See Fig. 2.36.)

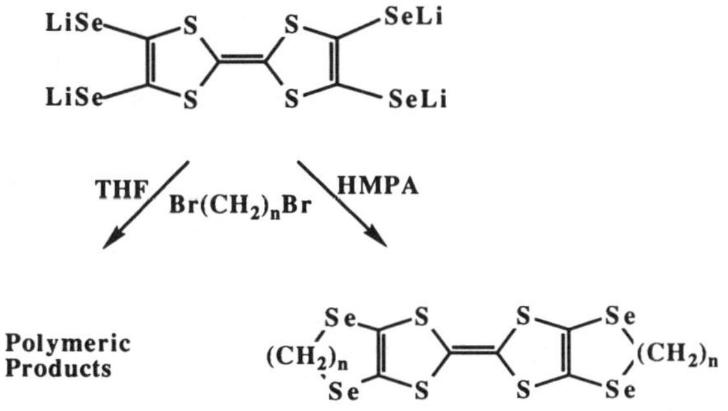

Figure 2.36

Tetralithiation of tetraselenafulvalene can be effected only by use of LDA,[83, 84] and not with n-BuLi. The latter reagent reportedly leads to ring-opened products resulting from anionic attack at carbon or selenium of TseF, rather than lithium-hydrogen exchange.[84] The tetralithio-TSeF can also be functionalized by (a) chalcogen insertion followed by alkylation,[83] (b) treatment with di(alkyl or aryl) di(sulfides or selenides),[83, 84] and (c) treatment with methyl chloroformate and carbon dioxide.[84] (See Fig. 2.37.)

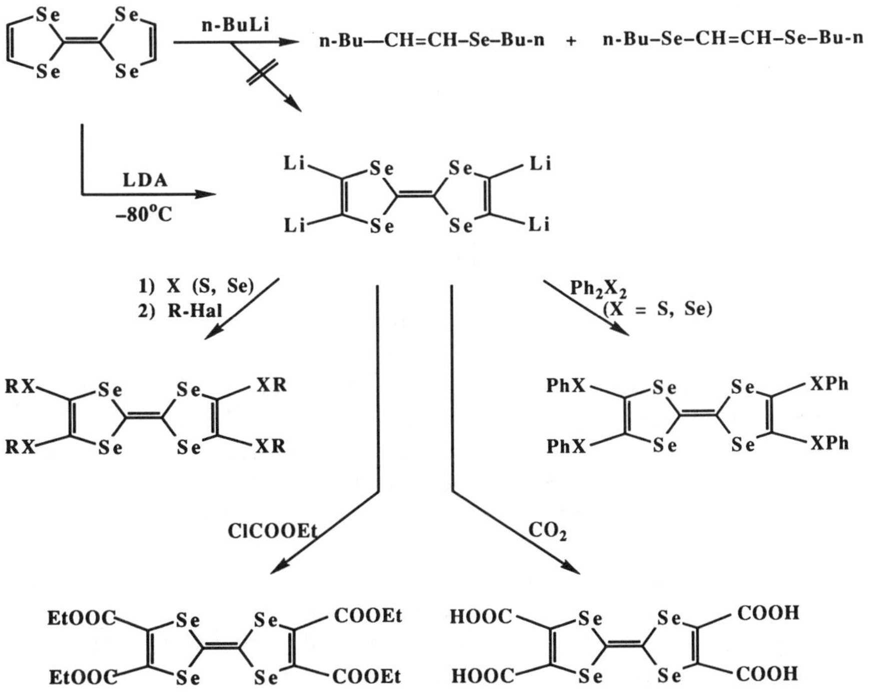

Figure 2.37

Another useful conversion of tetralithio-TTF involves its reaction with formaldehyde to obtain tetrakis(hydroxymethyl)TTF, which can be further converted to bis(oxydimethylene)TTF by the action of thionyl chloride.[85] (See Fig. 2.38.)

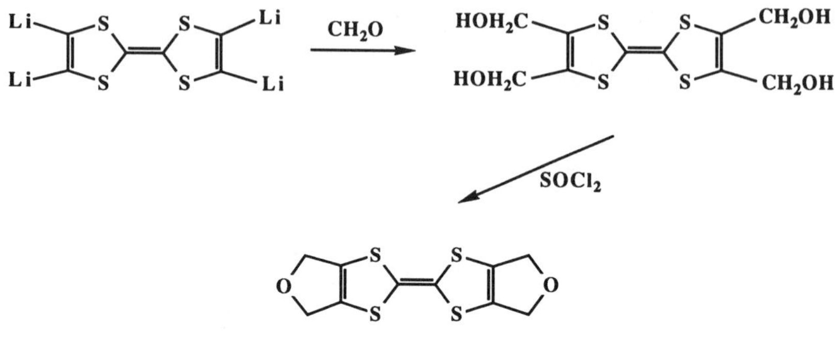

Figure 2.38

Halogen-substituted tetrathiafulvalenes are not readily obtained by electrophilic halogenation of tetrathiafulvalene, since stable cation-radical salts are easily formed.[28] These synthetically useful derivatives may be obtained from lithiated tetrathiafulvalene by the action of hexachloroethylene (for tetrachloroTTF) and 1,2-dibromotetrachloroethylene (for tetrabromoTTF).[86] (See Fig. 2.39.)

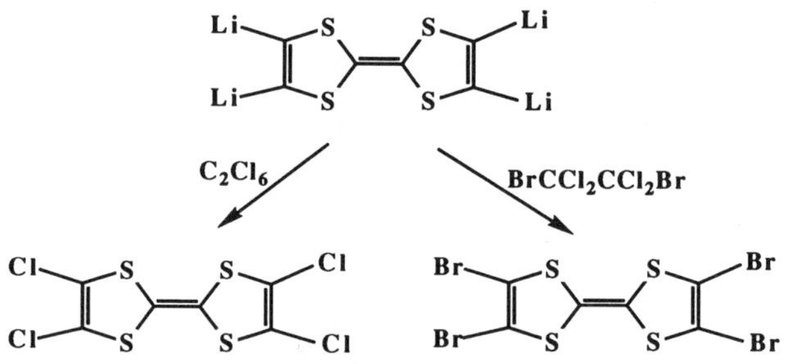

Figure 2.39

Unsymmetrical Tetrachalcogenafulvalenes

Unsymmetrical tetrachalcogenafulvalenes are readily synthesized by cross-coupling of two halves, either by trivalent phosphorous mediated coupling of two different 1,3-dichalcogenole-2-chalcogenones[87] or by deprotonation by tertiary amines of two different 1,3-dichalcogenolium ions.[88, 89] Alternately, two different 2-thioalkyl-1,3-dithiolium salts can be coupled electrochemically to obtain a mixture of three possible hexathioorthooxalates, which when thermally decomposed yield the corresponding tetrathiafulvalene derivatives as a mixture.[90] The product consists of a mixture caused by random coupling, and the separation of the desired unsymmetrical donor from the two other symmetrical byproducts is essential. The separation usually consists of fractional crystallization provided there are solubility differences among the products, or chromatographic methods. A large number of unsymmetrical donors has been synthesized this way, including DMET and MDT-TTF, which have given rise to superconducting solids. (See Fig. 2.40.)

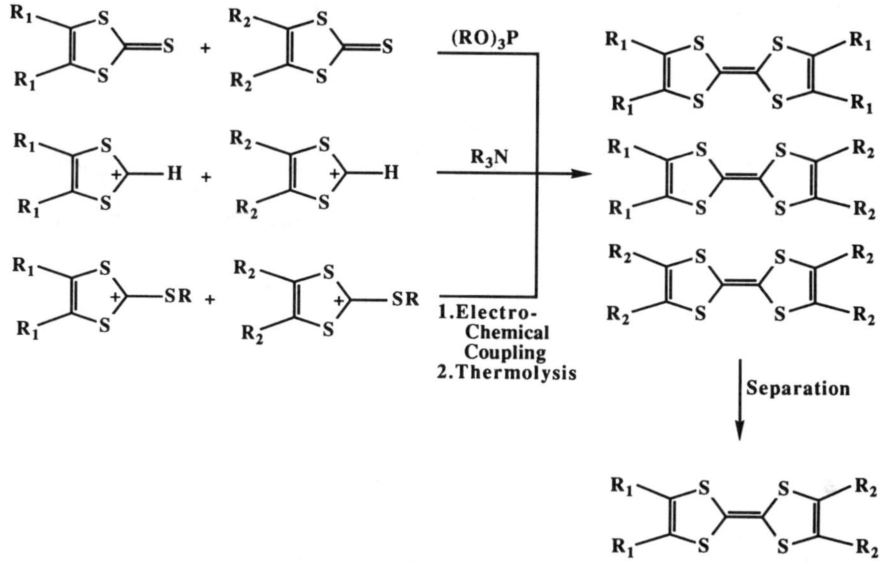

Figure 2.40

However, sometimes the separation of the mixture poses serious difficulties, and synthetic methodologies that are capable of giving the desired unsymmetrical donor without the formation of symmetrical byproducts are needed. Several strategies have been reported in the last few years, and these will be discussed in this section.

1. Cross-coupling of 1,3-dithiole-2-phosphoranes with 1,3-dithiolium ions

    Gonnella and Cava reported this general method for the directed syntheses of unsymmetrical TTF derivatives.[91] The reaction of a 1,3-dithiolium ion with triphenyl phosphine results in an adduct, which can be deprotonated by a strong base such as n-butyllithium to generate a phosphorane intermediate. The phosphorane is then reacted with a different 1,3-dithiolium ion to obtain an addition product, which on treatment with triethylamine brings about the elimination of triphenyl phosphine and furnishes the unsymmetrical TTF derivative. This methodology has been employed subsequently by several groups.[92, 93] (See Fig. 2.41.)

2. Cross-coupling of 1,3-dithiole-2-phosphonates with 1,3-dithiole-2-iminium salts

    In the previously described method involving phosphoranes, the formation of small amounts of symmetrical TTF derivatives has been noted in some cases.[93] It has been attributed to the instability of the phosphorane or caused by the dissociation of the phosphonium salt intermediate in certain cases.[93] An improved version of the phosphorane methodology, developed by Lerstrup and coworkers,[72] involves the reaction of 1,3-dithiolium ions with tri-

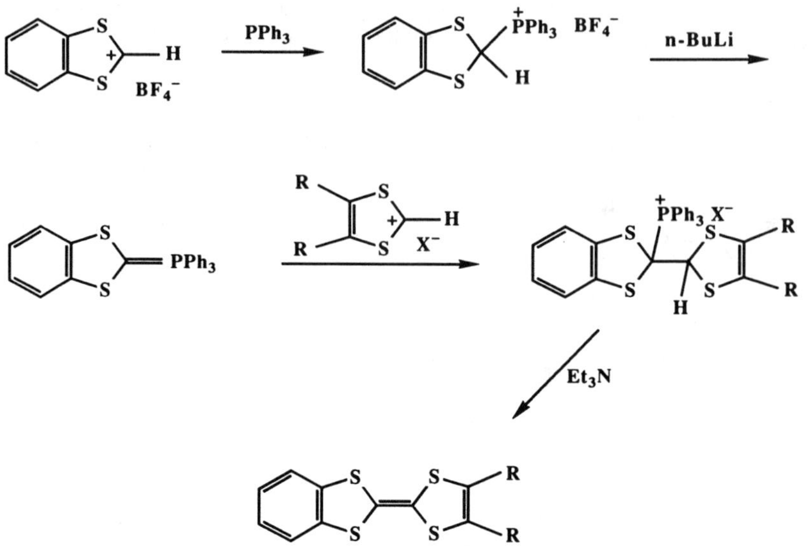

Figure 2.41

alkylphosphites to give 1,3-dithiole-2-phosphonate ester, which is then treated with a 1,3-dithiole-2-iminium salt (different 1,3-dithiole derivative) in the presence of potassium tert-butoxide. The product of this reaction is then treated with silica to obtain the unsymmetrical TTF derivative, without any detectable amounts of symmetrical TTF.[72] Neither the function of silica, nor the mechanism, is clear at present. (See Fig. 2.42.)

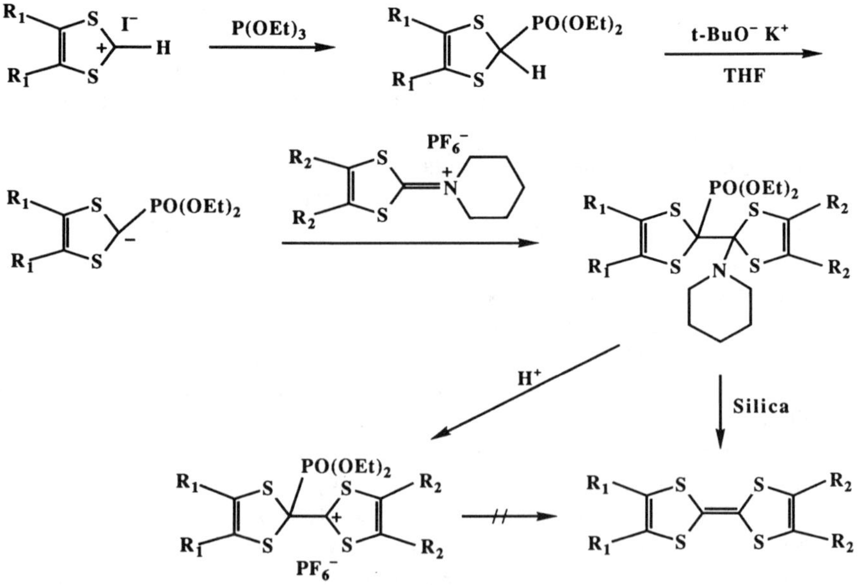

Figure 2.42

3. The hexathioorthooxalate route

Unsymmetrical hexathioorthooxalates, as precursors to unsymmetrical tetrathiafulvalenes, can be easily prepared by the reaction of the conjugate base of 2-thiomethyl-1,3-dithioles with a second 1,3-dithiole-2-thione, followed by methylation.[39] The former can be prepared easily by the thiophilic addition of MeLi to 1,3-dithiole-2-thiones[22, 39] or via $NaBH_4$ reduction of 2-thiomethyl-1,3-dithiolium ions. The unsymmetrical hexathioorthooxalates are then readily converted to the corresponding unsymmetrical TTF derivatives by decomposition in high-boiling halogenated solvents (CCl_4, 1,2-dichloroethane, 1,1,2-trichloroethane).[21, 22] Acid catalysis (p-toluenesulfonic acid) has been reported to be beneficial in this decomposition,[40] although in one instance it was found to be detrimental.[22] (See Fig. 2.43.)

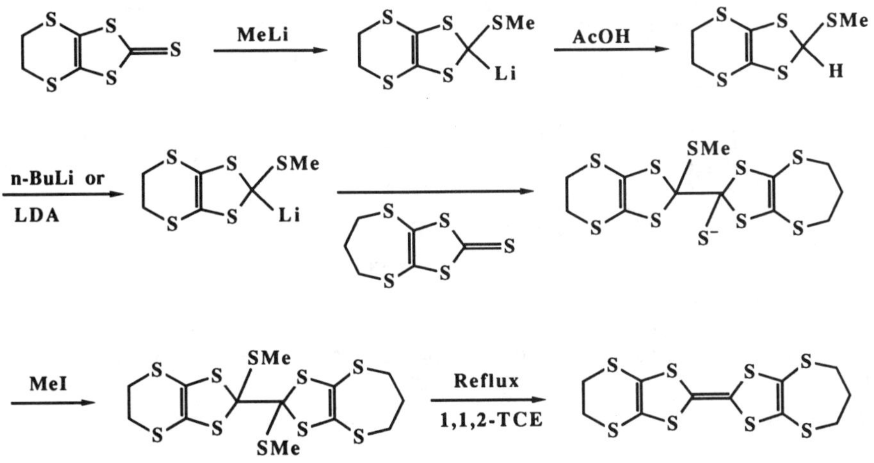

Figure 2.43

TETRACHALCOGENOACENES

In addition to the tetrachalcogenafulvalenes, the tetrachalcogenoacenes, Fig. 2.44, have given rise to a large number of organic conductors. These compounds contain two 1,2-dichalcogenole rings fused to the aromatic hydrocarbon skeleton, and because of the chalcogens on the periphery of the molecules, form efficient intermolecular chalcogen···chalcogen contacts in the solid state. No superconducting salts have been discovered in this class of materials yet, although many good organic metals are known.

Tetrathiotetracene (TTT), also called tetrathionaphthacene in the early literature, was first reported in 1948,[94] and highly conducting charge-transfer salts (with o-chloranil, o-bromanil, tetracyanoethylene) of this material were first reported in 1965.[95] Subsequently, several conducting salts of TTT with halides and TCNQ, which are quasi-one-dimensional metals, were reported.[96–99] These results have prompted the syntheses of a variety of compounds belonging to this class, wherein both the chalcogen and the aromatic hydrocarbon backbone have been modified.[100–111] (See Fig. 2.44.)

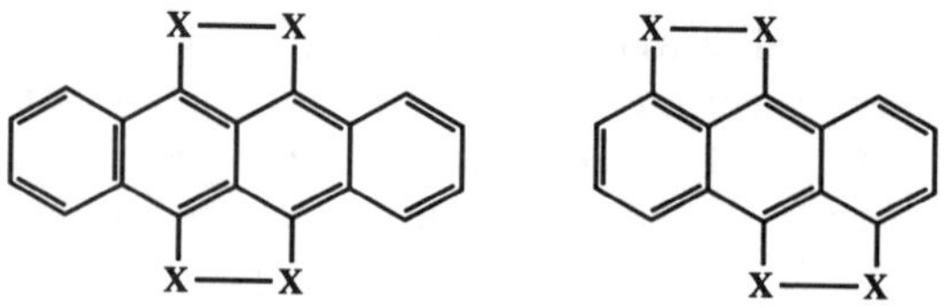

X = S TTT
X = Se TSeT
X = Te TTeT

X = S TTA
X = Se TSeA

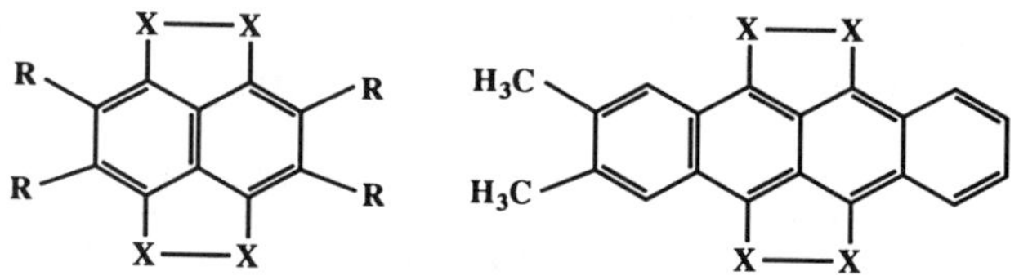

R = H, X = S   TTN
R = H, X = Se TSeN
R = H, X = Te TTeN
R = Me, X = S TMTTN
R = Me, X = Se TMTSeN
R = Me, X = Te TMTTeN

X = S DMTTT
X = Se DMTSeT

Figure 2.44 Structures of TTT compounds.

All the tetrachalcogenoacenes synthesized to date are prepared by one common method. The reaction of an appropriate tetrachloroacene with alkali metal dichalcogenides (e.g., Na_2S_2, Na_2Se_2, Na_2Te_2) in a dipolar aprotic solvent (DMF, HMPA, etc.) results in the formation of the tetrachalcogenoacenes in fair to good yields. (See Fig. 2.45.)

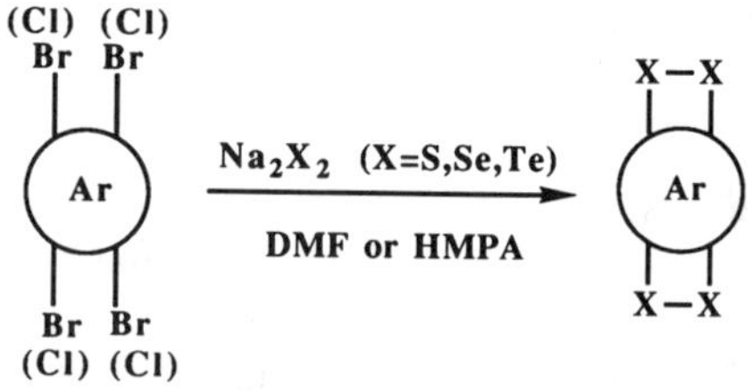

Na_2X_2 (X=S,Se,Te)
DMF or HMPA

Figure 2.45

In a slightly modified version of this method, TSeT was also synthesized by the reaction of elemental selenium with 5,11-dichlorotetracene in boiling Dowtherm (a eutectic of biphenyl and diphenyl ether, b.p. ca. 260°C).[106] (See Fig. 2.46.)

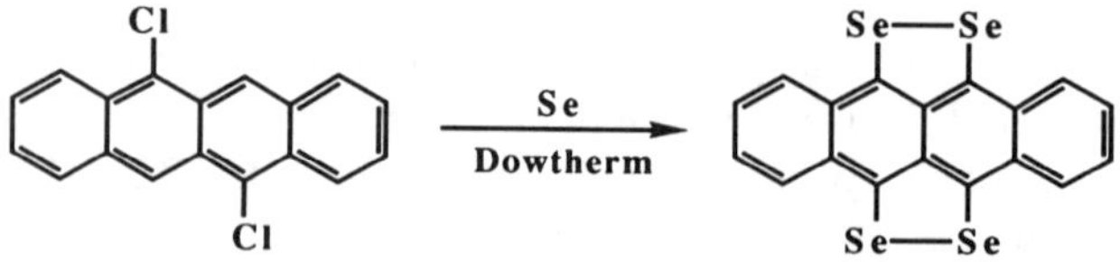

Figure 2.46

By a similar methodology, dichalcogenoacenes have also been synthesized as potential electron donors.[112-116] These compounds may be viewed as a subclass of the tetrachalcogenoacene family. (See Fig. 2.47.)

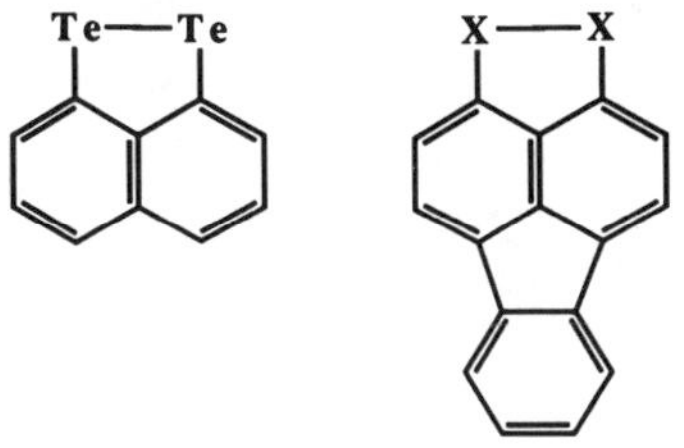

Figure 2.47 Structures of dichalcogenoacenes.

MISCELLANEOUS DONOR MOLECULES

TTF-type

A highly distorted, cage-type molecule with a bent TTF moeity has recently been described by Röhrich et al.[117] The synthetic method consists of alkylation of 4,5-dimercapto-1,3-dithiole-2-thione dianion with tetrabromodurene, followed by phosphite-induced intramolecular coupling. The X-ray structure of this unusual donor molecule reveals a highly distorted nature of the tetrathiafulvalene moiety, and the result of such distortion is manifested in high oxidation potentials and the ready protonation of this compound. (See Fig. 2.48.)

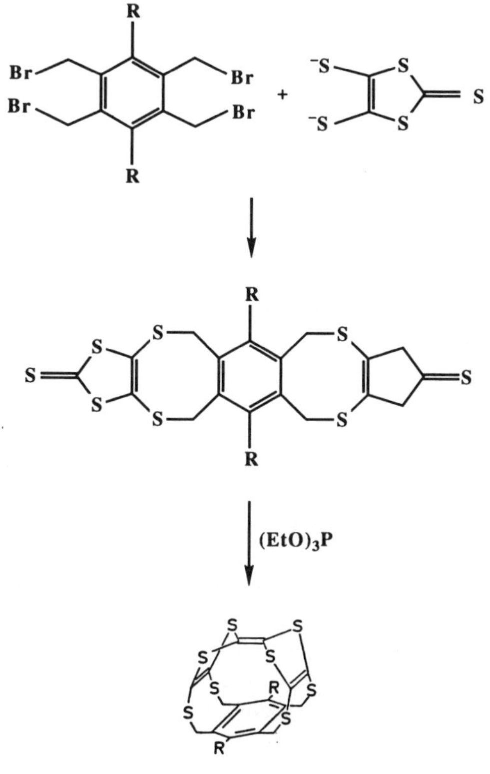

Figure 2.48

Chap. 2   Miscellaneous Donor Molecules

Another example of such a highly distorted TTF derivative has been reported by Girmay et al.[118] The synthetic methodology is very similar to the above and involves the reaction of 4,5-dimercapto-1,3-dithiole-2-thione dianion with two equivalents of 2,6-bis(bromomethyl)pyridine, followed by an intramolecular coupling. (See Fig. 2.49.)

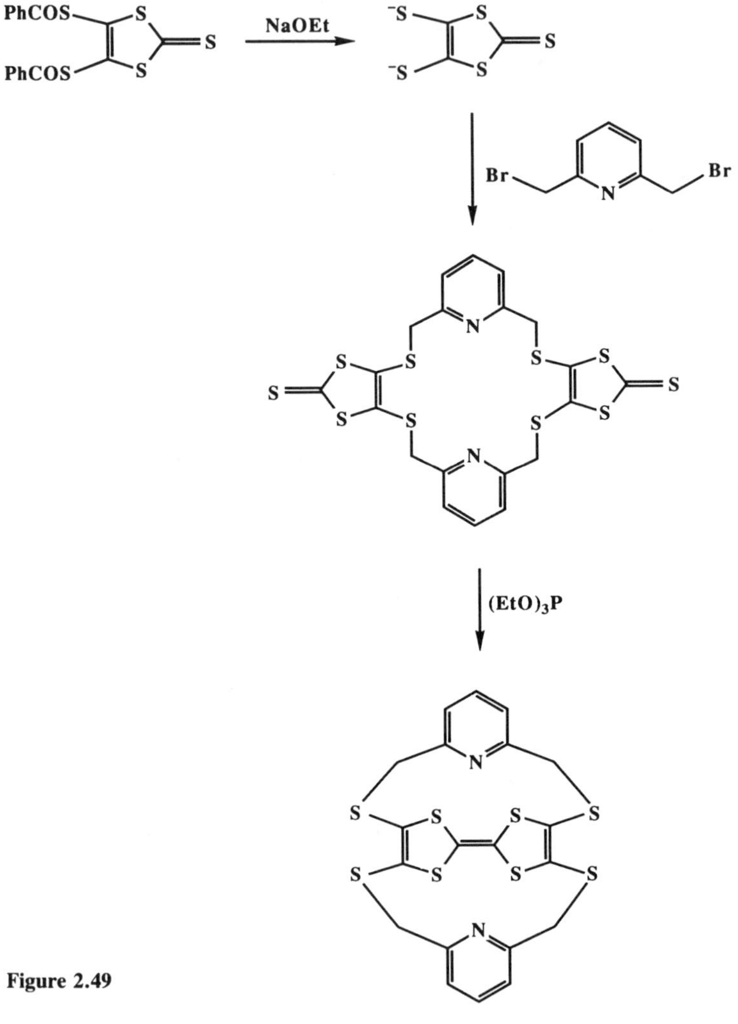

Figure 2.49

Another unusual TTF-type molecule is the so-called TTF-dimer, reported by Nishikawa et al.[119] Here two TTF units are fused to a bicyclic spiro-ring, and as a result are held at a fixed dihedral angle. The motivation for this work was to form or establish two- and three-dimensional S· · ·S networks, and hence, possibly higher dimensional conductors. The synthetic procedure is outlined in Fig. 2.50.

There are several examples of donor molecules in which two TTF (or TSF) moeities are connected via a bridging atom, for example, Te,[120] a bridging group, for example, ethylene,[121, 122] or directly.[123] (See Fig. 2.51.)

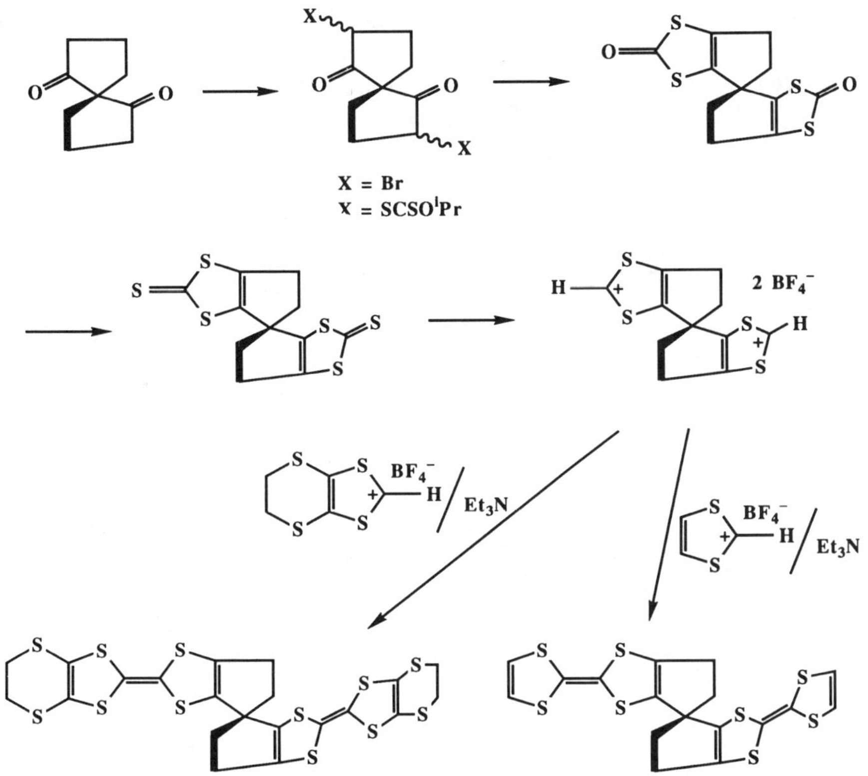

Figure 2.50

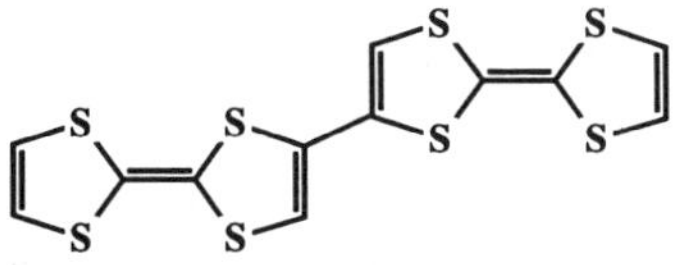

Bi-TTF

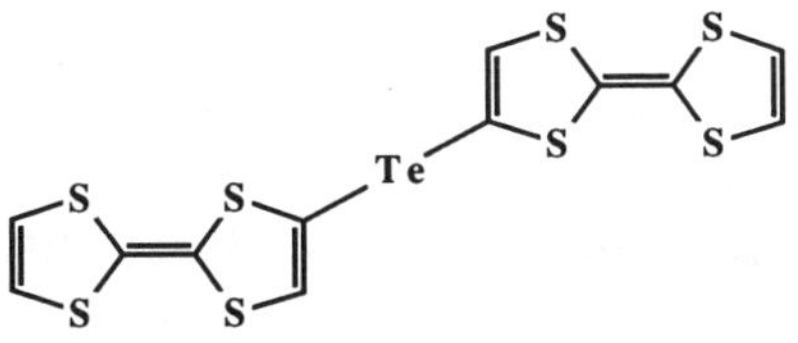

Bis(TTF)telluride

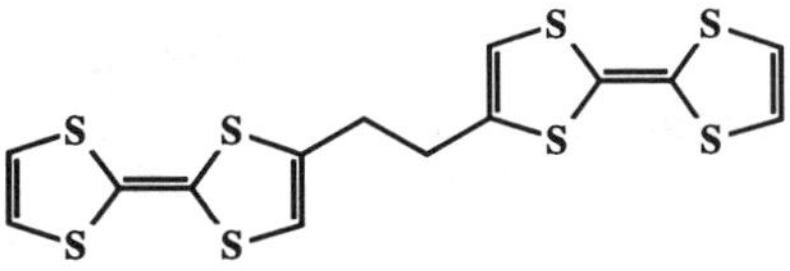

Bis(TTF)ethane

Figure 2.51 Structures of bridged TTFs.

Chap. 2 Miscellaneous Donor Molecules

Yui et al., have reported binaphtho[1,6-d,e]-1,3-dithiin-2-ylidene (C) and its selenium analog (D).[124] These compounds are formally tetrachalcogenoethylenes, but the heteroatoms are part of six-membered rings rather than five-membered rings as in tetrachalcogenafulvalenes. (See Fig. 2.52.)

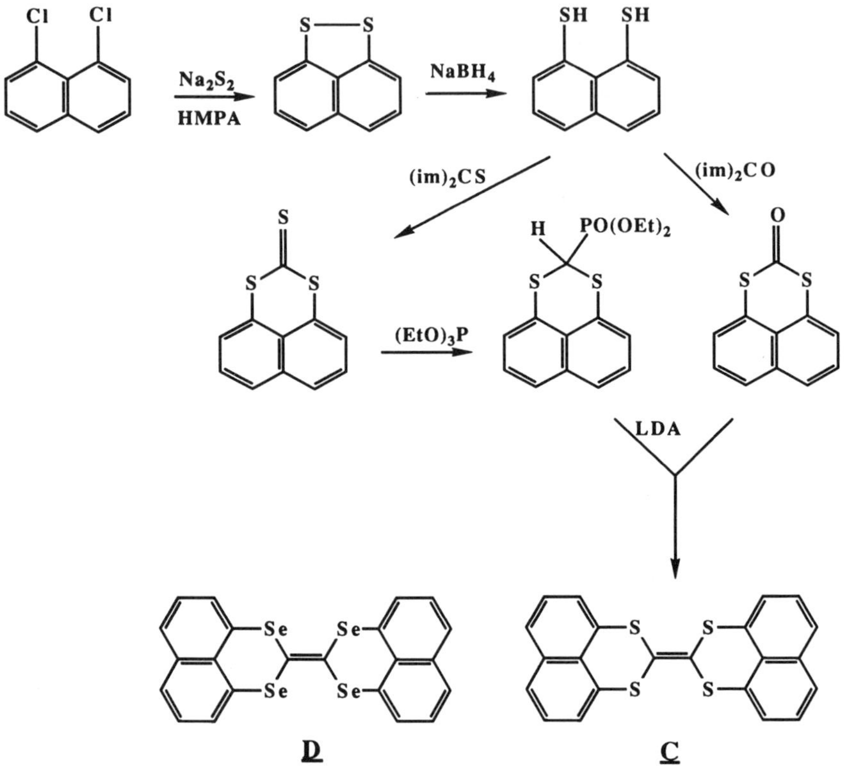

Figure 2.52

1,3-dithiole-2-ylidene Compounds

There is a considerable interest in the design and syntheses of compounds of this class, primarily because they are "chemical models" for the preparation of conducting solids with reduced on-site Coulomb repulsion energies (U). The importance of reduced U in relation to achieving high-electrical conductivities and, more recently, in relation to superconductivity, in organic materials whose electronic bandwidths are similar in magnitude to U, has been discussed.[125] Therefore, various model compounds have been synthesized, wherein the 1,3-chalcogenole-2-ylidene moieties of tetrachalcogenafulvalenes have been separated by different spacer groups. Some of the spacer groups include ethanediylidene,[126] 2-butene-1,4-diylidene,[126] aromatic quinodiylidene,[2, 127, 128] and 2,5-dihydrothiophene-2,5-diylidene.[129] (See Fig. 2.53.)

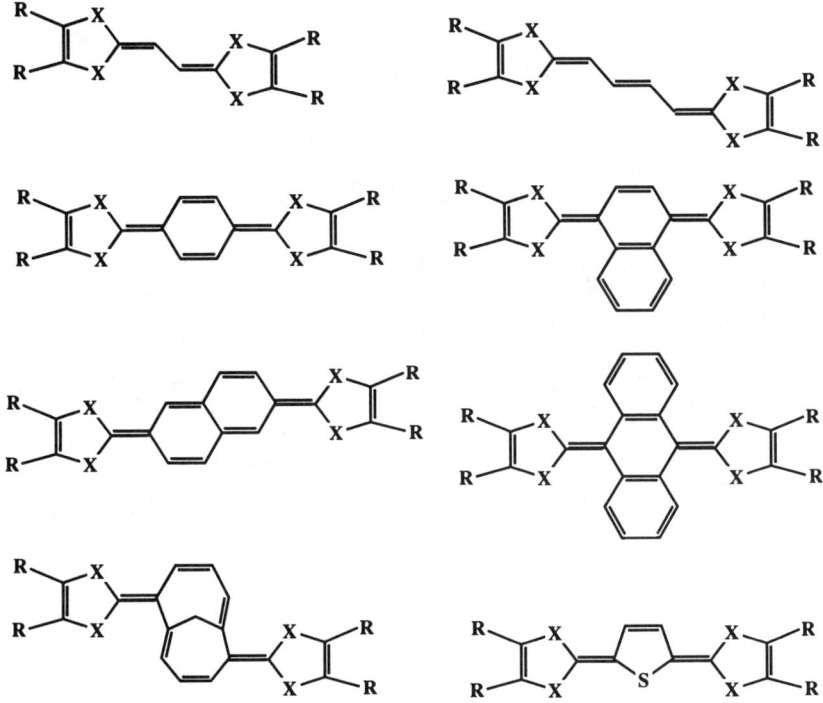

Figure 2.53 Structures of 1,3-dithiole-2-ylidene compounds.

In addition to the compounds described, where only two 1,3-dichalcogenole-2-ylidene moeities are connected through a spacer group, there are several other related compounds in which more than two 1,3-dichalcogenole-2-ylidene moeities are attached to a cyclic ring system.[130–132] These compounds are shown in Fig. 2.54. These compounds are synthesized by Ni(0) catalyzed oligomerization of dibromoethanediylidene-2,2′-bis(1,3-dithiole)s. The products depend on the nature of the solvent as well as the presence of carbon monoxide as a reactant (for synthesis of the cyclopentanone derivative).

Other examples of such compounds are the monofulvathiane, difulvathiane, and trifulvathiane systems recently described by Schumaker et al.,[133] and 1,3,5-(cyclohexane-2,4,6-trione)-tris(1,3-dithiole-2-ylidene).[134] (See Fig. 2.55.)

Polychalcogenaarenes

High-electrical conductivities in cation-radical salts of aromatic hydrocarbons, for example, perylene,[135–138] have prompted several workers to synthesize compounds that incorporate chalcogen atoms in large aromatic hydrocarbon frameworks. Some of the examples include 1,4,5,8-tetrathiaanthracene,[139] 1,6-dithiapyrene,[140] 3,4,9,10-tetrathiaperylene,[141] trithiaphenanthrene,[142, 143] triselenaphenanthrene,[143] 3,10-dithiaperylene,[144] 1,7-dithiaperyene,[145] 3,9-dithiaperylene,[145] 2,7-bis(methylthio)-1,6-dithiapyrene,[146] and 2,3:7,8-bis-(ethylenedithio)-1,6-dithiapyrene,[146] and 1,6-dioxapyrene.[122] (See Fig. 2.56.)

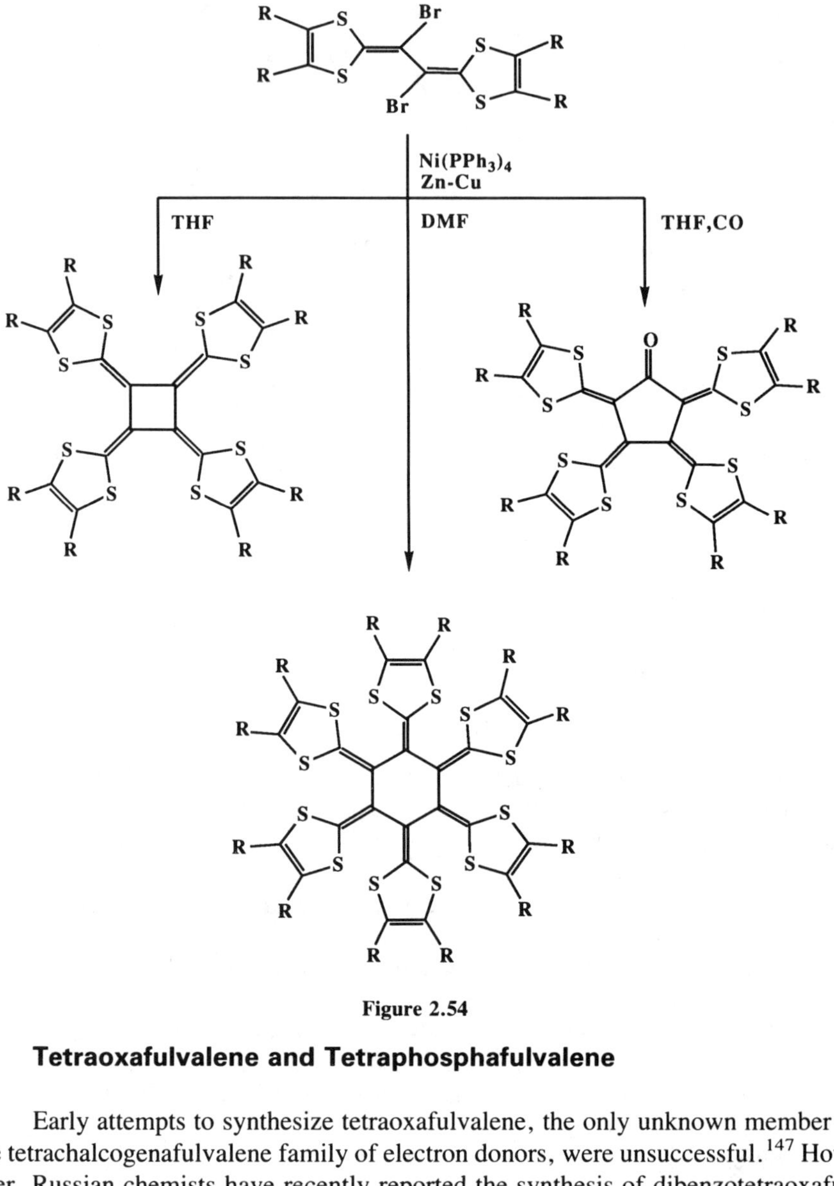

Figure 2.54

Tetraoxafulvalene and Tetraphosphafulvalene

Early attempts to synthesize tetraoxafulvalene, the only unknown member of the tetrachalcogenafulvalene family of electron donors, were unsuccessful.[147] However, Russian chemists have recently reported the synthesis of dibenzotetraoxafulvalene by acid catalyzed reaction of 2-alkoxy-4,5-benzo-1,3-dioxolane.[148] The reaction is believed to involve a cyclic dialkoxycarbene intermediate, which has also been trapped by triphenylphosphine. Cyclic dialkoxycarbenes are normally known to lose carbon dioxide and form olefines (Corey-Winter reation), but in this case, the presence of the benzo substituent is believed to have altered the course of the reaction to give rise to dimerization. (See Fig. 2.57.)

Another set of interesting and related compounds which have been recently reported are the tetraphosphafulvalene derivatives.[149] (See Fig. 2.58.)

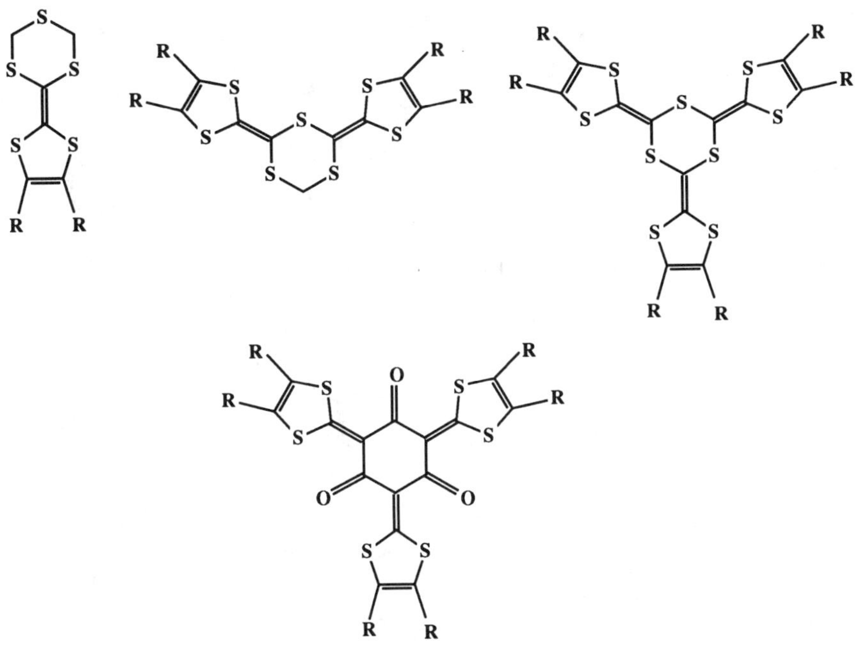

Figure 2.55 Structures of polyfulvathianes.

DESIGN AND SYNTHESIS OF ORGANIC CONDUCTING SALTS

Now that we have discussed the synthesis of organic electron-donor molecules, we can turn to the preparation of the organic conductors themselves. The first well-documented report of an organic conductor can be traced back to 1954 when (perylene)(Br)$_x$ was prepared by doping perylene donor molecules with bromine vapor.[150] Bromine is a strong oxidant, so that charge transfer occurred between the donor molecule and the acceptor. The resulting product was nonstoichiometric, unstable, and it is a semiconductor. Various charge-transfer complexes have been prepared since then with significantly improved stability and conductive properties. The general synthetic procedure involves controlled mixing of neutral donor and acceptor molecules. The most representative example is the first organic metal, TTF-TCNQ, discovered in 1973, where TTF is tetrathiafulvalene and TCNQ is tetracyanoquinodimethane.[151, 152] The TCNQ molecule with its four cyano groups was designed to stabilize negative charges. The TTF molecule with its four sulfur atoms, on the other hand, tends to donate electrons (1 or 2 electrons) in order to fulfill the (4n + 2) Huckel's rule. Many important guidelines for the design of organic metals have been set based on this class of material.[153] Two major criteria for obtaining high-electrical conductivity salts involve segregated molecular stacks in the crystal packing and a noninteger amount of charge transfer between donor and acceptor species. While crystal quality can be controlled by standard experimental conditions, such as diffusion rate, concentration, temperature, solvent, and

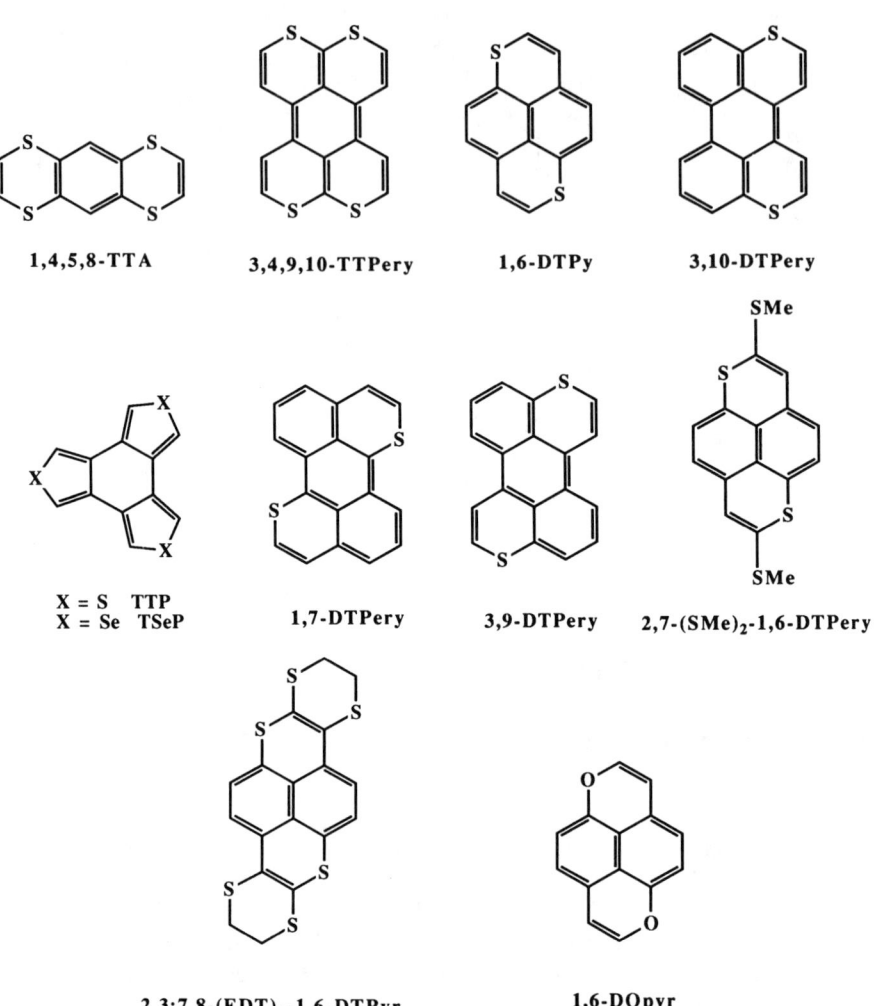

Figure 2.56 Structures of polychalcogenaarenes.

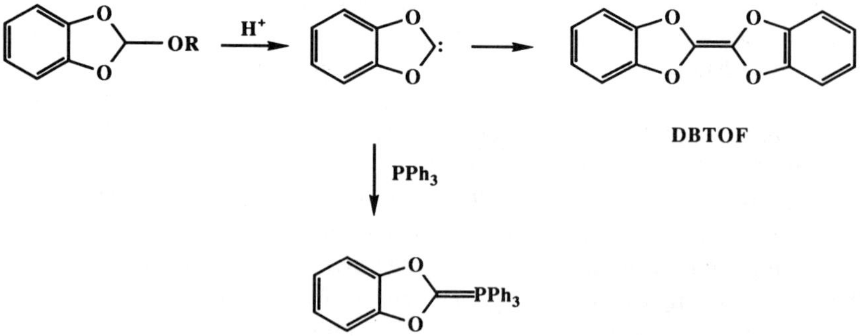

Figure 2.57

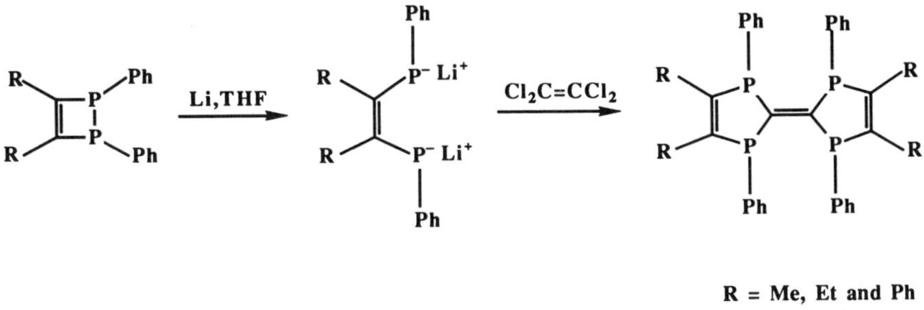

Figure 2.58

R = Me, Et and Ph

so on, the two aforementioned criteria are usually governed by the chemical nature of the donor and acceptor species. In order to satisfy the two previous criteria, and to synthesize new organic metals, many conducting derivatives based on TTF and TCNQ have been prepared. At the same time, a slight modification was made to the simple diffusion procedure, that is, the donor molecules were initially oxidized with a chemical oxidant followed by the addition of acceptor anions, so that a variety of simple nonoxidizing anions such as BF_4^-, SCN^-, Br^-, could also be introduced.[154] The procedure also worked well when the oxidation of the donor molecule was carried out by the use of the electrochemical method instead of chemical oxidation, as exemplified by the preparation of $(TSeF)(Br)_{0.8}$.[155, 156] The electrochemical method has the additional advantage of providing a finely controlled oxidation potential. The two-step synthesis was further refined to a one-step process called "electrocrystallization," wherein the donor molecules are oxidized electrochemically in the presence of a supporting electrolyte prepared from the desired anions. The use of electrocrystallization led to the preparation of high quality single crystals of the first organic superconductor, $(TMTSF)_2X$ in 1979, where TMTSF is tetramethyltetraselenafulvalene and X is a simple monovalent anion such as PF_6^-, ClO_4^-.[157] Electrocrystallization has now become the standard procedure worldwide for the synthesis of new organic conductors. In this section, a detailed procedure for constant current electrocrystallization, which frequently results in the formation of high-quality crystals, is provided. The known BEDT-TTF based organic conductors and superconductors, that are usually structurally two-dimensional in nature in contrast to the previous TTF-TCNQ type complexes, are grouped according to the different kind of anions used, where BEDT-TTF is bis(ethylenedithio)tetrathiafulvalene (abbreviated ET). In each category, the synthesized complexes are discussed briefly according to their physical properties and band electronic structures which form the basis for the rational design of future organic conductors as well as superconductors.

EXPERIMENTAL PROCEDURES

The crystal growing apparatus for electrocrystallization shown in Fig. 2.59 consists of a glass H-cell (15 mL or 45 mL capacity) with an ultrafine porosity glass frit and two platinum wire electrodes.[158] One end of the platinum wire electrode (0.6 to 2

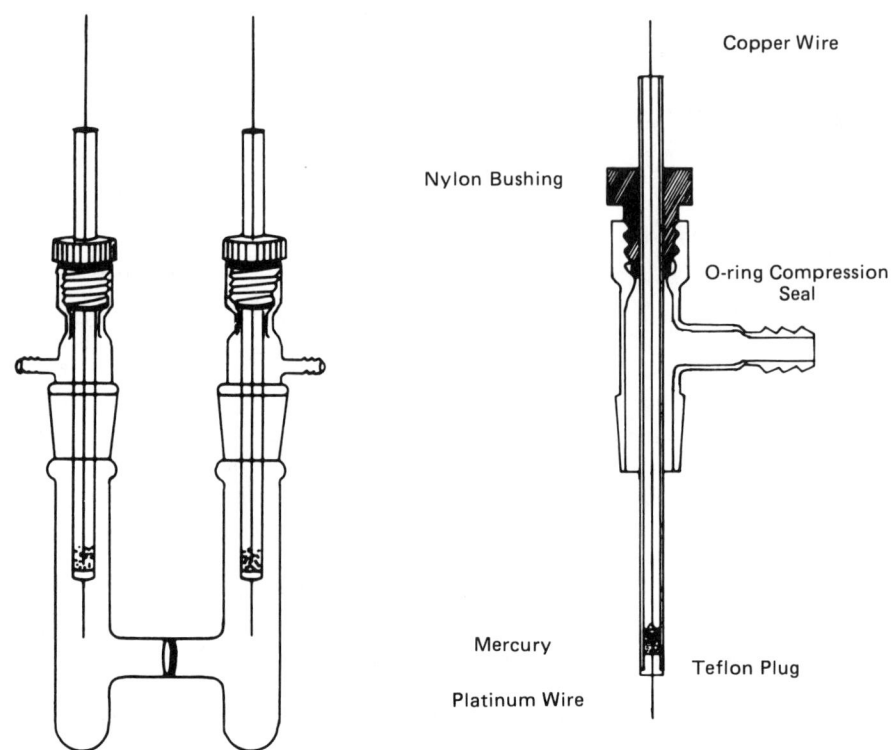

Figure 2.59

mm diameter, 3 cm long) is sealed in a glass tube and connected to an external power supply through a copper wire, and a few drops of mercury are used for the contact medium. A two electrode cell, separated by a frit, is commonly used for synthetic purposes in order to avoid any possible chemical cross contamination from the reference electrode during long term crystal growth. The H-cell is cleaned after each use by soaking in a fresh aqua regia bath overnight. To rinse the cell, distilled water is drawn through the frit several times from both directions by use of a vacuum pump, and the procedure is repeated with absolute methanol. The cell is then stored in an oven to completely remove water.

The key factors to the growth of large single crystals include purifying and drying the solvents and chemicals to the highest possible degree, the preparation of the electrodes prior to electrocrystallization (vide infra),[159, 160] and careful control of the electrolytic growth conditions. The anode used for crystal growth by anodic oxidation is connected to the negative end of a 3V battery, and the cathode is connected to a timer switch, then to the positive lead after the electrodes are immersed in a 1 M H_2SO_4 solution. A potential is applied for 4 minutes, and gases should evolve from both electrodes, after which time the polarity is reversed for 4 minutes. Finally, the polarity is switched back to that originally used for an additional 8-minute period. The electrodes are then rinsed and soaked in distilled water, and then absolute methanol for 10 minutes each followed by drying with a heat gun.

Typical solution concentrations for crystal growth are as follows: a 5mL solution containing 8 to 10 mg of the donor molecule ET (2mM) is added to the anode compartment of the cell, and a 10 mL 0.1 M NBu$_4$$X$ (X = desired anion) solution is added to the cathode compartment and the anode compartment, to level both sides. The cell is assembled either in a dry box, if the supporting anions are air-sensitive, or purged with argon before the electrodes are inserted. The optimum concentration for crystal growth varies from donor molecule to donor molecule. A concentrated anion or electrolyte solution, and nearly saturated donor molecule solution, is desirable so that the concentration is larger than the solubility product (K_{sp}) of the resulting charge-transfer complex, and this is often easily attainable. Initial nucleation and subsequent growth of high quality crystals is usually favorable under these conditions.

Finally, the anode and cathode of the H-cell assembly are connected to an external constant current power supply.[161] The current density, which is calculated as current divided by the surface area of the platinum anode, is set as low as possible to initiate crystal growth, e.g., at 0.05 to 1.5 μA/cm^2. Initial seed crystals usually appear after 1 to 3 days. The current density is kept at the lowest possible level once seed crystals begin growing in order to maintain very slow crystal growth. Low-current densities are desirable for large single crystal growth. The low-electrochemical potential associated with the low-current density is also desirable in order to avoid generating divalent donor cations during electrocrystallization. The crystals are harvested on depletion of the organic donor molecule, or at the sight of notable solution color change, or voltage drop. The growth period varies from a week or two weeks to as long as many months.

SURVEY OF ET-BASED CONDUCTORS

The majority of the (ET)$_m$$X_n$ salts are prepared by use of electrocrystallization techniques. In a few cases in which oxidizing anions such as polyhalides, cupric halides, mercuric halides, and so on, are involved, a growth procedure based on chemical oxidation has been applied. Mechanistic studies of electrocrystallization reactions have been carried out with videomicroscopy and cyclic voltammetry for (TMTSF)$_2$PF$_6$ and (FA)$_2$PF$_6$ (FA is fluoranthene), respectively.[162, 163] In the case of fluoranthene crystal growth, the mechanism involves the initial oxidation of the donor molecule to generate a radical-cation (FA$^{\cdot+}$), which is followed by the formation of radical dimer-cation (FA$_2^{\cdot+}$). The dimer cation is further complexed with the anion to form the final seed crystals. It was also noted in the same study that the mechanism involved in the electrocrystallization of perylene complexes may be different.[163] The mechanism for the formation of (ET)$_m$$X_n$ salts is expected to be more complex since different stoichiometries, or crystal packing motifs, are often observed in the final products (vide infra).

ET Salts with Octahedral, Tetrahedral, and Simple Anions

The crystallographic unit cell volumes and electrical properties of ET-based materials with octahedral, tetrahedral, and simple anions are summarized in Table 2.1. The α- and β-(ET)$_2$PF$_6$ complexes are among the earliest reported ET-based materials due to the previous discovery of organic superconductivity in (TMTSF)$_2$PF$_6$. These complexes are prepared by use of the electrocrystallization procedure from ET and the corresponding tetrabutylammonium salts. The α- and β-phases denote two different structural phases and do not imply the same crystal packing as that of the well-known superconducting β-(ET)$_2$X (X is a linear anion class of materials). The β-(ET)$_2$PF$_6$ and (ET)$_2$X (X$^-$ = AsF$_6^-$, SbF$_6^-$) salts have very similar donor molecule packing motifs and all have metal-to-insulator (MI) transitions near room temperature. Band electronic structure calculations based on the crystal structure of the β-(ET)$_2$PF$_6$ salt reveal one-dimensional character which usually leads to either charge-density wave (CDW) or spin-density wave (SDW) transitions.[181] Because of the high MI transition temperature of these salts, no other ET salts containing octahedral anions have been reported since 1984.

Multiple stoichiometries of the derived ET salts with tetrahedral anions are obtained simultaneously during electrocrystallization. Three major types, that is, 3:2, 2:1, and 2:1:0.5, are listed in Table 2.1. In order to help understand the

TABLE 2.1 ET-Based Complexes with Octahedral, Tetrahedral, and Simple Anions

Compound	Unit Cell Vol (Å^3)	T_{MI}(K)	T_c(K)	Reference
α-(ET)$_2$PF$_6$	794.3	>300		164
β-(ET)$_2$PF$_6$	3255.4	297		165
(ET)$_2$AsF$_6$	3274	264		166
(ET)$_2$SbF$_6$	3349	273		166
(ET)$_2$ReO$_4$	1565	81	2(4 kbar)	167
(ET)$_2$BrO$_4$	1589	180		168
(ET)$_2$InBr$_4$	1819.9	>300		169
(ET)$_2$GaI$_4$	1907	>300		170
(ET)$_2$InI$_4$	1941.8	>300		170
(ET)$_2$BrO$_4$(TCE)$_{0.5}$	1668.4			171
(ET)$_2$ClO$_4$(TCE)$_{0.5}$	1684	16		172
(ET)$_2$FSO$_3$(TCE)$_{0.5}$	1699.8	90		173
(ET)$_2$ClO$_4$(Dioxane)	1814	>300		174
γ-(ET)$_3$(FSO$_3$)$_2$	1174.1			175
(ET)$_3$Br$_2$·2(H$_2$O)	1176.0	185		176
γ-(ET)$_3$(HSO$_4$)$_2$	1181.3	130		177
γ-(ET)$_3$(ClO$_4$)$_2$	1182	170		174
γ-(ET)$_3$(BF$_4$)$_2$	1183.9			175
γ-(ET)$_3$(BrO$_4$)$_2$	1213.6	210		171
γ-(ET)$_3$(ReO$_4$)$_2$	1221.7	10		175
γ-(ET)$_3$(IO$_4$)$_2$	1231.4			175
(ET)$_3$Cl$_2$·2(H$_2$O)	2304.3	100	2(16 kbar)	179,180

multiphasic growth behavior, we must consider the relative tetrahedral anion volumes which can be estimated by use of the following equation:

$$\text{Anion volume} = (R_i + 2R_o)^3$$

where R_i and R_o are the ionic radii (in Å) of the inner (central) and outer atom, respectively. The estimated anion volumes for tetrahedral anions are listed in Table 2.2, and the result is a nearly linear correlation with respect to the unit cell volumes of the isostructural $(ET)_3X_2$ salts, where X is a tetrahedral anion.[182] The nearly linear correlation suggests that the estimated anion volumes are good indicators of unit cell volume when placed on a relative scale.

When comparing Tables 2.1 and 2.2, a simple observation can be made, that is, small anions such as FSO_3^-, HSO_4^-, BF_4^-, and ClO_4^-, lead to either 3:2 or 2:1:0.5 stoichiometries. Larger anions ($GaCl_4^-$ to InI_4^-) lead to a 2:1 stoichiometry exclusively, while the TlI_4^- anion leads to a 2:1:1, $(ET)_2(TlI_4)(I_3)$, complex. Anions with intermediate volumes, such as BrO_4^-, tend to form all three different stoichiometries. The observation can be understood in terms of space filling or molecular packing considerations. In other words, for the aforementioned mixed-valent complexes, the number of donor molecules per anion tends to be larger for the larger anions. For anions with intermediate volumes, for example, ClO_4^-, BrO_4^-, and ReO_4^-, the controlling factors which favor a particular stoichiometry, are not known.

The electrical properties of $(ET)_3X_2$ salts are basically semimetallic. The temperature dependent conductivity is usually rather flat at high temperature and drops quickly below metal-insulator transitions that occur approximately between 100 to 200 K. Results from the band structure calculations on these salts are rather complicated and consist of multiple Fermi surfaces (see Chap. 8 for a discussion). The conductivity arises from partially empty hole pockets and partially filled electron pockets. In a structurally different 3:2 salt, $(ET)_3Cl_2 \cdot 2H_2O$, the conductivity arises from a similar origin; however, superconductivity is observed at 2 K under high pressure. This is the first example that indicates that a 3:2 stoichiometry can also lead to superconductivity.[183] The $(ET)_3Br_2 \cdot 2H_2O$ salt, which is isostructural to all other 3:2 salts in Table 2.1, has also been synthesized but did not exhibit superconductivity. At present, no simple correlation exists between the occurrence of superconductivity and the structure packing motif for the 3:2 salts. The conduc-

TABLE 2.2 The Estimated Relative Anion Volumes for Tetrahedral Anions

Anion	Volume (Å^3)	Anion	Volume (Å^3)
FSO_3^-	14.7	$InCl_4^-$	68.9
HSO_4^-	15.3	$TlCl_4^-$	75.7
BF_4^-	16.4	$GaBr_4^-$	76.8
ClO_4^-	18.4	$InBr_4^-$	85.2
BrO_4^-	22.2	$TlBr_4^-$	93.0
ReO_4^-	25.4	GaI_4^-	105.8
IO_4^-	26.5	InI_4^-	116.2
$GaCl_4^-$	61.6	TlI_4^-	125.8

tivities of 2:1 or 2:1:0.5 salts with intermediate size anions, are metallic. The related electronic structures of these salts as derived from band electronic structure calculations again consist of multiple Fermi surfaces and indicate both one- and two-dimensional character in the derived system. The $(ET)_2ReO_4$ salt was the first ET complex that was reported to be superconductive under pressure. Many other ET salts containing tetrahedral anions have been prepared, but none have exhibited superconductivity. The MX_4^- (M^{3+} = Ga^{3+}, In^{3+}, Tl^{3+}) type anion volumes are estimated to be very large, and the electrical properties of these $(ET)_2MX_4$ salts are all reported to be semiconductive.[182] In summary, smaller tetrahedral anions are more promising in terms of obtaining metallic salts. However, multiple stoichiometries of the derived salts are to be expected, and no clear correlation between unit cell volumes and MI transition temperatures can be drawn at this time.

ET Salts with Linear Anions

ET-based materials with linear polyhalide anions have been the focus of synmetal research since the first report of ambient pressure superconductivity in β-$(ET)_2I_3$ at 1.4 K.[184] To date, there are more than a dozen ET/I_3^- salts that have been reported with superconducting critical temperatures (T_c's) ranging from 1.4 to 8 K. The salts containing triiodide (I_3^-) anions are listed in Table 2.3 along with the methods of preparation, unit cell volumes, T_c's, and room temperature electron spin resonance (ESR) peak-to-peak linewidths. (The latter being a common means of distinguishing the various different crystal phases.)

As shown in Table 2.3, five phases $(ET)_2I_3$ have been reported under electrocrystallization growth conditions. The α and β-phases are the most commonly observed. It has been reported that when highly purified tetrahydrofuran (THF) or chlorobenzene are used as solvents, under low-current density (on the order of 1 $\mu A/cm^2$), the pure superconducting β phase can be selectively prepared. When the solvent is purposely doped with a trace amount (approximately 5% with respect to

TABLE 2.3 Summary of $(ET)_m(I_3)_n$ Salts

Phase	Synthesis	Unit all Vol ($Å^3$)	$T_c(k)$	ΔHp-$p(G)$	Reference
α-$(ET)_2I_3$	EC[a]	1698	T_{MI}135K	70–110	185
β-$(ET)_2I_3$	EC	855.9	1.4	18–24	184, 186
θ-$(ET)_2I_3$	EC	3386	3.6	60–80	187, 188
κ-$(ET)_2I_3$	EC	1687.6	3.6		189
β_D-$(ET)_2I_3$	EC	854.9	T_{MI}140K	22–32	190
δ-$(ET)I_3$	CO[b]	12736	T_{MI}130K		191
λ_D-$(ET)_2I_3$	CO	3404	T_{MI}230K		190
ϵ-$(ET)_2(I_3)(I_8)_{0.5}$	CO	4211	Metal[c]		192
γ-$(ET)_3(I_3)_{2.5}$	CO	6812	2.5		191
η-$(ET)_2(I_3)(I_5)$	CO	4309	Metal[c]		193
ζ-$(ET)_2(I_2)(I_8)$	CO	2315	—		194

[a] Electrocrystallization.
[b] Chemical Oxidation.
[c] Metallic to 1.5 K.

ET) of water, acid, or chemical oxidants, the α phase can be synthesized with better than 95 percent phase selectivity.[195] The α phase can also be prepared by use of high-current density (e.g., 10 μA/cm^2). These experimental observations can be understood in terms of thermodynamic versus kinetic stability. The β phase is known to be thermodynamically more stable[196] as is evidenced by the novel α (T_{MI} = 135 K) to β*-(ET)$_2$I$_3$ (T_c = 8 K) thermal conversion (vide infra). The α phase, on the other hand, is the kinetically favored product. This has been demonstrated by dissolving a single crystal of β-(ET)$_2$I$_3$ in boiling trichloroethane (TCE), and quenching the hot solution to room temperature after filtration. The resulting (ET)$_2$I$_3$ microcrystalline product when analyzed by use of ESR spectroscopy is found to contain 93 percent of the α phase.[195]

The θ-(ET)$_2$I$_3$ phase crystals were prepared from the electrocrystallization of ET and (n-Bu)$_4$NI$_3$ with 5 percent of (n-Bu)$_4$NAuI$_2$ in THF.[187] The θ phase was originally formulated as θ-(ET)$_2$(I$_3$)$_{1-x}$(AuI$_2$)$_x$ (x < 0.02). It was soon discovered that the dopant can also be replaced by (n-Bu)$_4$NI$_2$Br. Occasionally, the θ-phase can even be detected from pure (n-Bu)$_4$NI$_3$ without dopant. It was also reported that not all the θ phase crystals show superconductivity.[197] We have prepared the θ phase crystals quite unexpectedly: when the electrocrystallization of ET with CdI$_2$ and (n-Bu)$_4$NI was carried out in TCE solvent, a semiconductive compound (ET)$_4$(Cd$_2$I$_6$) was prepared. However, when the electrocrystallization was carried out in THF, it led to both θ and α-(ET)$_2$I$_3$ without CdI$_2$ incorporation. Apparently, the dopants in the solution mixture play certain roles during electrocrystallization, because the θ phase salt is not commonly observed. Our θ phase crystals are metallic to temperatures below 10 K, but do not show superconductivity in a rf penetration depth measurement (see Chap. 4) carried out at 0.6 K. In addition, the κ-(ET)$_2$I$_3$ crystals were also prepared by electrocrystallization from a mixture of (n-Bu)$_4$NI$_3$ and (n-Bu)$_4$NAuI$_2$ in THF.[189] It was reported to coexist with α and θ-phases.

It should be noted at this point that crystals of (ET)$_2$I$_3$ prepared by electrocrystallization, which are α and β mixtures *in themselves*, have been identified.[198] These crystals were found growing on the bottom of the anodic portion of the H-cell. They can be identified as α and β mixtures by use of ESR lineshape analysis of each "single crystal." One interesting phenomenon associated with these crystals are their abnormal T_c's as determined from rf penetration depth measurements, which range from 2.5 to 6.9 K, in contrast to 1.4 K normally found in β-(ET)$_2$I$_3$. The high onset temperatures are usually accompanied by fairly broad transition widths that are reminiscent of the thermally converted α$_t$-(ET)$_2$I$_3$. It appears that the α, β, and β*(or α$_t$)-(ET)$_2$I$_3$ phases may coexist in the same "single" crystal, thereby showing a range of T_c's. This behavior is also observed when the electrocrystallization of ET with the I$_3^-$ anion is carried out in the presence of small amount of bis(propylenedithio)tetrathiafulvalene (PT) or methylenedithioethylenedithio-tetrathiafulvalene (MET) donor molecules.[199]

A typical procedure for employing the chemical oxidation technique is to expose an ET solution to iodine vapor.[191–194] The resulting charge-transfer complexes simply crystallize in the container because of their highly insoluble nature. In such experiments, many parameters can be varied, for instance, solvents, concentration of the donor, concentration of the oxidant, reaction temperature, cooling rate, with or without I_3^- anion added, and so on. Two most commonly used solvents for chemical oxidations are trichloroethane (TCE) and benzonitrile. Six different phases in the ET:I system have been structurally characterized and several phases are usually observed simultaneously even under one set of experimental chemical oxidation conditions. Systematic studies of chemical oxidation reactions are needed in order to determine how to prepare these phases selectively. The ϵ-$(ET)_2(I_3)(I_8)_{0.5}$ and ζ-$(ET)_2(I_2)(I_8)$ phases have been reported to lose iodine readily at elevated temperature to form the $(ET)_2I_3$ phase in its high-T_c state ($T_c = 6 \sim 7$ K by resistive measurements).[194, 200]

The ET salts with linear anions, their unit cell volumes, different structural phases and the corresponding physical properties are listed in Table 2.4. As shown in Table 2.4, the longer anions (I_3^- to IBr_2^-) tend to favor the α and β-$(ET)_2X$ packing motifs, while the shorter anions ($BrICl^-$ to $AuCl_2^-$) favor the α', β', and β'' structural motifs. Among all the aforementioned packing motifs, the β type appears to be the most desirable one for finding superconductivity. Four β-type superconductors have been reported, that is, β-$(ET)_2X$ ($X = I_3^-$, low-T_c state 1.4 K,[184] the high-T_c state at 8 K;[214–216] IBr_2^-, 2.8 K;[205] AuI_2^-, 4.98 K[204]). For the case of the I_3^- anion, the growth of the α or β morphology can be controlled by choosing the appropriate solvent during electrocrystallization (vide supra). While the same strategy works well with the AuI_2^- anion, it simply does not work with the IBr_2^- anion. In other words, electrocrystallization of ET with the IBr_2^- anion in THF solvent invariably leads to a mixture of α and β phases even under very low current density. The α-$(ET)_2IBr_2$ undergoes thermal conversion at 416 K to β-$(ET)_2IBr_2$.[217] To our knowledge, the only way to prepare pure β-$(ET)_2IBr_2$ is to thermally convert an α and β mixture.

Among the α', β', and β''-phases, the electrical properties of the first two are semiconductive. The β'-phase also shows a magnetic phase transition at temperatures below 30 K. The β''-phases, which exhibit metallic behavior, seems to be a minor component among the product mixture as judged from their relatively recent

TABLE 2.4 Summary of $(ET)_2X$ Salts (X = Linear Anion)

X	Anion Length (Å)	Unit Cell Volumes (Å³) and Electrical Properties (in parentheses if known)						Reference
		$\alpha(T_{MI}K)$	$\beta(T_cK)$	α'	β'		$\beta''(T_{MI}K)$	
I_3^-	10.14	1698(135)	855.9(1.4)					184–186,201
I_2Br^-	9.72	1687.9	842.3[a]					202,203
AuI_2^-	9.42		845.2(4.98)					204
IBr_2^-	9.30	1652[b]	828.7(2.8)					205
$BrICl^-$	9.01	1647[b]			821.3[b]			206,207
$IAuBr^-$	8.98		832.3		1689.2[b]		826.3	208,209
ICl_2^-	8.72				814.3[b]		814.3(10)	207,210
$AuBr_2^-$	8.70				1662[b]		813.7(10)	211,212
$AuCl_2^-$	8.14				800.7[b]			213

[a] Metallic to 0.5K but not superconductive.
[b] Semiconductor.

discovery. (The β″ salts are potential candidates for superconductors under high pressure.) One of the β″ salts, $(ET)_2AuBr_2$, is metallic to 1.4 K under 5.6 kbar pressure.

Based on the known $(ET)_2X$ salts, where X is a linear anion, new linear anions longer than IBr_2^- are promising in terms of growing new β phase materials. Two new linear anions have been prepared and characterized by the use of crown ethers. The first is $(K \cdot 18\text{-crown-}6)^+ CuI_2^-$ which is prepared from CuI and crown ether in saturated aqueous KI solution.[218] Under similar reaction conditions, AgI leads to $[(K \cdot 18\text{-crown-}6)]_2Ag_4I_6$ as the major product as well as $(K \cdot 18\text{-crown-}6)AgI_2$. Both the CuI_2^- and AgI_2^- anions have been characterized with single crystal X-ray diffraction techniques and found to be linear. Electrocrystallization of ET by use of these two new linear anions, however, led to the first "polymeric anion" ET salts, $(ET)_2Cu_5I_6$ and $(ET)_3Ag_{6.4}I_8$, respectively.[219, 220] Both of these ET salts are metallic and have metal-insulator transitions, and contain a polymeric metal-halide anion network. In this case, although the less common linear CuI_2^- and AgI_2^- anions can be stabilized by use of the $(K \cdot 18\text{-crown-}6)^+$ cation, they rearrange in solution during electrocrystallization to form polymeric anion networks with tetrahedral interstitial sites which appear to be the most favorable configurations for Cu(I) and Ag(I).

Two slightly shorter linear anions, $Ag(CN)_2^-$ and $Au(CN)_2^-$, have also been used to prepare new ET salts.[211] They both lead to the semiconductive α′-$(ET)_2X$ packing motif, which fits in Table 2.4 as one would expect for the short linear anions. The $Ag(CN)_2^-$ anion further rearranges during electrocrystallization to form $(ET)Ag_4(CN)_5$, another ET salt with a polymeric anion.[221] It should be pointed out that although short linear anions result in formation of mostly semiconductive α′ or β′-$(ET)_2X$ salts, they work very well with the small unsymmetric donor DMET, dimethyl(ethylenedithio)diselenadithiafulvalene. As shown in Table 2.5, there are seven DMET based superconductors reported.[222] All of these materials contain linear anions. Four of them with the shorter linear anions ($AuBr_2^-$, $Au(CN)_2^-$, and $AuCl_2^-$) actually show slightly higher superconducting T_c's than those containing the longer anions.

ET Salts with Planar Anions

Another approach to preparing new, and hopefully β-type ET salts, has been to pursue the use of planar anions. Since the linear $AuCl_2^-$ anion generates a

TABLE 2.5   Properties of $(DMET)_2X$ Superconductors

Anion Unit	Cell Vol($Å^3$)	T_c(K)	Pressure (kbar)
I_3^-	794.8	0.47	
AuI_2^-	799.5	0.55	5.0
IBr_2^-	791.7	0.58	
$AuBr_2^-$	769	1.9	
$AuBr_2^-$	759	1.0	1.5
$Au(CN)_2^-$	766.0	0.80	5.0
$AuCl_2^-$	761.5	0.83	

β'-type packing motif, a larger and planar $AuCl_4^-$ anion was attempted in electrocrystallizations. When freshly prepared $AuCl_4^-$ anion in THF was used, a microcrystalline product, $(ET)_2AuCl_4$, was obtained and structurally characterized.[223] The crystal structure was β-like in packing motif, and it appeared to be a metal from room temperature to 120 K based on ESR measurements. However, a destructive and irreversible phase transition occurred below 120 K, and crystals became polycrystalline. When the electrocrystallization was carried out in benzonitrile, a semiconductive product, $(ET)(AuCl_2)(AuCl_4)$, was isolated because of a spontaneous redox reaction between ET and $AuCl_4^-$ anion.[224] The $AuBr_2Cl_2^-$ anion also produces a semiconductive 1:1 salt, $(ET)AuBr_2Cl_2$.[225] The $Au(CN)_2Cl_2^-$ anion gives rise to $(ET)_2Au(CN)_2Cl_2$[223] which structurally resembles $(ET)_2SbF_6$. Its conductive properties are also similar to those of the octahedral anion containing ET salts, that is, showing a metal-to-insulator transition near room temperature.

ET Salts with Polymeric Anions

As mentioned previously (vide supra), the first ET salts with polymeric anions were $(ET)_3Ag_{6.4}I_8$, $(ET)Ag_4(CN)_5$, and $(ET)_2Cu_5I_6$, which are derived from the linear anions AgI_2^-, $Ag(CN)_2^-$, and CuI_2^- or CuI/LiI mixtures, respectively.[219-222] Although the synthesis of these compounds is generally quite reproducible, and in the cases of the $(ET)_3Ag_{6.4}I_8$ and $(ET)_2Cu_5I_6$ salts only one major product has been reported for each salt, the discovery of these salts was totally unexpected. We do not know the mechanism behind the anion rearrangement and how the polymeric anions were formed. Thus, we have very limited predictability as far as what will be the stoichiometry or packing motif of the final product whenever polymeric anions are formed. This situation is a great contrast to that of the ET salts with polyhalide linear anions. The further possibility of preparing new polymeric anions is enormous, and the conductive properties of the products prepared to date are usually good as evidenced by the large number of metals and superconductors that now exist. We will focus first on the κ-phase salts presently known.

As shown in Table 2.6, among the eight ET based κ-phase materials, seven

TABLE 2.6 Summary of κ-Phase Salts

Compound	Unit Cell Vol($Å^3$)	T_c(K)	Reference
$(ET)_2Ag(CN)_2·H_2O$	1651.2	5.0	227,150
$(ET)_2I_3$	1687.6	3.6	189
$(ET)_2Cu(NCS)_2$	1694.8	10.4	226
$(ET)_2Cu[N(CN)_2]Cl$	3299	12.8 (0.3 kbar)	245
$(ET)_2Cu[N(CN)_2]Br$	3317	11.6	246
$(ET)_2Cu[N(CN)_2]I$	3408	T_{MI} 150	247
$(ET)_4Hg_3Cl_8$	3478	1.8(12 kbar) 5.3(29 kbar)	228,229
$(ET)_4Hg_{2.89}Br_8$	3624	4.3(ambient) 6.7(3.5 kbar)	230,178
$(MDT-TTF)_2AuI_2$	2438.1	4.5	231
$(DMET)_2AuBr_2$	769	1.9	222
$(MT)_2Au(CN)_2$	1390	T_{MI}76	232

are derived from polymeric anions. A common procedure is to generate these polymeric anions *in situ,* as described with the following equation:

$$MX_n + X^- \rightarrow MX_{n+1}^- \qquad X^- = Cl^-, Br^-, CN^-, SCN^-,$$

where MX_n is a coordinatively unsaturated neutral metal halide and X is a halide or pseudohalide species. The κ-$(ET)_2Cu(NCS)_2$ salt was prepared from a novel technique[220, 226] by electrocrystallization of ET in the presence of CuSCN, KSCN, and 18-crown-6 in TCE. The CuSCN is extremely insoluble, and a crown ether is used to encapsulate the potassium cation in order to generate a "$Cu(NCS)_2^-$" anion which is soluble. Several factors are now known that improve the crystal quality, that is, very low-current density (0.05 μA/cm^2), 10 percent ethanol added to the solvent, and crystal growth under argon. The role of ethanol is not clear but may be related to the enhanced solubility of CuSCN, or species derived from it. Two other minor phases have also been reported, $(ET)Cu_2(NCS)_3$[233] and α-$(ET)_2Cu(NCS)_2$.[234] The former impurity phase is an insulator and was observed occasionally growing on the bottom of the anode compartment during electrocrystallization. The latter phase has not been fully characterized and undergoes a MI transition at 200 K. Hydrostatic pressure applied to κ-$(ET)_2Cu(NCS)_2$ greatly depresses its T_c.[235] The implication of this finding is that if one can expand the unit cell volume, the T_c should thereby increase. Simple replacement of CuSCN with AgSCN has led to a different phase, $(ET)Ag_{1.6}(SCN)_2$ which is a semiconductor.[233] Electrocrystallization with the $Au(SCN)_2^-$ anion has only generated an intractable material. Various efforts have been made to substitute SCN^- with $SeCN^-$ in the hope of further lengthening the anion, and thereby increasing the unit cell volume and raising T_c in the κ-phase materials, but all such attempts have failed. Electrocrystallization of ET with CuX (X = Cl, Br, and I) and $(PPh_4)N(CN)_2$ gives rise to three isostructural salts, κ-$(ET)_2Cu[N(CN)_2]X$ (X = Cl, Br, and I).[245-247] The $Cu[N(CN)_2]Br^-$ salt is an ambient pressure superconductor (T_c = 11.6 K). The $Cu[N(CN)_2]Cl^-$ salt superconducts at 12.8 K under 0.3 kbar pressure and it is the highest T_c organic superconductor known.

Electrocrystallization of ET with either $HgCl_2$ and $(n-Bu)_4NCl$ or $(n-Bu)_4NHgCl_3$ produces two phases, $(ET)_3Hg_2Cl_6$[236] and κ-$(ET)_4Hg_3Cl_8$.[228, 229] The former salt consists of a segregated $Hg_2Cl_6^{2-}$ dimer dianion and the latter a polymeric linear $Hg_3Cl_8^{2-}$ chain. The procedure for selective preparation of these two phases has not been reported. The chemistry of ET with the $HgBr_2/Br^-$ or $HgBr_3^-$ systems is far more complicated. Up to seven phases have been characterized, as listed in Table 2.7. The experimental conditions utilized for the synthesis of these compounds are similar except for slight variations in the supporting electrolytes and solvents. The resulting anions vary from the planar $HgBr_3^-$, tetrahedral $HgBr_4^{2-}$, tetrahedral dimer $Hg_2Br_6^{2-}$, the combination of $HgBr_3^-$ and $HgBr_4^{2-}$, and polymeric linear-chain $Hg_3Br_8^{2-}$. The first five salts are semiconductive, $(ET)_5(HgBr_4)_2(HgBr_3)$ has an MI transition at 120 K, and κ-$(ET)_4Hg_{2.89}Br_8$ is an ambient pressure superconductor with a T_c of 4.3 K. The latter κ-phase salt is a very minor reaction product, and its synthesis is not totally reproducible. However, the inclusion of 10 percent ethanol in the synthesis improves the yield slightly. Addi-

TABLE 2.7 Electrocrystallization of ET-Hg-Br salts

Compound	Electrolyte	Solvent	$CD^a(\mu A/cm^2)$	Reference
(ET)HgBr$_3$	Me$_4$NHgBr$_3$	PhCN		237
(ET)$_2$HgBr$_3$(TCE)	(n-Bu)$_4$NHgBr$_3$	TCE	0.5	238
(ET)$_2$HgBr$_4$(TCE)	(Et$_4$N)$_2$HgBr$_4$/ (n-Bu)$_4$NPF$_6$	TCE	0.8	239
(ET)$_3$(HgBr$_3$)$_2$	(n-Bu)$_4$NHgBr$_3$	THF/CH$_3$CN 1:2	0.5	240
(ET)$_4$(Hg$_2$Br$_6$)(TCE)	HgBr$_2$,(n-Bu)$_4$NBr	TCE	0.5	241
(ET)$_4$Hg$_{2.89}$Br$_8$	(n-Bu)$_4$NHgBr$_3$	TCE	0.5	230
(ET)$_5$(HgBr$_4$)$_2$(HgBr$_3$)	(Et$_4$N)$_2$HgBr$_4$	PhCN		237

tional studies are needed in order to optimize the preparation of κ-(ET)$_4$Hg$_{2.89}$Br$_8$.

In summary, more than half of the existing κ-phase salts are based on polymeric anions, and the others are derived from linear anions. Although these polymeric anions may not be easily applied to other organic donors, the synthetic strategy is certainly applicable. Much effort is still needed in order to develop a controlled procedure to lead to the κ-phase.

Finally, a few other thiocyanate salts have been prepared since the discovery of the κ-(ET)$_2$Cu(NCS)$_2$ superconductor. When the electrocrystallization of ET was carried out with (n-Bu)$_4$NHg(SCN)$_3$ in TCE, a β-like salt, (ET)Hg$_{0.776}$(SCN)$_2$, was obtained.[242] This compound remained metallic to very low temperature but did not show superconductivity at 0.6 K. However, when Hg(SCN)$_2$, KSCN, and 18-crown-6 were used as the electrolyte in TCE/10 percent ethanol, a different material α-(ET)$_2$KHg(SCN)$_4$ was obtained.[243] The incorporation of the potassium cation in the presence of the encapsulating agent, 18-crown-6, was again totally unexpected. The corresponding α-(ET)$_2$(NH$_4$)Hg(SCN)$_4$ salt can be prepared with NH$_4$SCN replacing KSCN. The potassium salts exhibited a resistive hump at low temperature similar to that of the κ-(ET)$_2$Cu(NCS)$_2$, but remained metallic to 1.5 K. The ammonium salt, on the other hand, showed metallic conductivity all the way to 1.5 K and became an ambient pressure superconductor at 1.15 K.[244] The organic donor molecule layers in these two compounds belong to the α-type packing motif, and the anion layer is a polymeric two-dimensional sheet. These two salts offer the possibility of developing structural-property correlations if additional superconductive and isostructural α-(ET)$_2$(M)Hg(SCN)$_4$ salts can be synthesized, where M is a monovalent cation such as Na$^+$, Rb$^+$, Cs$^+$, Tl$^+$, and so on.

REFERENCES

1. Narita, M.; Pittman, Jr. C. U. *Synthesis* **1976**, 489.
2. Bryce, M. R. *Aldrichimica Acta* **1985**, *18*, 73.
3. Krief, A. *Tetrahedron* **1986**, *42*, 1209.
4. Cowan D.; Kini, A. In *The Chemistry of Organic Selenium and Tellurium Compounds*, Volume 2, Edited by S. Patai, Wiley, Chichester, **1987**, p. 463.
5. Schukat, G.; Richter, A. M.; Fanghänel, E. *Sulfur Reports*, **1987**, *7*, 155.
6. Corey, E. J.; Carey, F. A.; Winter, R. A. E. *J. Am. Chem. Soc.* **1965**, *87*, 934.

7. Yoshida, Z.; Kawase, T.; Yoneda, S. *Tetrahedron Lett.* **1975**, 331.
8. Yoneda, S.; Kawase, T.; Inaba, M.; Yoshida, Z. *J. Org. Chem.* **1978**, *43*, 595.
9. Miles, M. G.; Wager, J. S.; Wilson, J. D.; Seidle, A. R. *J. Org. Chem.* **1975**, *40*, 2577.
10. Scherowsky, G.; Weiland, J. *Chem Ber.* **1974**, *107*, 3155.
11. Fanghanel, E.; Richter, A. M.; Schukat, G. *J. Prakt. Chemie* **1984**, *326*, 479.
12. Ueno, Y.; Nakayama, A.; Okawara, M. *J. Am. Chem. Soc.* **1976**, *98*, 7440.
13. Chen, W.; Cava, M. P.; Takassi, M. A.; Metzger, R. M. *ibid*, **1988**, *110*, 7903.
14. Tsujimoto, K.; Okeda, Y.; Ohashi, M. *J. Chem. Soc., Chem. Commun.* **1985**, 1803.
15. Storm, M.; Nespurek, S.; Ryba, O.; Kunbanek, V. *ibid*, **1987**, 696.
16. Le Costumer, J; Mollier, Y. *ibid*, **1980**, 38.
17. Sallé, M.; Gorgues, A.; Fabre, J.-M.; Bechgaard, K.; Jubault, M.; Texier, F. *ibid*, **1989**, 1520.
18. Gorgues, A.; Batail, P.; Le Coq, A. *J. Chem. Soc., Chem. Commun.* **1983**, 405.
19. Alper, H.; Paik, H.-N. *ibid*, **1977**, *42*, 3522.
20. Saeva, F. D.; Morgan, B. P.; Fichtner, M. W.; Haley, N. F. *J. Org. Chem.* **1984**, *49*, 390.
21. Moses, P. R.; Chambers, J. Q. *J. Am. Chem. Soc.* **1974**, *96*, 945.
22. Kini, A. M.; Tytko, S. F.; Hunt, J. E.; Williams, J. M. *Tetrahedron Lett.* **1987**, *28*, 4153.
23. Klingsberg, E.; *J. Am. Chem. Soc.* **1962**, *84*, 3410; *ibid*, **1964**, *86*, 5290.
24. Prinzbach, H.; Berger, H.; Luttringhaus, A. *Angew. Chem. Internat. Edit. Engl.* **1965**, *4*, 435.
25. G. Kiesslich, Dissertation, Universitat Wurzburg, **1968**.
26. Hünig, S.; Kiesslich, G.; Quast, H.; Scheutzov, D. *Liebigs Ann. Chem.* **1973**, 310.
27. Coffen, D. L.; Chambers, J. Q.; Williams, D. R.; Garrett, P. E.; Canfield, N. D. *J. Am. Chem. Soc.* **1971**, *93*, 2258.
28. Wudl, F.; Smith, G. M.; Hufnagel, E. J. *J. Chem. Soc. Chem. Commun.* **1970**, 1453.
29. Engler, E. M.; Patel, V. V. *Tetrahedron Lett.* **1975**, 1259.
30. Scherowsky, G.; Weiland, J. *Justus Liebigs Ann. Chem.* **1974**, 403.
31. Bittner, S.; Moradpour, A.; Kreif, P. *Synthesis* **1989**, 132.
32. Nakayama, J. *Synthesis* **1975**, 38 and references therein.
33. Harzler, H. D. *J. Am. Chem. Soc.* **1973**, *95*, 4379 and references therein.
34. Nakayama, J. *Synthesis* **1975**, 168.
35. Souizi, A.; Robert, A. *Tetrahedron* **1984**, *40*, 1817.
36. Bertho, F.; Robert, A.; Batail, P.; Robin, P. *Tetrahedron* **1990**, *46*, 433.
37. Fanghanel, E.; van Hinh, L.; Schukat, G. *Z. Chem.* **1976**, *16*, 317.
38. Jackson, Y. A.; White, C. L.; Lakshmikantham, M. V.; Cava, M. P. *Tetrahedron Lett.* **1987**, 5635.
39. Brown, C. A.; Miller, R. D.; Lindsay, C. M.; Smith, K. *Tetrahedron Lett.* **1984**, 991.
40. Lindsay, C. M.; Smith, K.; Brown, C. A.; Betterton-Cruz, K. *Tetrahedron Lett.* **1984**, 995.
41. Hurtley, W. R. H.; Smiles, S. *J. Chem. Soc.* **1926**, 2263.
42. Bajwa, G. S.; Berlin, K. D.; Pohl, H. A. *J. Org. Chem.* **1976**, *41*, 145.

43. Mizuno, M.; Cava, M. P. *ibid,* **1978,** *43,* 416.
44. Wudl, F.; Aharon-Shalom, E. *J. Am. Chem. Soc.* **1982,** *104,* 1154.
45. Lerstrup, K.; Talham, D.; Bloch, A. N.; Poehler, T. O.; Cowan, D. O. *J. Chem. Soc., Chem. Commun.* **1982,** 336.
46. Lerstrup, K. A.; Cowan, D. O. *J de Phys.* **1983,** *44,* C3-1247.
47. Lerstrup, K.; Cowan, D. O.; Kistenmacher, T. J. *J. Am. Chem. Soc.* **1984,** *106,* 8303.
48. Lerstrup, K.; Bailey, A.; McCullough, R.; Mays, M.; Cowan, D.; Kistenmacher, T. J. *Synth. Met.* **1987,** *1981,* 647.
49. McCullough, R. D.; Kok, G. B.; Lerstrup, K. A.; Cowan, D. O. *J. Am. Chem. Soc.* **1987,** *109,* 4115.
50. McCullough, R. D.; Mays, M. D.; Bailey, A. B.; Cowan, D. O. *Synth. Met.* **1988,** *27,* B487.
51. McCullough, R. D.; Cowan, D. O. *J. Org. Chem.* **1985,** *50,* 4646.
52. Okada, N.; Saito, G.; Mori, T. *Chem. Lett.* **1986,** 311.
53. Lambert, C.; Christiaens, L.; *Tetrahedron Lett.* **1984,** 833.
54. Hartzler, H. D. *J. Amer. Chem. Soc.* **1970,** *92,* 1412.
55. Rice, J. E.; Okamoto, Y. *J. Org. Chem.* **1981,** *46,* 446.
56. Rice, J. E.; Wojciechowski, P.; Okamoto, Y. *Heterocycles* **1982,** *18,* 191.
57. Nagawa, T.; Zama, Y.; Okamoto, Y. *Bull. Chem. Soc. Jpn.* **1984,** *57,* 2035.
58. Okamoto, Y.; Lee, H. S.; Attarwala, S. T. *J. Org. Chem.* **1985,** *50,* 2788.
59. Melby, L. R.; Hartzler, H. D.; Sheppard, W. A. *J. Org. Chem.* **1974,** *39,* 2456.
60. Hartzler, H. D. *J. Am. Chem. Soc.* **1971,** *93,* 4691.
61. Le Bozec, H.; Gorgues, A.; Dixneuf, P. H. *J. Am. Chem. Soc.* **1978,** *100,* 3946.
62. Ngounda, M.; Le Bozec, H.; Dixneuf, P. *J. Org. Chem.* **1982,** *47,* 4000.
63. Bozec, H. Le; Dixneuf, P. H., *J. Chem. Soc. Chem. Commun.* **1983,** 1462.
64. Bianchini, C.; Meli, A. *J. Chem. Soc. Chem. Commun.* **1983.** 1309.
65. Moinet, A.; Bozec, H. Le; Dixneuf, P. H. *Organometallics* **1989,** *8,* 1493.
66. Schumaker, R. R.; Lee, V. Y., Engler, E. M. *J. Org. Chem.* **1984,** *49,* 564.
67. Krebs, A.; Kimling, H. *Angew. Chem., Int. Ed. Engl.* **1971,** *10,* 509.
68. Fields, E. K.; Meyerrson, S. *Tetrahedron Lett.* **1970,** 629.
69. Nakayama, J. *J. Chem. Soc. Chem. Commun.* **1974,** 166.
70. Available from Aldrich, Strem, and Fluka
71. Wudl, J.; Kaplan, M. L.; Hufnagel, E. J.; Southwick, Jr., E. W. *J. Org. Chem.* **1974,** *39,* 3608.
72. Lerstrup, K.; Johannsen, I.; Jørgensen, M. *Synth. Met.* **1988,** *27,* B9.
73. Green, D. C. *J. Am. Chem., Chem. Commun.* **1977,** 161.
74. Green, D. C.; Allen, R. W. *J. Am. Chem., Chem. Commun.* **1978,** 832.
75. Green, D. C. *J. Org. Chem.* **1979,** *44,* 1476.
76. Dhindsa, A. S.; Bryce, M. R.; Lloyd J. P.; Petty, M. C. *Synth. Met.* **1988,** *27,* B563.
77. Aharon-Shalom, E.; Becker, J. Y.; Bernstein, J.; Bittner, S.; Shaik, S. *Tetrahedron Lett.* **1985,** 2783.
78. Okada, N.; Yamochi, H.; Shinozaki, F.; Oshima, K.; Saito, G. *Chem. Lett* **1986,** 1861.

79. Hsu, S-Y. Chiang, L. Y. *J. Org. Chem.* **1987,** *52,* 3444.
80. Lee, V. Y.; *Synth. Met.* **1987** *20,* 161.
81. Yamochi, H.; Iwasaka, N.; Urayama, H.; Saito, G. *Chem. Lett.* **1987,** 2265.
82. Kini, A. M.; Gates, B. D.; Beno, M. A.; Williams, J. M. *J. Chem. Soc. Chem. Commun.* **1989,** 169.
83. Iwasaka, N.; Saito, G.; Imaeda, K.; Mori, T.; Inokuchi, H. *Chem. Lett.* **1987,** 2399.
84. Rajeswari, S.; Jackson, Y. A.; Cava, M. P. *J. Chem. Soc., Chem. Commun.* **1988,** 1089.
85. Hsu, S-Y.; Chiang, L. Y. *Synth. Met.* **1988,** *27,* B651.
86. Bechgaard, K. presented at ICSM '88, Santa Fe, NM, June 26–July 2, 1988.
87. Fabre, J. M.; Giral, L.; Dupart, E.; Coulon, C.; Manceau, J. P.; Delhaes, P. *J. Chem. Soc., Chem. Commun.* **1983,** 1477.
88. Fabre, J. M.; Torreilles, E.; Gilbert, J. P.; Chanaa, M.; Giral, L. *Tetrahedron Lett.* **1977,** 4033.
89. Fabre, J. M.; Giral, L.; Dupart, E.; Coulon, C.; Delhaes, P. *J. Chem. Soc., Chem. Commun.* **1983,** 426.
90. Morand, J. P.; Brzezinski L.; Manigand, C. *J. Am. Chem. Chem. Commun.* **1986,** 1050.
91. Gonnella, N. C.; Cava, M. P. *J. Org. Chem.* **1978,** *43,* 369.
92. Tatemitsu, H.; Nishikawa, E.; Sakata, Y.; Misumi, S. *J. Am. Chem. Chem. Commun.* **1985,** 106.
93. Giral, L.; Fabre, J. M.; Gouasmia, A. *Tetrahedron Lett.* **1986,** 4315.
94. Marschalk C.; Stumm, C. *Bull. Chem. Soc. France* **1948,** 418.
95. Matsunaga, Y. *J. Chem. Phys.* **1965,** *42,* 2248.
96. Perlstein, J. H.; Ferraris, J. P.; Walatka, Jr., V. V.; Cowan, D. O.; Candela, G. A. *AIP Conf. Proc.* **1973,** *10,* 1494.
97. Wheland, R. C.; Gillson, J. L. *J. Am. Chem. Soc.* **1976,** *98,* 3916.
98. Mihaly, G.; Janossy, A.; Gruner, G. *Solid State Commun.* **1977,** *22,* 771.
99. Isett, L. C.; Perez-Albuerne, E. A. *Solid State Commun.* **1977,** *21,* 433.
100. Wudl, F.; Schafer, D. E.; Miller, B. *J. Am. Chem. Soc.* **1976,** *98,* 252.
101. Otsubo, T.; Sukenobe, N.; Aso, Y.; Ogura, F. *Synth. Met.* **1988,** *27,* B509.
102. Yamahira, A.; Nogami, T.; Mikawa, H. *J. Chem. Soc. Chem. Commun.* **1983,** 904.
103. Stark, J. C.; Reed, R.; Acampora, L. A.; Sandman, D. J.; Jensen, S.; Jones, M. T.; Foxman, B. M. *Organometallics* **1984,** *3,* 732.
104. Endres, H.; Keller, H. J.; Queckbörner, J.; Veigel, J.; Schweitzer, D. *Mol. Cryst. Liq. Cryst.* **1982,** *86,* 111.
105. Balodis, K. A.; Livdane, A. D.; Medne, R. S.; Neiland, O. Ya. *Z. Org. Chem. (Russian),* *15,* **1979,** 391.
106. Hilti, B.; Mayer, C. W.; Rihs, G. *Helv. Chim. Acta* **1978,** *61,* 1462.
107. Sandman, D. J.; Stark, J. C.; Foxman, B. M. *Organometallics* **1982,** *1,* 739.
108. Shibaeva, R. P.; Kaminski, V. F. *Cryst. Struct. Commun.* **1981,** *10,* 663.
109. Goodings, E. P.; Mitchard D. A.; Owen, G. *J. Chem. Soc. Perkin I* **1972,** 1310.
110. Maruo, T.; Singh, M.; Jones, M. T.; Rath, N. P.; Min, D. *MRS Proceedings* **1990,** *173,* 149.

111. Wegman, A.; Tieke, B.; Mayer, C. W.; Hilti, B. *J. Chem. Soc. Chem. Commun.* **1989**, 716.
112. Meinwald, J.; Dauplaise, D.; Wudl, F.; Hauser, J. J. *J. Am. Chem. Soc.* **1977**, *99*, 255.
113. Meinwald, J.; Dauplaise, D.; Clardy, J. *ibid*, **1977**, *99*, 7743.
114. Chaing, L.; Meinwald, J. *Tetrahedron Lett.* **1981**, 4565.
115. Bock, H.; Brahler, G.; Dauplaise, D.; Meinwald, J. *Chem. Ber.* **1981**, *114*, 2622.
116. Miyamoto, H.; Yui, K.; Aso, Y.; Otsubo, T.; Ogura, F. *Tetrahedron Lett.* **1986**, 2011.
117. Röhrich, J.; Wolf, P.; Enkelmann, V.; Mullen, K. *Angew. Chem. Int. Ed. Engl.* **1988**, *27*, 1377.
118. Girmay, B.; Kilburn, J. D.; Underhill, A. E.; Varma, K. S.; Hursthouse, M. B.; Harman, M. E.; Becher, J.; Bojesen, G. *J. Chem. Soc., Chem. Commun.* **1989**, 1406.
119. Nishikawa, E.; Tatemitsu, H.; Sakata, Y.; Misumi, S. *Chem. Lett.* **1986**, 2131.
120. Becker, J. Y.; Bernstein, J.; Bittner, S.; Sharma, J. A. R. P.; Shahal, L. *Tetrahedron Lett.* **1988**, 6177.
121. Lerstrup, K.; Jørgensen, M.; Johannsen, I.; Christensen, J.; Bechgaard, K. in ISSP-ISOS Proceedings, *The Physics and Chemistry of Organic Superconductors*, Edited by G. Saito and S. Kagoshima, Springer-Verlag, Berlin, **1990**, p. 383.
122. Bechgaard, K.; Lerstrup, K.; Jørgensen, M.; Johannsen, I.; Christensen, J.; Larsen, J. *Mol. Cryst. Liq. Cryst.* **1990**, *181*, 161.
123. Lee, V. Y; Schumker, R. R.; Engler, E. M.; Meyerle, J. J. *Mol. Cryst. Liq. Cryst.* **1982**, *86*, 317.
124. Yui, K.; Aso, Y.; Otsubo, T.; Ogura, F. *Chem. Lett.* **1986**, 551.
125. Mazumdar S.; Ramasesha, S. *Synth. Met.* **1988**, *27*, A105 and references therein.
126. Sugimoto, T.; Awaji, H.; Sugimoto, I.; Misaki, Y.; Kawase, T.; Yoneda, S.; Yoshida, Z.; Kobayashi, T.; Anzai, H. *Chem. Mater.* **1989**, *1*, 535 and references therein.
127. Bryce M. R.; Moore, A. J. *Pure and Appl. Chem.* **1990**, *62*, 473 and references therein.
128. Yamashita, Y.; Kobayashi, Y.; Miyashi, T. *Angew Chem. Int. Ed. Engl.* **1989**, *28*, 1052.
129. Takahashi, K.; Nihira, T.; Takase, K.; Shibata, K. *Tetrahedron Lett.* **1989**, 2091.
130. Sugimoto, T.; Awaji, H.; Misakai, Y.; Yoshida, Z.; Kai, Y.; Nakagawa, H.; Kasai, N. *J. Am. Chem. Soc.* **1985**, *107*, 5792.
131. Sugimoto, T.; Misaki, Y.; Kajita, T.; Yoshida, Z.; Kai, Y.; Kasai, N.; *ibid*, **1987**, *109*, 4106.
132. Sugimoto, T.; Misaki, Y.; Arai, Y.; Yamamoto, Y.; Yoshida, Z.; Kai, Y.; Kasai, N. *ibid*, **1988**, *110*, 628.
133. Schumaker, R. R.; Rajesswari, S.; Joshi, M. V.; Cava, M. P.; Takassi, M. A.; Metzger, R. M. *ibid*, **1989**, *111*, 308.
134. Kimura, M.; Watson, W. H.; Nakayama, J. *J. Org. Chem.* **1980**, *45*, 3719.
135. Akamatu, H.; Inokuchi, H.; Matsunaga, Y. *Nature* **1954**, *173*, 168.
136. Keller, H. J.; Nothe, D.; Pritzkow, H.; Wehe, D.; Werner, M.; Koch, P.; Schweitzer, D. *Mol. Cryst. Liq. Cryst.* **1980**, *62*, 181.
137. Keller, H. J.; Nothe, D.; Pritzkow, H.; Wehe, D.; Werner, M.; Harms, R. H.; Koch, P.; Schweitzer, D. *Chemica Scripta* **1981**, *17*, 101.
138. Alcacer, L.; Chasseau, D.; Gaultier, J. *Solid State Commun.* **1980**, *35*, 945.

139. Nabeshima, T; Iwata, S.; Furukawa, N.; Morihashi, K.; Kikuchi, O. *Chem. Lett.* **1988**, 1325.
140. Bechgaard, K. *Mol. Cryst. Liq. Cryst.* **1985**, *125*, 81.
141. Kono, Y.; Miyamoto, H.; Aso, Y.; Otsubo, T.; Ogura, F.; Tanaka, T.; Sawada, M. *Angew Chem. Int. Ed. Engl.* **1989**, *28*, 1222.
142. Hart, H.; Sasaoka, S. *J. Am. Chem. Soc.* **1978**, *100*, 4326.
143. Cowan, D. O.; Kini, A.; Chiang, L.-Y.; Lerstrup, K.; Talham, D. R.; Poehler, T. O.; Bloch, A. N. *Mol. Cryst. Liq. Cryst.* **1982**, *86*, 1.
144. Nakasuji, K.; Kubota, H.; Kotani, T.; Murata, I.; Saito, G.; Enoki, T.; Imaeda, K.; Inokuchi,H.; Honda, M.; Katayama, C.; Tanaka, J. *J. Am. Chem. Soc.* **1986**, *108*, 3460.
145. Nakasuji, K. ISSP-ISOS Proceedings, *The Physics and Chemistry of Organic Superconductors*, Edited by G. Saito and S. Kagoshima, Springer-Verlag, Berlin, **1990**, p. 399.
146. Nakasuji, K.; Sasaki, M.; Kotani, T; Murata, I.; Enoki, T.; Imaeda, K.; Inokuchi, H.; Kawamoto, A.; Tanaka, J. *J. Am. Chem. Soc.* **1987**, *109*, 6970.
147. Unpublished results of Wudl, F. cited in Suzuki, T.; Yamochi, H.; Srdanov, G.; Hinkelmann, K.; and Wudl, F., *ibid*, **1989**, *111*, 3108.
148. Safiev, O. G.; Nazarov, D. V.; Zorin, V. V.; Rakhmankulov, D. L. *Khim. Geterosiklishe. Soedin.* **1988**, 852 Engl. Transl. **1988**, 702.
149. Maigrot, N; Ricard, L.; Charrier, C.; Mathey, F. *Angew Chem. Int. Ed. Engl.* **1988**, *27*, 950.
150. Akamatu, H.; Inokuchi, H.; Matsunaga, Y. *Nature*, **1954**, 173, 168.
151. Ferraris, J.; Cowan, D. O.; Walatka, V., Jr.; Perlstein, J. H. *J. Am. Chem. Soc.* **1973**, *95*, 948.
152. Coleman, L. B.; Cohen, M. J.; Sandman, D. J.; Yamagishi, F. G.; Garito, A. F.; Heeger, A. J. *Solid State Commun.* **1973**, *12*, 1125.
153. Cowan, D. O.; *Proceedings of the 4th International Kyoto Conference on New Aspects of Organic Chemistry*, Yoshida, Z.; Shiba, T.; Ohshiro, Y., Eds., Kodansha Ltd.: Tokyo and VCH Verlagsgesellschaft: FRG, 1989.
154. Wudl, F. *J. Am. Chem. Soc.* **1975**, *97*, 1962.
155. Kaufman, F. B.; Engler, E. M.; Green, D. C.; Chambers, J. Q. *J. Am. Chem. Soc.* **1976**, *98*, 1596.
156. Scott, B. A.; Placa, S. J. La; Torrance, J. B.; Silverman, B. D.; Welker, B. *J. Am. Chem. Soc.* **1977**, *99*, 6631.
157. Bechgaard, K.; Carneiro, K.; Rasmussen, F. B.; Olsen, M.; Rindorf, G.; Jacobsen, C.S.; Pedersen, H. J.; Scott, J. C. *J. Am. Chem. Soc.* **1981**, *103*, 2440.
158. Stephens, D.A.; Rehan, A. E.; Compton, S. J.; Barkhau, R. A.; Williams, J. M. *Inorg. Synth.* **1986**, *24*, 135. An excellent electronic device for constant current or voltage electrocrystallization can be purchased from Custom Research Instruments, 39 Chieftain Dr., St. Louis, Missouri 63146 (314-991-9626).
159. Williams, J. M.; Wang, H. H.; Emge, T. J.; Geiser, U.; Beno, M. A.; Leung, P. C. W.; Carlson, K. D.; Thorn, R. J.; Schultz, A. J.; Whangbo, M.-H. *Progress in Inorg. Chem.* **1987**, *35*, 51.
160. Emge, T. J.; Wang, H. H.; Beno, M. A.; Williams, J. M.; Whangbo, M.-H.; Evain, M. *J. Am. Chem. Soc.* **1986**, *108*, 8215.

161. Enquiries to: Sambrook Engineering, Department of Chemistry, University College of North Wales, Bangor Gwynedd LL57 2UW, Great Britain.
162. Ward, M. D. *Electroanalytical Chemistry* **1989**, *16*, 181, Bard, A. J., Ed., Marcel Dekker, Inc.: New York.
163. Enkelmann, V.; Morra, B. S.; Kröhnke, Ch.; Wegner, G.; Heinze, J. *Chem. Phys.* **1982**, *66*, 303.
164. Kobayashi, H.; Kato, R.; Mori, T.; Kobayashi, A.; Sasaki, Y.; Saito, G.; Inokuchi, H. *Chem. Lett.* **1983**, 759.
165. Kobayashi, H.; Mori, T.; Kato, R.; Kobayashi, A.; Sasaki, Y.; Saito, G.; Inokuchi, H. *Chem. Lett.* **1983**, 581.
166. Laversanne, R.; Amiell, J.; Delhaes, P.; Chasseau, D; Hauw, C. *Solid State Commun.* **1984**, *52*, 177.
167. Parkin, S. S. P.; Engler, E. M.; Schumaker, R. R.; Lagier, R.; Lee, V. Y.; Scott, J. C.; Greene, R. L. *Mol. Cryst. Liq. Cryst.* **1983**, *23*, 1790.
168. Williams, J. M.; Beno, M. A.; Wang, H. H.; Reed, P. E.; Azevedo, L. J.; Schirber, J. E. *Inorg. Chem.* **1984**, *23*, 1790.
169. Beno, M. A.; Cox, D. D.; Williams, J. M.; Kwak, J. F. *Acta Cryst. Allogr.* **1984**, *C40*, 1334.
170. Geiser, U.; Wang, H. H.; Schlueter, J. A.; Hallenbeck, S. L.; Allen, T. J.; Chen, M. Y.; Kao, H.-C. I.; Carlson, K. D.; Gerdom, L. E.; Williams, J. M. *Acta Cryst. Allogr.* **1988**, *C44*, 1544.
171. Beno, M. A.; Blackman, G. S.; Leung, P. C. W.; Carlson, K. D.; Copps, P. T. and Williams, J. M. *Mol. Cryst. Liq. Cryst.* **1985**, *119*, 409.
172. Kobayashi, H.; Kobayashi, A.; Sasaki, Y.; Saito, G.; Enoki, T.; Inokuchi, H. *J. Am. Chem. Soc.* **1983**, *105*, 297.
173. Cox, D. D.; Ball, G. A.; Alonso, A. S.; Williams, J. M. *Inorg. Synth.* **1989**, *26*, 393.
174. Kobayashi, H.; Kato, R.; Mori, T.; Kobayashi, A.; Sasaki, Y.; Saito, G.; Enoki, T. and Inokuchi, H. *Chem. Lett.* **1984**, 179.
175. Parkin. S. S. P.; Engler, E. M.; Lee, V. Y.; Schumaker, R. R. *Mol. Cryst. Liq. Cryst.* **1985**, *119*, 375.
176. Urayama, H.; Saito, G.; Kawamoto, A.; Tanaka, J. *Chem. Lett.* **1987**, 1753.
177. Porter, L. C.; Wang, H. H.; Miller, M. M.; Williams, J. M. *Acta Cryst. Allogr.* **1987**, *C43*, 2201.
178. Schirber, J. E.; Overmyer, D. L.; Venturini, E. L.; Wang, H. H.; Carlson, K. D.; Kwok, W. K.; Kleinjan, S. and Williams, J. M. *Physica C.* **1989**, *161*, 412.
179. Mori, T.; Inokuchi, H. *Chem. Lett.* **1987**, 1657.
180. Rosseinsky, M. J.; Kurmoo, M.; Talham, D. R.; Day, P.; Chasseau, D.; Watkin, D. *J. Chem. Soc., Chem. Commun.* **1988**, 88.
181. Mori, T.; Kobayashi, A.; Sasaki, Y.; Kato, R.; Kobayashi, H. *Solid State Commun.* **1985**, *53*, 627.
182. Wang, H. H.; Allen, T. J.; Schlueter, J. A.; Hallenbeck S. L.; Stupka, D. L.; Chen, M. Y.; Despotes, A. M.; Kao, H.-C. I.; Carlson, K. D.; Geiser, U.; Williams, J. M. *Phosphorus and Sulfur*, **1988**, *38*, 329.
183. Mori, T.; Inokuchi, H. *Solid State Commun.* **1987**, *64*, 335.
184. Yagubskii, É. B.; Shchegolev, I. F.; Laukhin, V. N.; Kononovich, P. A.; Kartsovnik, M. V.; Zvarykina, A. V.; Buravov, L. I. *JETP Lett.* **1984**, *39*, 12.

185. Bender, K.; Hennig, I.; Schweitzer, D.; Dietz, K.; Endres, H.; Keller, H. J. *Mol. Cryst. Liq. Cryst.* **1984**, *108*, 359.

186. Shibaeva, R. P.; Kaminskii, V. F.; Beliskii, V. K.; *Kristallografiya* **1984**, *29*, 1089; *Sov. Phys. Crystallog.* **1984**, *29*, 638.

187. Kobayashi, H.; Kato, R.; Kobayashi, A.; Nishio, Y.; Kajita, K.; Sasaki,W. *Chem. Lett.* **1986**, 833.

188. Wang, H. H.; Vogt, B. A.; Geiser, U.; Beno, M. A.; Carlson, K. D.; Kleinjan, S.; Thorup, N.; Williams, J. M. *Mol. Cryst. Liq. Cryst.* **1990**, *181*, 135.

189. Kobayashi, A.; Kato, R.; Kobayashi, H.; Moriyama, S.; Nishio, Y.; Kajita, K.; Sasaki, W. *Chem. Lett.* **1987**, 459.

190. Qian, M.-X.; Wang, X.-H.; Zhu, Y.-L.; Zhu, D.; Li, L.; Ma, B.-H.; Duan, H.-M.; Zhang, D.-L. *Synth. Met.* **1988**, *27*, A277

191. Shibaeva, R. P.; Kaminskii, V. F.; Yagubskii, É. B. *Mol. Cryst. Liq. Cryst.* **1985**, *119*, 361.

192. Shibaeva, R. P.; Lobkovskaya, R. M.; Yagubskii, É. B.; Kostyuchenko, E. E.; *Kristallografiya* **1986**, *31*, 455; *Sov. Phys. Crystallogr.* **1986**, *31*, 267.

193. Beno, M. A.; Geiser, U.; Kostka, K. L.; Wang, H. H.; Webb, K. S.; Firestone, M. A.; Carlson, K. D.; Nuñez, L.; Williams, J. M.; Whangbo, M.-H. *Inorg. Chem.* **1987**, *26*, 1912.

194. Shibaeva, R. P.; Lobkovskaya, R. M.; Yagubskii, E. B.; Kostyuchenko, E. E. *Kristallografiya* **1986**, *31*, 1110; *Sov. Phys. Crystallogr.* **1986**, *31*, 657.

195. Wang, H. H.; Montgomery, L. K.; Husting, C. A.; Vogt, B. A.; Williams, J. M.; Budz, S. M.; Lowry, M. J.; Carlson, K. D.; Kwok, W. K.; Mikheyev, V. *Chem. of Mater.* **1989**, *1*, 484.

196. Wang, H. H.; Ferraro, J. R.; Carlson, K. D.; Montgomery, L. K.; Geiser, U.; Williams, J. M.; Whitworth, J. R.; Hill, S.; Whangbo, M.-H.; Evain, M.; Novoa, J. *J. Inorg. Chem.* **1989**, *28*, 2267.

197. Kobayashi, H.; Kato, R.; Kobayashi, A.; Moriyama, S.; Nishio, Y.; Kajita, K.; Sasaki, W. *Synth. Met.* **1988**, *27*, A283.

198. Montgomery, L. K.; Geiser, U.; Wang, H. H.; Beno, M. A.; Schultz, A. J.; Kini, A. M.; Carlson, K. D.; Williams, J. M.; Whitworth, J. R. Gates, B. D.; Cariss, C. S.; Pipan, C. M.; Donega, K. M.; Wenz, C.; Kwok, W. K.; Crabtree, G. W. *Synth. Met.* **1988**, *27*, A195.

199. Beno, M. A.; Kini, A. M.; Montgomery, L. K.; Whitworth, J. R.; Carlson, K. D.; Williams, J. M. *Synth. Met.* **1988**, *27*, A219.

200. Merzhanov, V. A.; Kostyuchenko, E. É.; Laukhin, V. N.; Lobkovskaya, R. M.; Makova, M. K.; Shibaeva, R. P.; Shchegolev, I. F.; Yagubskii, É. B. *JETP Lett.* **1985**, *41*, 179.

201. Williams, J. M.; Emge, T. J.; Wang, H. H.; Beno, M. A.; Copps, P. T.; Hall, L. N.; Carlson, K. D.; Crabtree, G. W. *Inorg. Chem.* **1984**, *23*, 2558.

202. Daoben, Z.; Ping, W.; Meixiang, W.; Zhaolou, Y.; Naiijue, Z. *Solid State Commun.* **1986**, *57*, 843.

203. Emge, T. J.; Wang, H. H.; Beno, M. A.; Leung, P. C. W.; Firestone, M. A.; Jenkins. H. C.; Cook, J. D.; Carlson, K. D.; Williams, J. M.; Venturini, E. L.; Azevedo, L. J.; Schirber, J. E. *Inorg. Chem.* **1985**, *24*, 1736.

204. Wang, H. H.; Beno, M. A.; Geiser, U.; Firestone, M. A.; Webb, K. S.; Nuñez, L.; Crabtree, G. W.; Carlson, K. D.; Williams, J. M.; Azevedo, L. J.; Kwak, J. F.; Schirber, J. E. *Inorg. Chem.* **1985**, *24*, 2465.

205. Williams, J. M.; Wang, H. H.; Beno, M. A.; Emge, T. J.; Sowa, L. M.; Copps, P. T.; Behroozi, F.; Hall, L. N.; Carlson, K. D.; Crabtree, G. W. *Inorg. Chem.* **1984**, *23*, 3839.

206. Kobayashi, H.; Kato, R.; Kobayashi, A.; Saito, G.; Tokumoto, M.; Anzai, H.; Ishiguro, T. *Chem. Lett.* **1986**, 93.

207. Emge, T. J.; Wang, H. H.; Leung, P. C. W.; Rust, P. R.; Cook, J. D.; Jackson, P. L.; Carlson, K. D.; Williams, J. M.; Whangbo, M.-H.; Venturini, E. L.; Schirber, J.E.; Azevedo, L. J.; Ferraro, J. R. *J. Am. Chem. Soc.* **1986**, *108*, 695.

208. Ugawa, A.; Yakushi, K.; Kuroda, H.; Kawamoto, A.; Tanaka, J. *Chem. Lett.* **1986**, 1875.

209. Ugawa, A.; Yakushi, K.; Kuroda, H.; Kawamoto, A.; Tanaka, J. *Synth. Met.* **1988**, *22*, 305.

210. Ugawa, A.; Okawa, Y.; Yakushi, K.; Kuroda, H.; Kawamoto, A.; Tanaka, J.; Tanaka, M.; Nogami, Y.; Kagoshima, S.; Murata, K.; Ishiguro, T. *Synth. Met.* **1988**, *27*, A407.

211. Beno, M. A.; Firestone, M. A.; Leung, P. C. W.; Sowa, L. M.; Wang, H. H.; Williams, J. M.; Whangbo, M.-H. *Solid State Commun.* **1986**, *57*, 735.

212. Kurmoo, M.; Talham, D. R.; Day, P.; Parker, I. D.; Friend, R. H.; Stringer, A. M.; Howard, J. A. K. *Solid State Commun.* **1987**, *61*, 459.

213. Emge, T. J.; Wang, H. H.; Bowman, M. K.; Pipan, C. M.; Carlson, K. D.; Beno, M. A.; Hall, L. N.; Anderson, B. A.; Williams, J. M.; Whangbo, M.-H. *J. Am. Chem. Soc.* **1987**, *109*, 2016.

214. Laukhin, V. N.; Kostyuchenko, E. É.; Sushko, Yu. V.; Shchegolev, I. F.; Yagubskii, É. B. *JETP Lett.* **1985**, *41*, 81.

215. Murata, K.; Tokumoto, M.; Anzai, H.; Bando, H.; Saito, G.; Kajimura, K.; Ishiguro, T. *J. Phys. Soc. Jpn.* **1985**, *54*, 1236.

216. Schirber, J. E.; Azevedo, L. J.; Kwak, J. F.; Venturini, E. L.; Leung, P. C. W.; Beno, M. A.; Wang, H. H.; Williams, J. M. *Phys. Rev. B: Condens. Matter* **1986**, *33*, 1987.

217. Wang, H. H.; Carlson, K. D.; Montgomery, L. K.; Schlueter, J. A.; Cariss, C. S.; Kwok, W. K.; Geiser, U.; Crabtree, G. W.; Williams, J. M. *Solid State Commun.* **1988**, *66*, 1113.

218. Rath, N. P.; Holt, E. M. *J. Chem. Soc., Chem. Commun.* **1986**, 311.

219. Buravov, L. I.; Zvarykina, A. V.; Kartsovnik, M. V.; Kushch, N. D.; Laukhin, V. N.; Lobkovskaya, R. M.; Merzhanov, V.A.; Fedutin, L. N.; Shibaeva, R. P.; Yagubskii, É. B. *Sov. Phys. JETP* **1987**, *65*, 336.

220. Geiser, U.; Wang, H. H.; Donega, K. M.; Anderson, B. A.; Williams, J. M.; Kwak, J. F. *Inorg. Chem.* **1986**, *25*, 401.

221. Geiser, U.; Wang, H. H.; Gerdom, L. E.; Firestone, M. A.; Sowa, L.M.; Williams, J. M.; Whangbo, M.-H. *J. Am. Chem. Soc.* **1985**, *107*, 8305.

222. Kikuchi, K.; Saito, K.; Ikemoto, I.; Murata, K.; Ishiguro, T.; Kobayashi, K. *Synth. Met.* **1988**, *27*, B269.

223. Geiser, U.; Anderson, B. A.; Murray, A.; Pipan, C. M.; Rohl, C. A.; Vogt, B.A.;

Wang, H. H.; Williams, J. M.; Kang, D. B.; Whangbo, M.-H. *Mol. Cryst. Liq. Cryst.* **1990**, *181*, 105.

224. Mori, T.; Inokuchi, H. *Chem. Lett.* **1986**, 2069.
225. Porter, L. C.; Wang, H. H.; Beno, M. A.; Carlson, K. D.; Pipan, C. M.; Proksch, R. B.; Williams, J. M. *Solid State Commun.* **1987**, *64*, 387.
226. Urayama, H.; Yamochi, H.; Saito, G.; Sato, S.; Kawamoto, A.; Tanaka, J.; Mori, T.; Maruyama, Y.; Inokuchi, H. *Chem. Lett.* **1988**, 463.
227. Kurmoo, M.; Talham, D. R.; Pritchard, K. L.; Day, P.; Stringer, A. M.; Howard, J. A. K. *Synth. Met.* **1988**, *27*, A177.
228. Lyubovskaya, R. N.; Lyubovskii, R. B.; Shibaeva, R. P.; Aldoshina, M. Z.; Gol'denberg, L. M.; Rozenberg, L. P.; Khidekel', M. L.; Shul'pyakov, Yu. F. *JETP Lett.* **1985**, *42*, 468.
229. Shibaeva, R. P.; Rozenberg, L. P. *Kristallografiya* **1988**, *33*, 1402; *Sov. Phys. Crystallogr.* **1988**, *33*, 834.
230. Lyubovskaya, R. N.; Zhilyaeva, E. I.; Pesotskii, S. I.; Lyubovskii, R. B.; Atovmyan, L. O.; D'yachenko, O. A.; Takhirov, T. J. *JETP Lett.* **1987**, *46*, 188.
231. Papavassiliou, G. C.; Mousdis, G. A.; Zambounis, J. S.; Terzis, A.; Hountas, A.; Hilti B.; Mayer, C. W.; Pfeiffer, J. *Synth. Met.* **1988**, *27*, B379.
232. Nigrey, P. J.; Morosin, B.; Kwak, J. F.; Venturini, E. L.; Baughman, R. J. *Synth. Met.* **1986**, *16*, 1.
233. Geiser, U.; Beno, M. A.; Kini, A. M.; Wang, H. H.; Schultz, A. J.; Gates, B. D.; Cariss, C. S.; Carlson, K. D.; Williams, J. M. *Synth. Met.* **1988**, *27*, A235.
234. Kinoshita, N.; Takahashi, K.; Murata, K.; Tokumoto, M.; Anzai, H. *Solid State Commun.* **1988**, *67*, 465.
235. Schirber, J. E.; Venturini, E. L.; Kini, A. M.; Wang, H. H.; Whitworth, J. R.; Williams J. M. *Physica C* **1988**, *152*, 157.
236. Shibaeva, R. P.; Rozenberg, L. P.; Aldoshina, M. Z.; Lobkovskaya, R. N. *Kristallografiya* **1988**, *33*, 125; *Sov. Phys. Crystallogr.* **1988**, *33*, 71.
237. Mori, T.; Wang, P.; Imaeda, K.; Enoki, T.; Inokuchi, H. *Solid State Commun.* **1987**, *64*, 733.
238. Geiser, U.; Wang, H. H.; Williams, J. M. (unpublished results).
239. Bu, X.; Coppens, P.; Naughton, M. J., *Acta Cryst.* **1990**, *C46*, 1609.
240. Müller, H.; Fritz, H. P.; Heidmann, C.-P.; Gross, F.; Veith, H.; Lerf, A.; Andres, K.; Fuchs, H.; Polborn, K.; Abriel, W. *Synth. Met.* **1988**, *27*, A257.
241. Geiser, U.; Wang, H. H.; Kleinjan, S.; Williams, J. M. *Mol. Cryst. Liq. Cryst.* **1990**, *181*, 125.
242. Wang, H. H.; Beno, M. A.; Carlson, K. D.; Thorup, N.; Murray, A.; Porter, L. C.; Williams, J. M.; Maly, K.; Bu, X.; Petricek, V.; Cisarova, I.; Coppens, P.; Jung, D.; Whangbo, M.-H., Schirber, J. E.; Overmyer, D. L.; *Chem. Mater.*, **1991**, in press.
243. Oshima, M.; Mori, H.; Saito, G.; Oshima, K. *Chem. Lett.* **1989**, 1159.
244. Wang, H. H.; Carlson, K. D.; Geiser, U.; Kwok, W. K.; Vashon, M. D.; Thompson, J. E.; Larsen, N. F.; McCabe, G. D.; Hulscher, R. S.; Williams, J. M. *Physica. C* **1990**, *166*, 57.
245. Williams, J. M.; Kini, A. M.; Wang, H. H.; Carlson, K. D.; Geiser, U.; Montgomery, L. K.; Pyrka, G. J.; Watkins, D. M.; Kommers, J. M.; Boryschuk, S. J.; Strieby

Crouch, A. V.; Kwok, W. K.; Schirber, J. E.; Overmyer, D. L.; Jung, D.; Whangbo, M.-H. *Inorg. Chem.* **1990,** *29,* 3262.
246. Kini, A. M.; Geiser, U.; Wang, H. H.; Carlson, K. D.; Williams, J. M.; Kwok, W. K.; Vandervoort, K. G.; Thompson, J. E.; Stupka, D. L.; Jung, D.; Whangbo, M.-H. *Inorg. Chem.* **1990,** *29,* 2555.
247. Wang, H. H.; Carlson, K. D.; Geiser, U.; Kini, A. M.; Schultz, A. J.; Williams, J. M.; Montgomery, L. K.; Kwok, W.K.; Welp, U.; Vandervoort, K. G.; Boryschuk, S. J.; Strieby Crouch, A. V.; Kommers, J. M.; Watkins, D. M.; Schirber, J. E.; Overmyer, D. L.; Jung, D.; Novoa, J. J.; Whangbo, M.-H. Proceedings, International Conference on Science and Technology of Synthetic Metals, Tübingen, FRG, **1990,** *Synth. Met.,* in press.

3

Structural Aspects

INTRODUCTION

It has been mentioned in the previous chapter that ET salts exhibit a wide variety of electronic characteristics, depending on the anion present in the material. In addition, multiple phases exist, sometimes even forming simultaneously in the same preparation, with vastly differing physical properties. These properties derive primarily from the *packing* of the donor-radical cations in their crystal structure. The electronic interactions between the cations are determined not only by the distance between them, but also by the spatial arrangement: for example, face-to-face interactions between the (usually rather flat) donor molecules differ greatly from side-to-side interactions. In the molecules of interest, the heteroatoms, sulfur or selenium, are located near the periphery of the molecule, and in combination with their larger size compared with carbon atoms, it is mostly the heteroatom···heteroatom contacts that determine the intermolecular electronic interactions. In this chapter, we present the molecular structures of selected electron-donor molecules, give a survey of all the crystal structures reported for ET, and related donor-radical salts, present a description of some of the more common structural types observed, and describe the structures of the superconducting salts in more detail.

The structures of many donor molecules related to ET have been determined, and their inclusion in this monograph is somewhat arbitrary. The discussion will include donor molecules containing the tetra(chalcogeno)fulvalene moiety, and, in addition, either further fused rings or substituents with a chalcogen atom in the first position of the substituent. Examples of the first kind include ring-forming alkylenedithio substituents, but also groups like pyrazine or benzene capable of fusing to the fulvalene core. The second group will be represented by substituents of the CH$_3$S-type. This distinction excludes most simply substituted tetrathia- and tetraselenafulvalenes which have been extensively reviewed elsewhere.

MOLECULAR STRUCTURE OF THE DONOR MOLECULES
Bis(ethylenedithio)tetrathiafulvalene, ET

The crystal structure of neutral ET was reported by Kobayashi et al.[1] The molecule consists of two five-membered and two six-membered heterocyclic ring systems, each containing two sulfur atoms. On each side of the molecule, a five- and a six-membered ring are fused together along a C=C double bond, and the five-membered rings are connected by a C=C double bond.

The molecule is rather flat, due to the presence of an extended π-electron system, approaching D_{2h} symmetry in the planar limit. However, noticeable distortions from planar geometry are always present, especially in the ethylene end groups which can take on a number of conformations (vide infra). In the neutral molecule, even the central portion of the molecule is bent: folds through the inner sulfur atoms perpendicular to the long molecular axis exist, leading to dihedral angles of 165.3° and 167.7°, across S1—S2 and S3—S4, respectively.[1] In the charge-transfer salts of ET, the central, conjugated portion of the molecule usually becomes more planar with increasing charge.

The conformation of the terminal ethylene groups with respect to each other and the central part of the molecule, their influence on the crystal packing, and ultimately on the physical properties has been the topic of extensive discussions. Some of the more common conformations found in ET charge-transfer salts are shown in Fig. 3.1. The six-membered outer rings of the ET molecule usually

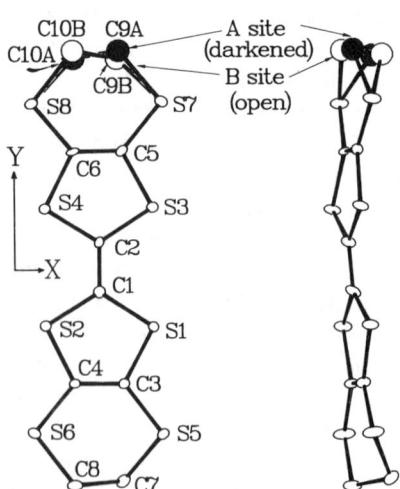

Figure 3.1 Two views of the ET molecule showing the possible crystallographically disordered ethylene group sites A and B, corresponding to the staggered and eclipsed conformations, respectively. Reprinted with permission from: Williams, J. M.; Wang, H. H.; Emge, T. J.; Geiser, U.; Beno, M. A.; Leung, P. C. W.; Carlson, K. D.; Thorn, R. J.; Schultz, A. J.; Whangbo, M.-H. In *Prog. Inorg. Chem.*; Lippard, S. J., Ed.; Vol. 35, p. 51. Copyright © 1987 John Wiley & Sons (ref.2).

contain the inner carbon and the sulfur atoms in a common plane, and at least one of the ethylene group carbon atoms displaced from the plane. In the "twist" conformation, the second ethylene group carbon atom is found on the other side of the plane, but usually at a smaller distance. There are, of course, two enantiomorphic "twist"-conformations, corresponding to a positive or negative dihedral angle between the inner C=C and the ethylene C–C bond directions. In addition, a "boat" conformation is occasionally found, wherein both ethylene carbon atoms are located on the same side of the plane. However, the occurrence of a "boat" conformation is generally accompanied by large atomic thermal parameters and an artificially shortened C–C distance between the carbon atomic positions obtained from the crystallographic analysis, indicating some degree of disorder. The "boat" position is probably not intrinsically stable and is only found in the presence of a special arrangement of the crystal packing forces exerted by the anions on the ethylene group hydrogen atoms. In those cases where two "twist" conformations are present in the same molecule, two diastereomeric configurations are possible: (a) the eclipsed configuration with both ethylene groups parallel to each other, giving the molecule approximate C_{2h} symmetry (two-fold axis along the long direction of the molecule), and (b) the staggered configuration, where in a projection of the molecule along the long axis the ethylene group C–C bonds appear crossed, and with opposite orientation with respect to the plane of the molecule. This latter configuration has approximate D_2 symmetry. In the crystal structure of neutral ET, one end group is found in the "twist" conformation, whereas the other is in the "boat" conformation. Both enantiomers are found since the crystals are inversion-symmetric (space group $P2_1/c$).[1]

Other Bis(alkylenedithio)tetrathiafulvalenes

The simplest modifications of ET are made by the replacement of the ethylene end groups with similar constituents, for example, methylene or 1,3-propylene. All six possible combinations have been synthesized, that is, in addition to ET with two ethylene groups, there are the symmetric molecules MT (two methylene groups) and PT (two propylene groups), as well as the unsymmetrical combinations, MET, MPT, and EPT (with end groups according to the acronym letters). These substitutions do not affect the geometry of the central portion of the molecule significantly, but the end groups differ considerably from ET. As an example, the molecular structure of methylenedithiopropylenedithiotetrathiafulvalene (MPT) is shown in Fig. 3.2. The methylene carbon atom is found essentially in the plane of the molecule, thus imposing a rather rigid structure. The propylene end group is invariably found in an ordered conformation containing all three carbon atoms on the same side of the molecular plane. Furthermore, the plane formed by the three propylene carbon atoms is usually approximately parallel to the molecular plane.

Another small modification of ET is the oxidation (dehydrogenation) of the end group ethylene groups to vinylene groups leading to π-conjugation throughout the bis(vinyldithio)tetrathiafulvalene (BVDT-TTF or VT) molecule. Despite the extended conjugation, however, the end groups are far out of the molecular plane (boat conformation) in both neutral and charged molecules, leading to folding

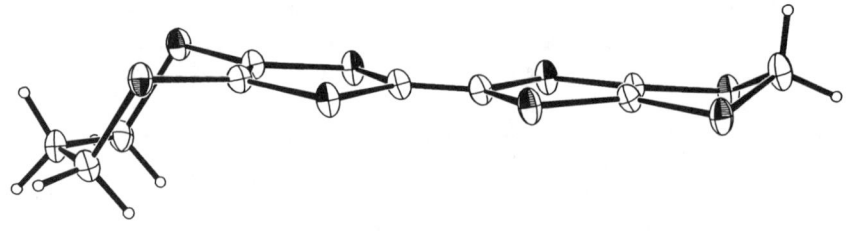

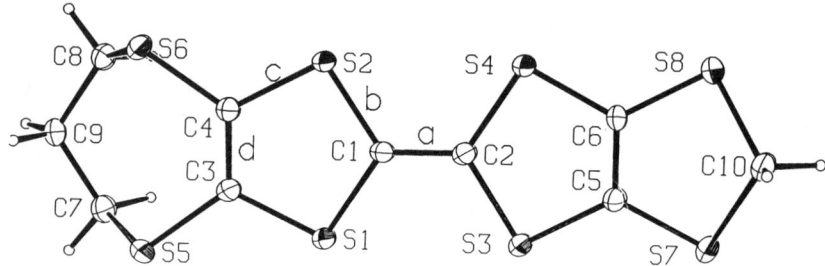

Figure 3.2 Side and top views of the molecule methylenedithiopropylene-dithiotetrathiafulvalene (MPT) in (MPT)$_2$ClO$_4$(THF) (from ref. 3, with permission).

dihedral angles around the outer sulfur-to-sulfur vector of 48.7° (neutral VT) and 34.6° and 47.8° [two nonequivalent ends, respectively, of VT in (VT)$_2$PF$_6$].[4] The fact that the six-membered rings in VT are less planar than in ET, and that the boat conformation is preferred, is caused by a mismatch of bond angles within the ring; that is, it is not possible to build a planar ring if all C–S–C and C–C–S angles are required to be ~100° and ~120°, respectively. Therefore, the ring overcomes the strain on the bond angles by folding, against the electronic forces that favor a planar arrangement for the extended π-electronic system.

Finally, in a further expansion of the outer rings, a *cis*-1,4-dithio-2-butenyl (or *cis*-acetylenedimethyldithio) group was successfully inserted to give the donor molecule BADT-TTF.[5] That molecule (only the structure of the neutral molecule is known) is much less planar than the previously mentioned examples. The folding angles across the inner sulfur atoms are 159° and 151.5°, respectively, or approximately 10° smaller than in neutral ET. Furthermore, the dihedral angles between planes through the end group carbon atoms and the adjacent C$_2$S$_4$ entities are only 66° and 70.5°; that is, the end groups wrap around backwards into the molecule. The bending of the end groups is in the opposite direction of the bending of the conjugated portion of the molecule, giving a wave-like appearance in a side view.

Chalcogen Substitutions

Another simple modification is the replacement of some (or all) of the sulfur atoms with other chalcogens, for example, oxygen, selenium, or tellurium. Selenium-containing donor molecules have been known for quite some time, and among the tetraselena-fulvalenes, the tetra-methyl derivative, TMTSF, has yielded superconductors with a number of anions, all with low-transition temperatures (around 1 K), and all but (TMTSF)$_2$ClO$_4$ require applied pressure to suppress metal-to-insulator transitions which prevent them from becoming superconductors at ambient pressure. However, we have excluded TMTSF and its salts from the scope of this text and refer to older reviews.[6-10] Donor molecules based on dithiodiselenafulvalene are listed in the section on unsymmetrical donors. No telluriumcontaining analogue of ET is known, but tetratellurafulvalene has been synthesized recently.[11] The donor molecules hexamethylenetetratellurafulvalene[12] and dibenzotetratellurafulvalene also have been prepared.

Seleno-analogs of ET included in this discussion are bis(ethylenediseleno)tetraselenafulvalene (BEDSe-TSeF), bis(ethylenediseleno)tetrathiafulvalene (BEDSe-TTF), and bis(ethylenedithio)tetraselenafulvalene (BEDT-TSeF), that is, substitution of all (BEDSe-TSeF), outer (BEDSe-TTF), and inner (BEDT-TSeF) sulfur atoms of ET. BEDT-TSeF has been mentioned in the literature but once,[13] and no structural information on either the parent molecule or any of its charge-transfer salts is available. Structures of the other two donor molecules are known for both neutral molecules and charge-transfer salts. The principal difference from ET arises from the increased bond lengths in the outer rings, where sulfur is replaced by selenium: C-Se distances are approximately 0.1 Å longer than the corresponding C-S distances. A secondary effect is the slightly smaller optimal C-Se-C angle compared to corresponding C-S-C angle, as a result of reduced shielding of the chalcogen atom lone pair electrons. Both effects cause the six-membered ring to become less planar than in ET, and the boat conformation becomes more common.[14] Figure 3.3 shows the ethylene group conformations found in the two independent BEDSe-TTF molecules present in (BEDSe-TTF)$_2$AuI$_2$.

Oxygen-substituted derivatives have become available recently. So far, only the outer ring sulfur atoms of ET have been replaced by oxygen atoms to form BEDO-TTF.[15] The structure of BEDO-TTF is known both in its neutral form[15] and in a few charge transfer salts,[16] including a single superconductor, (BEDO-TTF)$_3$Cu$_2$(NCS)$_3$.[17] Differences in the packing of BEDO-TTF as compared to ET

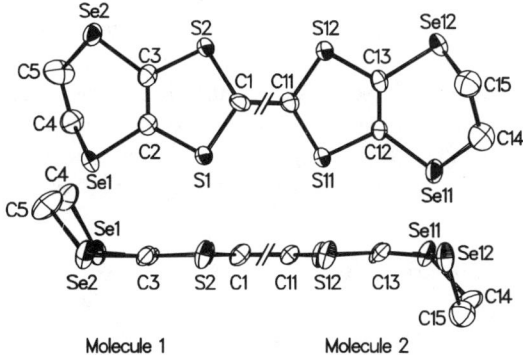

Figure 3.3 Top and side views of the two crystallographically independent BEDSe-TTF molecules (each located on an inversion center; one half of each shown) found in (BEDSe-TTF)$_2$AuI$_2$. Reprinted with permission from *Chem. Mater.* **1989**, *1*, 143 (ref. 14). Copyright © 1989 American Chemical Society.

are due primarily to the shorter C–O bond lengths (ca. 1.4 Å vs. 1.8 Å for C–S). Thus the outer chalcogen atoms of the molecule protrude much less from the side of the donor molecule, allowing the inner sulfur atoms to participate in additional chalcogen···chalcogen contacts. In fact, neither the neutral donor molecule nor the charge-transfer salts known exhibit any O···O contacts shorter than the sum of the van der Waals radii, but numerous such short S···S and S···O contacts exist. In the oxygen-containing donor molecules, the ethylene group hydrogen atoms are involved in hydrogen bonding to the oxygen atoms of neighboring molecules. In addition to BEDO-TTF, the unsymmetrical ethylenedioxo-ethylenedithio-tetrathiafulvalene, EDOEDT-TTF or "EOET,"[18] has been synthesized. Its crystal structure has not been completed yet, but it appears to be isostructural to ET. Bis(2-oxapropylenedithio)tetrathiafulvalene, BOPDT-TTF, an oxygen-containing derivative of PT, has been reported recently, and the structure of the 2:1 salt with hexafluorophosphate anion has been determined.[19] The oxygen atoms in this case do not replace the sulfur atoms of PT, but rather the central CH_2-group of the propylene moiety is replaced by an ether-like oxygen atom.

Unsymmetrical Donor Molecules

Recently, a number of donor molecules containing an outer ring on only one side of the molecule have been synthesized and characterized structurally: MDT-TTF and EDT-TTF contain only one methylenedithio and ethylenedithio group, respectively, per molecule. MDT-TTF even forms a superconductor, κ-(MDT-TTF)$_2$AuI$_2$ (T_c = 4.5 K)[20, 21] whose crystal structure[22] will be discussed. The conformations of these molecules are as expected: the methylene or ethylene end groups behave as in MT or ET, respectively, and the fulvalene end is planar. Furthermore, the absence of any out-of-plane hydrogen atoms at the fulvalene end allows for close intermolecular contacts. Centrosymmetric dimers, as well as chains where the alkylene end groups protrude from alternating sides, are common packing motifs observed in MDT-TTF and EDT-TTF. In both cases, the major face-to-face contacts are found on the fulvalene end of the molecules.

The molecules 3,4-dimethyl-3',4'-ethylenedithio-2,5-diselena-2',5'-dithiafulvalene, DMET, and 3,4-dimethyl-3',4'-ethylenedithio-tetrathiafulvalene, DIMET, are other important examples of unsymmetrical donor molecules. The former is a hybrid between ET and TMTSF, incorporating half a molecule of each, whereas DIMET is the all-sulfur analogue. Several superconducting salts of DMET have been reported (vide infra). Both molecules have a propensity to form cen-

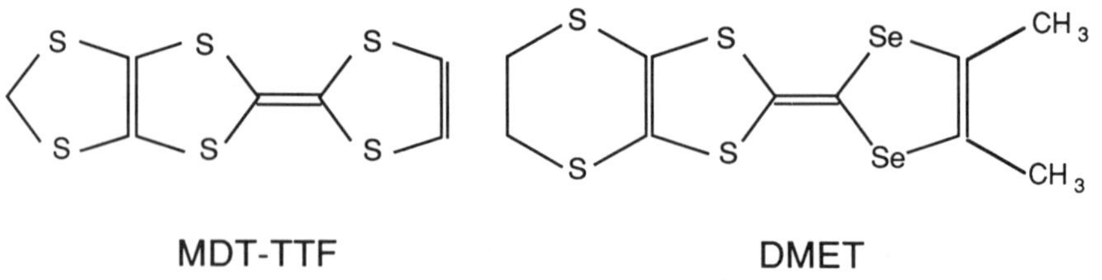

MDT-TTF   DMET

Figure 3.4

trosymmetric dimers, and their salts are often isostructural, for example, the perchlorate salts $(DMET)_2ClO_4$[23] and $(DIMET)_2ClO_4$.[24, 25] Of the neutral molecules, only the structure of DMET has been reported,[26] but many charge-transfer salts are known for both.

Further Modifications of the Outer Rings

It is possible to replace the outer rings in the prototype ET molecules with a number of aromatic ring systems, for example, benzene, pyridine, or pyrazine. The possible combinations are of course numerous: only a few salts have been prepared for each donor (even fewer structurally characterized), and much is still open for further exploration. Historically important, since their salts with the electron acceptor TCNQ (tetracyanoquinodimethane) are metallic to very low temperatures,[27] are the donor molecules hexamethylene-tetrachalcogenafulvalene (HMTSF and HMTTeF, for the selenium and tellurium derivatives, respectively), which contain saturated five-membered outer rings.

Alkylthio(seleno-,telluro-)-Substituted Tetrathiafulvalenes

The most studied alkylthio-substituted tetrathiafulvalene is tetrakis(methylthio)tetrathiafulvalene, TTM-TTF. Because of intermolecular steric interactions, the methylthio groups are usually rotated such that the methyl groups are pointing away from the molecule, sometimes in the molecular plane and sometimes out of the plane. The methyl groups thus block the inner sulfur atoms from effective intermolecular interactions, except for the direction perpendicular to the molecular plane. On the other hand, the neutral molecule is capable of coordinating as a bidentate ligand to metal centers, as in $(TTM-TTF)(HgI_2)_2$,[28] shown in Fig. 3.5. TTM-TTF[29] and its methylseleno- and methyltelluro-analogues[30] form mixed stacks with the organic acceptor molecule TCNQ (tetracyanoquinodimethane).

Longer-chain tetrakis(alkylthio)tetrathiafulvalenes are also known. They are usually denoted TTC_n-TTF, where n stands for the number of carbon atoms in the alkyl group. Compounds with $n > 10$ have been investigated for incorporation into films and polymers; for a review, see Saito.[31]

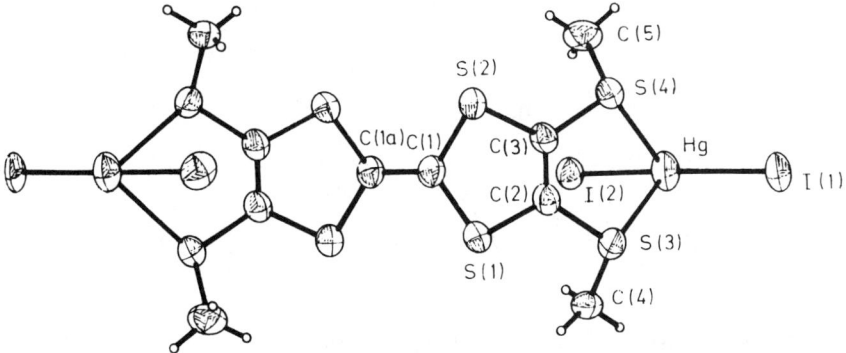

Figure 3.5 Tetrakis(methylthio)tetrathiafulvalene as a complexing ligand in $(TTM-TTF)(HgI_2)_2$ (from ref. 28, with permission).

Other donor molecules have been derivatized with methylthio groups, for example, VT.[32] In $(CH_3S)_4VT$, as in unsubstituted VT, the molecule is folded across a line through the outer sulfur atoms (the molecule lies on a center of inversion). The CH_3S-groups are bent even further, with the methyl groups pointing back towards the center of the molecule. In a side-view, this gives the molecule an S-shape appearance.[32]

The most sophisticated donor molecule in this category is probably the macrocycle formed from a tetrathiafulvalene and two 2,6-bis(methylthiolo)pyridine units, connecting to the 3,3' and the 4,4'-carbon positions, respectively, of the TTF molecule.[33] Thus the tetrathiafulvalene is forced to bend over backwards and is extremely nonplanar, as shown in Fig. 3.6. No charge-transfer salts have been reported for this donor molecule.

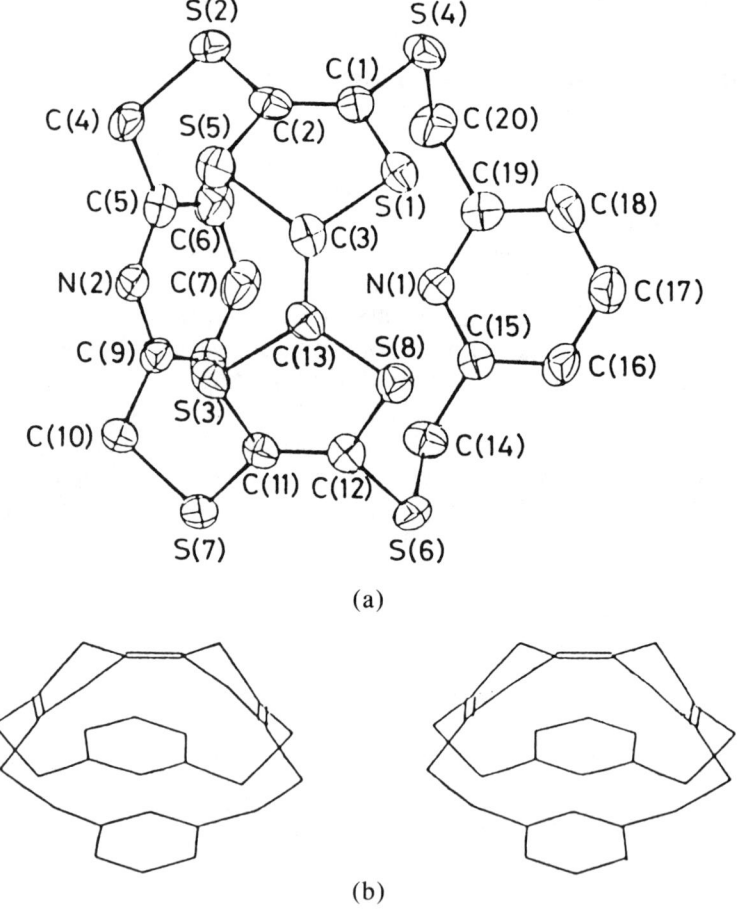

Figure 3.6 Crystal structure of the macrocycle formed from tetrathiafulvalene and 2,6-bis(methylthiolo)pyridine. (a) ORTEP plot; (b) stereoscopic stick figure (from ref. 33, with permission).

OVERVIEW OF CRYSTAL STRUCTURES

BEDT-TTF Salts

More than 130 crystal structures of ET-containing salts have been determined. Their unit cells (room temperature) and space groups are summarized in Table 3.1. In some cases, the unit cell axes were switched with respect to the original report, in order to facilitate comparison between related structures. Furthermore, it has become evident that triclinic unit cells determined on some diffractometers are left-handed, requiring an *odd* number of axis transpositions in order to bring them into standard form. Where that situation was recognized, as when the unit cells were determined independently by other workers, using other diffractometers, the handedness was inverted (usually by taking the complement to 180° for one of the unit cell angles). The table is grouped according to the shape of the anion, except that all polyhalides (especially poly-iodides) are listed in the "linear" category. There are examples where more than one anion exists in the crystal structure, for example, $(ET)_2(TlI_4)(I_3)$[42] or $(ET)_5(HgBr_4)_2(HgBr_3)$.[86] In those cases, the larger of the anions determines the category. In some cases, the unit cell data are obviously incorrect, as indicated by a unit cell volume incompatible with the reported cell constants. These cases are flagged in the table.

In an earlier review[2] a classification scheme (M,S,L) for listing ET crystal structures was presented. First, the number of ET donor layers is counted, to yield the variable L. Within each layer (if more than one occurs per unit cell, they are usually related to each other by some symmetry operation) it is usually possible to discern a stacking direction. The number of stacks per layer is thus indicated by S, and within a stack, the number of ET donor molecules is denoted by M. As an example, α'-$(ET)_2AuBr_2$ is classified as (4,1,2), indicating the presence of two layers, each containing one stack, and four ET molecules make up the stacking repeat unit. The product MSL should equal the number of ET molecules per formula unit times Z, the number of formula units per unit cell, thus the number of ET donor molecules in the entire cell. This classification scheme has been applied to other donor systems as well,[141] although some of the other donor molecules (and occasionally even ET) adopt packing patterns which are not easily expressed by the (M,S,L)-scheme.

Salts Derived from Other Donor Molecules

Table 3.2 lists unit cell crystallographic data for many salts based on electron-donor molecules other than ET. Since there rarely are many different salts for a given donor molecule, the table is not grouped by anion, but rather by donor molecules. Too many nonstandard packing patterns are observed with these salts, and therefore the (M,S,L)-scheme is omitted from the table although it could be applied to a number of table entries. Most of the donor molecules listed have been described in the previous sections of this chapter. We have attempted to be thorough, but no claim is made for completeness of this list, especially for the older donors such as HMTSF and DBTTF.

TABLE 3.1 Summary of Room Temperature Crystallographic Data for $(ET)_nX_m$ Salts

Compound	Space group	a(Å)	b(Å)	c(Å)	$\alpha(°)$	$\beta(°)$	$\gamma(°)$	Unit cell vol (Å^3)	Z	(M,S,L)	Reference
ET	$P2_1/c$	6.614	13.985	16.646	90	109.55	90	1449.6[a]	4	—	1
				Linear anions							
α-(ET)$_2$I$_3$	$P\bar{1}$	9.183	10.804	17.422	96.96	97.93	90.85	1698	2	(2,2,1)	34
β-(ET)$_2$I$_3$	$P\bar{1}$	6.615	9.100	15.286	94.38	95.59	109.78	855.9	1	(2,1,1)	35, 36
β_D-(ET)$_2$I$_3$	$P\bar{1}$	6.466	9.257	15.273	98.28	89.75	108.91	854.9	1	(2,1,1)	37
γ-(ET)$_3$(I$_3$)$_{2.5}$	$Pbnm$	13.76	14.73	33.61	90	90	90	6812	4	(3,2,2)	38
δ-(ET)I$_3$(TCE)$_{0.333}$[b]	$C2/c$	10.728	34.14	34.92	95.00	90	90	12736	24	—[c]	39
ε-(ET)$_2$(I$_3$)(I$_8$)$_{0.5}$	$P2_1/c$	13.974	18.77	17.40	67.3	90	90	4211	4	(4,2,1)	40
ζ-(ET)$_2$(I$_2$)(I$_8$)	$P2/c$	7.827	17.02	19.16	65.08	90	90	2315	2	—[c]	41
η-(ET)$_2$(I$_3$)(I$_5$)	$P2_1/a$	15.113	15.993	18.159	90	100.99	90	4309	4	(4,2,1)	42
η'-(ET)I$_3$	$P2_1/n$	12.792	10.745	14.768	90	98.92	90	2005.3	4	(2,2,1)	43
θ-(ET)$_2$(I$_3$)$_{1-x}$(AuI$_2$)$_x$	$Pnma$	10.076	33.853	4.994	90	90	90	1693	2	(1,2,2)	44
superstructure	$P2_1/c$	9.928	10.076	34.220	90	98.39	90	3386	4	(2,2,2)	45
θ-(ET)$_2$(I$_3$)$_{1-x}$(I$_2$Br)$_x$[e]	$Pnma$	10.068	33.851	4.970	90	90	90	1694	2	(1,2,2)	46
λ_D-(ET)$_2$I$_3$	monoclinic	10.086	9.947	34.31	90	98.51	90	3404	4	(2,2,2)	37
κ-(ET)$_2$I$_3$	$P2_1/c$	16.387	8.466	12.832	90	108.56	90	1687.6	2	(-,-,1)[f]	47
α-(ET)$_2$I$_2$Br	$P\bar{1}$	9.142	10.818	17.370	96.98	97.97	90.81	1687.9	2	(2,2,1)	48
β-(ET)I$_2$Br	$P\bar{1}$	6.612	9.024	15.192	94.16	95.23	110.12	842.3	1	(2,1,1)	49
α-(ET)$_2$IBr$_2$	$P\bar{1}$	8.905	12.031	16.402	85.16	88.66	70.86	1652	2	(2,2,1)	50
β-(ET)$_2$IBr$_2$	$P\bar{1}$	6.593	8.975	15.093	93.79	94.97	110.54	828.7	1	(2,1,1)	50
ε-(ET)$_2$(IBr$_2$)$_2$(TCE)$_{0.5}$[b]	$P2_1/c$	18.767	13.859	17.133	90	103.7	90	4329.3	4	(4,2,1)	51
α-(ET)$_2$BrICl	$P\bar{1}$	8.879	12.035	16.361	85.24	88.62	70.94	1647	2	(2,2,1)	52
β'-(ET)$_2$BrICl	$P\bar{1}$	6.642	9.816	12.975	87.29	101.11	98.28	821.3	1	(2,1,1)	53
β'-(ET)$_2$ICl$_2$	$P\bar{1}$	6.645	9.771	12.921	87.19	100.91	98.63	814.3	1	(2,1,1)	53
β''-(ET)$_2$ICl$_2$	$P\bar{1}$	5.769	9.003	16.290	93.07	96.58	103.64	814.3	1	(2,1,1)	54
β'''-(ET)$_2$ICl$_2$	$P\bar{1}$	5.734	8.979	16.313	93.24	96.58	103.57	808	1	(2,1,1)	55
β-(ET)$_2$AuI$_2$	$P\bar{1}$	6.603	9.015	15.403	94.95	96.19	110.66	845.2	1	(2,1,1)	56
γ'-(ET)$_2$AuI$_2$	$Acam$	12.833	7.914	34.16	90	90	90	3392	4	(2,2,2)	57
δ-(ET)$_2$AuI$_2$	$Pbcm$	6.849	14.914	33.586	90	90	90	3431	4	(4,1,2)	58, 59
α'-(ET)$_2$IAuBr	$P2$	32.044	6.735	7.838	90	92.96	90	1689.2	2	(2,1,2)	60
β-(ET)$_2$IAuBr	$P\bar{1}$	6.590	8.993	15.236	94.46	95.80	110.99	832.3	1	(2,1,1)	60
β'-(ET)$_2$IAuBr	$P\bar{1}$	9.071	16.406	5.770	97.61	103.43	91.70	826.3	1	(2,1,1)	61

TABLE 3.1 Summary of Room Temperature Crystallographic Data for $(ET)_nX_m$ Salts (continued)

Compound	Space group	$a(\text{Å})$	$b(\text{Å})$	$c(\text{Å})$	$\alpha(°)$	$\beta(°)$	$\gamma(°)$	Unit cell vol (Å^3)	Z	(M,S,L)	Reference
δ-(ET)$_2$IAuBr	$Pbcm$	6.823	14.842	33.091	90	90	90	3351.1	4	(4,1,2)	60
α'-(ET)$_2$AuBr$_2$	$P2/n$	7.799	6.723	31.756	90	93.46	90	1662.0	2	(2,1,2)	62
β''-(ET)$_2$AuBr$_2$	$P\bar{1}$	9.027	16.372	5.712	97.60	102.94	92.09	813.7	1	(2,1,1)	63
δ-(ET)$_2$AuBr$_2$	$Pbcm$	6.798	14.837	33.624	90	90	90	3290	4	(4,1,2)	64
β'-ET$_2$AuCl$_2$	$P\bar{1}$	6.651	9.761	12.734	86.12	100.70	99.41	800.7	1	(2,1,1)	65
(ET)AuCl$_2$	$P\bar{1}$	10.033	17.412	5.916	95.76	93.63	93.48	1023.8	2	(2,1,1)	66
α'-(ET)$_2$Au(CN)$_2$	$P2/n$	7.932	6.735	31.018	90	91.14	90	1657.7	2	(2,1,2)	62
η-(ET)$_2$Au(CN)$_2$	$C2/c$	36.517	4.254	22.117	90	118.38	90	3199.3	4	(1,4,2)	67
α'-(ET)$_2$Ag(CN)$_2$	$P2/n$	7.956	6.732	30.738	90	90.05	90	1646.3	2	(2,1,2)	62
(ET)$_2$Ag(CN)$_2$	$P2_1/n$	10.996	4.281	34.093	90	90.02	90	1631	2	(1,2,2)	68
(ET)$_3$Ag(CN)$_2$	$P2_12_12$	9.519	4.952	33.927	90	90	90	1599.3	2	(1,2,2)	68
(ET)CuCl$_2$	$P\bar{1}$	6.577	12.154	5.781	100.87	94.74	108.58	424.4	1	(1,1,1)	69
(ET)$_2$CuCl$_2$	$P2/c$	7.941	6.676	30.586	90	97.47	90	1607.7	2	(2,1,2)	70
α-(ET)$_2$Cu(NCS)$_2$	$P\bar{1}$ (?)	9.052	10.854	17.471	100.95	97.34	90.37	1672	2	(2,2,1)	71
				Planar anions							
α-(ET)$_3$(NO$_3$)$_2$	$P2_1/n$	5.890	31.125	12.915	90	103.73	90	2300	2	(3,1,2)	72
β-(ET)$_n$(NO$_3$)$_m$	$P2_1/c$	6.529	12.379	28.487	90	95.01	90	2294	—	—[g]	72
γ-(ET)$_n$(NO$_3$)$_m$	$P2_1/n$	15.035	6.584	30.900	90	101.64	90	2996	2	—[g]	72
(ET)$_2$[C(CN)$_3$]	$P2/a$	14.979	6.700	16.395	90	94.80	90	1639.7	2	(4,1,1)	73
(ET)$_2$[C$_4$(CN)$_6$]	$P2/a$	34.212	6.725	7.958	90	94.18	90	1825.9	2	(2,1,2)	74
(ET)$_2$[C$_4$(CN)$_6$](Ø-CN)^b	$P\bar{1}$	12.919	18.643	7.855	93.63	107.81	91.43	1795.5	—	—[g]	74
(ET)$_2$[C$_5$(CN)$_3$](TCE)$_x$	$Pncm$	6.774	14.657	41.471	90	90	90	4075	4	(4,1,2)	75
(ET)(TCNQ)	$P\bar{1}$	23.915	7.817	6.650	105.10	94.33	89.18	1197.0	2	(2,1,1)	76
(ET)(TCNQ)	$P2_1/n$	23.897	7.327	7.053	90	98.37	90	1221.8	2	—[h]	77
(ET)$_4$Pt(C$_2$O$_4$)$_2$	$P\bar{1}$	8.678	11.878	15.757	105.49	91.05	91.96	1563.6	1	(2,2,1)	78
(ET)$_4$Ni(CN)$_4$	$P\bar{1}$	9.699	10.959	16.430	95.99	97.87	115.07	1541	1	(2,2,1)	79, 80
(ET)$_4$Pt(CN)$_4$	$P\bar{1}$	9.721	10.940	16.552	95.71	98.48	115.14	1550	1	(2,2,1)	79
(ET)$_4$Pt(CN)$_4$	$P\bar{1}$	9.762	11.002	16.625	95.83	98.32	115.01	1574.9	1	(2,2,1)	81
(ET)$_4$Pt(CN)$_4$(TCE)											81
(ET)AuBr$_2$Cl$_2$	$Pnnm$	4.467	18.670	11.901	90	90	90	992.5	2	—[i]	82
(ET)$_2$AuCl$_4$	$P\bar{1}$	6.590	8.561	18.319	93.19	105.52	119.47	845.1	1	(2,1,1)	83
(ET)$_2$AuCN$_2$Cl$_2$	$P2/c$	17.727	6.690	15.284	90	108.19	90	1722	2	(4,1,1)	83
(ET)(AuCl$_4$)(AuCl$_2$)	$C2/m$	7.683	9.375	16.079	90	98.55	90	1145.3	2	—[c]	84

TABLE 3.1 Summary of Room Temperature Crystallographic Data for $(ET)_nX_m$ Salts (continued)

Compound	Space group	$a(\text{Å})$	$b(\text{Å})$	$c(\text{Å})$	$\alpha(°)$	$\beta(°)$	$\gamma(°)$	Unit cell vol (Å^3)	Z	(M,S,L)	Reference
$(ET)_6(CuBr_4)(CuBr_2)$	$P2_1/c$	16.937	10.123	14.178	90	102.59	90	2372.2	1	(3,2,1)	85
$(ET)HgBr_3$	$P\bar{1}$	10.278	15.730	6.437	97.93	96.01	106.1	978.8	2	(2,1,1)	86
$\zeta\text{-}(ET)_3(HgBr_3)_2$	$P\bar{1}$?	7.758	10.555	18.371	90.45	94.16	101.31	1471.0	1	(3,1,1)	87
$(ET)_2HgI_3$	$C2/c$	19.881	12.690	22.523	90	103.02	90	5536.7	—	—[g]	87
$(ET)Ni(dmit)_2$[j]	$P\bar{1}$	7.355	6.667	15.031	93.43	109.91	97.69	682.3	1	—[h]	88
$(ET)Ni(mnt)_2$[j]	$P\bar{1}$	6.273	7.833	13.127	96.17	92.59	96.22	636.4	1	—[h]	89
$(ET)[C_8S_2(CN)_4]$[k]	$P\bar{1}$	7.205	13.654	6.898	91.76	103.85	101.32	643.9	1	—[h]	90
Tetrahedral anions										
$(ET)_2ReO_4$	$P\bar{1}$	7.802	12.596	17.117	73.46	80.45	89.15	1589	2	(2,2,1)	91
$(ET)_2BrO_4$	$P\bar{1}$	7.795	12.613	17.102	72.97	80.44	88.74	1589.0	2	(2,2,1)	92
$\alpha\text{-}(ET)_3(ReO_4)_2$	$P2_1/n$	8.498	30.566	9.413	90	89.57	90	2417.9	2	(2,1,2)	93
$\beta\text{-}(ET)_3(ReO_4)_2$	$P2_1/c$	16.298	12.013	12.416	90	91.24	90	2430.2	2	—[g]	94
$\gamma\text{-}(ET)_3(ReO_4)_2$	$P\bar{1}$	8.409	9.295	16.124	101.97	88.39	82.95	1221.7	1	(1,3,1)	94
$\gamma\text{-}(ET)_3(IO_4)_2$	$P\bar{1}$	8.418	9.303	16.231	101.93	88.23	82.67	1231.4	1	(1,3,1)	94
$\gamma\text{-}(ET)_3(BrO_4)_2$	$P\bar{1}$	7.670	9.550	16.686	89.38	87.02	83.87	1213.6	1	(1,3,1)	95
$\gamma\text{-}(ET)_3(ClO_4)_2$	$P\bar{1}$	7.613	9.498	16.463	89.16	87.17	95.91	1182	1	(1,3,1)	96
$\gamma\text{-}(ET)_3(HSO_4)_2$	$P\bar{1}$	7.633	9.440	16.607	87.72	85.04	83.13	1181.3	1	(1,3,1)	97
$\gamma\text{-}(ET)_3(FSO_3)_2$	$P\bar{1}$	7.605	9.421	16.578	87.74	95.91	96.89	1174.1	1	(1,3,1)	94
$\gamma\text{-}(ET)_3(BF_4)_2$	$P\bar{1}$	7.654	9.496	16.398	88.99	92.42	96.09	1183.9	1	(1,3,1)	94
$(ET)(ReO_4)(THF)_{0.5}$[b]	$P2/c$	12.679	8.073	19.292	90	97.69	90	1948.2	4	(2,2,1)	98
$(ET)(IO_4)(THF)_{0.5}$	$P2/c$	12.692	8.036	19.285	90	97.66	90	1950	4	(2,2,1)	98
$(ET)_2(BrO_4)(TCE)_{0.5}$	$P\bar{1}$	7.656	12.957	18.590	109.6	90.2	105.1	1668.4	2	(2,2,1)	95
$(ET)_2(ClO_4)(TCE)_{0.5}$	$P\bar{1}$	7.740	12.966	18.620	110.85	79.32	104.80	1684	2	(2,2,1)	99
$(ET)_2(FSO_3)(TCE)_{0.5}$	$P\bar{1}$	7.786	13.033	18.590	100.09	90.21	105.27	1699.8	1	(2,2,1)	100
$(ET)_2(ClO_4)(C_4H_8O_2)_2$[b]	$P2/c$	8.242	6.677	32.998	90	92.71	90	1814	2	(2,1,2)	96
$(ET)_2(TlI_4)(I_3)$	$P2_1/n$	8.348	15.294	34.217	90	93.61	90	4360	4	(2,2,2)	42
$(ET)_2InI_4$	$P\bar{1}$	6.726	16.313	19.496	97.91	90.26	113.39	1941.8	2	(4,1,1)	101
$(ET)_2InBr_4$	$P\bar{1}$	6.618	16.040	17.470	95.29	92.43	99.12	1819.9	2	(4,1,1)	102
$(ET)_2InCl_4$[l]	triclinic	6.529	(8.347)	16.249	92.29	99.70	93.25	(870.3)	(1)	—	103
$(ET)_2GaI_4$	$P\bar{1}$	6.719	14.950	19.216	81.27	90.47	91.34	1907	2	(4,1,1)	101
$(ET)_2CuCl_4$	$P\bar{1}$	16.238	8.978	19.369	120.84	91.61	90.69	2406.5	3	—[g]	104
$(ET)_2FeCl_4$	$P\bar{1}$	6.626	15.025	17.805	82.80	89.53	88.15	1757.5	2	(4,1,1)	105

TABLE 3.1 Summary of Room Temperature Crystallographic Data for $(ET)_n X_m$ Salts (*continued*)

Compound	Space group	a (Å)	b (Å)	c (Å)	α (°)	β (°)	γ (°)	Unit cell vol (Å^3)	Z	(M,S,L)	Reference
(ET)FeBr$_4$	$P\bar{1}$	8.634	10.980	11.773	91.91	102.84	93.73	1084.5	2	—^{m}	105
(ET)$_3$(ZnCl$_4$)$_2$	$P\bar{1}$	6.800	9.64	20.39	93.5	90.0	101.4	1307	1	—^{n}	81, 106
(ET)$_3$MnCl$_4$)$_2$	$P\bar{1}$	6.802	9.714	20.595	93.75	89.70	101.66	1329.8	1	—^{n}	107
(ET)$_5$(HgBr$_4$)$_2$(HgBr$_3$)	$P\bar{1}$	14.454	26.404	12.977	111.83	92.44	92.23	4585	2	—^{n}	86
(ET)$_2$(HgBr$_4$)(TCE)^o	$C2/c$	56.13	4.202(x2)	22.306	90	100.88	90	5166(x2)	4(8)	(1,4,2)	108
Octahedral Anions											
(ET)$_2$AsF$_6$	$A2/a$	14.890	6.666	35.379	90	111.19	90	3274	4	(4,1,2)	109
(ET)$_2$SbF$_6$	$I2/c$	33.56	14.93	6.70	90	93.88	90	3349	4	(4,1,2)	110
α-(ET)$_2$PF$_6$	$P\bar{1}$	14.771	8.597	6.462	95.71	97.64	98.87	794.3	1	(2,1,1)	111
β-(ET)$_2$PF$_6$	$Pnna$	14.960	32.643	6.664	90	90	90	3255.4	4	(4,1,2)	112
γ-(ET)$_2$PF$_6$	$P\bar{1}$										
(ET)$_4$PtCl$_6$(Ø-CN)^b	$P\bar{1}$	8.665	11.975	17.35	83.0	87.1	94.9	1778	1	(2,2,1)	113
Monoatomic anions											
(ET)$_3$Cl$_2$(H$_2$O)$_2$	$P\bar{1}$	11.214	13.894	15.924	94.71	109.00	97.03	2304.3	2	(3,2,1)	114, 115, 116
(ET)$_4$Cl$_2$(H$_2$O)$_4$	$P2/b$	6.684	14.38	17.20	109.1	90	90	1616	1	(4,1,1)	116
(ET)$_3$Br$_2$(H$_2$O)$_2$	$P\bar{1}$	7.718	9.587	16.167	85.42	85.91	81.09	1176.0	1	(1,3,1)	117
γ-(ET)$_3$Br$_2$	$P\bar{1}$	7.743	9.606	16.194	84.92	95.00	99.05	1181.5	1	(1,3,1)	94
Polynuclear and other molecular (discrete) anions											
(ET)$_3$Hg$_2$Cl$_6$	$P2_1/c$	11.24	6.645	36.36	90	105.45	90	2617	2	(3,1,2)	118
(ET)$_4$Hg$_2$Br$_6$(TCE)	$P2_1/c$	19.344	13.401	29.418	90	103.87	90	7404	4	(8,2,1)	119
(ET)$_4$Hg$_2$Br$_6$(THF)	$P2_1/c$	17.095	13.470	16.001	90	91.71	90	3683	2	(4,2,1)	120
(ET)$_4$Hg$_2$Br$_6$(Ø-Cl)$_2^b$	$P\bar{1}$	19.131	16.133	12.706	105.57	78.45	98.43	3683.9	2	—^{g}	121
(ET)$_x$Hg$_{2.5}$Br$_2$(Ø-CN)$_2$	$P\bar{1}$	20.911	17.106	12.043	110.17	94.71	66.18	3691	2	—^{g}	121
(ET)$_2$Cd$_2$I$_6$	$P\bar{1}$	9.067	10.515	12.441	97.12	103.43	106.13	1085	1	—^{m}	122
(ET)$_4$Cd$_2$I$_6$	$P\bar{1}$	8.856	11.961	18.092	87.72	84.80	74.04	1835	1	(4,1,1)	120
(ET)$_3$Ta$_2$F$_{11}$	$P2_1/m^p$	16.683	11.928	12.550	90.23	90	90	2497	2	(3,2,1)	123
(ET)$_4$[Mo$_6$Cl$_8$)I$_6$](THF)$_2$	$P\bar{1}$	10.384	12.372	19.696	95.85	91.45	110.18	2325	1	(-,-,1)^q	124
(ET)$_4$((Mo$_6$Cl$_8$)I$_6$)(THF)	$C2/c$	28.57	13.44	17.60	90	108.04	90	8679	4	—^{g}	124
ζ-(ET)$_4$[(Mo$_6$Cl$_8$)Cl$_6$](THF)$_2$	$P\bar{1}$	9.561	12.400	19.325	92.86	91.33	110.18	2146	1	(2,2,1)	67
(ET)$_3$[V(dmit)$_3$]$_2^j$	$P\bar{1}$	6.390	17.536	19.167	103.37	85.86	90.28	2084	1	—^{i}	125
(ET)$_2$CF$_3$SO$_3$	$P2_1/n$	6.649	33.050	15.018	90	90.98	90	3300	4	(4,1,2)	126
(ET)$_2$(p-CH$_3$-Ø-SO$_3$)	$P2$	7.785	6.697	34.402	90	91.10	90	1793.2	2	(2,1,2)	126

TABLE 3.1 Summary of Room Temperature Crystallographic Data for $(ET)_nX_m$ Salts (continued)

Compound	Space group	a(Å)	b(Å)	c(Å)	$\alpha(°)$	$\beta(°)$	$\gamma(°)$	Unit cell vol (Å³)	Z	(M,S,L)	Reference
$(ET)(Re_6Se_5Cl_9)(DMF)_2^b$	$P\bar{1}$	8.823	9.920	12.527	98.48	91.48	97.87	1073.0	1	$(2,-,-)^f$	127
$(ET)_8SiW_{12}O_{14}$	$I2$	14.017	43.259	14.065	90	107.26	90	8144	2	(4,2,2)	128
Polymeric anions											
$(ET)_3Ag_6.4I_8$	$P\bar{1}$	4.430	16.874	21.10	96.55	91.56	93.44	1566.4	1	(1,3,1)	129
$(ET)Ag_4(CN)_5$	$Fddd$	13.24	19.48	19.62	90	90	90	5060	8	—^{i}	130
κ-$(ET)_2Ag(CN)_2(H_2O)$	$P2_1$	12.593	8.642	16.080	90	109.33	90	1651.2	2	$(-,-,1)^f$	68
$(ET)Ag_{1.6}(SCN)_2$	$Pbca$	4.247	11.588	40.18	90	90	90	1982	4	(1,2,2)	131
$(ET)Ag_{2.4}Br_3$	$P2_12_12_1$	4.232	11.699	40.29	90	90	90	1994	4	(1,2,2)	132
κ-$(ET)_2Cu(NCS)_2$	$P2_1$	16.256	8.456	13.143	90	110.28	90	1694.8	2	$(-,-,1)^f$	133
$(ET)Cu_2(NCS)_3$	$P2_1/a$	12.688	10.444	16.697	90	90.89	90	2215	4	—^{m}	131
$(ET)_2Cu_5I_6$	$P2_1/c$	4.326	21.29	22.41	91.31	90	90	2063	2	(1,4,1)	134
$(ET)BiI_4$	$P\bar{1}$	8.265	11.118	14.424	110.76	96.41	103.57	1176.8	2	(2,1,1)	135
$(ET)_2HgBr_5(TCE)$	$P2_1/c$	40.583	4.183	22.76	90	104.32	90	3745	4	(1,4,2)	120
$(ET)_4Hg_3I_8^s$	—	—	—	—	—	—	—	—		—	87
$(ET)_4Hg_3I_8$	$C2/c$	11.920	40.740	8.430	90	110.21	90	3842	2	(2,2,2)	136
$(ET)_2Hg_3I_5$	$C2/c$	11.926	40.795	8.369	90	109.79	90	3821.2		—^{g}	87
$(ET)_4Hg_{2.89}Br_8$	$I2/c$	11.219	8.706	37.105	90	90.97	90	3624	2	$(-,-,2)^f$	137
(Hg sublattice)	$I2$	3.877	8.706	37.141	90	87.30	90	(1252)			
$(ET)_4Hg_{3-x}Cl_8^t$	$I2/c$	11.062	8.754	35.92	90	91.01	90	3478	2	$(-,-,2)^f$	138
(Hg sublattice)	$I2$	ca. 12	8.754	?	90	?	90)				
$(ET)_2KHgSCN)_4$	$P\bar{1}$	9.933	10.082	20.565	93.06	103.70	90.91	1997	2	(2,2,1)	139
$(ET)_2(NH_4)Hg(SCN)_4$	$P\bar{1}$	9.963	10.091	20.595	93.30	103.65	90.53	2008	2	(2,2,1)	140
κ-$(ET)_2Cu[N(CN)_2]Br$	$Pnma$	12.942	30.016	8.539	90	90	90	3317	4	$(-,-,2)^f$	261
κ-$(ET)_2Cu[N(CN)_2]Cl$	$Pnma$	12.977	29.979	8.480	90	90	90	3299	4	$(-,-,2)^f$	263

[a] Reported unit cell volume incompatible with cell constants. $\sigma(a)$ is reported five times larger than $\sigma(b)$ or $\sigma(c)$.

[b] Solvent molecule abbreviations: TCE = 1,1,2-trichloroethane; Ø-CN = benzonitrile; THF = tetrahydrofuran; $C_4H_8O_2$ = 1,4-dioxane; Ø-Cl = chlorobenzene; DMF = dimethylformamide.

[c] Mixed layers.

[d] $x \approx 0.02$; "average" structure.

[e] $x \approx 0$.

[f] Twisted dimers.
[g] Unit cell only reported.
[h] Mixed stacks.
[i] 3-dimensional network.
[j] dmit = 1,3-dithia-2-thione-4,5-dithiolate, $C_3S_5^{2-}$; mnt = maleonitrile-2,3-dithiolate, $C_4N_2S_2^{2-}$.
[k] $C_8S_2(CN)_4$ = 2,5-bis(dicyanomethylene)-2,5-dihydrothieno[3,2-b]thiophene:

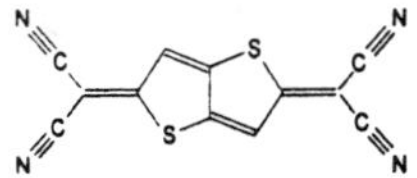

[l] Unit cell obtained from a twinned crystal. b-axis could be a multiple of indicated value.
[m] Isolated dimers.
[n] Two distinct donor molecule layers, one containing some ET^{2+} cations, the other lesser-charged species.
[o] ET lattice; b = 4.202 Å; anion lattice: b = 8.404 Å. M = 1 refers to ET lattice.
[p] Space group not given in original, but inferred from unit cell drawing.
[q] Double ribbons.
[r] Isolated side-by-side chains.
[s] Reported "in progress." However, it is strange that the unit cell reported by Müller et al.,[87] for $(ET)_2Hg_2I_5$ matches that indicated by Aldoshina et al.,[136] for $(ET)_4Hg_3I_8$.
[t] Incommensurate composite structure. The ET molecule and the halogen atoms are associated with a different unit cell from that of the mercury atoms.

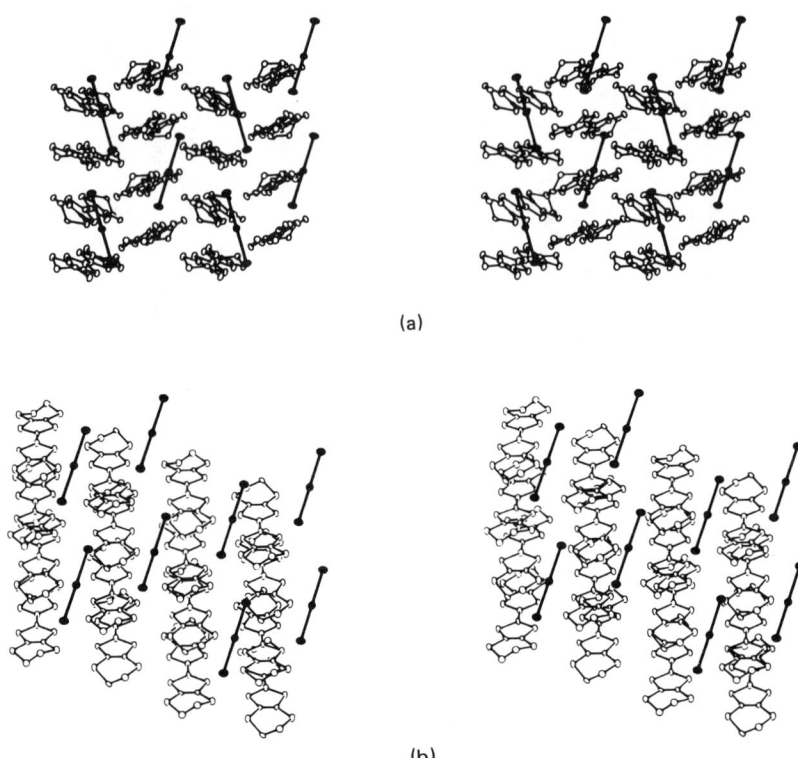

Figure 3.7 Stereoviews of (a) the two-dimensional network of ET molecules and the adjacent anion sheet of α-(ET)$_2$I$_3$; (b) the ET network and adjacent anion sheet of β-(ET)$_2$I$_3$. Both of these phases of (ET)$_2$I$_3$ contain a layered composition of alternating sheets of ET donors and I$_3^-$ anions. Reprinted with permission from Williams, J. M.; Wang, H. H.; Emge, T. J.; Geiser, U.; Beno, M. A.; Leung, P. C. W.; Carlson, K. D.; Thorn, R. J.; Schultz, A. J.; Whangbo, M.-H. In *Prog. Inorg. Chem.*; Lippard, S. J., Ed.; Vol. 35, p. 51. Copyright © 1987 John Wiley & Sons (ref. 2).

CRYSTAL STRUCTURES OF SUPERCONDUCTING ET SALTS

β-(ET)$_2$X Family

The β-(ET)$_2$X family of compounds contains several members, among them the superconducting salts with $X^- = $ I$_3^-$,[35, 36] AuI$_2^-$,[56] and IBr$_2^-$,[50] as well as nonsuperconducting β-(ET)$_2$I$_2$Br.[49] Like most ET compounds, these salts contain layers of ET donor molecules separated by anion layers. The β-type structures are triclinic, with $Z = 1$ (i.e., two-donor molecules), containing one donor molecule layer per unit cell. The molecules are stacked along the $a + b$-diagonal direction, with two molecules per repeat unit. Subsequent stacks are related by translations. All molecules are coplanar by symmetry. Along the stacks, the molecules are slightly dimerized in their lateral displacements when viewed along the molecular plane normal. The short (less than the van der Waals distance of 3.6 Å) S···S contacts are found between stacks, rather than within the stacks, and their two-dimensional connectivity gives the donor layer the appearance of a honeycomb. The anions are located on inversion centers, even unsymmetrical I_2Br^-, which is disordered in the crystal structure of β-$(ET)_2I_2Br$.[49] It is thought that this disorder, and the associated random electrostatic potential, prevents β-$(ET)_2I_2Br$ from becoming a superconductor.

The β-structural type is often compared with the α-type, as found in α-$(ET)_2I_3$,[34] that is, the first ET salt reported with a linear, triatomic anion. Both types are shown in Fig. 3.7. They are both triclinic, with one donor layer per unit cell. However, the α-cell volume is twice the size of the β-cell volume; thus not all the ET molecules nor the anions are crystallographically equivalent. The anions occupy two distinct inversion centers, and their orientations are different. Of the ET molecules, one set is located on a general position, forming one kind of stack (I). The other stack (II) is formed by ET molecules located on two sets of inversion centers and almost parallel to each other (11° dihedral angle). However, the molecules on stack I form a marked dihedral angle with the molecules on stack II (59.4° and 70.4° with respect to the two different molecules on stack II). A much more detailed discussion of the α- and β-structural types has been published in an earlier review article.[2]

β-$(ET)_2I_3$ at Low Temperature or High Pressure

Unlike the other β-$(ET)_2X$ salts, β-$(ET)_2I_3$ undergoes a crystallographic phase transition at temperatures below 175 K (originally[225, 226] estimated in the vicinity of 200 K, and later measured more accurately as 175 K at ambient pressure[227]) to an incommensurate modulated structure.[225, 226] Incommensurate diffraction satellites [$\mathbf{q} = (0.075(3)a^*, 0.275(5)b^*, 0.205(5)c^*)$][228] to the main Bragg reflections were observed both with X-rays[226] and with neutrons.[225] A histogram of neutron diffraction data obtained at the Argonne Intense Pulsed Neutron Source (IPNS) by use of an area detector is presented in Fig. 3.8(a), showing incommensurate satellite reflections. A full structural analysis of the first-order satellite intensities, collected at 120 K on an X-ray diffractometer, was carried out[230] by use of a rigid-body displacement/rotation method.[231] In this model, the sinusoidal modulation components of displacement and librational orientation are simultaneously applied to all atoms comprising a molecule, in this case separately for the ET donor cation and the individual atoms of the triiodide anion (under the constraints of centro-symmetry at the crystal origin; no orientational components for individual atoms), while the basic-structure parameters (positional and thermal) were varied in the usual way. It was found that both the triiodide units and the ET molecules retain their orientation within approximately 1°, but they exhibit considerable displacement amplitudes, that is, 0.29 Å and 0.12 Å, respectively. The major displacement components are along the a-axis for I_3^-, and within the molecular plane perpendicular to the long molecular axis

TABLE 3.2 Summary of Room Temperature Crystallographic Data for (Donor)$_n$X$_m$ Salts Other Than ET Salts

Compound	Space group	a(Å)	b(Å)	c(Å)	α(°)	β(°)	γ(°)	Unit cell vol (Å^3)	Z	Reference
				MT[a]						
(MT)SbF$_6$	P$\bar{1}$	9.573	7.500	5.633	100.20	91.61	90.96	397.7	1	142, 143
(MT)AsF$_6$[b]										143
MT	P2$_1$/c	4.172	24.203	6.581	90	112.59	90	613.6	2	144
θ-(MT)$_2$AuBr$_2$	Pnma	10.406	31.453	4.163	90	90	90	1362.5	2	145
θ-(MT)$_2$AuI$_2$[b]	Pnma									145
θ-(MT)$_2$AsF$_6$	C2/c	34.661	4.027	10.894	90	95.94	90	1529.3	2	145
(MT)$_2$Au(CN)$_2$	P2$_1$/a	10.827	8.420	15.614	90	102.46	90	1390	2	146
(MT)$_3$PF$_6$(DCE)[c]	P$\bar{1}$	16.841	9.998	7.776	113.64	99.72	100.03	1138.6	1	147
(MT)$_3$ClO$_4$(DCE)[c]	Pnma	12.728	33.302	10.204	90	90	90	4325.1	4	148
(MT)I$_3$	P$\bar{1}$	13.185	9.433	7.830	102.99	104.64	67.56	861.8	2	149
				MET[d]						
MET	P2$_1$/n	6.422	13.972	15.479	90	92.45	90	1387.6	4	150
(MET)ClO$_4$	C2/m	13.842	10.255	5.842	90	104.31	90	803.6	2	151
(MET)PF$_6$	C2/m	14.146	10.521	5.901	90	105.02	90	848.2	2	151
(MET)$_3$(ReO$_4$)$_2$	P$\bar{1}$	8.007	9.814	15.526	90.90	90.85	110.58	1142	1	151
				MPT[e]						
(MPT)$_2$ClO$_4$(THF)[c]	P$\bar{1}$	9.110	12.501	16.611	104.46	88.47	102.42	1788	2	3
(MPT)$_2$PF$_6$(THF)[c]	P$\bar{1}$	9.215	12.528	17.110	105.03	91.14	103.56	1848	2	3
				EPT[f]						
t-(EPT)$_2$ICl$_2$	P$\bar{1}$	6.880	10.925	12.048	87.32	100.30	100.54	875.9	1	152
m-(EPT)$_2$ICl$_2$	P2/c	15.752	6.780	16.910	90	101.34	90	1771	2	152
(EPT)Au(CN)$_2$	P$\bar{1}$	9.796	12.728	16.457	103.77	93.07	102.72	1932	4	153
(EPT)InI$_4$	P2$_1$/c	6.920	12.176	30.421	90	92.17	90	2561	4	154
(EPT)$_2$PF$_6$	P2/c	15.452	6.767	16.539	90	97.24	90	1716	2	155
(EPT)$_2$GaBr$_4$	P$\bar{1}$	11.277	13.253	15.426	92.66	95.30	92.26	2291	4	156
				PT[g]						
(PT)I$_3$	P$\bar{1}$	6.843	9.246	16.350	91.39	93.21	111.11	962.3	1	157
(PT)$_2$IBr$_2$	P$\bar{1}$	6.802	9.259	16.102	91.15	92.46	111.73	941.2	1	158
PT	P2$_1$/c	10.484	12.308	13.040	90	96.55	90	1671.7	4	159
α-(PT)$_2$ICl$_2$	I2/c	16.83	6.974	31.92	90	94.99	90	3732	4	160
(PT)$_2$ICl$_2$	P2/c	15.950	6.930	17.170	90	100.84	90	1864	2	161
(PT)$_3$(InI$_4$)$_2$	P$\bar{1}$	7.403	9.170	25.883	89.29	96.34	92.83	1774.2	1	162
(PT)$_3$(TlI$_4$)$_2$	P$\bar{1}$	7.400	9.166	25.858	89.43	96.51	92.67	1740.6	1	162
				16.865	90	93.94	90	2833.7	2	163

Compound	Space group	$a(\text{Å})$	$b(\text{Å})$	$c(\text{Å})$	$\alpha(°)$	$\beta(°)$	$\gamma(°)$	Unit cell vol (Å^3)	Z	Reference
(PT)HgBr$_3$	$P\bar{1}$	15.411	10.998	6.778	103.51	93.26	101.03	1090	2	121
α-(PT)[Ni(dmit)$_2$]$_2$[h]	$P2_1/c$	4.897	25.512	17.231	90	101.20	90	2111.6	2	164
BE(P)DSe-TSeF[i]										
BEDSe-TSeF	$P2_1/c$	6.783	14.235	17.555	90	109.20	90	1600.7	4	144
(BEDSe-TSeF)PF$_6$	$C2/m$	14.868	11.003	6.114	90	104.07	90	970.2	2	165
(BEDSe-TSeF)$_2$AuBr$_2$	$P\bar{1}$	6.815	8.101	17.049	97.76	99.44	106.39	874.3	1	166
(BEDSe-TSeF)AuBr$_2$	$P\bar{1}$	17.625	16.244	12.466	116.64	107.84	92.85	2962	6	167
BPDSe-TSeF	$P2_1/c$	10.971	12.681	13.353	90	94.59	90	1835	4	168
BEDSe-TTF[j]										
(BEDSe-TTF)$_2$I$_3$	$P\bar{1}$	6.781	8.767	16.023	89.68	94.76	110.66	887.9	1	14
(BEDSe-TTF)$_2$AuI$_2$	$P\bar{1}$	7.614	8.341	15.538	77.56	98.31	112.71	886.9	1	14
(BEDSe-TTF)$_2$IBr$_2$	$P\bar{1}$	6.863	10.065	13.183	87.94	100.49	98.84	884.8	1	169
BEDSe-TTF	$P2_1/n$	6.842	14.047	16.094	90	94.49	90	1542	4	170, 171
(BEDSe-TTF)I$_3$	$P2_1/n$	12.503	10.592	15.842	90	100.12	90	2065.4	4	170
DMET (DMEDT-DSeDTF, EDTDM-DTDSF)[k]										
DMET	$P2_1/c$	6.724	12.144	18.228	90	111.21	90	1387.7	4	26
(DMET)$_2$Au(CN)$_2$	$P\bar{1}$	6.763	7.710	15.314	91.29	96.81	75.04	766.0	1	172
(DMET)$_2$AuCl$_2$	$P\bar{1}$	7.021	7.742	15.622	108.60	98.27	70.82	759.5	1	173
(DMET)$_2$AuI$_2$	$P\bar{1}$	6.772	7.724	15.776	90.02	98.35	75.71	784.7	1	174
(DMET)$_2$I$_3$[l]	$P\bar{1}$	6.703	7.752	16.254	94.85	105.92	78.25	794.8	1	175
(DMET)$_2$IBr$_2$	$P\bar{1}$	6.695	7.748	16.220	94.80	105.92	78.21	791.7	1	176
(DMET)$_2$AuBr$_2$	$P\bar{1}$	6.732	7.687	15.775	97.37	103.13	75.99	768	1	177
κ$_1$-(DMET)$_2$AuBr$_2$	$P2_1/a$	11.487	8.207	16.107	90	90.94	90	1518	2	177
κ$_2$-(DMET)$_2$AuBr$_2$	$P2_1/a$	11.542	8.237	16.168	90	90.99	90	1537	2	177
(DMET)$_2$BF$_4$	$P\bar{1}$	7.032	7.812	27.502	90.86	90.28	106.21	1450.5	2	176
(DMET)$_2$ClO$_4$	$P\bar{1}$	7.041	7.839	27.515	88.67	90.49	105.79	1461.0	2	23
(DMET)$_2$PF$_6$	$P\bar{1}$	7.700	7.903	13.096	101.93	97.72	94.57	767.9	1	178
(DMET)$_2$AsF$_6$	$P\bar{1}$	7.710	7.935	13.244	102.05	97.76	94.97	779.6	1	179
DIMET (DMEDT-TTF)[m]										
(DIMET)$_2$PF$_6$	$P\bar{1}$	7.644	7.814	12.966	78.84	82.17	84.58	747.0	1	24, 180
(DIMET)$_2$SbF$_6$	$P\bar{1}$	8.360	6.645	14.555	76.88	80.53	75.60	757.7	1	181
(DIMET)$_2$ClO$_4$	$P\bar{1}$	7.034	7.850	27.024	87.92	88.92	74.65	1438.1	2	24
(DIMET)$_2$ClO$_4$	$P\bar{1}$	7.000	7.824	27.010	88.10	89.02	74.58	1425	2	25
(DIMET)$_2$ClO$_4$(THF)	$P\bar{1}$	6.731	7.632	33.32	96.53	92.99	103.13	1650.7	2	182
(DIMET)$_2$Br	$P2_1/c$	15.412	7.082	17.082	90	99.78	90	1977.6	?	24

TABLE 3.2 Summary of Room Temperature Crystallographic Data for $(Donor)_nX_m$ Salts Other Than ET Salts (continued)

Compound	Space group	$a(\text{Å})$	$b(\text{Å})$	$c(\text{Å})$	$\alpha(°)$	$\beta(°)$	$\gamma(°)$	Unit cell vol (Å^3)	Z	Reference
				EDT-TTF[a]						
(EDT-TTF)I_{3+x}	$P2_1/c$	15.307	9.622	12.013	90	110.69	90	1655.7	4	183
(EDT-TTF)$_2$AuI$_2$	$F222$	12.473	7.238	29.650	90	90	90	2676	4	22
(EDT-TTF)$_2$AuBr$_2$	$C2/m$	7.200	28.974	6.647	90	111.68	90	1288.7	2	141, 184
(EDT-TTF)$_2$IBr$_2$	$P\bar{1}$	6.242	8.353	13.443	91.27	94.48	110.06	655.6	1	141, 185
(EDT-TTF)$_2$Au(CN)$_2$	$P\bar{1}$	6.388	7.225	14.752	90.50	101.56	106.15	639.2	1	186
(EDT-TTF)$_2$TaF$_6$	$P\bar{1}$	6.463	7.049	15.213	98.77	102.57	102.22	646.7	1	186
α-(EDT-TTF)$_2$AsF$_6$	$P\bar{1}$	6.457	7.043	14.855	101.68	96.15	102.38	638.1	1	186
β-(EDT-TTF)$_2$AsF$_6$	$Pccn$	28.303	7.120	12.706	90	90	90	2560	4	186
β-(EDT-TTF)$_2$PF$_6$	$Pccn$	28.014	7.122	12.650	90	90	90	2524	4	186
(EDT-TTF)$_2$ReO$_4$	$C2/c$	29.720	7.185	12.416	90	110.56	90	2482	4	186
(EDT-TTF)$_2$ClO$_4$	$C2/c$	29.283	7.145	12.417	90	111.54	90	2416	4	186
(EDT-TTF)$_2$BF$_4$	$C2/c$	29.133	7.134	12.379	90	111.85	90	2388	4	186
α-(EDT-TTF)[Ni(dmit)$_2$][h]	$P\bar{1}$	6.658	7.627	27.385	86.77	91.43	119.29	1208.6	2	187
β-(EDT-TTF)[Ni(dmit)$_2$][h]	$P2_1/c$	27.685	7.845	11.508	90	101.33	90	2450.7	4	187
			Oxygen-containing Donor Molecules							
BEDO-TTF[g]	$P2_12_12_1$	7.515	7.627	21.794	90	90	90	1249	4	15
(BEDO-TTF)$_2$AuBr$_2$	$P2_1/m$	5.308	32.47	8.165	90	98.47	90	1392	2	16
(BEDO-TTF)$_2$ClO$_4$	$P2_1/m$	5.340	32.43	8.069	90	98.25	90	1426	2	16
(BEDO-TTF)$_2$PF$_6$	$P\bar{1}$	5.358	17.04	8.080	92.68	97.91	94.76	727	1	16
(BEDO-TTF)$_3$Cu$_2$(SCN)$_3$	$P2_1/m$	7.202	33.476	9.340	90	99.56	90	2221	2	17
(Superstructure)	$P2_1/m$	15.812	33.476	12.706	90	97.92	90	6662	6	17)
(BEDO-TTF)$_2$I$_3$	$P\bar{1}$	5.327	4.029	16.885	88.29	83.45	81.21	355.8[q]		188
EDOEDT-TTF	$P\bar{1}$	5.840	9.620	17.115	99.55	80.19	115.67	849.4[r]		188
(BOPDT-TTF)$_2$PF$_6$[s]	$P2_1/c$	6.473	14.231	16.239	90	110.71	90	1399	4	18
	$P\bar{1}$	9.024	6.921	14.440	103.32	98.27	96.45	858	1	19
				DBTTF[t]						
(DBTTF)I$_3$	$C2/m$	19.888	9.294	14.551	90	90.03	90	2690	6	189
(DBTTF)$_2$Cu$_2$Cl$_6$	$C2/m$	9.224	24.979	7.317	90	108.28	90	1600.7	2	190
(DBTTF)$_2$Cu$_2$Br$_6$	$P\bar{1}$	9.982	12.122	7.893	83.70	112.80	104.97	850.4	1	191
(DBTTF)$_3$[Sn$_3$(CH$_3$)$_6$Cl$_8$](ØCN)[c]	$P\bar{1}$	12.931	20.992	12.485	90.07	99.18	79.41	3287.4	2	192
(DBTTF)[Sn(C$_2$H$_5$)$_2$Cl$_3$](CH$_3$CN)	$P\bar{1}$	12.795	12.960	7.463	95.86	108.62	71.44	1111.8	2	193
(DBTTF)$_8$(SnCl$_6$)$_3$	$P4_2/c$	14.991	14.991	27.885	90	90	90	6267	2	194, 195
(DBTTF)$_3$(SnBr$_6$)	$Pnn2$	15.013	15.417	10.398	90	90	90	2407	2	195
(DBTTF)[PtBr$_5$S(CH$_3$)$_2$]	$P\bar{1}$	26.074	12.899	7.674	68.91	83.76	89.65	2392	4	196

TABLE 3.2 Summary of Room Temperature Crystallographic Data for (Donor)$_n$X$_m$ Salts Other Than ET Salts (*continued*)

Compound	Space group	a(Å)	b(Å)	c(Å)	α(°)	β(°)	γ(°)	Unit cell vol (Å^3)	Z	Reference
[(CH$_3$)$_2$DBTTF]BF$_4$	*Ibam*	14.637	17.625	13.614	90	90	90	3512	8	197
(DBTTF)[Ni(dmit)$_2$]^h	$P\bar{1}$	14.29	3.830	12.256	90.03	106.26	89.84	644.0	1	198
TTM-TTF, TSeM-TTF, TTeM-TTF										
α-[(CH$_3$S)$_4$TTF]	$P2_1/n$	15.668	7.804	14.010	90	106.16	90	1645.5	4	199
β-[(CH$_3$S)$_4$TTF]	$P\bar{1}$	5.141	7.565	10.950	72.46	79.03	84.87	398.4	1	28
[(CH$_3$S)$_4$TTF](AuCl$_4$)$_2$	*C2/m*	14.799	11.513	8.034	90	97.05	90	1358	2	200, 201
[(CH$_3$S)$_4$TTF]Br$_3$	$P2_1/n$	7.587	11.424	23.145	90	95.80	90	1997	4	202
[(CH$_3$S)$_4$TTF](Br$_3$)$_2$	*C2/m*	11.510	17.039	9.770	90	141.62	90	1190	2	202
[(CH$_3$S)$_4$TTF]IBr$_2$	$P2_1/n$	8.207	11.537	21.804	90	91.88	90	2064	4	203
[(CH$_3$S)$_4$TTF]I$_{2.47}$	*C2/c*	24.78	3.978	22.65	90	115.50	90	2015	4	204
[(CH$_3$S)$_4$TTF](HgI$_2$)$_2$^{u}	$P\bar{1}$	7.667	8.308	11.608	81.42	75.15	67.06	657.2	1	28
α-[(CH$_3$S)$_4$TTF]$_2$Cu$_2$Cl$_6$	$P\bar{1}$	8.591	11.314	11.868	63.44	88.13	74.28	988.0	1	205
β-[(CH$_3$S)$_4$TTF]$_2$Cu$_2$Cl$_6$	$P\bar{1}$	8.871	10.652	12.505	109.70	106.10	102.63	1003.2	1	205
[(CH$_3$S)$_4$TTF][C$_4$(CN)$_6$]	*C2/c*	30.161	4.046	23.414	90	117.48	90	2534.7	4	199
[(CH$_3$S)$_4$TTF]$_2$TCNQ^v	$P\bar{1}$	9.747	14.378	7.892	103.39	90.55	96.02	1069.4	2	29
[(CH$_3$S)$_4$TTF]TCNQ	$P\bar{1}$	9.534	17.418	7.983	93.40	91.27	99.06	1306.1	2	29
[(CH$_3$Se)$_4$TTF]TCNQ	$P\bar{1}$	9.538	17.732	8.114	93.64	91.06	99.54	1350	2	30
(CH$_3$Te)$_4$TTF	$P2_1/n$	13.769	5.480	12.245	90	90.49	90	923.8	2	206
[(CH$_3$Te)$_4$TTF]TCNQ	*C2/c*	15.659	8.207	22.234	90	91.98	90	2855	4	30
Other Donor Molecules										
(MDT-TTF)I$_{3+x}$^{w}	$P2_1/c$	14.583	9.350	11.995	90	107.76	90	1557.5	4	183
(MDT-TTF)$_2$AuI$_2$	*Pbnm*	10.797	7.789	28.991	90	90	90	2438.1	4	20, 22
(PzEDT-TTF)$_2$PF$_6$^{x}	$P\bar{1}$	5.573	8.173	16.332	94.70	89.39	75.74	718.2	1	207
(PzEDT-TTF)$_2$BF$_4$(TCE)$_{0.5}$^{c}	*C2/c*	34.655	12.122	14.212	90	97.13	90	5925	8	207
(PzEDT-TTF)$_2$IBr$_2$	$P\bar{1}$	8.847	5.671	15.818	98.44	89.66	107.13	749.6	1	208
(PzEDT-TTF)$_2$(I$_3$)$_{2/3}$	$P\bar{1}$	8.179	5.588	16.735	104.75	89.02	104.82	714.1	1	209
(Pz$_2$-TTF)$_2$BF$_4$^{y}	*Fddd*	11.547	13.127	34.182	90	90	90	5181	8	210
(Pz$_2$-TTF)$_2$PF$_6$	*Fddd*	11.642	13.122	35.339	90	90	90	5398	8	210
[(CH$_3$)$_2$PzEDT-TTF]$_2$(I$_3$)$_{2/3}$^{z}	*C2*	34.943	6.610	7.335	90	92.13	90	1693.0	2	209
(PzEDT-DSeDTF)$_2$I$_3$^{aa}	$A2_1am$	13.139	7.491	34.688	90	90	90	3254.8	4	211
(EDT-DTDSeF)$_2$IBr$_2$^{bb}	$P\bar{1}$	6.298	8.383	13.528	91.46	94.48	109.60	669.9	1	141
(PyEDT-TTF)$_2$Br$_2$^{cc}	$P\bar{1}$	7.308	6.475	16.895	92.68	92.31	105.83	767.2	1	208
[(CH$_3$)$_2$EDT-TTF]I$_{3+x}$^{dd}	$P\bar{1}$	7.721	15.352	31.356	93.65	92.74	91.68	3724	8	211
[(CH$_3$)$_2$EDT-TTF]$_m$ClO$_4$)$_n$^{b}										211
[(CH$_3$)$_2$EDT-TTF]$_m$(PF$_4$)$_n$^{b}										211

TABLE 3.2 Summary of Room Temperature Crystallographic Data for (Donor)$_n$X$_m$ Salts Other Than ET Salts (*continued*)

Compound	Space group	a(Å)	b(Å)	c(Å)	α(°)	β(°)	γ(°)	Unit cell vol (Å^3)	Z	Reference
[(CH$_3$)$_2$EDT-TTF]$_m$(TCNQ)$_n$[b,v]	$P\bar{1}$	6.915	8.085	19.384	82.66	84.40	70.98	1014	1	211
[(CH$_3$)$_4$ET]$_2$PF$_6$	$P2_1/c$	6.468	11.277	12.879	90	134.42	90	671.0	2	212
VT[ee]	$P\bar{1}$	16.652	7.408	6.448	100.05	103.44	96.74	751.2	1	4
(VT)$_2$PF$_6$	orthorhombic	33.3	12.6	3.94	90	90	90	1653	4?	4
(VT)(I$_3$)$_{0.4}$[u]	$C2/c$	15.061	7.7832	24.930	90	101.28	90	2866	8	213
(DMCTTF)$_2$ClO$_4$[ff]	$P\bar{1}$	7.710	7.821	25.718	83.76	81.77	68.58	1426	2	214
(DMCTTF)$_2$BF$_4$	$P2_1/n$	7.782	26.129	9.067	90	105.73	90	1775	4	214
BADT-TTF[gg]	$P2_1/n$	7.939	28.302	5.067	90	95.74	90	1132.8	2	5
(CH$_3$S)$_4$VT	$C2/m$	21.999	12.573	3.890	90	90.29	90	1075.9	2	32
(HMTSF)(TCNQ)[hh,v]	$P2_1/n$	9.242	9.039	14.787	90	112.92	90	1137.8	2	215
(C$_2$H$_5$Te)$_4$TTF	$P2_1/c$	6.623	17.760	12.859	90	99.05	90	1493.7	4	206, 216
HMTTeF[ii]	$P\bar{1}$	11.336	14.318	11.027	84.65	101.87	98.90	1727	1	217
α-(HMTTeF)$_4$(PF$_6$)$_2$	$C2/c$	17.355	13.777	28.105	90	99.18	90	6634	4	218
β-(HMTTeF)$_4$(PF$_6$)$_2$	$P\bar{1}$	15.466	13.525	10.586	92.85	102.50	75.91	2096.8	2	219
(HMTTeF)$_2$[Pt(dmit)$_2$][h]	$P\bar{1}$	8.566	11.750	7.926	108.68	92.52	89.06	750.6	1	220
(HMTTeF)(TTeDCN)[jj]	$P\bar{1}$	9.589	11.232	8.635	94.08	102.12	89.41	907.0	2	221
BDMT-TTeF[kk]	$P\bar{1}$	9.590	9.847	8.967	115.61	113.10	74.07	696.7	2	222
DBTSF[ll]	$C2/c$	10.115	12.402	9.602	90	109.16	90	1074	2	223
(DTTTF)(TCNQ)[v,mm]	$C2/m$	21.867	12.794	3.891	90	91.48	90	1088	2	224
α-(DTTTF)(TCNQF$_4$)[nn]	$P2_1/a$	13.059	24.213	7.135	90	103.38	90	2195	4	224
β-(DTTTF)(TCNQF$_4$)										224

[a] MT = bis(methylenedithio)tetrathiafulvalene
[b] No unit cell given
[c] Solvent molecule abbreviations: DCE = 1,2-dichloroethane; THF = tetrahydrofuran; ØCN = benzonitrile; TCE = 1,1,2-trichloroethane
[d] MET = methylenedithio-ethylenedithio-tetrathiafulvalene
[e] MPT = methylenedithio-propylenedithio-tetrathiafulvalene
[f] EPT = ethylenedithio-propylenedithio-tetrathiafulvalene
[g] PT = bis(propylenedithio)tetrathiafulvalene
[h] dmit = 1,3-dithia-2-thione-4,5-dithiolate, C$_3$S$_5^{2-}$
[i] BEDSe-TSeF = bis(ethylenediseleno)tetraselenafulvalene; BPDSe-TSeF = bis(propylenediseleno)tetraselenafulvalene
[j] BEDSe-TTF = bis(ethylenediseleno)tetrathiafulvalene

[k] DMET, DMEDT-DSeDTF, or EDTDM-DTDSF = 3,4-ethylenedithio-3',4'-dimethyl-2,5-dithia-2',5'-diselenafulvalene

[l] A typographical error in the original reads $c = 17.791$ Å instead of $c = 15.791$ Å. The unit cell reported in this table has been transformed for comparison purposes ($\mathbf{b} \leftarrow -\mathbf{b}$, $\mathbf{c} \leftarrow -\mathbf{a} - \mathbf{c}$). Another typographical error in the original (both Russian and English translation) is the space group symbol, $P\bar{1}$. From the context of the paper it is clear that the centrosymmetric space group $P\bar{1}$ is correct.

[m] DIMET or DMEDT-TTF = 3,4-ethylenedithio-3',4'-dimethyl-tetrathiafulvalene

[n] EDT-TTF = ethylenedithio-tetrathiafulvalene

[o] Average structure

[p] BEDO-TTF = bis(ethylenedioxo)tetrathiafulvalene

[q] BEDO-TTF subcell of composite structure;. 1 BEDO-TTF molecule per unit cell

[r] Iodide subcell of composite structure; 2 I atoms per unit cell

[s] BOPDT-TTF = bis(2-oxapropylenedithio)tetrathiafulvalene

[t] DBTTF = dibenzo-tetrathiafulvalene

[u] Donor molecule subcell

[v] TCNQ = tetracyanoquinodimethane

[w] MDT-TTF = methylenedithio-tetrathiafulvalene

[x] PzEDT-TTF = (2,3-pyrazino)-ethylenedithio-tetrathiafulvalene

[y] Pz$_2$-TTF = bis(2,3-pyrazino)tetrathiafulvalene

[z] (CH$_3$)$_2$PzEDT-TTF = 3,4-(5,6-dimethyl-2,3-pyrazino)-3',4'-ethylenedithio-tetrathiafulvalene

[aa] PzEDT-DSeDTF = 3,4-(2,3-pyrazino)-3',4'-ethylenedithio-2,5-diselena-2',5'-dithiafulvalene

[bb] EDT-DTDSeF = 3,4-ethylenedithio-2,5-dithia-2',5'-diselenafulvalene

[cc] PyEDT-TTF = 3,4-(2,3-pyridino)-3',4'-ethylenedithio-tetrathiafulvalene

[dd] (CH$_3$)$_2$EDT-TTF = 3,4-(1,2-dimethyl-ethylenedithio)tetrathiafulvalene (configuration unspecified)

[ee] VT = bis(vinylenedithio)tetrathiafulvalene

[ff] DMCTTF = dimethyltetramethylene-tetrathiafulvalene

[gg] BADT-TTF = bis(cis-1,4-dithia-2-butenyl)tetrathiafulvalene or 3,4,3',4'-bis(1,3-propylene)tetraselenafulvalene or bis(cis-acetylenedimethyldithio)tetrathiafulvalene

[hh] HMTSF = hexamethylenetetraselenafulvalene or 3,4,3',4'-bis(1,3-propylene)tetraselenafulvalene

[ii] HMTTeF = hexamethylenetetratellurafulvalene or 3,4,3',4'-bis(1,3-propylene)tetratellurafulvalene

[jj] TTeDCN = tetratelluradicyclopenta[b,g]naphtalene, $C_{12}H_{12}Te_4$; an isomer of HMTTeF

[kk] BDMT-TTeF = bis(dimethylthieno)tetratellurafulvalene

[ll] DBTSF = dibenzotetraselenafulvalene

[mm] DTTTF = $\Delta^{2,2'}$-bithieno[3,4-d]-1,3-dithiole or bis(thiopheno)tetrathiafulvalene

[nn] TCNQF$_4$ = Tetrafluoro-tetracyanoquinodimethane

(thus perpendicular to the stacking direction, **a** + **b**). Furthermore, it was found that the one ethylene group which is disordered at room temperature encounters some H · · · I contacts that are too short in some phase areas along the modulation period, unless the ethylene group conformations are adjusted according to the phase angle. A preliminary and simplified model taking into account this complication was included in the structural analysis,[230] and a more sophisticated analysis is currently in progress.[103]

The modulation vector remains approximately constant on cooling to temperatures no lower than 110 K, at which temperature it decreases from $|Q| = 0.27 - 0.28$ Å^{-1} to 0.24 Å^{-1} at 100 K, but the change is only noticeable if one equilibrates the crystal for many hours at 109 K.[232] Otherwise, no change is observed at 110 K, and $|Q|$ remains constant to the lowest temperatures measured. All components of the modulation vector are reduced, and there is a possibility that the b^*-component becomes commensurate with b^*, that is, $q_2 = b^*/4$. A discontinuity at 110 K[232] or 125 K[228] was also seen from an analysis of satellite peak intensities. The application of moderate pressure (400 bar) depresses the onset temperature of the modulated structure (to 150 K at 400 bar) and significantly facilitates the transition at 110 K. Under applied pressure, some reduction of the modulation vector length is already observable above 110 K.[232]

A further increase in applied pressure completely suppresses the onset of the modulated structure, and an ordered structure (denoted[237] β*) with a superconducting transition temperature, $T_c \approx 8$ K is formed.[233-235] See also Fig. 3.8(b). Provided that the samples are not heated above 125 K, the "high-T_c" state is preserved even after the pressure is reduced back to 1 atm.[229, 236] A full structure analysis[237] at 4.5 K and 1.5 kbar by use of neutron diffraction data revealed that under these conditions the structure is ordered, and that the terminal ethylene groups of the ET molecule assume the staggered conformation, in contrast to the other β-(ET)$_2$X salts, $X^- = IBr_2^-$, I_2Br^-, and AuI_2^-, where the eclipsed conformation is found at low temperature. At room temperature and 9.5 kbar pressure, β-(ET)$_2$I$_3$ also assumes the staggered conformation, indicating that the β*-structural phase region extends to that area of the phase diagram.[238] No phase transition analogous to that found in β-(ET)$_2$I$_3$ is found on cooling the other β-(ET)$_2$X salts. It could be said that the latter always condense in their high-T_c state.

A comparison of transition temperatures with structural parameters (e.g., anion length, see Fig. 3.9, or unit cell volume) in the β-(ET)$_2$X salts reveals that the larger anions lead to higher T_c's. This correlation[2] assumes that for the triiodide salt, the high-T_c form represents the true T_c, and that the value measured in the modulated phase is suppressed, likely by the quasi-random potential introduced by the structural modulation. The correlation, which of course only holds within an isostructural series, and the observation (see Chap. 4) that applied pressure (and associated lattice compression) usually diminishes T_c, have stimulated the search for ET salts with larger and larger anions. In a previous review,[2] it was suggested that bis(thiocyanato)-metal complex anions were likely candidates, and as the next section shows, that strategy was successful, albeit in an unexpected manner.

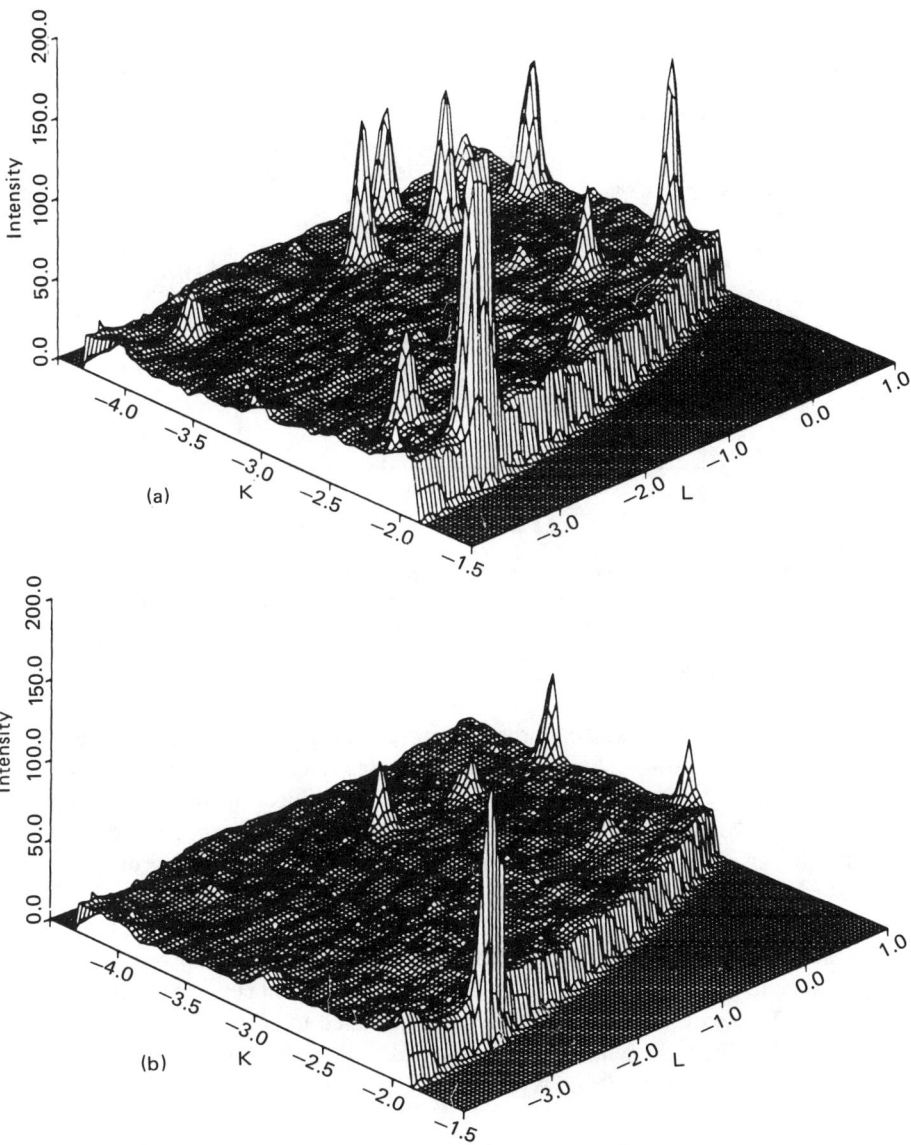

Figure 3.8 (a) Plot of the neutron-diffraction intensity distribution in the $h = 4.92$ reciprocal lattice plane of β-$(ET)_2I_3$ at 20 K and with zero pressure. Satellite peaks at (5, -2, -3)$-\mathbf{q}$, and (5, -4, 1)$-\mathbf{q}$ are clearly observable. (b) The same $h = 4.92$ reciprocal lattice plane after applying a pressure of 1.4 kbar at 20 K, warming to room temperature, and cooling back down to 20 K. There are no observable satellite reflections (from ref. 229, with permission).

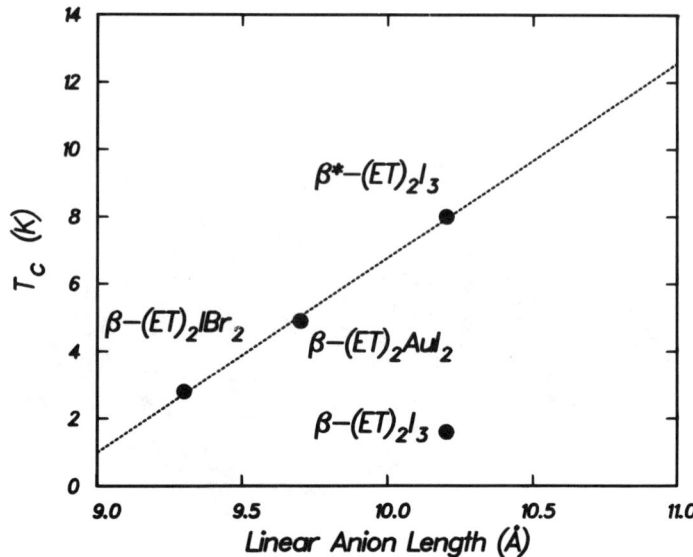

Figure 3.9 Structure-properties correlation[2] in β-type (ET)$_2$X salts with linear, triatomic anions. T_c increases with increasing anion length (sum of bond lengths and twice the van der Waals radius of the terminal atom).

κ-(ET)$_2$Cu(NCS)$_2$

κ-(ET)$_2$Cu(NCS)$_2$ was the organic superconductor with the highest transition temperature, $T_c \approx 10.4$ K, until very recently (see Sec. Latest Developments). It was discovered by a Japanese group,[239] and was the subject of well over 50 papers over the last two years. Its structure, which is distinctly different from that of the β-phase linear-anion systems, was subsequently determined by two groups,[133, 240] both at room temperature and at 104 K.[133] It contains one ET donor molecule layer per unit cell, but no stacks are distinguishable within the layer. The donor molecules form face-to-face orthogonal dimers; that is, they are rotated by approximately 90° with respect to each other, or ca. 45° with respect to the unit cell axes (b and c), see Fig. 3.10. There are two dimers per unit cell, related to each other by a 2$_1$ screw rotation. Taken by itself, without the anion, the ET donor layer in κ-(ET)$_2$Cu(NCS)$_2$ is essentially centrosymmetric. A centrosymmetric variant is found in κ-(ET)$_2$I$_3$ (vide infra).

The anion, shown in Fig. 3.11, forms zig-zag chains of alternating copper(I) atoms and bridging, linear NCS$^-$ ions. In addition, each copper atom is coordinated to a terminal, N-bound NCS$^-$ unit. In both bridging and terminal thiocyanate units, the angle formed by the Cu–N bond and the axis of the NCS$^-$ unit is nearly 180°, whereas the Cu–S–C angle of the bridging thiocyanate unit is 108°, reflecting the presence of nonbonding lone-pair electrons.

The presence of "twisted dimers" is the most distinguishing feature of the κ-family of compounds, other members being κ-(ET)$_2$I$_3$,[47] κ-(ET)$_2$-Ag(CN)$_2$·H$_2$O,[68] κ-(MDT-TTF)$_2$AuI$_2$ (see below), (MT)$_2$Au(CN)$_2$,[146] κ-

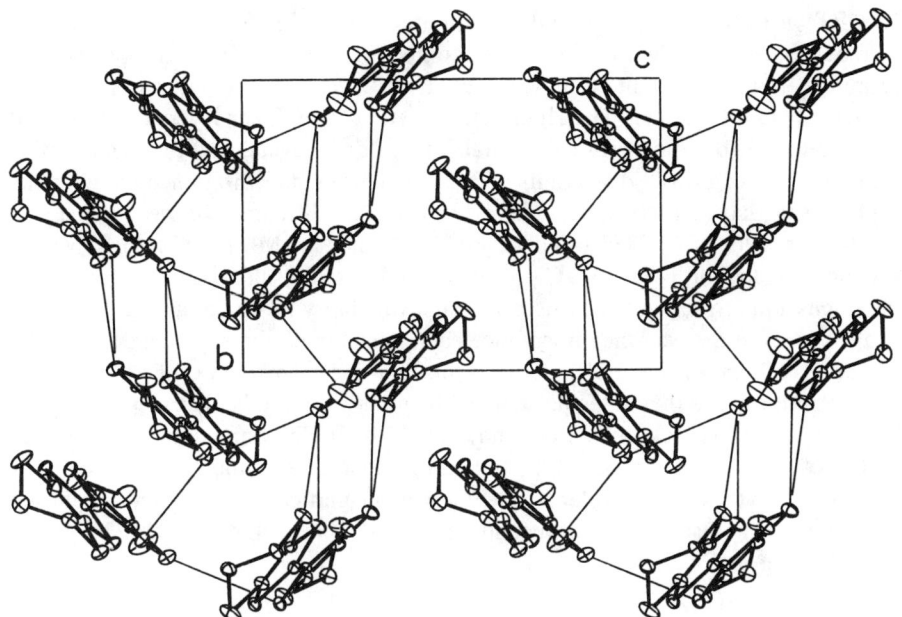

Figure 3.10 The ET donor molecule layer in κ-(ET)$_2$Cu(NCS)$_2$ at 118 K. S⋯S contacts shorter than 3.6 Å are indicated by thin lines.

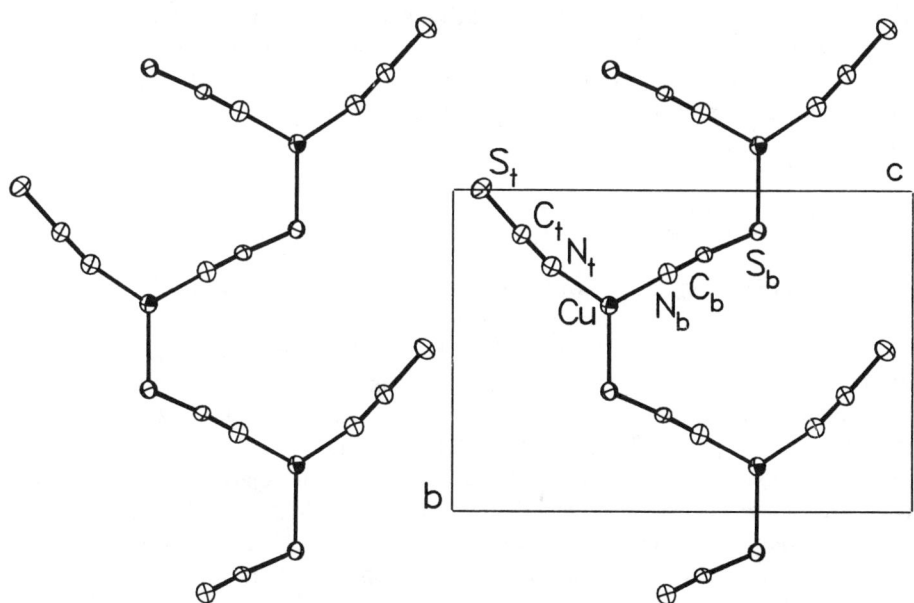

Figure 3.11 The polymeric anion in κ-(ET)$_2$Cu(NCS)$_2$. The subscripts b and t refer to bridging and terminal thiocyanate units, respectively.

(DMET)$_2$AuBr$_2$ (see below), and (to some extent) (ET)$_4$Hg$_{3-\delta}$X$_8$ (X = Cl, Br) (see below). All but the mercury halide compounds and κ-(MDT-TTF)$_2$AuI$_2$ have primitive, monoclinic unit cells, with one donor layer per unit cell, and they comprise the κ-family proper. The Hg:X salts contain κ-like donor layers (with twisted donor dimers), but their unit cell contains two donor-layers, and the symmetry is different. An extensive discussion of the similarities and differences among the κ-phase salts has recently appeared.[241] The principal difference between the superconducting salts {κ-(ET)$_2$Cu(NCS)$_2$, κ-(ET)$_2$I$_3$, and κ-(MDT-TTF)$_2$AuI$_2$} versus nonsuperconducting (MT)$_2$Au(CN)$_2$ (the silver cyanide salt[68] was excluded from the discussion, presumably because of the unavailability of structural details to Jung et al.[241]) lies in the way the dimers face each other. In all the superconductors, the central C=C bond of one molecule is on top of a five-membered ring when viewed along the normal to the molecular planes, thus indicating a displacement along the long molecular axis. On the other hand, in (MT)$_2$Au(CN)$_2$, the dimers are exactly on top of each other. The slipped-pair arrangement of the superconductors allows for a somewhat shorter interplanar spacing which significantly affects the electronic properties. This discussion was recently extended[242] to include κ-(DMET)$_2$AuBr$_2$ and (ET)$_4$Hg$_{3-\delta}$Cl$_8$.

κ-(ET)$_2$I$_3$

Mixed electrolytes (n-C$_4$H$_9$)$_4$NI$_3$ and (n-C$_4$H$_9$)$_4$NAuI$_2$, yield, in addition to α- and θ-phase salts (see below), yet another superconductor, κ-(ET)$_2$I$_3$,[47, 243] with T_c = 3.6 K. The ET-donor molecule packing is very similar to that described in the copper thiocyanate salt. However, the space group is centrosymmetric, allowing for the I$_3^-$ anions to be located on inversion centers. The presence of inversion symmetry also renders all the ET molecules equivalent, as there are now centers of inversion located in the middle of each dimer.

θ-(ET)$_2$(I$_3$)$_{1-x}$(AuI$_2$)$_x$

θ-(ET)$_2$(I$_3$)$_{1-x}$(AuI$_2$)$_x$ (with $x < 0.02$) was reported to be a superconductor at ambient pressure with a transition temperature, T_c = 3.6 K.[46, 244] In this compound, the basic structure[44] comprising the ET-donor molecule network is orthorhombic, with space group *Pnma*. It contains two-donor molecule layers, perpendicular to the longest unit cell axis (b), as seen in Fig. 3.12. Each layer in turn consists of two uniform stacks of ET-donor molecules, with c as the stacking axis. Within each layer the long molecular axes of adjacent ET-donor molecules are approximately parallel to each other, but tilted from the plate normal (b-axis) by approximately 20° (estimated from the published[44] *ab*-projection). Neighboring layers are tilted in opposite directions, allowing for a mirror reflection plane between layers. The plane normal of the molecules is also inclined with respect to the stacking direction, again in such a manner that molecules located on adjacent stacks (related by a screw rotation operation) are inclined

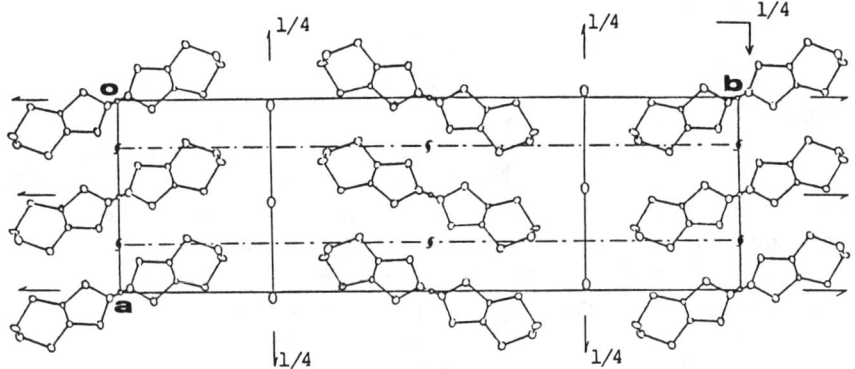

Figure 3.12 Unit cell packing of θ-$(ET)_2(I_3)_{1-x}(AuI_2)_x$ ($x < 0.02$) (from ref. 44, with permission).

with opposite signs. The dihedral angle between the planes formed by ET donors on neighboring planes is 79.6°. The donor molecule packing within the layer is very similar to that found in α-$(ET)_2I_3$[34] and similar salts (see Table 3.1), even though the symmetry in the θ-phase layers is higher, and all ET molecules are centrosymmetric and equivalent to each other (at least in the average structure). Alpha-phase crystals are triclinic, and of the four donor molecules per layer (only one per unit cell), two are located on nonequivalent inversion centers, and the other two (equivalent to each other) are in a general position.

It was also noted[44] that the symmetry of the overall structure is lower, and the unit cell is twice as large as that described. In the smaller orthorhombic unit cell, the iodine atomic positions cannot be described in a chemically acceptable model, except for their placement on the crystallographic mirror plane. In a subsequent paper, Kobayashi et al.,[45] reported a model for the anion positions based on the larger, monoclinic unit cell, where the intensities of the extra (type-II) Bragg reflections could be reproduced satisfactorily, if not exactly (crystallographic agreement factor $R = 0.17$), by an ordered arrangement of I_3^- anions. It was found[45] that the anion positions described by such a model coincided with the positions of minimum H···I energy.

In none of the crystallographic analyses, was the presence of AuI_2^- explicitly included, but rather the anions were treated as being composed solely of I_3^-. Nevertheless, the presence of gold in these crystals was proven by microanalysis.[46] Furthermore, in the absence of either AuI_2^- or I_2Br^- no θ-phase crystals of θ-$(ET)_2I_3$ formed. The fraction of impurity anion in the crystals was found to be much smaller than in the starting electrolysis solutions. For example, in a 1:1 mixture of $[(n\text{-}C_4H_9)_4N]I_3$ and $[(n\text{-}C_4H_9)_4N]AuI_2$, only 3 percent of the AuI_2^- is incorporated into the crystals. In the case of I_2Br^- contamination, θ-$(ET)_2I_3$ crystals with bromine content below the detection limit of X-ray microanalysis were grown.[46] The average structure of those crystals was determined[46] to be essentially identical to that of the AuI_2^--containing crystals.

γ-(ET)$_3$(I$_3$)$_{2.5}$

The reports on γ-(ET)$_3$(I$_3$)$_{2.5}$ are sparse: the compound was reported to be a superconductor with $T_c = 2.5$ K,[245] and later the unit cell, composition, and a structure drawing were published in a volume of conference proceedings.[38] From the latter picture, two layers of ET-donor molecules are discernible, separated by mirror planes which contain some of the I$_3^-$ anions. More anions are seen interspersed among the donor molecule cavities, similar to those found in δ-(ET)I$_3$,[38] ϵ-(ET)$_2$(I$_3$)(I$_8$),[40] η-(ET)$_2$(I$_3$)(I$_5$) (also called ζ-phase[42]), and (ET)$_2$(I$_3$)(TlI$_4$).[42] However, a count of I$_3^-$ anions in the unit cell drawing[38] reveals 12 anions per unit cell, whereas the compositional formula indicates only ten. Hence, one must assume that some of the I$_3^-$ positions are only partially occupied. Two stacks of three ET-donor molecules each make up every layer in the unit cell. Space group considerations require that the ET molecules located on adjacent stacks are tilted with opposite signs with respect to the stacking axis, a. This latter aspect of the structure is also found in the θ-phases, for example, θ-(ET)$_2$(I$_3$)$_{1-x}$(AuI$_2$)$_x$,[44] (vide supra). Further similarities include the presence of two stacks and two layers per unit cell, the latter related by a mirror reflection plane. In contrast to the γ-type, the θ-structures do not show anions within the donor molecular layer, and the packing along the stacking axis is more uniform. Shibaeva et al.,[38] also mention that X-ray diffraction photographs exhibit additional $00l$ reflections and diffuse streaks with an incommensurate k-index in the $0kl$ net, possibly because of the presence of an additional phase. It would be highly desirable to see more details on this unusual crystal structure.

κ-(ET)$_4$Hg$_{3-\delta}$X$_8$ Family (X = Cl, Br)

Superconductivity was first reported in κ-(ET)$_4$Hg$_{3-\delta}$Cl$_8$ by Lyubovskaya et al.,[246] with a transition temperature $T_c = 1.8$ K at 12 kbar. Later, T_c was found to increase to 5.3 K when a higher pressure, 39 kbar, was applied.[247] It was recognized immediately that the crystal structure of this compound contains two incommensurate penetrating lattices with different periodicity along the crystallographic a-axis, a lattice with sharp Bragg reflections corresponding to $a' \approx 11$ Å, and a lattice with diffuse rows according to $a'' \approx 12$ Å (most of the intensity on every third row, indicating an approximate periodicity of $a''' \approx 4$ Å). Details of the structure were later published by Shibaeva et al.:[138] the ET-donor molecules and the chlorine atoms are associated with one lattice, whereas the mercury atoms possess the periodicity of the other sublattice. Thus, the chlorine atoms form infinite channels, inside which the mercury atoms are distributed according to their own sublattice, a''. The position of the chlorine atoms is such that the coordination around the Hg^{2+} ions can be distorted octahedral or tetrahedral, depending on the exact location of the metal center. The stoichiometry, in particular the exact value of δ in the empirical formula, has not been determined independently. Assuming eight chlorine atoms per 11 Å of channel length and three Hg^{2+} ions per 12 Å, a Hg:Cl ratio of $0.34 \equiv 2.75:8$ is deduced, thus $\delta \approx 0.25$.

As characteristic for a κ-structure, the ET-donor molecules are dimerized in a face-to-face fashion, and adjacent dimers are rotated by approximately 90° with respect to each other. In a variant to the proper κ-structures, κ-$(ET)_2Cu(NCS)_2$ and κ-$(ET)_2I_3$, the unit cell is composed of two donor molecule layers instead of one, hence the centered space group $I2/c$, rather than the primitive $P2_1$ and $P2_1/c$, depending on the absence or presence of a center of inversion within the anion layer. It is not currently known what structural changes are caused by the applied pressure which induces superconductivity, at 1.8 K (12 kbar) and 5.3 K (39 kbar).

More recently, it was pointed out[121] that the crystals used in the original investigation[138, 246] were grown from a different solvent (tetrahydrofuran, THF) than the crystals used in the further investigations (1,1,2-trichloroethane, TCE), and that slight crystallographic differences exist: in the former crystals the interplanar distance between molecules within a dimer is shorter (3.59 Å)[138] than in the latter (3.614 Å).[121] Differences in the resistivity behavior of the two systems have been found, and in particular it is not known to date if the crystals grown from TCE are superconducting under pressure.

The corresponding bromide compound, κ-$(ET)_4Hg_{3-\delta}Br_8$, is a superconductor at ambient pressure with $T_c = 4.3$ K.[137] Its structure[137, 121] is essentially that of the chloride analogue (TCE-grown phase), except that the interplanar spacing within dimers is even larger, that is, 3.696 Å. Surprisingly, in the bromide compound, the repeat distance of the Hg-subcell (which apparently shows no tripling) is approximately 0.1 Å shorter than in κ-$(ET)_4Hg_{3-\delta}Cl_8$, leading to a larger mercury content in the compositional formula: $(ET)_4Hg_{2.89}Br_8$.

It is interesting to note that the corresponding iodide compound, nonsuperconducting $(ET)_4Hg_3I_8$ is stoichiometric, and that the structure[136] is different from the chloride and bromide derivatives, despite comparable unit cell lengths and the same space group ($I2/c$ and $C2/c$ are equivalent). One major difference is that the monoclinic unique axis in the iodide analog is out of the donor molecule plane, thus along the longest unit cell directions, whereas in the other two salts, the monoclinic axis is contained in the plane. Müller et al.[87] have reported a compound $(ET)_4Hg_2I_5$ with essentially identical lattice parameters as those of $(ET)_4Hg_3I_8$,[136] but they mention a different salt of composition $(ET)_4Hg_3I_8$ whose structure analysis is "in progress."[87]

$(ET)_3Cl_2 \cdot 2H_2O$

$(ET)_3Cl_2 \cdot 2H_2O$, with $T_c = 2$ K at 16 kbar,[248] is unique among the organic superconductors for several reasons: (a) It is the only salt with 3:2 stoichiometry, whereas most of the others have a 2:1 cation to (monovalent) anion ratio. (b) It is the only superconducting hydrate, and one of only a few charge-transfer salts containing water molecules. Among the structurally characterized ET salts, only $(ET)_3Cl_2 \cdot 2H_2O$,[114–116] $(ET)_4Cl_2 \cdot 2H_2O$,[116] $(ET)_3Br_2 \cdot 2H_2O$,[117] and κ-$(ET)_2$-$Ag(CN)_2 \cdot H_2O$[68] are hydrates. Structural information on $(ET)_3Cl_2 \cdot 2H_2O$ has been published by at least three independent groups.[114–116] Interestingly, in each case, the crystals formed from electrolyte mixtures that were supposed to crystallize completely unrelated ET salts: a $CoCl_4^{2-}$ salt was expected by the Japanese,[114] a substituted dithiocarbamate by the English[115] (the source of chlorine in this case is the solvent, CH_2Cl_2), and a palmitate ($n\text{-}C_{15}H_{31}CO_2^-$) in the presence of trichloroethane by the Soviet groups.[116] In our own, unpublished experiments on the same compound, anionic manganese(II) fluoride aquo-complexes and chlorinated solvent were present.

The structure of $(ET)_3Cl_2 \cdot 2H_2O$[114-116] is shown in Fig. 3.13. It contains slipped stacks of three crystallographically independent ET molecules per repeat unit (along b, in the coordinate system used in Table 3.1), two stacks per layer, and one layer (perpendicular to c^*) per unit cell. All molecules are oriented approximately parallel to each other, and in the c-axis projection (c being roughly parallel to the long ET molecular axis) the ET molecules appear to be periodic with repeat distances of $a/2$ and $b/3$. However, the translational pseudosymmetry is broken along the stacking b-direction by out of plane displacements in a pattern ... (z_1, z_2, z_3)(z_1, z_2, z_3) ... , where z_i denotes the z-coordinate of molecule i, and $z_1 < z_2 < z_3$. In the transverse a-direction, the pseudosymmetry is broken by the inversion (the center of symmetry has the y-coordinate of molecule 3) of the above stacking patterns, that is, ... $(-z_2, -z_1, -z_3)(-z_2, -z_1, -z_3)$ ... .

The chloride anions are connected by hydrogen bonds to $Cl_4(H_2O)_4$ units, where two chlorine atoms and two water molecules form a rhomboid-shaped ring around an inversion center, and on each side a further water molecule bridges the oxygen atom of a ring water unit to a terminal chloride ion.

The bromide analogue, $(ET)_3Br_2 \cdot 2H_2O$, has a related, but slightly simpler structure.[117] In this salt, all the stacks, which are similar to those in the chloride salt, are equivalent, and the hydrogen bonding network forms rhomboid-shaped clusters

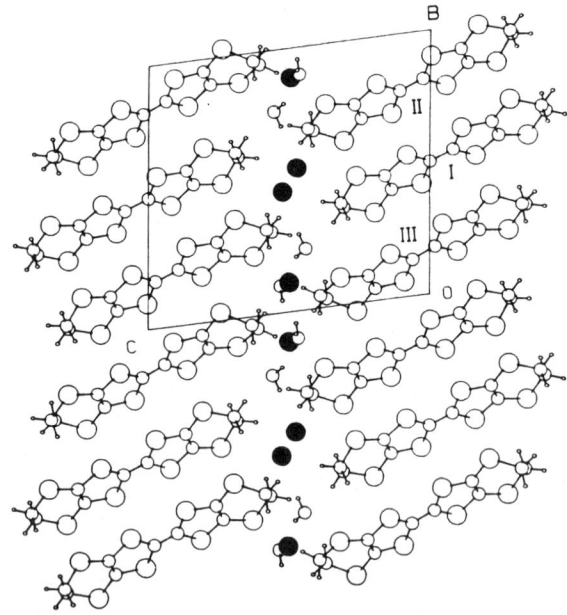

Figure 3.13 Unit cell view of $(ET)_3Cl_2 \cdot 2H_2O$ along the a axis. The chlorine atoms shown as dark circles (from ref. 115, with permission).

Br$_2$(H$_2$O)$_2$ without the extra, terminal (H$_2$O)-halide groups. The ET-donor molecule network in (ET)$_3$Br$_2$·2H$_2$O is that of γ-(ET)$_3$(X)$_2$, where X^- is a tetrahedral anion, for example, BrO$_4^-$,[95] ClO$_4^-$,[96] FSO$_3^-$,[94] HSO$_4^-$,[97] or BF$_4^-$.[94] The salt γ-(ET)$_3$Br$_2$ briefly mentioned by Parkin et al.,[94] is probably the dihydrate, since the lattice constants are essentially the same. No superconductivity has been reported to date in (ET)$_3$Br$_2$·2H$_2$O, nor in the other chloride-hydrate, (ET)$_4$Cl$_2$·2H$_2$O.[116]

(ET)$_2$(NH$_4$)Hg(SCN)$_4$

A recently discovered ET-based superconductor is (ET)$_2$(NH$_4$)Hg(SCN)$_4$, with T_c = 1.15 K (onset).[249] The compound was first prepared as an analogue of (ET)$_2$KHg(SCN)$_4$,[139] which undergoes a metal-to-insulator transition at 180 K. Based on unit cell data, the structure of the ammonium salt was reported to be isostructural to the potassium derivative,[139, 249] and a full crystallographic analysis subsequently confirmed this assumption.[140]

The organic part of the crystal structure is closely related to that of α-(ET)$_2$I$_3$[34] (see Fig. 3.14). It contains one ET donor molecule layer per unit cell,

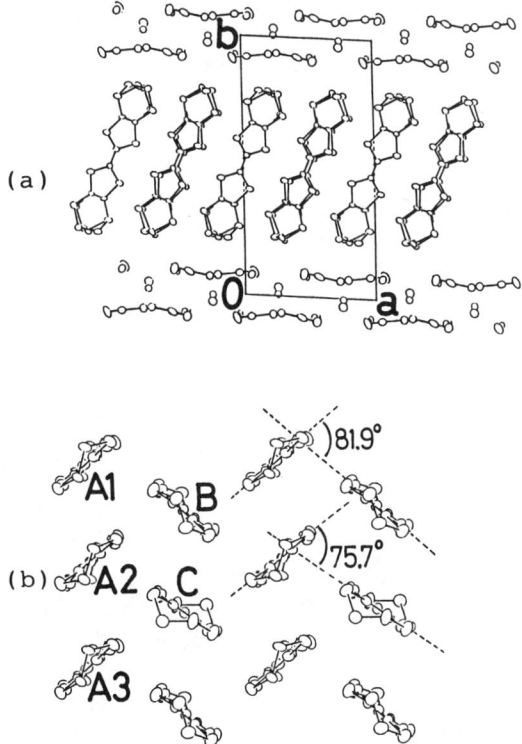

Figure 3.14 The crystal structure of (ET)$_2$KHg(SCN)$_4$: (a) projected along the c-axis (a in Table 3.1); (b) projected along the average long molecular axis of the ET molecules. Molecules B and C are located on inversion centers. The structure of the NH$_4^+$ salt is essentially the same (from ref. 139, with permission).

each layer having two stacks of two molecules each per repeat unit. The molecules on one stack are located on two nonequivalent inversion centers (labeled B and C by the Japanese group[140]), whereas the donor molecules of the other stack are equivalent to each other and occupying a general position (A). Furthermore, the normals to the molecular planes of donor molecules located on one stack are inclined with opposite direction with respect to the stacking axis, c, forming a kind of "herring-bone" arrangement as in the α-phases. The dihedral angles between the molecules are 82.3°, 75.6°, and 6.6°, for pairs A–B, A–C, and B–C, respectively; thus they are closer to 90° than in α-$(ET)_2I_3$ and of similar magnitude as those in θ-$(ET)_2(I_3)_{1-x}(AuI_2)_x$.[44]

The complex, polymeric anion is formed by a double layer of SCN^- units and Hg^{2+} and NH_4^+ ions in between. The mercury ions are tetrahedrally coordinated to the sulfur ends of four SCN^- ions, and the ammonium-nitrogen atoms (or K^+ in the potassium derivative) form the apex of a square pyramid containing SCN-nitrogen atoms in the basal plane. All four thiocyanate units are located in the same plane, and contacts between the NH_4^+ or K^+ ions and thiocyanate ions in the other layer are substantially longer. In the case of the ammonium compound, hydrogen atoms are presumably located between nitrogen atoms, forming hydrogen bonds. However, in order to satisfy the tetrahedral geometry of the NH_4^+ ion, only two out of the four hydrogen bonds can be simultaneously formed, leading to disorder. No hydrogen atoms were located in the X-ray diffraction analysis,[140] and no information about the nature of the hydrogen atom disorder is available at present.

CRYSTAL STRUCTURES OF SUPERCONDUCTORS BASED ON OTHER DONOR MOLECULES

κ-(MDT-TTF)$_2$AuI$_2$

$(MDT-TTF)_2AuI_2$ was discovered by a Greek group[20] to be a superconductor with $T_c = 4.5$ K.[21] It is the organic superconductor with the highest transition temperature based on a donor molecule other than ET, except for some salts of the dithiolate complexes $Ni(dmit)_2^{\delta-}$ [250, 251] and $Pd(dmit)_2^{\delta-}$ [252, 253] (see Table 1.1). We have excluded molecular conductors built from transition metal dithiolate complexes from the current discussion, except where they form the counter-ion in conjunction with an electron donor molecule.

The crystal structure of κ-$(MDT-TTF)_2AuI_2$ was solved independently by the Athens[22] and Argonne[21] groups. It contains donor molecule layers with dimers that are rotated with respect to each other (by 79.3°), typical of a κ-phase material (the prototype being κ-$(ET)_2I_3$, see above); see Fig. 3.15. However, the compound is not isomorphic with κ-$(ET)_2I_3$ or κ-$(ET)_2Cu(NCS)_2$, as it possesses two MDT-TTF layers per unit cell, related by a mirror reflection operation. This is indicated by the higher symmetry (orthorhombic) as compared to the other, monoclinic κ-phase salts. The centrosymmetric molecular dimers overlap mainly with the fulvalene ends, the ethylene groups protruding away from the dimer.

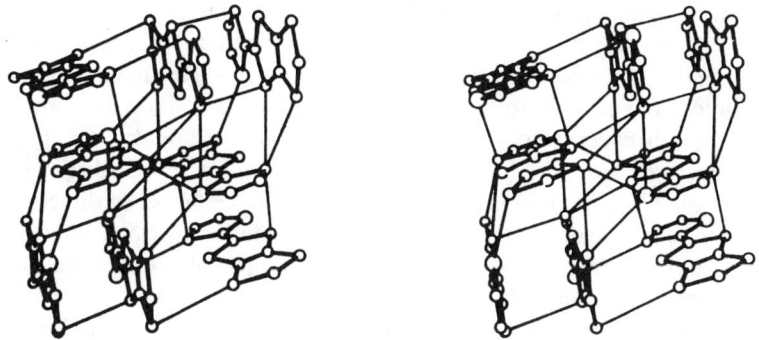

Figure 3.15 Stereoview of the dimeric structure in κ-(MDT-TTF)$_2$AuI$_2$. S⋯S contacts (<3.6 Å) are indicated by thin lines (from ref. 22, with permission).

(DMET)$_2$X Family

Seven salts of the donor molecule DMET, a hybrid of ET and TMTSF, with six different anions have been found to be superconductors with transition temperatures between 0.47 and 1.9 K; see Table 1.1. The DMET salts have been grouped by Ikemoto et al.[176, 179] into five structural categories. No superconductivity is found in the first and second groups which consist of 2:1 salts with tetrahedral (BF$_4^-$ and ClO$_4^-$) and octahedral (PF$_6^-$, AsF$_6^-$) anions, respectively. The different groups are also distinguished by their normal-state conductive properties.[179, 254]

Group 3 salts (see Fig. 3.16) contain linear gold(I) complex anions and include the first DMET-based superconductor (T_c = 0.8 K under 5 kbar applied pressure), (DMET)$_2$Au(CN)$_2$,[256, 257] whose crystal structure was described in detail by Kikuchi et al.[172] That salt and the isostructural superconductors (DMET)$_2$AuCl$_2$ (T_c = 0.83 K at ambient pressure; little structural detail known to date)[258] and (DMET)$_2$AuI$_2$ (T_c = 0.55 K at 5 kbar;[258] crystal structure by Ishikawa et al.[174])

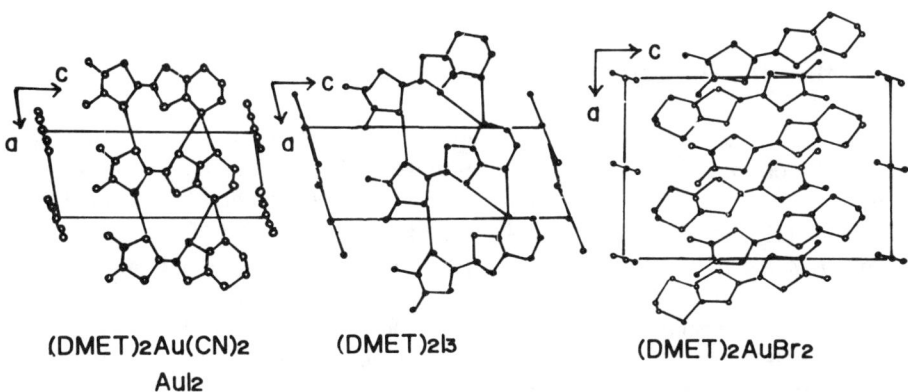

(DMET)2Au(CN)2   (DMET)2I3   (DMET)2AuBr2
AuI2

Figure 3.16 Unit cell views of several structural families of DMET salts: Group 3–5, represented by (DMET)$_2$Au(CN)$_2$, (DMET)$_2$I$_3$, and (DMET)$_2$AuBr$_2$, respectively (from ref. 255, with permission).

contain parallel stacks of slightly dimerized DMET donor molecules grouped around inversion centers such that the ethylene end groups protrude from opposing sides of the stack. In the b-axis projection (along the stacking axis), the long direction of the DMET molecule is parallel to the c-axis (inter-layer repeat direction), and it is directed at a gap between anions in the adjacent anion layer. Within the dimers, the S···Se contacts are around 3.8 Å, whereas between dimers Se···Se distances longer than 3.9 Å exist. Furthermore, the molecules are slipped between dimers, but approximately face-to-face within the dimer. The shorter intermolecular contacts (~3.55 Å) are found in a side-by-side manner between molecules of adjacent stacks.

The group 4 compounds contain typically larger linear triatomic anions than group 3, with the superconductors $(DMET)_2I_3$ ($T_c = 0.47$ K at ambient pressure),[259] $(DMET)_2IBr_2$ ($T_c = 0.59$ K at ambient pressure),[259] as well as non-superconducting $(DMET)_2I_2Br$,[259] $(DMET)_2SCN$,[179] and one modification (sometimes named "p" for the characteristic plate-like shape of the crystals) of $(DMET)_2AuBr_2$.[177] The DMET stacks in the group 4 salts are very similar to those in group 3, but the overall packing and the arrangement of the stacks with respect to the anions show a marked difference. In the group 4 salts, when viewed along the stacking (**b**) direction, the long direction of the DMET molecules is along the **a** + **c** diagonal, instead of along **c**, and the molecular axis points directly at the anion center, rather than in between anions, as in group 3.

Finally, in group 5, the other modification (called "r" for the rhombus-shaped groups) of $(DMET)_2AuBr_2$, at least one κ-phase like κ-$(ET)_2Cu(NCS)_2$ and κ-$(MDT-TTF)_2AuI_2$ is found. Actually, two kinds of crystals can be found, originally denoted r_1 and r_2 ($κ_1$ and $κ_2$, respectively, in Table 3.2), with slightly different unit cells and somewhat different electrical properties: $κ_1$-crystals require 1.5 kbar of applied pressure to turn superconducting at $T_c = 1.0$ K,[260] whereas $κ_2$-crystals are ambient-pressure superconductors with $T_c = 1.9$ K, the highest for any DMET salt.[177] As is normal for a κ-phase salt, $(DMET)_2AuBr_2$ exhibits face-to-face dimers which are rotated by approximately 80° (details of the structure have not been published yet). A comparison of κ-$(DMET)_2AuBr_2$ with other κ-phase charge-transfer salts has recently been published.[242]

$(BEDO-TTF)_3Cu_2(NCS)_3$

One of the most recently discovered organic superconductors, and the only one based on an oxygen-containing donor molecule, is $(BEDO-TTF)_3Cu_2(NCS)_3$,[17] with $T_c \approx 1.06$ K at ambient pressure. Its structure is a superstructure of some other BEDO-TTF salts, that is, $(BEDO-TTF)_2AuBr_2$ and $(BEDO-TTF)_2ClO_4$,[16] with a tripling of the $a + c$ diagonal (in some crystals, reflections corresponding to a superstructure with tripling of both a and c-axis, independently); see Fig. 3.17. The donor molecule packing is essentially that of the basic substructure, and a structural model of the BEDO-TTF donor molecules could be refined from the strong sublattice reflections alone. However, the anion assumes the periodicity of the larger cell. It is a two-dimensional polymer, related to the $[Cu(NCS)_2^-]_\infty$ chains of κ-$(ET)_2Cu(NCS)_2$ (vide supra) by elimination of half of the terminal SCN-groups, and condensation of the other terminal thiocyanate anions with the copper atoms

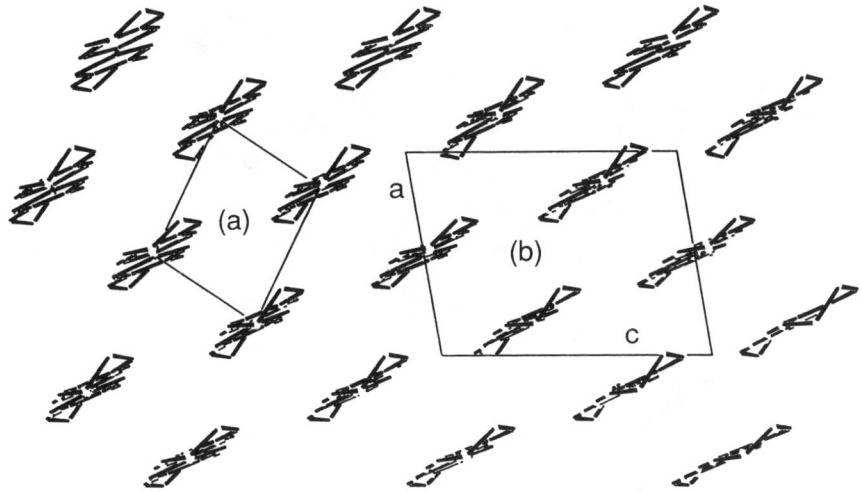

Figure 3.17 Donor molecule layer in (BEDO-TTF)$_3$Cu$_2$(NCS)$_3$. Unit cells (a) and (b) indicate, respectively, the subcell of the organic constituents and the threefold supercell imposed by the polymeric anion. Many BEDO-TTF salts with simple anions crystallize in the unit cell indicated by (a), thus essentially conserving the donor molecular network.

whose SCN-groups were removed. Thus, two inequivalent copper sites with approximately triangular coordination are created, one surrounded by two nitrogen and one sulfur atoms, and the other by one N and two S atoms.[17]

LATEST DEVELOPMENTS

κ-(ET)$_2$Cu[N(CN)$_2$]X, X = Br, Cl

Just as this book was about to go to press, the organic superconductor with the highest transition temperature reported to date was discovered: κ-(ET)$_2$Cu[N(CN)$_2$]Br, with T_c = 11.6 K (inductive onset) or 12.8 K (resistive onset).[261, 262] As shown in Fig. 3.18, its crystal structure contains donor molecule layers similar to those in κ-(ET)$_2$Cu(NCS)$_2$, with mutually orthogonal molecular dimers arranged in a typical κ-fashion (see above). However, close comparison (see Fig. 3.10 for comparison) of the topology of close S · · · S contacts in the layer reveals some differences between the two salts: In κ-(ET)$_2$Cu(NCS)$_2$, the contacts connect the ET-donor molecules into one-dimensional ribbons along the short unit cell direction, whereas in (ET)$_2$Cu[N(CN)$_2$]Br they form a two-dimensionally connected network. It should be noted that the two salts are not isostructural, with different unit cell sizes and symmetries. As a result, all ET molecules in the latter salt are crystallographically equivalent, and two, rather than one, donor molecule layers constitute a unit cell along the interlayer cell direction (b, in accordance with the standard setting of the space group symbol). The dicyanamido-bromo-cuprate(I) anion is also depicted in Figure 3.18. It consists of zig-zag chains of alternating

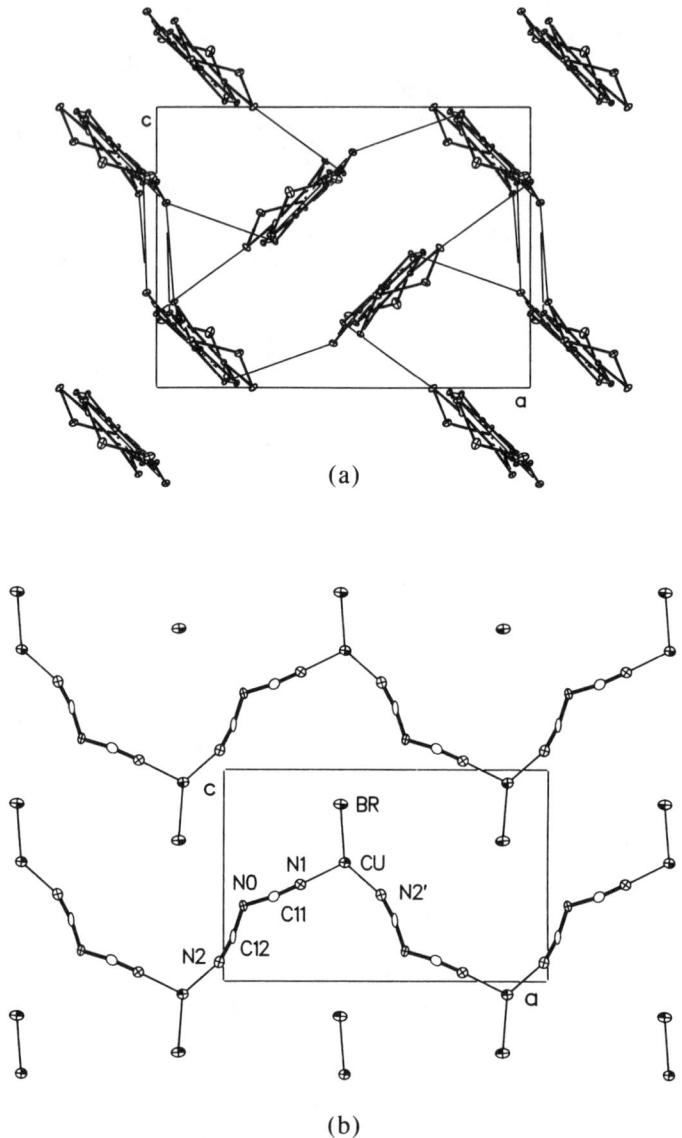

Figure 3.18 (a) ET donor molecule network in κ-(ET)$_2$Cu[N(CN)$_2$]Br at 127 K. Short S···S contacts are indicated by thin lines. Compare with the very similar κ-(ET)$_2$Cu(NCS)$_2$ shown in Figure 3.10. (b) Anion layer in κ-(ET)$_2$Cu[N(CN)$_2$]Br.

copper(I) and dicyanamide ions. The copper atoms complete their approximately trigonal coordination environment with a bond to a terminal bromide atom. In contrast to the κ-(ET)$_2$Cu(NCS)$_2$ salt, which also contains infinite-chain polymeric anions, the chains in the dicyanamide salt run along the longer of the in-plane unit cell directions, thus enhancing the two-dimensional nature of the donor-molecule network *via* hydrogen-anion contacts.

Yet more recently, the chloride analogue, κ-(ET)$_2$Cu[N(CN)$_2$]Cl, was also found[263] to be a superconductor, with an even higher transition temperature, T_c = 12.8 K (inductive onset), but requiring a modest applied pressure of 0.3 kbar. The chloride salt is isostructural to the bromide salt. Thus for the first time in the κ-family of ET compounds more than one member of an isostructural series have been found. Structure-property correlations similar to those described for the β-(ET)$_2$X class should be within reach for the κ-salts.

REFERENCES

1. Kobayashi, H.; Kobayashi, A.; Sasaki, Y.; Saito, G.; Inokuchi, H. *Bull. Chem. Soc. Jpn.* **1986**, *59*, 301.
2. Williams, J. M.; Wang, H. H.; Emge, T. J.; Geiser, U.; Beno, M. A.; Leung, P. C. W.; Carlson, K. D.; Thorn, R. J.; Schultz, A. J.; Whangbo, M.-H. In *Prog. Inorg. Chem.;* Lippard, S. J., Ed.; John Wiley: New York, **1987**, Vol. 35, p. 51.
3. Beno, M. A.; Kini, A. M.; Wang, H. H.; Tytko, S. F.; Carlson, K. D.; Williams, J. M. *Acta Cryst.* **1988**, *C44*, 1223.
4. Kobayashi, H.; Kobayashi, A.; Nakamura, T.; Nogami, T.; Shirota, Y. *Chem. Lett.* **1987**, 559.
5. Shibaeva, R. P.; Rozenberg, L. P.; Abramov, M. A.; Rubtsova, I. K.; Petrov, M. L. *Kristallografiya* **1989**, *34*, 117; *Sov. Phys. Crystallogr.* (Engl. Transl.) **1989**, *34*, 68.
6. Bechgaard, K. *Mol. Cryst. Liq. Cryst.* **1982**, *79*, 1.
7. Jérome, D.; Schultz, H. J. *Adv. Phys.* **1982**, *31*, 299.
8. Williams, J. M.; Beno, M. A.; Wang, H. H.; Leung, P. C. W.; Emge, T. J.; Geiser, U.; Carlson, K. D.; *Acc. Chem. Res.* **1985**, *18*, 261.
9. Williams, J. M. *Prog. Inorg. Chem.* **1985**, *33*, 183.
10. Williams, J. M.; Carneiro, K. *Adv. Inorg. Chem. Radiochem.* **1985**, *29*, 249.
11. McCullough, R. D.; Kok, G. B.; Lerstrup, K. A.; Cowan, D.O. *J. Am. Chem. Soc.* **1987**, *109*, 4115.
12. Wudl, F.; Aharon-Shalom, E. *J. Am. Chem. Soc.* **1982**, *104*, 1154.
13. Schumaker, R. R.; Lee, V. Y.; Engler, E. M. *J. Phys. (Les Ulis, France)* **1983**, *Colloq. C3*, 1139.
14. Wang, H. H.; Montgomery, L. K.; Geiser, U.; Porter, L. C.; Carlson, K. D.; Ferraro, J. R.; Williams, J. M.; Cariss, C. S.; Rubinstein, R. L.; Whitworth, J. R.; Evain, M.; Novoa, J. J.; Whangbo, M.-H. *Chem. Mater.* **1989**, *1*, 140.
15. Suzuki, T.; Yamochi, H.; Srdanov, G.; Hinkelmann, K.; Wudl, F. *J. Am. Chem. Soc.* **1989**, *111*, 3108.
16. Beno, M. A.; Wang, H. H.; Carlson, K. D.; Kini, A. M.; Frankenbach, G. M.; Ferraro, J. R.; Larson, N.; McCabe, G. D.; Thompson, J.; Purnama, C.; Vashon, M.; Williams, J. M.; Jung, D.; Whangbo, M.-H. *Mol. Cryst. Liq. Cryst.* **1990**, *181*, 145.
17. Beno, M. A.; Wang, H. H.; Kini, A. M.; Carlson, K. D.; Geiser, U.; Kwok, W. K.; Thompson, J. E.; Williams, J. M.; Ren, J.; Whangbo, M.-H. *Inorg. Chem.* **1990**, *29*, 1599.
18. Kini, A. M.; Mori, T.; Geiser, U.; Budz, S. M.; Williams, J. M. *J. Chem. Soc. Chem. Commun.* **1990**, 647.

19. Medne, R. S.; Khodokovskii, V. Y.; Neiland, O. Y.; Aldoshina, M. Z.; Gol'denberg, L. M.; Lyubovskaya, R. N.; Takhirov, T. G.; D'yachenko, O. A.; Atovmyan, L. O. *Izv. Akad. Nauk SSSR, Ser. Khim.* **1989,** 174; *Bull. Acad. Sci. USSR* (Engl. Transl.) **1989,** *38,* 161.
20. Papavassiliou, G. C.; Mousdis, G. A.; Zambounis, J. S.; Terzis, A.; Hountas, A.; Hilti, B.; Mayer, C. W.; Pfeiffer, J. *Synth. Met.* **1988,** *27,* B379.
21. Kini, A. M.; Beno, M. A.; Son, D.; Wang, H. H.; Carlson, K. D.; Porter, L. C.; Welp, U.; Vogt, B. A.; Williams, J. M.; Jung, D.; Evain, M.; Whangbo, M.-H; Overmyer, D. L.; Schirber, J. E. *Solid State Commun.* **1989,** *69,* 503.
22. Hountas, A.; Terzis, A.; Papavassiliou, G. C.; Hilti, B.; Bürkle, M.; Mayer, C. W.; Zambounis, J. *Acta Cryst.* **1990,** *C46,* 228.
23. Takhirov, T. G.; Krasochka, O. N.; D'yachenko, O. A.; Atovmyan, L. O.; Petrov, M. L.; Rubtsova, I. K.; Lyubovskaya, R. N. *Zh. Strukt. Khim.* **1989,** *30,* 114; *J. Struct. Chem.* (Engl. Transl.) **1989,** *30,* 455.
24. Aldoshina, M. Z.; Atovmyan, L. O.; Gol'denberg, L. M.; Krasochka, O. N.; Lyubovskaya, R. N.; Lubovskii, R. B.; Merzhanov, V. A.; Khidekel', M. L. *J. Chem. Soc., Chem. Commun.* **1985,** 1658.
25. Endres, H.; Heid, R.; Keller, H. J.; Heinen, I.; Schweitzer, D. *Acta Cryst.* **1987,** *C43,* 115.
26. Saito, K.; Ishikawa, Y.; Kikuchi, K.; Ikemoto, I.; Kobayashi, K. *Acta Cryst.* **1989,** *C45,* 1403.
27. Bloch, A. N.; Cowan, D. O.; Bechgaard, K.; Pyle, R. E.; Banks, R. H.; Poehler, T. O. *Phys. Rev. Lett.* **1975,** *34,* 1561.
28. Endres, H. Z. *Naturforsch.* **1986,** *41,* b1351.
29. Mori, T.; Wu, P.; Imaeda, K.; Enoki, T.; Inokuchi, H.; Saito, G. *Synth. Met.* **1987,** *19,* 545.
30. Iwasawa, N.; Shinozaki, F.; Saito, G.; Oshima, K.; Mori, T.; Inokuchi, H. *Chem. Lett.* **1988,** 215.
31. Saito, G. *Pure Appl. Chem.* **1987,** *59,* 999.
32. Nakano, H.; Nogami, T.; Shirota, Y.; Harada, S.; Kasai, N. *Bull. Chem. Soc. Jpn.* **1989,** *62,* 2382.
33. Girmay, B.; Kilburn, J. D.; Underhill, A. E.; Varma, K. S.; Hursthouse, M. B.; Harman, M. E.; Becher, J.; Bojesen, G. *J. Chem. Soc., Chem. Commun.* **1989,** 1406.
34. Bender, K.; Hennig, I.; Schweitzer, D.; Dietz, K.; Endres, H.; Keller, H. J. *Mol. Cryst. Liq. Cryst.* **1984,** *108,* 359.
35. Shibaeva, R. P.; Kaminskii, V. F.; Bel'skii, V. K. *Kristallografiya* **1984,** *29,* 1089; *Sov. Phys. Crystallogr.* (Engl. Transl.) **1984,** *29,* 638.
36. Williams, J. M.; Emge, T. J.; Wang, H. H.; Beno, M. A.; Copps, P. T.; Hall, L. N.; Carlson, K. D.; Crabtree, G. W. *Inorg. Chem.* **1984,** *23,* 2558.
37. Qian, M.-X.; Wang, X.-H.; Zhu, Y.-L.; Zhu, D.; Li, L.; Ma, B.-H.; Duan, H.-M.; Zhang, D.-L. *Synth. Met.* **1988,** *27,* A277.
38. Shibaeva, R. P.; Kaminskii, V. F.; Yagubskii, É. B. *Mol. Cryst. Liq. Cryst.* **1985,** *119,* 361.
39. Shibaeva, R. P.; Lobkovskaya, R. M.; Kaminskii, V. F.; Lindeman, S. V.; Yagubskii, É. B. *Kristallografiya* **1986,** *31,* 920; *Sov. Phys. Crystallogr.* (Engl. Transl.) **1986,** 31, 546.

40. Shibaeva, R. P.; Lobkovskaya, R. M.; Yagubskii, É. B.; Kostyuchenko, E. É. *Kristallografiya* **1986**, *31*, 455; *Sov. Phys. Crystallogr.* (Engl. Transl.) **1986**, *31*, 267.
41. Shibaeva, R. P.; Lobkovskaya, R. M.; Yagubskii, É. B. Kostyuchenko, E. É. *Kristallografiya* **1986**, *31*, 1110; *Sov. Phys. Crystallogr.* (Engl. Transl.) **1986**, *31*, 657.
42. Beno, M. A.; Geiser, U.; Kostka, K. L.; Wang, H. H.; Webb, K. S.; Firestone, M. A.; Carlson, K. D.; Nuñez, L.; Williams, J. M.; Whangbo, M.-H. *Inorg. Chem.* **1987**, *26*, 1912.
43. Shibaeva, R. P.; Lobkovskaya, R. M.; Yagubskii, É. B.; Laukhina, E. É. *Kristallografiya* **1987**, *32*, 901; *Sov. Phys. Crystallogr.* (Engl. Transl.) **1987**, *32*, 530.
44. Kobayashi, H.; Kato, R.; Kobayashi, A.; Nishio, Y.; Kajita, K; Sasaki, W. *Chem. Lett.* **1986**, 833.
45. Kobayashi, A.; Kato, R.; Kobayashi, H.; Moriyama, S.; Nishio, Y.; Kajita, K.; Sasaki, W. *Chem. Lett.* **1986**, 2017.
46. Kato, R.; Kobayashi, H.; Kobayashi, A.; Nishio, Y.; Kajita, K.; Sasaki, W. *Chem. Lett.* **1986**, 957.
47. Kobayashi, A.; Kato, R.; Kobayashi, H.; Moriyama, S.; Nishio, Y.; Kajita, K.; Sasaki, W. *Chem. Lett.* **1987**, 459.
48. Daoben, Z.; Ping, W.; Meixiang, W.; Zhaolou, Y.; Naijue, Z. *Solid State Commun.* **1986**, *57*, 843.
49. Emge, T. J.; Wang, H. H.; Beno, M. A.; Leung, P. C. W.; Firestone, M. A.; Jenkins, H. C.; Cook, J. D.; Carlson, K. D.; Williams, J. M.; Venturini, E. L.; Azevedo, L. J.; Schirber, J. E. *Inorg. Chem.* **1985**, *24*, 1736.
50. Williams, J. M.; Wang, H. H.; Beno, M. A.; Emge, T. J.; Sowa, L. M.; Copps, P. T.; Behroozi, F.; Hall, L. N.; Carlson, K. D.; Crabtree, G. W. *Inorg. Chem.* **1984**, *23*, 3839.
51. Yamochi, H.; Urayama, H.; Saito, G.; Oshima, K.; Kawamoto, A.; Tanaka, J. *Chem. Lett.* **1988**, 1211.
52. Kobayashi, H.; Kato, R.; Kobayashi, A.; Saito, G.; Tokumoto, M.; Anzai, H.; Ishiguro, T. *Chem. Lett.* **1986**, 93.
53. Emge, T. J.; Wang, H. H.; Leung, P. C. W.; Rust, P. R.; Cook, J. D.; Jackson, P. L.; Carlson, K. D.; Williams, J. M.; Whangbo, M.-H.; Venturini, E. L.; Schirber, J. E.; Azevedo, L. J.; Ferraro, J. R. *J. Am. Chem. Soc.* **1986**, *108*, 695.
54. Ugawa, A.; Okawa, Y.; Yakushi, K.; Kuroda, H.; Kawamoto, A.; Tanaka, J.; Tanaka, M.; Nogami, Y.; Kagoshima, S.; Murata, K.; Ishiguro, T. *Synth. Met.* **1988**, *27*, A407.
55. Buravov, L. I.; Zvarykina, A. V.; Ignat'ev, A. A.; Kotov, A. I.; Laukhin, V. N.; Makova, M. K.; Merzhanov, V. A.; Rozenberg, L. P.; Shibaeva, R. P.; Yagubskii, É. B. *Izv. Akad. Nauk SSSR, Ser. Khim.* **1988**, 2027; *Bull. Acad. Sci. USSR* (Engl. Transl.) **1988**, 1815.
56. Wang, H. H.; Beno, M. A.; Geiser, U.; Firestone, M. A.; Webb, K. S.; Nuñez, L.; Crabtree, G. W.; Carlson, K. D.; Williams, J. M.; Azevedo, L. J.; Kwak, J. F.; Schirber, J. E. *Inorg. Chem.* **1985**, *24*, 2465.
57. Geiser, U.; Wang, H. H.; Beno, M. A.; Firestone, M. A.; Webb, K. S.; Williams, J. M.; Whangbo, M.-H. *Solid State Commun.* **1986**, *57*, 741.
58. Kobayashi, A.; Kato, R.; Kobayashi, H.; Tokumoto, M.; Anzai, H.; Ishiguro, T. *Chem. Lett.* **1986**, 1117.

59. Whangbo, M.-H.; Evain, M.; Beno, M. A.; Wang, H. H.; Webb, K. S.; Williams, J. M. *Solid State Commun.* **1988**, *68*, 421.
60. Ugawa, A.; Yakushi, K.; Kuroda, H.; Kawamoto, A.; Tanaka, J. *Synth. Met.* **1988**, *22*, 305.
61. Ugawa, A.; Yakushi, K.; Kuroda, H.; Kawamoto, A.; Tanaka, J. *Chem. Lett.* **1986**, 1875.
62. Beno, M. A.; Firestone, M. A.; Leung, P. C. W.; Sowa, L. M.; Wang, H. H.; Williams, J. M.; Whangbo, M.-H. *Solid State Commun.* **1986**, *57*, 735.
63. Mori, T.; Sakai, F.; Saito, G.; Inokuchi, H. *Chem. Lett.* **1986**, 1037.
64. Mori, T.; Sakai, F.; Saito, G.; Inokuchi, H. *Chem. Lett.* **1986**, 1589.
65. Emge, T. J.; Wang, H. H.; Bowman, M. K.; Pipan, C. M.; Carlson, K. D.; Beno, M. A.; Hall, L. N.; Anderson, B. A.; Williams, J. M.; Whangbo, M.-H. *J. Am. Chem. Soc.* **1987**, *109*, 2016.
66. Mori, T.; Inokuchi, H. *Solid State Commun.* **1987**, *62*, 525.
67. Amberger, E.; Fuchs, H.; Polborn, K. *Synth. Met.* **1987**, *19*, 605.
68. Kurmoo, M.; Talham, D. R.; Pritchard, K. L.; Day, P.; Stringer, A. M.; Howard, J. A. K. *Synth. Met.* **1988**, *27*, A177.
69. Kawamoto, A.; Tanaka, J.; Tanaka, M. *Acta Cryst.* **1987**, *C43*, 205.
70. Geiser, U.; Wang, H. H.; Hammond, C. E.; Firestone, M. A.; Beno, M. A.; Carlson, K. D.; Nuñez, L.; Williams, J. M. *Acta Cryst.* **1987**, *C43*, 656.
71. Kinoshita, N.; Takahashi, K.; Murata, K.; Tokumoto, M.; Anzai, H. *Solid State Commun.* **1988**, *67*, 465.
72. Weber, A.; Endres, H.; Keller, H. J.; Gogu, E.; Heinen, I.; Bender, K.; Schweitzer, D. *Z. Naturforsch.* **1985**, *40b*, 1658.
73. Beno, M. A.; Wang, H. H.; Soderholm, L.; Carlson, K. D.; Hall, L. N.; Nuñez, L.; Rummens, H.; Anderson, B.; Schlueter, J. A.; Williams, J. M.; Whangbo, M.-H.; Evain, M. *Inorg. Chem.* **1989**, *28*, 150.
74. Yamochi, H.; Tsuji, T.; Saito, G.; Suzuki, T.; Miyashi, T.; Kabuto, C. *Synth. Met.* **1988**, *27*, A479.
75. Watson, W. H.; Kini, A. M.; Beno, M. A.; Montgomery, L. K.; Wang, H. H.; Carlson, K. D.; Gates, B. D.; Tytko, S. F.; DeRose, J.; Cariss, C.; Rohl, C. A.; Williams, J. M. *Synth. Met.* **1989**, *33*, 1.
76. Mori, T.; Inokuchi, H. *Solid State Commun.* **1986**, *59*, 355.
77. Mori, T.; Inokuchi, H. *Bull. Chem. Soc. Jpn.* **1987**, *60*, 402.
78. Gärtner, S.; Heinen, I.; Schweitzer, D.; Nuber, B.; Keller, H. J. *Synth. Met.* **1989**, *31*, 199.
79. Ouahab, L.; Padiou, J.; Grandjean, D.; Garrigou-Lagrange, C.; Delhaes, P.; Bencharif, M. *J. Chem. Soc., Chem. Commun.* **1989**, 1038.
80. Tanaka, M.; Takeuchi, H.; Sano, M.; Enoki, T.; Suzuki, K.; Imaeda, K. *Bull. Chem. Soc. Jpn.* **1989**, *62*, 1432.
81. (a) Shibaeva, R. P.; Lobkovskaya, R. M.; Korotkov, V. E.; Kushch, N. D.; Yagubskii, É. B.; Makova, M. K. *Synth. Met.* **1988**, *27*, A457. (b) Lobkovskaya, R. M.; Kushch, N. D.; Shibaeva, R. P.; Yagubskii, É. B.; Simonov, M. A. *Kristallografiya* **1989**, *34*, 1158; *Sov. Phys. Crystallogr.* (Engl. Transl.) **1989**, *34*, 698.

82. Porter, L. C.; Wang, H. H.; Beno, M. A.; Carlson, K. D.; Pipan, C. M.; Proksch, R. B.; Williams, J. M. *Solid State Commun.* **1987,** *64,* 387.
83. Geiser, U.; Anderson, B. A.; Murray, A.; Pipan, C. M.; Rohl, C. A.; Vogt, B. A.; Wang, H. H.; Williams, J. M.; Kang, D. B.; Whangbo, M.-H. *Mol. Cryst. Liq. Cryst.* **1990,** *181,* 105.
84. Mori, T.; Inokuchi, H. *Chem. Lett.* **1986,** 2069.
85. Mori. T.; Sakai, F.; Saito, G.; Inokuchi, H. *Chem. Lett.* **1987,** 927.
86. Mori, T.; Wang, P.; Imaeda, K.; Enoki,T.; Inokuchi, H. *Solid State Commun.* **1987,** *64,* 733.
87. Müller, H.; Fritz, H. P.; Heidmann, C.-P.; Gross, F.; Veith, H.; Lerf, A.; Andres, K.; Fuchs, H.; Polborn, K.; Abriel, W. *Synth. Met.* **1988,** *27,* A257.
88. Kobayashi, H.; Kato, R.; Kobayashi, A.; Sasaki, Y. *Chem. Lett.* **1985,** 191.
89. Reith, W.; Polborn, K.; Amberger, E. *Angew. Chem. Int. Ed. Engl.* **1988,** *27,* 699.
90. Aso, Y.; Yui, K.; Ishida, H.; Otsubo, T.; Ogura, F.; Kawamoto, A.; Tanaka, J. *Chem. Lett.* **1988,** 1069.
91. Parkin, S. S. P.; Engler, E. M.; Schumaker, R. R.; Lagier, R.; Lee, V. Y.; Scott, J. C.; Greene, R. L. *Phys. Rev. Lett.* **1983,** *50,* 270.
92. Williams, J. M.; Beno, M. A.; Wang, H. H.; Reed, P. E.; Azevedo, L. J.; Schirber, J. E. *Inorg. Chem.* **1984,** *23,* 1790.
93. Carneiro, K.; Scott, J. C.; Engler, E. M. *Solid State Commun.* **1984,** *50,* 477.
94. Parkin, S. S. P.; Engler, E. M.; Lee, V. Y.; Schumaker, R. R. *Mol. Cryst. Liq. Cryst.* **1985,** *119,* 375.
95. Beno, M. A.; Blackman, G. S.; Leung, P. C. W.; Carlson, K. D.; Copps, P. T.; Williams, J. M. *Mol. Cryst. Liq. Cryst.* **1985,** *119,* 409.
96. Kobayashi, H.; Kato, R.; Mori, T.; Kobayashi, A.; Sasaki, Y.; Saito, G.; Enoki, T.; Inokuchi, H. *Chem. Lett.* **1984,** 179.
97. Porter, L. C.; Wang, H. H.; Miller, M. M.; Williams, J. M. *Acta Cryst.* **1987,** *C43,* 2201.
98. Kobayashi, H.; Kobayashi, A.; Sasaki, Y.; Saito, G.; Inokuchi, H. *Chem. Lett.* **1984,** 183.
99. Kobayashi, H.; Kobayashi, A.; Sasaki, Y.; Saito, G.; Enoki, T.; Inokuchi, H. *J. Am. Chem. Soc.* **1983,** *105,* 297.
100. Cox, D. D.; Ball, G. A.; Alonso, A. S.; Williams, J. M. *Inorg. Synth.* **1989,** *26,* 393.
101. Geiser, U.; Wang, H. H.; Schlueter, J. A.; Hallenbeck, S. L.; Allen, T. J.; Chen, M. Y.; Kao, H.-C. I.; Carlson, K. D.; Gerdom, L. E.; Williams, J. M. *Acta Cryst.* **1988,** *C44,* 1544.
102. Beno, M. A.; Cox, D. D.; Williams, J. M.; Kwak, J. F. *Acta Cryst.* **1984,** *C40,* 1334.
103. Geiser, U. unpublished results.
104. Mallah, T.; Hollis, C.; Bott, S.; Day, P.; Kurmoo, M. *Synth. Met.* **1988,** *27,* A381.
105. Mallah, T.; Hollis, C.; Bott, S.; Kurmoo, M.; Day, P.; Allan, M.; Friend, R. H. *J. Chem. Soc. Dalton, Trans.* **1990,** 859.
106. (a) Korotkov, V. E.; Kushch, N. D.; Makova, M. K.; Shibaeva, R. P.; Yagubskii, É. B. *Izv. Akad. Nauk SSSR, Ser. Khim.* **1988,** *37,* 1686; *Bull. Acad. Sci. USSR* (Engl. Transl.) **1988,** *37,* 1500. (b) Korotkov, V. E.; Shibaeva, R. P. *Kristallografiya* **1989,** *34,* 1442; *Sov. Phys. Crystallogr.* (Engl. Transl.) **1989,** *34,* 865.

107. Mori, T.; Inokuchi, H. *Bull. Chem. Soc. Jpn.* **1988,** *61,* 591.
108. Bu, X.; Coppens, P.; Naughton, M. J. *Acta Cryst.* **1990,** *C46,* 1609.
109. Leung, P. C. W.; Beno, M. A.; Blackman, G. S.; Coughlin, B. R.; Miderski, C. A.; Joss, W.; Crabtree, G. W.; Williams, J. M. *Acta Cryst.* **1984,** *C40,* 1331.
110. Laversanne, R.; Amiell, J.; Delhaes, P.; Chasseau, D.; Hauw, C. *Solid State Commun.* **1984,** *52,* 177.
111. Kobayashi, H.; Kato, R.; Mori, T.; Kobayashi, A.; Sasaki, Y.; Saito, G.; Inokuchi, H. *Chem. Lett.* **1983,** 759.
112. Kobayashi, H.; Mori, T.; Kato, R.; Kobayashi, A.; Sasaki, Y.; Saito, G.; Inokuchi, H. *Chem. Lett.* **1983,** 581.
113. Galimzyanov, A. A.; Ignat'ev, A. A.; Kushch, N. D.; Laukhin, V. N.; Makova, M. K.; Merzhanov, V. A.; Rozenberg, L. P.; Shibaeva, R. P.; Yagubskii, É. B. *Synth. Met.* **1989,** *33,* 81.
114. Mori, T.; Inokuchi, H. *Chem. Lett.* **1987,** 1657.
115. Rosseinsky, M. J.; Kurmoo, M.; Talham, D. R.; Day, P.; Chasseau, D.; Watkin, D. *J. Chem. Soc., Chem. Commun.* **1988,** 88.
116. Shibaeva, R. P.; Lobkovskaya, R. M.; Rozenberg, L. P.; Buravov, L. I.; Ignatiev, A. A.; Kushch, N. D.; Laukhina, E. É.; Makova, M. K.; Yagubskii, É. B.; Zvarykina, A. V. *Synth. Met.* **1988,** *27,* A189.
117. Urayama, H.; Saito, G.; Kawamoto, A.; Tanaka, J. *Chem. Lett.* **1987,** 1753.
118. Shibaeva, R. P.; Rozenberg, L. P.; Aldoshina, M. Z.; Lobkovskaya, R. M. *Kristallografiya* **1988,** *33,* 125; *Sov. Phys. Crystallogr.* (Engl. Transl.) **1988,** *33,* 71.
119. Geiser, U.; Wang, H. H.; Kleinjan, S.; Williams, J. M. *Mol. Cryst. Liq. Cryst.* **1990,** *181,* 125.
120. Geiser, U.; Budz, S.; Kleinjan, S.; Wang, H. H.; Williams, J. M.; Kang, D. B.; Whangbo, M.-H. to be submitted to *Chem. Mater.*
121. Aldoshina, M. Z.; Goldenberg, L. M.; Zhilyaeva, E. I.; Lyubovskaya, R. N.; Takhirov, T. G.; Dyachenko, O. A.; Atovmyan, L. O.; Lyubovskii, R. B. *Mater. Sci.* **1988,** *14,* 45.
122. Bu, X.; Su, Z.; Coppens, P. *Acta Cryst.* **1990,** *c47,* 279.
123. Acrivos, J. V.; Hughes, H. P.; Parkin, S. S. P. *J. Chem. Phys.* **1987,** *86,* 1780.
124. Fuchs, H.; Fuchs, S.; Polborn, K.; Lehnert, T.; Heidmann, C.-P.; Müller, H. *Synth. Met.* **1988,** *27,* A271.
125. Broderick, W. E.; McGhee, E. M.; Godfrey, M. R.; Hoffman, B. M.; Ibers, J. A. *Inorg. Chem.* **1989,** *28,* 2902.
126. Chasseau, D.; Watkin, D.; Rosseinsky, M. J.; Kurmoo, M.; Talham, D. R.; Day, P. *Synth. Met.* **1988,** *24,* 117.
127. Pénicaud, A.; Lenoir, C.; Batail, P.; Coulon, C.; Perrin, A. *Synth. Met.* **1989,** *32,* 25.
128. Davidson, A.; Boubekeur, K.; Pénicaud, A.; Auban, P.; Lenoir, C.; Batail, P.; Hervé, G. *J. Chem. Soc., Chem. Commun.* **1989,** 1373.
129. Geiser, U.; Wang, H. H.; Donega, K. M.; Anderson, B. A.; Williams, J. M.; Kwak, J. F. *Inorg. Chem.* **1986,** *25,* 401.
130. Geiser, U.; Wang, H. H.; Gerdom, L. E.; Firestone, M. A.; Sowa, L. M.; Williams, J. M.; Whangbo, M.-H. *J. Am. Chem. Soc.* **1985,** *107,* 8305.

131. Geiser, U.; Beno, M. A.; Kini, A. M.; Wang, H. H.; Schultz, A. J.; Gates, B. D.; Cariss, C. S.; Carlson, K. D.; Williams, J. M. *Synth. Met.* **1988**, *27*, A235.

132. Geiser, U.; Wang, H. H.; Rust, P. R.; Tonge, L. M.; Williams, J. M. *Mol. Cryst. Liq. Cryst.* **1990**, *181*, 117.

133. Urayama, H.; Yamochi, H.; Saito, G.; Sato, S.; Kawamoto, A.; Tanaka, J.; Mori, T.; Maruyama, Y.; Inokuchi, H. *Chem. Lett.* **1988**, 463.

134. Shibaeva, R. P.; Lobkovskaya, R. M. *Kristallografiya* **1988**, *33*, 408; *Sov. Phys. Crystallogr.* (Engl. Transl.) **1988**, *33*, 241.

135. Geiser, U.; Wang, H. H.; Budz, S. M.; Lowry, M. J.; Williams, J. M.; Ren, J.; Whangbo, M.-H. *Inorg. Chem.* **1990**, *29*, 1611.

136. Aldoshina, M. Z.; Goldenberg, L. M.; Lyubovskaya, R. N.; Takhirov, T. G.; Dyachenko, O. A.; Atovmyan, L. O.; Merzhanov, V. A.; Lyubovskii, R. B. *Mater. Sci.* **1988**, *14*, 53.

137. Lyubovskaya, R. N.; Zhilyaeva, E. I.; Pesotskii, S. I.; Lyubovskii, R. B.; Atovmyan, L. O.; D'yachenko, O. A.; Takhirov, T. J. *Pis'ma Zh. Eksp. Teor. Fiz.* **1987**, *46*, 149; *JETP Lett.* (Engl. Transl.) **1987**, *46*, 188.

138. Shibaeva, R. P.; Rozenberg, L. P. *Kristallografiya* **1988**, *33*, 1402; *Sov. Phys. Crystallogr.* (Engl. Transl.) **1988**, *33*, 834.

139. Oshima, M.; Mori, H.; Saito, G.; Oshima, K. *Chem. Lett.* **1989**, 1159.

140. Mori, H.; Tanaka, S.; Oshima, M.; Saito, G.; Mori, T.; Maruyama, Y.; Inokuchi, H. *Bull. Chem. Soc. Jpn.* **1990**, *63*, 2183.

141. Terzis, A.; Hountas, A.; Papavassiliou, G. C.; Hilti, B.; Pfeiffer, J. *Acta Cryst.* **1990**, *C46*, 224.

142. Kato, R.; Kobayashi, H.; Kobayashi, A. *Chem. Lett.* **1986**, 2013.

143. Yoshitake, M.; Yakushi, K.; Kuroda, H.; Kobayashi, A.; Kato, R.; Kobayashi, H. *Bull. Chem. Soc. Jpn.* **1988**, *61*, 1115.

144. Kato, R.; Kobayashi, H.; Kobayashi, A.; Sasaki, Y. *Chem. Lett.* **1985**, 1231.

145. Kato, R.; Kobayashi, H.; Kobayashi, A. *Chem. Lett.* **1987**, 567.

146. Nigrey, P. J.; Morosin, B.; Kwak, J. F.; Venturini, E. L.; Baughman, R. J. *Synth. Met.* **1986**, *16*, 1.

147. Kato, R.; Kobayashi, A.; Sasaki, Y.; Kobayashi, H. *Chem. Lett.* **1984**, 993.

148. Kato, R.; Kobayashi, H.; Kobayashi, A.; Sasaki, Y. *Chem. Lett.* **1984**, 1693.

149. Sugano, T.; Sato, S.; Konno, M.; Kinoshita, M. *Acta Cryst.* **1988**, *C44*, 1764.

150. Kini, A. M.; Beno, M. A.; Williams, J. M. *J. Chem. Soc., Chem. Commun.* **1987**, 335.

151. Beno, M. A.; Geiser, U.; Kini, A. M.; Wang, H. H.; Carlson, K. D.; Miller, M.; Allen, T.; Schlueter, J.; Proksch, R.; Williams, J. M. *Synth. Met.* **1988**, *27*, A209.

152. Schultz, A. J.; Geiser, U.; Kini, A. M.; Wang, H. H.; Schlueter, J.; Cariss, C. S.; Williams, J. M. *Synth. Met.* **1988**, *27*, A229.

153. Geiser, U.; Gates, B. D.; Kini, A. M.; Wang, H. H.; Williams, J. M.; Kang, D. B.; Whangbo, M.-H. unpublished results.

154. Porter, L. C.; Allen, T. J.; Carlson, K. D.; Chen, M. Y.; Geiser, U.; Kao, H.-C. I.; Kini, A. M.; Schlueter, J. A.; Wang, H. H.; Williams, J. M. *Acta Cryst.* **1988**, *C44*, 1712.

155. Geiser, U.; Allen, T. J.; Kini, A. M.; Wang, H. H.; Schlueter, J.A.; Williams, J. M.; Evain, M.; Whangbo, M.-H. to be submitted to *Solid State Commun.*
156. Geiser, U.; Montgomery, L. K.; Wang, H. H.; Kini, A. M.; Rubinstein, R. L.; Williams, J. M.; Kang, D. B.; Whangbo, M.-H. unpublished results.
157. Kobayashi, H.; Takahashi, M.; Kato, R.; Kobayashi, A.; Sasaki, Y. *Chem. Lett.* **1984**, 1331.
158. Nigrey, P. J.; Morosin, B.; Venturini, E. L.; Azevedo, L. J.; Schirber, J.E.; Perschke, S. E.; Williams, J. M. *Physica B & C* **1986**, *143*, 290.
159. Porter, L. C.; Kini, A. M.; Williams, J. M. *Acta Cryst.* **1987**, *C43*, 998.
160. Shibaeva, R. P.; Rozenberg, L. P.; Simonov, M. A.; Kushch, N. D.; Yagubskii, É. B. *Kristallografiya* **1988**, *33*, 1156; *Sov. Phys. Crystallogr.* (Engl. Transl.) **1988**, *33*, 685.
161. Williams, J. M.; Emge, T. J.; Firestone, M. A.; Wang, H. H.; Beno, M. A.; Geiser, U.; Nuñez, L.; Carlson, K. D.; Nigrey, P. J.; Whangbo, M.-H. *Mol. Cryst. Liq. Cryst.* **1987**, *148*, 233.
162. Geiser, U.; Wang, H. H.; Schlueter, J.; Chen, M. Y.; Kini, A. M.; Kao, I. H.-C.; Williams, J. M.; Whangbo, M.-H.; Evain, M. *Inorg. Chem.* **1988**, *27*, 4284.
163. Kato, R.; Mori, T.; Kobayashi, A.; Sasaki, Y.; Kobayashi, A. *Chem. Lett.* **1984**, 781.
164. Kobayashi, A.; Kato, R.; Kobayashi, H.; Mori, T.; Inokuchi, H. *Physica B + C* **1986**, *143*, 562.
165. Kato, R.; Kobayashi, H.; Kobayashi, A.; Sasaki, Y. *Chem. Lett.* **1985**, 1943.
166. Kato, R.; Kobayashi, H.; Kobayashi, A. *Chem. Lett.* **1986**, 785.
167. Kato, R.; Kobayashi, H.; Kobayashi, A.; Mori, T.; Inokuchi, H. *Chem. Lett.* **1987**, 277.
168. Kato, R.; Kobayashi, H.; Kobayashi, A. *Synth. Met.* **1987**, *19*, 629.
169. Porter, L. C.; Cariss, C. S.; Carlson, K. D.; Geiser, U.; Kini, A. M.; Montgomery, L. K.; Rubenstein, R. L.; Wang, H. H.; Whitworth, J. R.; Williams, J. M. *Synth. Met.* **1988**, *27*, A223.
170. Nigrey, P. J.; Morosin, B.; Duesler, E. *Synth. Met.* **1988**, *27*, B481.
171. Kini, A. M.; Gates, B. D.; Beno, M. A.; Williams, J. M. *J. Chem. Soc., Chem. Commun.* **1989**, 169.
172. Kikuchi, K.; Ishikawa, Y.; Saito, K.; Ikemoto, I.; Kobayashi, K. *Acta Cryst.* **1988**, *C44*, 466.
173. Ishikawa, Y.; Saito, K.; Kikuchi, K.; Ikemoto, I.; Kobayashi, K.; Anzai, H. *Acta Cryst.* **1990**, *C46*, 1652.
174. Ishikawa, Y.; Kikuchi, K.; Saito, K.; Ikemoto, I.; Kobayashi, K. *Acta Cryst.* **1989**, *C45*, 572.
175. Aldoshina, M. Z.; Atovmyan, L. O.; Gol'denberg, L. M.; Krasochka, O. N.; Lyubovskaya, R. N.; Lyubovskii, R. B.; Khidekel', M. L.; *Dokl. Akad. Nauk SSSR* **1986**, *289*, 1140, *Proc. Acad. Sci. USSR; Phys. Chem.* (Engl. Transl.) **1986**, *289*, 697.
176. Kikuchi, K.; Ishikawa, Y.; Saito, K.; Ikemoto, I.; Kobayashi, K. *Synth. Met.* **1988**, *27*, B391.
177. Kikuchi, K.; Honda. Y.; Ishikawa, Y.; Saito, K.; Ikemoto, I.; Murata, K.; Anzai, H.; Ishiguro, T.; Kobayashi, K. *Solid State Commun.* **1988**, *66*, 405.
178. Kikuchi, K.; Ikemoto, I.; Kobayashi, K. *Synth. Met.* **1987**, *19*, 551.

179. Ikemoto, I. *Jpn. J. Appl. Phys. Ser. 1 (Superconducting Materials)* **1988**, 170.
180. Gallois, B.; Gaultier, J.; Bechtel, F.; Chasseau, D.; Hauw, C.; Ducasse, L. *Synth. Met.* **1987**, *19*, 419.
181. Gaultier, J.; Gallois, B.; Chasseau, D.; Bechtel, F.; Ducasse, L. *Acta Cryst.* (submitted for publication).
182. Heid, R.; Endres, H.; Keller, H. J.; Gogu, E.; Heinen, I.; Bender, K.; Schweitzer, D. *Z. Naturforsch B. Chem. Sci.* **1985**, *40*, 1703.
183. Hountas, A.; Terzis, A.; Papavassiliou, G. C.; Hilti, B.; Pfeiffer, J. *Acta Cryst.* **1990**, *C46*, 220.
184. Mori, T.; Inokuchi, H. *Solid State Commun.* **1989**, *70*, 823.
185. Terzis, A.; Hountas, A.; Papavassiliou, G. C. *Solid State Commun.* **1988**, *66*, 1161.
186. Kato, R.; Kobayashi, H.; Kobayashi, A. *Chem. Lett.* **1989**, 781.
187. Kato, R.; Kobayashi, H.; Kobayashi, A.; Naito, T.; Tamura, M.; Tajima, H.; Kuroda, H. *Chem. Lett.* **1989**, 1839.
188. Wudl, F.; Yamochi, H.; Suzuki, T.; Isotalo, H.; Fite, C.; Kasmai, H.; Liou, K.; Srdanov, G.; Coppens, P.; Maly, K.; Frost-Jensen, A. *J. Am. Chem. Soc.* **1990**, *112*, 2461.
189. Shibaeva, R. P.; Rozenberg, L. P.; Aldoshina, M. Z.; Lyubovskaya, R. N.; Khidekel', M. L. *Zh. Strukt. Khim.* **1979**, *20*, 485; *J. Struct. Chem.* (Engl. Transl.) **1979**, *20*, 409.
190. Honda, M.; Katayama, C.; Tanaka, J.; Tanaka, M. *Acta Cryst.* **1985**, *C41*, 197.
191. Honda, M.; Katayama, C.; Tanaka, J.; Tanaka, M. *Acta Cryst.* **1985**, *C41*, 688.
192. Matsubayashi, G.; Shimizu, R.; Tanaka, T. *Chem. Lett.* **1985**, 973.
193. Matsubayashi, G.; Shimizu, R.; Tanaka, T. *Synth. Met.* **1987**, *19*, 715.
194. Veretennikova, L. S.; Lyubovskaya, R. N.; Lyubovskii, R. B.; Rozenberg, L. P.; Simonov, M. A.; Shibaeva, R. P.; Khidekel', M. L. *Dokl. Akad. Nauk SSSR; Fiz. Khim.* **1978**, *241*, 862; *Proc. Acad. Sci. USSR; Phys. Chem.* (Engl. Transl.) **1978**, *241*, 693.
195. Shibaeva, R. P.; Rozenberg, L. P.; Lyubovskaya, R. N.; *Kristallografiya* **1980**, *25*, 507; *Sov. Phys. Crystallogr.* (Engl. Transl.) **1980**, *25*, 292.
196. Shibaeva, R. P.; Rozenberg, L. P. *Kristallografiya* **1980**, *25*, 268; *Sov. Phys. Crystallogr.* (Engl. Transl.) **1980**, *25*, 156.
197. Tanaka, C.; Tanaka, J.; Dietz, K.; Katayama, C.; Tanaka, M. *Bull. Chem. Soc. Jpn.* **1983**, *56*, 405.
198. Kato, R.; Kobayashi, H.; Kobayashi, A.; Sasaki, Y. *Chem. Lett.* **1985**, 131.
199. Katayama, C.; Honda, M.; Kumagai, H.; Tanaka, J.; Saito, G.; Inokuchi, H. *Bull. Chem. Soc. Jpn.* **1985**, *58*, 2272.
200. Brunn, K.; Endres, H.; Weiss, J. *Z. Naturforsch B. Chem. Sci.* **1988**, *43*, 224.
201. Jones, P. G. *Z. Naturforsch B. Chem. Sci.* **1989**, *44*, 243.
202. Endres, H. *Z. Naturforsch B. Chem. Sci.* **1986**, *41*, 1437.
203. Honda, K.; Goto, M.; Kurahashi, M.; Anzai, H.; Tokumoto, M.; Ishiguro, T. *Bull. Chem. Soc. Jpn.* **1988**, *61*, 588.
204. Wu, P.; Mori, T.; Enoki, T.; Imaeda, K.; Saito, G.; Inokuchi, H. *Bull. Chem. Soc. Jpn.* **1986**, *59*, 127.
205. Endres, H. *Z. Naturforsch B. Chem. Sci.* **1987**, *42*, 5.

206. Inokuchi, H.; Imaeda, K.; Enoki, T.; Mori, T.; Maruyama, Y.; Saito, G.; Okada, N.; Yamochi, H.; Seki, K.; Higuchi, Y.; Yasuoka, N. *Nature* **1987**, *329*, 39.

207. Mentzafos, D.; Psycharis, V.; Terzis, A. *Acta Cryst.* **1989**, *C45*, 1333.

208. Terzis, A.; Psycharis, V.; Hountas, A.; Papavassiliou, G. *Acta Cryst.* **1988**, *C44*, 128.

209. Psycharis, V.; Hountas, A.; Terzis, A.; Papavassiliou, G. *Acta Cryst.* **1988**, *C44*, 125.

210. Papavassiliou, G. C.; Terzis, A.; Underhill, A. E.; Geserich, H. P.; Kaye, B. *Synth. Met.* **1987**, *19*, 703.

211. Terzis, A.; Hountas, A.; Underhill, A. E.; Clark, A.; Kaye, B.; Hilti, B.; Mayer, C.; Pfeiffer, J.; Yiannopoulos, S. Y.; Mousdis, G.; Papavassiliou, G. C. *Synth. Met.* **1988**, *27*, B97.

212. Wallis, J. D.; Karrer, A. D.; Dunitz, J. D. *Helv. Chim. Acta* **1986**, *69*, 69.

213. Yagubskii, É. B.; Kotov, A. I.; Shibaeva, R. P.; Ignat'ev, A. A.; Neiland, O. Y.; Kreitsberga, Y. N. *Dokl. Akad. Nauk SSSR; Fiz. Khim.* **1986**, *289*, 676; *Proc. Acad. Sci. USSR; Phys. Chem.* (Engl. Transl.) **1986**, *289*, 673.

214. Granier, T.; Gallois, B.; Fabre, J. M. *Acta Cryst.* **1989**, *C45*, 1376.

215. Phillips, T. E.; Kistenmacher, T. J.; Bloch, A. N.; Cowan, D. O. *J. Chem. Soc., Chem. Commun.* **1976**, 334.

216. Becker, J. Y.; Bernstein, J.; Bittner, S.; Sarma, J. A. R. P.; Shahal, L.; Shaik, S. S. *Acta Cryst.* **1988**, *C44*, 1770.

217. Carroll, P. J.; Lakshmikantham, M. V.; Cava, M. P.; Wudl, F.; Aharon-Shalom, E.; Cox, S. D. *J. Chem. Soc., Chem. Commun.* **1982**, 1316.

218. Kikuchi, K.; Yakushi, K.; Kuroda, H.; Ikemoto, I.; Kobayashi, K.; Honda, M.; Katayama, C.; Tanaka, J. *Chem. Lett.* **1985**, 419.

219. Li, Z. S.; Matsuzaki, S.; Kato, R.; Kobayashi, H.; Kobayashi, A.; Sano, M. *Chem. Lett.* **1986**, 1105.

220. Kobayashi, A.; Sasaki, Y.; Kato, R.; Kobayashi, H. *Chem. Lett.* **1986**, 387.

221. Okada, N.; Saito, G.; Mori, T. *Chem. Lett.* **1986**, 311.

222. Lerstrup, K.; Cowan, D. O.; Kistenmacher, T. J. *J. Am. Chem. Soc.* **1984**, *106*, 8303.

223. Lerstrup, K.; Lee, M.; Wiygul, F. M.; Kistenmacher, T. J.; Cowan, D. O. *J. Chem. Soc., Chem. Commun.* **1983**, 294.

224. Shu, P.; Chiang, L.; Emge, T. J.; Holt, D.; Kistenmacher, T. J.; Lee, M.; Stokes, J.; Poehler, T.; Bloch, A.; Cowan, D. *J. Chem. Soc., Chem. Commun.* **1981**, 920.

225. Emge, T. J.; Leung, P. C. W.; Beno, M. A.; Schultz, A. J.; Wang, H. H.; Sowa, L. M.; Williams, J. M. *Phys. Rev. B: Condens. Matter* **1984**, *30*, 6780.

226. Leung, P. C. W.; Emge, T. J.; Beno, M. A.; Wang, H. H.; Williams, J. M.; Petricek, V.; Coppens, P. *J. Am. Chem. Soc.* **1984**, *106*, 7644.

227. Nogami, Y.; Kagoshima, S.; Sugano, T.; Saito, G. *Synth. Met.* **1986**, *16*, 367.

228. Ravy, S.; Pouget, J. P.; Moret, R.; Lenoir, C. *Phys. Rev. B: Condens. Matter* **1988**, *37*, 5113.

229. Schultz, A. J.; Beno, M. A.; Wang, H. H.; Williams, J. M. *Phys. Rev. B: Condens. Matter* **1986**, *33*, 7823.

230. Leung, P. C. W.; Emge, T. J.; Beno, M. A.; Wang, H. H.; Williams, J. M.; Petricek, V.; Coppens, P. *J. Am. Chem. Soc.* **1985**, *107*, 6184.

231. Petricek, V.; Coppens, P.; Becker, P. *Mol. Cryst. Liq. Cryst.* **1985**, *125*, 393.

232. Nogami, Y.; Kagoshima, S.; Anzai, H.; Tokumoto, M.; Mori, N.; Kinoshita, N.; Saito, G. *J. Phys. Soc. Jpn.* **1990,** *59,* 259.

233. Laukhin, V. N.; Kostyuchenko, E. É.; Sushko, Y. V.; Shchegolev, I. F.; Yagubskii, É. B. *Pis'ma Zh. Eksp. Teor. Fiz.* **1985,** *41,* 68; *JETP Lett.* (Engl. Transl.) **1985,** *41,* 81.

234. Murata, K.; Tokumoto, M.; Anzai, H.; Bando, H.; Saito, G.; Kajimura, K.; Ishiguro, T. *J. Phys. Soc. Jpn.* **1985,** *54,* 1236.

235. Veith, H.; Heidmann, C.-P.; Gross, F.; Lerf, A.; Andres, K.; Schweitzer, D. *Solid State Commun.* **1985,** *56,* 1015.

236. Creuzet, F.; Creuzet, G.; Jérome, D.; Schweitzer, D.; Keller, H. J. *J. Phys. Lett.* **1985,** *46,* L1079.

237. Schultz, A. J.; Wang, H. H.; Williams, J. M.; Filhol, A. *J. Am. Chem. Soc.* **1986,** *108,* 7853.

238. Molchanov, V. N.; Shibaeva, R. P.; Kachinskii, V. N.; Yagubskii, É. B.; Simonov, V. I.; Vainshtein, B. K. *Dokl. Akad. Nauk SSSR* **1986,** *286,* 637; *Sov. Phys. Dokl.* (Engl. Transl.) **1986,** *31,* 6.

239. Urayama, H.; Yamochi, H.; Saito, G.; Nozawa, K.; Sugano, T.; Kinoshita, M.; Sato, S.; Oshima, K.; Kawamoto, A.; Tanaka, J. *Chem. Lett.* **1988,** 55.

240. Carlson, K. D.; Geiser, U.; Kini, A. M.; Wang, H. H.; Montgomery, L. K.; Kwok, W. K.; Beno, M. A.; Williams, J. M.; Cariss, C. S.; Crabtree, G. W.; Whangbo, M.-H.; Evain, M. *Inorg. Chem.* **1988,** *27,* 965.

241. Jung, D.; Evain, M.; Novoa, J. J.; Whangbo, M.-H.; Beno, M. A.; Kini, A. M.; Schultz, A. J.; Williams, J. M.; Nigrey, P. J. *Inorg. Chem.* **1989,** *28,* 4516.

242. Whangbo, M.-H.; Jung, D.; Wang, H. H.; Beno, M. A.; Williams, J. M.; Kikuchi, K. *Mol. Cryst. Liq. Cryst.* **1990,** *181,* 1.

243. Kato, R.; Kobayashi, H.; Kobayashi, A.; Moriyama, S.; Nishio, Y.; Kajita, K.; Sasaki, W. *Chem. Lett.* **1987,** 507.

244. Kobayashi, H.; Kato, R.; Kobayashi, A.; Nishio, Y.; Kajita, K.; Sasaki, W. *Chem. Lett.* **1986,** 789.

245. Yagubskii, É. B.; Shchegolev, I. F.; Pesotskii, S. I.; Laukhin, V. N.; Kononovich, P. A.; Kartsovnik, M. V.; Zvarykina, A. V. *Pis'ma Zh. Eksp. Teor. Fiz.* **1984,** *39,* 275; *JETP Lett.* (Engl. Transl.) **1984,** *39,* 328.

246. Lyubovskaya, R. N.; Lyubovskii, R. B.; Shibaeva, R. P.; Aldoshina, M. Z.; Gol'denberg, L. M.; Rozenberg, L. P.; Khidekel', M. L.; Shul'pyakov, Y. F. *Pis'ma Zh. Eksp. Teor. Fiz.* **1985,** *42,* 380; *JETP Lett.* (Engl. Transl.) **1985,** *42,* 468.

247. Lyubovskii, R. B.; Lyubovskaya, R. N.; Kapustin, N. V. *Zh. Eksp. Teor. Fiz.* **1987,** *93,* 1863; *Sov. Phys. JETP* (Engl. Transl.) **1987,** *66,* 1063.

248. Mori, T.; Inokuchi, H. *Solid State Commun.* **1987,** *64,* 335.

249. Wang, H. H.; Carlson, K. D.; Geiser, U.; Kwok, W. K.; Vashon, M. D.; Thompson, J. E.; Larsen, N. F.; McCabe, G. D.; Hulscher, R. S.; Williams, J. M. *Physica C* **1990,** *166,* 57.

250. Brossard, L.; Ribault, M.; Bousseau, M.; Valade, L.; Cassoux, P. *C. R. Acad. Sci. Ser. 2,* **1986,** *302,* 205.

251. Kobayashi, A.; Kim, H.; Sasaki, Y.; Moriyama, S.; Nishio, Y.; Kajita, K.; Sasaki, W.; Kato, R.; Kobayashi, H. *Synth. Met.* **1988,** *27,* B339.

252. Brossard, L.; Hurdequint, H.; Ribault, M.; Valade, L.; Legros, J.-P.; Cassoux, P. *Synth. Met.* **1988,** *27,* B157.
253. Brossard, L.; Ribault, M.; Valade, L.; Cassoux, P. *J. Phys. (Fr.)* **1989,** *50,* 1521.
254. Kikuchi, K.; Saito, K.; Ikemoto, I.; Murata, K.; Ishiguro, T.; Kobayashi, K. *Synth. Met.* **1988,** *27,* B269.
255. Kagoshima, S.; Nogami, Y. *Synth. Met.* **1988,** *27,* A299.
256. Kikuchi, K.; Kikuchi, M.; Namiki, T.; Saito, K.; Ikemoto, I.; Murata, K.; Ishiguro, T.; Kobayashi, K. *Chem. Lett.* **1987,** 931.
257. Kikuchi, K.; Murata, K.; Kikuchi, M.; Honda, Y.; Takahashi, T.; Oyama, T.; Ikemoto, I.; Ishiguro, T.; Kobayashi, K. *Jpn. J. Appl. Phys.* **1987,** *26 (Suppl. 26-3),* 1369.
258. Kikuchi, K.; Murata, K.; Honda, Y.; Namiki, T.; Saito, K.; Anzai, H.; Kobayashi, K.; Ishiguro, T.; Ikemoto, I. *J. Phys. Soc. Jpn.* **1987,** *56,* 4241.
259. Kikuchi, K.; Murata, K.; Honda, Y.; Namiki, T.; Saito, K.; Ishiguro, T.; Kobayashi, K.; Ikemoto, I. *J. Phys. Soc. Jpn.* **1987,** *56,* 3436.
260. Kikuchi, K.; Murata, K.; Honda, Y.; Namiki, T.; Saito, K.; Kobayashi, K.; Ishiguro, T.; Ikemoto, I. *J. Phys. Soc. Jpn.* **1987,** *56,* 2627.
261. Kini, A. M.; Geiser, U.; Wang, H. H.; Carlson, K. D.; Williams, J. M.; Kwok, W. K.; Vandervoort, K. G.; Thompson, J. E.; Stupka, D. L.; Jung, D.; Whangbo, M.-H. *Inorg. Chem.* **1990,** *29,* 2555.
262. Williams, J. M.; Kini, A. M.; Geiser, U.; Wang, H. H.; Carlson, K. D.; Kwok, W, K.; Vandervoort, K. G.; Thompson, J. E.; Stupka, D. L.; Jung, D.; Whangbo, M.-H. *Proceedings of the International Conference on Organic Superconductors,* South Lake Tahoe, CA, May **1990;** Kresin, V.; Little, W. L., Eds.; Plenum Press: New York, **1990;** p. 33.
263. Williams, J. M.; Kini, A. M.; Wang, H. H.; Carlson, K. D.; Geiser, U.; Montgomery, L. K.; Pyrka, G. J.; Watkins, D. M.; Kommers, J. M.; Boryschuk, S. J.; Strieby Crouch, A. V.; Kwok, W. K.; Schirber, J. E.; Overmyer, D. L.; Jung, D.; Whangbo, M.-H. *Inorg. Chem.* **1990,** *29,* 3262.

4

Electrical Conductivities and Superconducting Properties

INTRODUCTION

In this chapter, we survey the normal-state electrical conductivities, or resistivities, and the superconducting properties of the synthetic organic metals which comprise the current group of organic superconductors beyond the $(TMTSF)_2X$ charge-transfer salts. For additional perspective, we discuss the properties of a variety of semiconducting organic salts, in particular those related to or contrasting with the metallic phases derived from a common anionic species. Most of the many known organic charge-transfer salts are, in fact, semiconductors. Furthermore, an appreciable number of the salts that are metallic-like near ambient temperatures undergo metal-to-semiconducting transitions as the temperature is lowered. In some cases these transitions can be suppressed by the application of high pressures, and the pressurized phase at lower temperatures may become superconducting. A few other organic metals are metallic to the lowest temperatures but never attain the superconducting state. However, a respectable number of organic metals remain metallic to low temperatures and undergo a superconducting transition at ambient pressure. Thus, there is a rich diversity of electrical conductive behavior among the organic charge-transfer salts. The features of these materials, such as the organic-donor molecule packing motifs and the anionic arrangements and coordination, in many cases readily explain the occurrence of metallic or semiconductive behavior, the relative ease in the transformation from the metallic to the semiconductive state, and the existence of strong conductive anisotropy, which is an ubiquitous feature of the organic conductors.

The presently known organic superconductors and their superconducting transition temperatures (T_c) are summarized in Table 1.1. Listed at the top of the table are the $(TMTSF)_2X$ superconductors,[1] which are historically significant as the original source of organic superconductivity, first discovered[2] in $(TMTSF)_2PF_6$ in 1979. In these salts, the selenium-based tetramethyltetraselenafulvalene molecule (TMTSF) serves as a radical-cation donor that provides the essential network for electrical conductivity, and X^- is a univalent, complex inorganic anion. All presently known organic superconductors are derived from charge-transfer salts similar in structure to the TMTSF salts, although the organic molecule in some cases may serve as a radical-anion acceptor species, and the organic molecule in all cases is the basic source of the electrical conductivity.

All of the $(TMTSF)_2X$ salts have superconducting transitions near 1 K, and six of the seven total require applied pressures of several kbar to suppress a metal-to-semiconducting transition at temperatures below 300 K. These salts are quasi one-dimensional metals at room temperature with electrical conductivity occurring principally along the stacking axis of the TMTSF radical-cation donor molecule; conductivity along directions normal to this axis is several orders-of-magnitude lower. Unlike a true one-dimensional metal, however, these materials do not undergo a Peierls[3] driven charge-density wave transition at lower temperatures, in part because the dimensionality increases with diminishing temperatures. The salts with octahedral anions, such as PF_6^-, at ambient pressure undergo a spin-density wave (SDW) transition to an insulating state with an antiferromagnetic arrangement of spins. Those with the tetrahedral anions (ReO_4^- and ClO_4^-) at ambient pressure undergo an anion-ordering structural rearrangement with a doubling of the unit cell length along one or more axes, which yields an insulating state for the ReO_4^- salt. The anion-ordered structure of the ClO_4^- salt, however, is metallic and becomes superconducting at ambient pressure.[4] An important feature of $(TMTSF)_2ClO_4$, because of its relevance to the properties of more recent organic superconductors, is that rapid cooling through the anion-ordering transition (near 24 K) leads to the freezing of partial crystallographic disorder in the anions and to a suppression of the superconducting transition temperature.[5] Although the TMTSF-based superconductors are interesting materials with yet unresolved questions for valuable future research, it is generally perceived that the TMTSF donor molecule is likely exhausted as a potential source of higher T_c materials.

The most important group of the current organic superconductors consists of salts derived from the sulfur-based BEDT-TTF (or ET) organic-donor molecule, bis(ethylenedithio)tetrathiafulvalene.[6] This group comprises more than a dozen different phases with T_c's ranging from 1 K to 13 K. In contrast to TMTSF, the ET-donor molecule yields salts of different stoichiometry and salts of the same anion having a variety of different structures, each with much different conductive properties. Furthermore, ET yields a substantially larger number of organic metals and a much larger number of ambient pressure superconductors. The ET metallic salts in a number of important cases are quasi two-dimensional metals in the sense that the conductivity is essentially isotropic in a plane coincident with the two-dimensional S···S network of the sulfur atoms of the donor molecules. Somewhat analogous to the case of crystallographic disorder in (TMTSF)$_2$ClO$_4$, disorder in the structures of the ET metallic salts is generally accompanied by a suppression of the superconducting transition temperature. The majority of the ET-based superconductors can be classified as belonging to one of the two structural groups denoted as the β-phase superconductors, which have linear triatomic anions (I$_3^-$, AuI$_2^-$, and IBr$_2^-$), and the κ-phase or κ-like superconductors, which include the Hg halide salts, κ-(ET)$_2$I$_3$, κ-(ET)$_2$Cu(NCS)$_2$, κ-(ET)$_2$Cu[N(CN)$_2$]Br, and κ-(ET)$_2$Cu[N(CN)$_2$]Cl. Until the discovery[204–206] in 1990 of superconductivity near 11.6 K in κ-(ET)$_2$Cu[N(CN)$_2$]Br and near 12.8 K under a minimal applied pressure of ~0.3 kbar in κ-(ET)$_2$-Cu[N(CN)$_2$]Cl, the κ-(ET)$_2$Cu(NCS)$_2$ salt, with T_c near 10 K, represented the maximum temperature for organic superconductivity.[7,8] It is not obvious that ET will yield even much higher T_c materials, but it is clear that further research on the unusual properties of the ET salts is important to the development of high-T_c organic superconductors.

Listed below the ET salts in Table 1.1 are four groups of organic superconductors involving salts of other donor or acceptor molecules. The salts of the donor MDT-TTF (methylenedithio-tetrathiafulvalene) and DMET [dimethyl(ethylenedithio)diselenafulvalene] are of considerable interest as examples of unsymmetric organic donor molecules which yield organic superconductors. Thus, symmetrical donor molecules like TMTSF or ET are not essential requirements for the avoidance of crystallographic disorder and the consequent suppression of superconductivity in organic charge-transfer salts. The superconducting salts of DMET at present are rather low-temperature superconductors with T_c's near 1 K, although the salt of this donor with the highest T_c has a κ-like structure. Only one superconducting salt, (MDT-TTF)$_2$AuI$_2$, is presently known for the MDT-TTF donor. This salt, which possesses a κ-like structure, is an ambient pressure superconductor with a relatively high superconducting transition temperature, $T_c = 4.5$ K.[9,10] It appears to be significant that the κ-phase and κ-like salts occur for at least three different types of donor molecules and that this structural type yields at present the highest T_c for an organic superconductor. Thus, the κ-phase salts hold considerable promise for the synthesis of yet higher T_c organic superconductors.

Four salts of the acceptor molecule Ni(dmit)$_2$, which is bis(4,5-dimercapto-1,3-dithiole-2-thione)nickel(II), or its analogue Pd(dmit)$_2$ are reported to be superconducting from 1.6 K to 6 K under high applied pressures. The structures of metal complexes of dmit such as Ni(dmit)$_2$ closely resemble the symmetrical structure of the ET molecule and, in addition, they include terminal sulfur atoms (see Chap. 3). Thus, they are appealing as candidates for higher dimensional organic superconductors. It is of interest to note that the structure of the TTF (tetrathiafulvalene) salt of Ni(dmit)$_2$ appears to have a three-dimensional network of sulfur atoms. So far, however, this electron-acceptor molecule has not yielded any ambient pressure superconductors.

Finally—but far from unimportant despite its bottom listing in Table 1.1—is the superconducting salt β$_m$-(BEDO-TTF)$_3$Cu$_2$(NCS)$_3$, in which BEDO-TTF is the electron-donor organic molecule bis(ethylenedioxy)tetrathiafulvalene, an oxygen-replacement analogue of BEDT-TTF or ET. This salt is important as the first oxygen-containing organic superconductor,[11] and it represents at least an initial step in uniting common features of the organic superconductors with those of the recently discovered high-T_c ceramic oxide superconductors. Although the T_c of β_m-(BEDO-TTF)$_3$Cu$_2$(NCS)$_3$ is modest at ~1 K, one has substantial historical precedent with many other kinds of superconducting materials to expect increasing T_c's in other salts of this donor or salts of other oxygen-containing organic donor or acceptor molecules.

The following survey of the normal-state conductivities of organic salts presents a brief discussion of the general nature of electrical transport in these materials and a reasonably comprehensive evaluation of the experimental conductivities of nearly all of the superconducting salts and several relevant nonsuperconducting metallic and semiconducting phases. The experimental aspects are not covered in equal detail for each salt because the primary objective is the delineation in a broad context of the similarities and contrasting features of electrical conductivities and the occurrence or absence of a superconducting transition. The survey of the superconducting properties is devoted largely to those of the β-phase and κ-phase, or κ-like superconductors, because these two classes each are composed of several salts having a common structural feature and well-documented volume superconducting properties, and because these two classes presently yield the highest superconducting transition temperatures, ~8 K and ~12.8 K, respectively. In addition to these phases, we include discussions of a variety of superconducting salts belonging to the ET/I system and the salts α-(ET)$_2$(NH$_4$)Hg(SCN)$_4$ and β_m-(BEDO-TTF)$_3$Cu$_2$((NCS)$_3$. The ET/I system, which includes the β-, β*-, and κ-(ET)$_2$I$_3$ superconducting phases, is remarkable in the variety of its salts and superconducting derivatives and is worthy of special considerations as a group. α-(ET)$_2$(NH$_4$)Hg(SCN)$_4$ and β_m-(BEDO-TTF)$_3$Cu$_2$(NCS)$_3$ are of interest because they both illustrate new opportunities and directions for research on organic superconductors.

NORMAL-STATE CONDUCTIVITIES

The Nature of Electrical Transport in Organic Salts

An organic metal is an organic compound that conducts electricity as does an ordinary metal. There are two classes of organic metals: the organic charge-transfer salts of interest here, and the so-called doped organic polymers such as polyacetylene doped with iodine.[12-14] These two classes have some common conductive properties but have important differences in their electrical transport mechanisms. Furthermore, the metallic organic polymers are not superconducting, although it is interesting to note that the inorganic polymeric material (SN)$_x$ is metallic and a low-temperature superconductor with $T_c = 0.3$ K.[15] Ordinary metals, consisting of metallic elements, alloys, and intermetallic compounds, are materials of generally high conductivity, σ, which monotonically increases with decreasing temperature T, $d\sigma/dT<0$. Organic metals are characterized by the same temperature dependence of conductivity, but the conductivities at comparable temperatures are very much smaller. A typical organic metal at room temperature, for example, has $\sigma < 10^3 (\Omega\text{cm})^{-1}$ compared to $\sigma \approx 10^6$ $(\Omega$ cm$)^{-1}$ for superior conductors such as copper, silver, and gold. Furthermore, organic metals are highly anisotropic conductors for which the conductivity may be larger along one axis or direction than another. Are organic metals, therefore, really metals in the conventional sense? For the organic charge-transfer salts, the answer is a qualified yes: they are metals much like ordinary metals, but there are special chemical and structural differences that generally account for the much lower conductivities and the highly anisotropic properties. Regarding the anistropic properties, as previously discussed, the organic salts are layered materials consisting, more or less, of rows and stacks of the organic donor (or acceptor) molecules separated by arrays of anions, along which direction the conductivity is generally very small compared to conductivities within the plane of the organic layers. Weak interactions along the stacking direction or along the interstacking direction can lead to conductivities that are relatively small along one axis or direction of the organic layer, thus yielding materials that may be described as one-dimensional conductors.

Some qualitative aspects of the electronic band theory of solids are useful here for describing the nature of metallic conductivity and the origin of semiconducting states in the organic charge-transfer salts. This subject is covered in greater detail with specific examples in Chap. 8. While there are relatively strong coulombic forces binding together the cationic and anionic constituents, the valence bands, which comprise the highest occupied energy levels analogous to the valence energy levels of a molecule, are derived from the overlapping of the orbitals of adjacent organic molecular ions, in particular, the orbitals of the chalcogen atoms S and Se for the salts that are of present interest. As is typical of all organic solids, the organic moieties of the charge-transfer salts retain their molecular integrity because interactions among them, of the order of van der Waals interactions, are much weaker than the intramolecular covalent forces.[16] For weak interactions, intermolecular orbital bonding and antibonding combinations are not widely spread in energy, so that conduction and valence bands are narrow, in contrast to the wide bands of conventional metals and insulators. The inorganic counter ion, in addition to serving as an electron charge-transfer acceptor (or donor), acts as a spacer between organic stacks and layers that is crucial, in terms of its size and coordination, in modifying the strengths of the interactions between the organic moieties and in giving the anisotropic dispersions in the energy bands.

If the organic moieties were neutral molecules with even numbers of electrons, the valence bands would be completely filled, and the compound would have an insulating electronic state. Charge transfer of an odd number of electrons per formula unit (usually one electron) partially depletes a valance band (or partially fills an unoccupied valence band), thus creating a conduction band and therefore a metallic electronic state. However, charge transfer does not guarantee a metallic state. A redistribution of the conduction electrons into an antiferromagnetic spin state (a Mott-Hubbard insulator) may occur to reduce strong coulomb repulsions of the conduction electron spin-pairs; or a dimerization of the molecular ions (a Peierls-like insulator) may occur to produce an energetically more favorable state.[17] In both cases, the valence bands are again completely filled, thus yielding an insulating electronic state. In a sense, it is the packing of the organic moieties into low-dimensional arrangements that often favors localization effects and the formation of insulating ground states or promotes a metal-to-insulator transition as the temperature is lowered.

Considerable empirical evidence exists that many of the metallic organic charge-transfer salts can be adequately interpreted in terms of a conduction band, which means that these materials are metallic in the conventional sense of ordinary metals, although the conduction band may be narrower and generally anisotropic. The doped organic polymers, however, are not simple conduction-band metals. These materials, depending on the level of doping, are metallic in terms of their increasing conductivity with decreasing temperature, but the charge carriers are thought to be associated with the existence of localized states, denoted as solitons or bipolarons, which are induced by local distortions and disorder in the oxidized or reduced polymer chains.[13,14,16] As presently understood, charge carriers other than conduction-band electrons are not required for the organic charge-transfer salts. However, as more and more organic salts have been synthesized and studied, it has become increasingly evident that the structures of the conduction bands for a number of these salts are quite complicated. For example, some of the apparent metallic salts, such as $(ET)_2ClO_4(1,1,2\text{-trichloroethane})_{0.5}$ and $\alpha\text{-}(ET)_2I_3$, may be semimetals or very small energy-gap semiconductors which behave as semimetals near room temperatures. A semimetal is a metallic state, similar to that of the divalent metals such as the alkaline earth metals, in which there are two (or more) partially filled (conduction) bands that overlap in energy at the Fermi level.[18] The electrical conductivities of such metals involve transport of electrons in both bands. In associating conductivities with the transport of conduction-band electrons, one must recognize that the transport is never a simple one-electron (band) phenomenon but a many-body process involving the complicated interactions of the conduction electrons with one another and with the core atoms of the lattice.

For conventional conduction-band electrons, one can express the electrical conductivity as

$$\sigma = \frac{ne^2 \tau}{m} \tag{4.1}$$

where n is the number density of conduction electrons of charge e and mass m, and τ is the relaxation time between scattering events.[18] In its simplest interpretation, Eq. (4.1) is the expression for isotropic electrical conductivity in a homogeneous electron gas, but it is applicable to conduction-band electrons if σ is interpreted as a conductivity tensor and the number density n is defined as an effective number of electrons that depends on the shape of the Fermi surface and direction of acceleration. For conduction-band conductivity, the charge carriers are the electrons at the Fermi surface, and this surface is highly anisotropic in metallic organic salts.

Band theory arguments show that the conduction number denisty n is generally small for narrow conduction bands and large for wide bands. On the intuitive basis of volume considerations, one expects the number density to be relatively small for the organic salts because the organic molecules furnish only about one conduction electron or less per molecule and the molecule is very much larger than the atomic donor of an ordinary metal.[19] For a metallic salt such as β-$(ET)_2I_3$, for example, with one conduction electron per unit cell of volume 855.9×10^{-24} cm^3, one calculates that $n = 1.17 \times 10^{21}$ cm^{-3}, which is several orders of magnitude smaller than that of an ordinary metal.[20] The relaxation time τ (or the mean free path $l = \tau v$, where v is the electron velocity) is another important factor in the conductivity—and the most difficult to evaluate—because $1/\tau$ is proportional to the electron scattering probability. Narrow band conductors have larger densities of (momentum) states at the Fermi level, thus increasing the available states into which electrons may be scattered. Because the dominant scattering at ambient temperatures arises from the thermal vibrations (phonons) of the crystal lattice, one intuitively expects a larger scattering probability, and thus a smaller τ, for the organic metals owing to the large number of contributing internal modes of vibration from the organic moiety. In the context of the electron gas theory, therefore, the metallic organic salts may be described as systems having conduction electrons with smaller number densities and lower mobilities, $e\tau/m$, than those of ordinary metals.

For conventional metals, the electron scattering probability or $1/\tau$, which is a measure of the electrical resistance, can be shown to be proportional to the mean-square amplitude of vibration of the metal atoms, which in turn is proportional to the absolute temperature T.[18] Thus, the usual temperature dependence of metallic conductivity is a linear relationship,

$$\sigma \propto \frac{1}{T} \tag{4.2}$$

Conductivities of the metallic organic salts, on the other hand, frequently exhibit a stronger temperature dependence, approximately that of an inverse quadratic relationship,[19–22] $\sigma \propto 1/T^2$. A quadratic temperature dependence occurs in inorganic layered compounds such as TiS_2, and it is usually associated with electron-electron scattering.[19,23] However, other considerations suggest that this quadratic dependence is due to scattering by librational (twisting) modes of the organic donor molecule, which is found to be a second-order (two-phonon) coupling interaction.[24,25] Another interesting feature of the electrical conductivities of metallic organic salts, as revealed by measurements of the thermoelectric power, is that hole (p-type) conduction frequently occurs, although the charge carriers are the conduction electrons. It is interesting to note that hole conduction also occurs in the layered copper oxide high-T_c superconductors in the normal metallic state.[26] The p-type conductivity is an example of the anomalous Hall effect, in which the conduction electrons behave like particles of negative mass or, more easily visualized, like positive particles (holes) of normal positive mass. This effect is a consequence of the curvature of the Fermi surface and indicative of a nearly filled conduction band.[18] Actually, both normal electron (n-type) and hole conduction occur simultaneously, but one or the other may dominate the transport.

The same weak interactions that lead to narrow energy bands also lead to narrow energy gaps E_g, approaching the magnitudes of thermal energies at room temperature ($kT \approx 26$ meV, k is Boltzmann's constant). Therefore, although the organic charge-transfer salts may have insulating ground states, they are often better described as small-gap (intrinsic) semiconductors. These semiconductors can have appreciable conductivity,

$$\sigma = \sigma_o \exp \frac{-Eg}{2kT} \quad (4.3)$$

at ambient temperature arising from thermal excitations of valence band electrons through the gap to an unoccupied band, with activation energy

$$E_a = \frac{Eg}{2} \quad (4.4)$$

Typical activation energies for the semiconducting organic salts are of the order of 100 meV. The proportionality factor σ_o in Eq. (4.3) is a sample-dependent parameter which also has some temperature dependence as a result of changes in electron scattering. The semiconducting organic salts frequently exhibit nonexponential conductivities, especially over large temperature ranges.

Experimental electrical conductivities are usually, and most reliably, measured by a technique called the "four-probe" method. In applications of this technique to the small, fragile single crystals of organic charge-transfer salts, the probes (or electrodes) consist of four thin gold wires (13–51 μm in diameter) attached by gold or silver conductive paste at short intervals along a crystallographic axis of interest, frequently a needle axis.[27,28] A constant current, I, is applied to the crystal from the terminal probes, and the voltage drop, V, developed between the central wire probes spaced at a distance d is measured as a function of temperature. Vanishingly small current flows through the voltage-measuring instrument, so that the resistance $R = V/I$ between the voltage probes is essentially independent of the contact and lead resistances. The conductivity σ, in practical units of $(\Omega \text{ cm})^{-1}$, or the resistivity ρ, is calculated as

$$\sigma = \frac{1}{\rho} = (I/V)(d/A) \quad (4.5)$$

where A is the cross-sectional area of the crystal. With probes attached along different axial directions, or with a special arrangement of probes which provides the components of a resistivity tensor,[27,29] one can determine the conductive anisotropy. For very thin, needle-shaped crystals, however, it is difficult to measure resistances along more than the needle axis.

For the description of the electrical conductive behavior, it is useful to denote differences in the electrical conductivities measured along different crystallographic axes. Conductivities within the plane of the organic donor molecules are designated here by $\sigma_{\parallel}$ and conductivities in the perpendicular direction, along which direction the anions often segregate the organic layers, are designated by $\sigma_{\perp}$. Conductivities along two different axes within the donor-molecule plane are denoted by $\sigma_{\parallel 1}$ and $\sigma_{\parallel 2}$ and identified where necessary by the specific unit cell axes. In general, $\sigma_{\parallel 1}$ designates the largest conductivity, and $\sigma_{\perp}$ designates the smallest. Similar definitions apply to the resistivities, $\rho_{\parallel 1}$, $\rho_{\parallel 2}$, and $\rho_{\perp}$. The conductivity ratios $\sigma_{\parallel 2}/\sigma_{\parallel 1}$ and $\sigma_{\perp}/\sigma_{\parallel 1}$ are useful indicators of the degree of anisotropy in the electrical conductive properties.

Electrical conductivity or resistivity is an intrinsic property of the crystal. However, it may differ from crystal-to-crystal for different specimens of the same material because of differences in impurity levels, crystal imperfections, and mechanical defects. Chemical impurities are presumed to be very low in concentration, but microcracks occur frequently and may be promoted by thermal stresses induced by the mechanical confinements of the resistance-measuring probes. Precise conductivities are difficult to determine because of inaccuracies in the measurements of the small, often irregular crystal dimensions (typically $1 \times 0.1 \times 0.05$ mm^3) and because of the occurrence of nonuniform electric fields. The temperature dependencies of the conductivity, however, are usually qualitatively reliable because the slopes depend only on relative changes in the resistance. A problem in measuring semiconductors is that the metal (probe)-semiconductor contacts give rectifying junctions. Despite the difficulties in measuring conductivities, however, there is sufficiently good qualitative agreement among measurements that one can discern the most important trends in the conductive properties.

ET Salts with Polyhedral Anions

The polyhedral anions, such as ClO_4^- and PF_6^-, that have provided the metallic salts and six superconductors of the TMTSF radical-cation donor molecule yield some interesting metallic salts with the ET donor but only one superconductor, $(ET)_2ReO_4$. These salts were the first derivatives of the ET-donor molecule to be synthesized and studied for the existence of new organic metals and superconductors. The electrical conductive properties of some of these salts are summarized in Table 4.1. These properties are generally quite different from those of the TMTSF salts.

The tetrahedral anion ClO_4^-, which provides the sole ambient-pressure superconducting salt of TMTSF, gives three salts with ET, two of which are metallic but not superconducting under presently known conditions. Two of these have 2:1 stoichiometry for the cation-to-anion mol ratio, but they incorporate neutral solvent molecules in their structures. The salt $(ET)_2ClO_4(TCE)_{0.5}$, where TCE denotes 1,1,2-trichloroethane, is important as the first example of a two-dimensional (2D) organic metal.[30] The conductivity is highest along the ET donor molecule interstacking direction, but it is nearly isotropic in the plane defined by the stacking and interstacking directions, which plane crystallographically shows an extended, approximately two-dimensional (2D) network of S$\cdots$S interactions.[31] At room temperature, $\sigma_{||1} \approx 26$ $(\Omega$ cm$)^{-1}$ and $\sigma_{||2}/\sigma_{||1} \approx 0.9$. This anisotropy ratio decreases slightly as the temperature is lowered. Along the direction of the arrays of anions and solvent molecules, the conductivity is more than two orders of magnitude smaller, $\sigma_\perp/\sigma_{||1} < 10^{-2}$. Figure 4.1 illustrates the temperature dependence of the conductivity of $(ET)_2ClO_4(TCE)_{0.5}$. With decreasing temperatures, the conductivity monotonically increases up to a value of ~ 1000 $(\Omega$ cm$)^{-1}$ at a temperature near 20 K, at which temperature the salt undergoes a somewhat broad MI transition associated with an ordering of the solvent molecules.[32] The temperature dependence of the conductivity above 80 K is approximately quadratic,[30] as is typical of many metallic organic salts. Reflectance spectra measurements,[33] which show the exist-

TABLE 4.1 Electrical Conductivities (σ) at Room Temperature, MI Transition Temperatures (T_{MI}) for the Metallic States, and Activation Energies (E_a) for the Semiconducting States of Some ET Salts with Tetrahedral and Octahedral Anions

Salt	State[a]	$\sigma_{\|\|1}$[b] (Ω cm)$^{-1}$	$\sigma_{\|\|2}/\sigma_{\|\|1}$	$\sigma_\perp/\sigma_{\|\|1}$	T_{MI} (K)	E_a (meV)	Reference[c]
$(ET)_2ClO_4(TCE)_{0.5}$	2D-M	26 (a)	0.9 (c)	$<10^{-2}$	20		30
$(ET)_2ClO_4(C_4H_8O_2)$	2D-S	1 (b)	1 (a)	10^{-5}		300	35
γ-$(ET)_3(ClO_4)_2$	2D-M	50 (c)[d]	0.5 (b)		170	80	35,37
$(ET)_2ReO_4$	1D-M	200 (a)	0.05 (b)		80	45	39
$(ET)_2BrO_4$	1D-M	600 (a)			180		42
α-$(ET)_2PF_6$	1D-S	0.1 (c)				50	45
β-$(ET)_2PF_6$	1D-M	10 (c)	0.02 (a)	10^{-4}	297	200	46
$(ET)_2AsF_6$	1D-M	2 (b)	0.05 (c)	10^{-5}	264	90	48
$(ET)_2SbF_6$	1D-M	4 (b)	0.05 (c)	10^{-5}	273	90	48

[a] State indicates dimensionality and metallic (M) or semiconducting (S) behavior at 300 K.

[b] Conductivities are approximate, with a range of values. $\sigma_{\|\|1}$ is an axis of high conductivity associated with (approximately) the unit cell axis given in parentheses, as defined by the cited references. $\sigma_{\|\|2}$ is an axis of lower conductivity associated with another unit cell axis. $\sigma_{\|\|1}$ and $\sigma_{\|\|2}$ lie in the plane of the stacking and interstacking direction of the organic donor molecules. $\sigma_\perp$ is the conductivity in a direction perpendicular to the organic layers, along which direction the layers are separated by the arrays of anions; conductivities are lowest along this direction.

[c] See text for additional references.

[d] Conductivity in the (100) plane; the molecules are stacked along the [0$\bar{1}$1] direction and have side-by-side interactions along the [012] and [021] directions.[37]

ence of an optical gap, and energy band calculations[34] suggest that the metallic state of $(ET)_2ClO_4(TCE)_{0.5}$ is a semimetal.

The second 2:1 salt of ET and the perchlorate anion is $(ET)_2ClO_4(C_4H_8O_2)$, where $C_4H_8O_2$ is a neutral solvent molecule of dioxane. This salt is a 2D semiconductor at room temperature and below with $\sigma_{\|\|1} \approx \sigma_{\|\|2} \approx 1$ (Ω cm)$^{-1}$ and thermal activation energy $E_a \approx 300$ meV.[35,36] It is interesting that the change in solvent molecule (and its stoichiometry) strongly affects the conductive properties. The third salt of this system is γ-$(ET)_3(ClO_4)$,[35,37] which with a 3:2 stoichiometry could be a semiconductor, semimetal, or a metal. This salt is metallic-like from room temperature down to 170 K, at which temperature it undergoes a sharp MI transition without appreciable change in crystal structure. γ-$(ET)_3(ClO_4)_2$ in the metallic state, which likely is a semi-metal, was initially thought to be a 1D metal with interactions primarily along the interstacking direction.[35] More recent measurements indicate that it is a 2D metal.[37] The room temperature conductivity, ~ 50 (Ω cm)$^{-1}$, is approximately isotropic within a factor of 2 in the plane of the organic donor molecules, and the anisotropy ratio of conductivities remains constant with change in temperature within the metallic region. The MI transition is suggested to arise from small rearrangements or displacements of the ET molecules or ClO_4^- anions.[38]

The ReO_4^- anion with ET produces five salts, of which $(ET)_2ReO_4$ is the most interesting because it is reported to be superconducting under applied pressures.[39] The observation of superconductivity in this salt represents the first

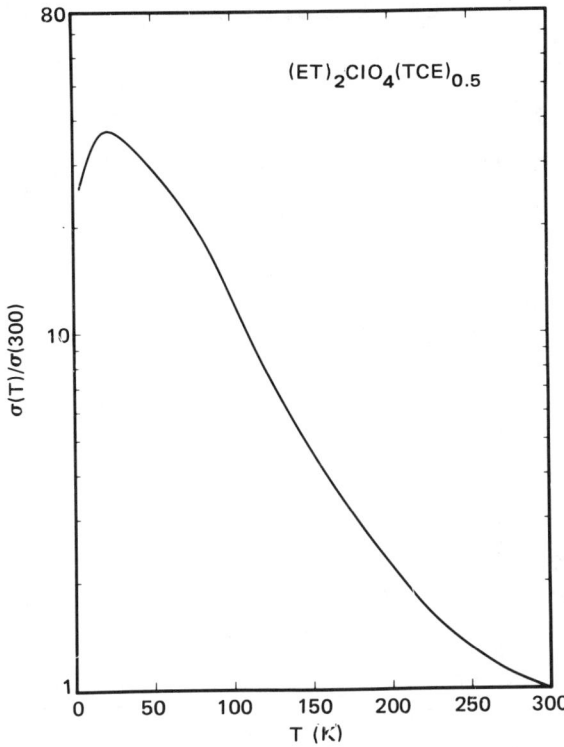

Figure 4.1 Relative electrical conductivity along the crystallographic b axis of $(ET)_2ClO_4(TCE)_{0.5}$ at ambient pressure and zero applied magnetic field as reported by Saito et al.[30] Reproduction of Fig. 21, ref. 6, with permission: *Progress in Inorganic Chemistry*, Vol. 35, pp. 51–218, Ed. by S. J. Lippard, an Interscience® Publication, Copyright © **1987** by John Wiley and Sons, Inc., NY.

discovery of superconductivity in an organic material beyond the TMTSF derivatives. $(ET)_2ReO_4$ is a curious material because it is isostructural with $(ET)_2BrO_4$, but the two salts have considerably different conductive properties. The conductivity of $(ET)_2ReO_4$ is largest along the stacking direction of the ET donor, $\sim 200\ (\Omega\ \text{cm})^{-1}$ at room temperature, and it is smaller by a factor of 1/20 along the interstacking direction.[40] Although not reported, the conductivity along the direction of the anion arrays is likely to be orders of magnitude smaller. Thus, $(ET)_2ReO_4$ is somewhat similar to the quasi-1D $(TMTSF)_2X$ salts with tetrahedral anions, except that the anions of this ET salt are ordered at room temperature.

Figure 4.2 illustrates the change in conductivity of $(ET)_2ReO_4$ along the stacking axis as a function of temperature. At ambient pressure (1 bar), the conductivity gradually increases as the temperature is lowered until at ~ 80 K there occurs a sharp MI transition. Below 80 K, the salt is semiconductive with an activation energy of ~ 45 meV. The MI transition is suggested to be a first-order structural phase transition associated with a displacing change in the positions of the ET-donor molecules or anions.[41] In $(TMTSF)_2ReO_4$ the semiconductive state oc-

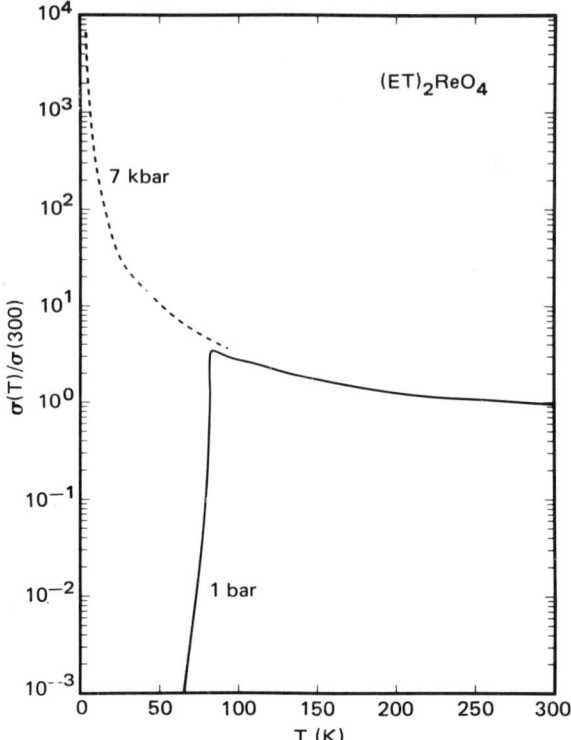

Figure 4.2 Relative electrical conductivity along the crystallographic a axis of $(ET)_2ReO_4$ at ambient pressure (1 bar) and at an applied pressure of 7 kbar, both at zero applied magnetic field, as reported by Parkin et al.[39] Reproduction of Fig. 22 of ref. 6 with permission: *Progress In Inorganic Chemistry*, Vol. 35, pp. 51–218, Ed. by S. J. Lippard, an Interscience® Publication, Copyright © **1987** by John Wiley and Sons, Inc., NY.

curs as the result of crystallographic anion ordering accompanied by a change in one of the unit cell lengths, but in $(ET)_2ReO_4$ the anions are ordered well above the transition temperature. Under applied pressure, the temperature of the MI transition can be suppressed, and it is totally suppressed by pressures above 4 kbar, at which pressures the salt becomes superconducting at a temperature of ~ 2 K.[39] The isostructural salt $(ET)_2BrO_4$ is metallic above 200 K and semiconductive below 100 K.[42-44] Unlike the perrhenate salt, however, there is no sharp MI transition between these temperatures but only a gradual change in the temperature dependence in which the conductivity reaches a broad maximum near 180 K. There is no evidence of superconductivity in $(ET)_2BrO_4$.

The octahedral anions PF_6^-, AsF_6^- and SbF_6^- form ET salts of stoichiometry $(ET)_2X$ which are 1D metals with MI transitions very near room temperature. The high-conductivity axis lies along the interstack directions of the ET-donor molecules, parallel to the plane of adjacent donor molecules. The PF_6^- anion forms two structural modifications, α-$(ET)_2PF_6$ and β-$(ET)_2PF_6$, of which the α modification is a 1D semiconductor at room temperature with $\sigma_{\parallel 1} \approx 0.1$ $(\Omega\ cm)^{-1}$ and an activation energy of ~ 50 meV.[45] β-(ET)$_2$PF$_6$[46,47] and the AsF$_6^-$ and SbF$_6^-$ salts[48–50] are metals near room temperature with MI transitions at 297, 264, and 273 K, respectively. The conductivities of these salts in the metallic state near room temperature are $\sigma_{\parallel 1} \approx$ 2–10 $(\Omega \text{ cm})^{-1}$ with $\sigma_{\parallel 2}/\sigma_{\parallel 1} \approx$ 0.02–0.05, and $\sigma_\perp/\sigma_{\parallel 1} \leq$ 10^{-4}. The MI transitions are apparently associated with charge-density wave (Peierls-type) distortions involving a doubling of the unit cell length along the interstack direction.[47,50,51] As determined from thermoelectric power measurements up to 340 K,[46] β-(ET)$_2$PF$_6$ exhibits hole (p-type) conduction in the metallic state and electron (n-type) conduction in the semiconducting state below 290 K. There are no reports of superconductivity in the ET salts of the octahedral anions.

ET Salts with Linear Anions

About 20 of the many reported salts composed of ET and the linear anions possess metallic conductive properties near room temperature, and a dozen of these are metallic to the lowest attainable temperatures. The conductive properties of the α- and β-type salts of this group are listed in Table 4.2. The most important of these materials are the β-phase salts, β-(ET)$_2$X, with X^- representing I$_3^-$, AuI$_2^-$, and IBr$_2^-$, because these comprise a fixed structural class of organic superconductors having T_c's which vary systematically with change in the length of the anion. β-(ET)$_2$I$_3$ was the first ambient pressure superconductor discovered[52] after (TMTSF)$_2$ClO$_4$ and, subsequently, was the incentive to a successful search for additional superconducting salts of ET derived from other linear anions.

The metallic salts β-(ET)$_2$I$_3$,[52] β-(ET)$_2$AuI$_2$,[53] and β-(ET)$_2$IBr$_2$[54] are quasi-2D metals that remain metallic to the lowest temperatures and undergo superconducting transitions at ambient pressure near 1.5 K, 5 K, and 3 K, respectively. The electrical conductivities[52,55,56] at room temperature are ~ 10–50 $(\Omega \text{ cm})^{-1}$ in the plane of the stacking and interstacking directions of the ET donor molecule (generally the crystallographic *ab* plane). Studies[57] of the conductive anisotropy of β-(ET)$_2$I$_3$ show that the conductivity is approximately isotropic in the plane of the donor molecules but three orders-of-magnitude smaller in the direction normal to this plane. The small differences in conductivity within the donor molecule plane can be attributed to slight differences in the scattering mechanisms.[58] Structural studies show that the plane of the organic donor molecules exhibits a 2D network of S···S interactions comparable to or slightly exceeding van der Waals interactions. This structural evidence, as well as energy band calculations[59] and other studies,[60] indicated that all three of these salts have similar electrical transport properties. Thus, these salts are somewhat similar to the quasi-2D metallic salt (ET)$_2$ClO$_4$(TCE)$_{0.5}$. Optical studies[61] indicate that the axis of highest electrical conductivity lies along the stacking axis of the ET-donor molecules, which is the direction [*a* + *b*] in the cited reference.

The salt β-(ET)$_2$I$_2$Br is isostructural with β-(ET)$_2$IBr$_2$ with the length of the anion I-I-Br$^-$ lying between that of the I-I-I$^-$ anion and the Br-I-Br$^-$ anion.[62,63] This suggests that β-(ET)$_2$I$_2$Br should be metallic and superconducting. This salt is indeed 2D metallic to the lowest temperatures, but it exhibits no superconductivity at ambient pressure down to 0.5 K or under applied pressures up to 5 kbar near

TABLE 4.2 Electrical Conductivities (σ) at Room Temperature and MI Transition Temperatures (T_{MI}) of Some α- and β-Type Salts of ET with Linear Anions

Salt	State[a]	$\sigma_{\|\|1}$[b] (Ω cm)$^{-1}$	$\sigma_{\|\|2}/\sigma_{\|\|1}$	$\sigma_{\perp}/\sigma_{\|\|1}$	T_{MI}(K)	Reference[c]
β-(ET)$_2$I$_3$	2D-M	50 (b')[d]	0.6 (a)	4×10^{-3}	—[e,f]	52,57
β^*-(ET)$_2$I$_3$	2D-M	$\sigma_{\|\|}$(1 kbar)/$\sigma_{\|\|}$(1 bar) = 1.3 (a axis)[g]			—[e,f]	57,68
β-(ET)$_2$AuI$_2$	2D-M	10 (a)[g]			—[e,f]	55
β-(ET)$_2$IBr$_2$	2D-M	20 (a)[g]			—[e,f]	56
β-(ET)$_2$I$_2$Br	2D-M	20 (a)[g]			—[e]	62,63
α-(ET)$_2$I$_3$	2D-M	200 (b)[h]	>0.5 (a)	10^{-3}	135	74
α-(ET)$_2$IBr$_2$	2D-S	1(a)[g]				56
α-(ET)$_2$I$_2$Br	2D-M	10 (a)[h]	$\geq$0.5 (b)	10^{-3}	265	75
β'-(ET)$_2$ICl$_2$	1D-S	0.01 (a)				82
β''-(ET)$_2$AuBr$_2$	2D-M	70 (c)[i]			—[e]	84,85
β''-(ET)$_2$ICl$_2$	2D-M	1000 (a)[i]			—[e]	86

[a] State indicates dimensionality and metallic (M) or semiconducting (S) behavior at 300 K.
[b] Conductivities are approximate. See footnote b of Table 4.1; $\sigma_{\perp}$ is along $c^* = a \times b$ for a unit cell in which the organic donor molecules are in the ab plane.
[c] See text for additional references.
[d] $b' = a \times c^*$, an axis normal to a in the ab plane. Reference 52 gives $\sigma = 30$ (Ω cm)$^{-1}$ for the needle axis (a); the value given here has been adjusted by the experimental anisotropy ratio.
[e] Salt is metallic to the lowest temperatures.
[f] This metallic salt is superconducting.
[g] Conductivity along the needle axis or the elongated direction of the crystal; this is not necessarily the axis of highest conductivity.
[h] For the unit cell of the cited references, a is the stacking direction and b is the interstacking direction (approximately) of the ET organic donor molecules.
[i] For the unit cell of the cited references, a is the stacking direction and c is the interstacking direction of the ET donor molecules.

1 K.[62,64] The absence of superconductivity has been attributed to crystallographic disorder in the orientations of the noncentrosymmetric I-I-Br$^-$ anion, which resides on a crystallographic center of inversion.[62] This suppression of superconductivity is somewhat similar to the case of disordered anions in the (TMTSF)$_2$ClO$_4$ superconductor, except that the disorder in the TMTSF salt leads to an MI transition. In the β-phase superconductors, the anions are symmetrical and perfectly ordered in all cases.

β-(ET)$_2$I$_3$, however, represents another example of the suppression of superconductivity caused by crystallographic disorder, but in this case the disorder arises from the arrangement of the terminal ethylene groups of the ET-donor molecules instead of the anions. At ambient pressure, the terminal ethylene group at one end of each ET-donor molecule in β-(ET)$_2$I$_3$ is disordered. As the temperature is lowered, the crystal exhibits an incommensurate structural modulation beginning near 200 K,[65] indicating some partial, long-range but incomplete ordering. This structural modulation persists to low temperatures where the salt becomes superconducting near 1.5 K. In the other two β-phase superconducting salts as well as in β-(ET)$_2$I$_2$Br the ethylene groups are perfectly ordered, and no structural modulation occurs. Under mild applied pressure, which has been established to be a minimum of 0.5 kbar,[66] β-(ET)$_2$I$_3$ undergoes a superconducting transition near 8 K.[67,68] Furthermore, the structure of the pressurized salt becomes crystallographically ordered with the ethylene groups arranged in a staggered conformation.[69–71] This structure is no longer isostructural with the other β-phase salts, and therefore this ordered salt has been denoted as β*-(ET)$_2$I$_3$.[69] The general literature, however, frequently denotes the low-temperature (1.4 K) and the high-temperature (8 K) superconducting salts as β$_L$-(ET)$_2$I$_3$ and β$_H$-(ET)$_2$I$_3$, respectively. The electrical conductivity and conductive anisotropy ratios of β*-(ET)$_2$I$_3$ do not differ significantly from those of the ambient-pressure salt.[57,68] Note that an increase in pressure to 1 kbar increases the conductivity along the needle axis of β-(ET)$_2$I$_3$ by only about 30 percent. There is no particular change in the normal-state conductive properties that can account for the remarkable increase in the superconducting transition temperature of β-(ET)$_2$I$_3$ under pressure. However, β*-(ET)$_2$I$_3$ [or β$_H$-(ET)$_2$I$_3$] is the key salt in the correlation of the structural and superconducting properties of the β-phase salts of ET. For when β*-(ET)$_2$I$_3$, β-(ET)$_2$AuI$_2$, and β-(ET)$_2$IBr$_2$ are considered together as a group of ordered structures, there is a straightforward correlation of decreasing T_c's with decreasing length of the anion.

Figure 4.3 illustrates the temperature dependence of the resistivity along the needle axis (a axis) of a β-(ET)$_2$I$_3$ crystal at ambient pressure.[22] Aside from some unusual features below 10 K, this figure shows typical metallic behavior to low temperatures, and despite the large conductive anisotropy between the directions parallel and perpendicular to the plane of the ET-donor molecules, the resistivity has been shown to be metallic in all directions for both the β-(ET)$_2$I$_3$ and β*-(ET)$_2$I$_3$ crystals.[57] The temperature dependence of the resistivity is $\rho \approx T^2$ below 100 K, again typical of metallic organic salts.[20] Above 100 K the resistivity is linear in T with a slight change in the slope near 200 K where the structural modulation occurs.[22] Thermoelectric power studies[58,72] indicate that electrical transport within the plane of the organic donor molecules is dominated by hole conduction above 200 K but involves both electron and hole conduction at lower temperatures.

Figure 4.3 shows at low temperatures an abrupt drop in the resistivity near 8 K, another noticeable change in slope near 4 K, and finally a large drop in the resistivity near 1.6 K, below which temperature the crystal becomes a volume superconductor at ambient pressure. The resistive anomaly near 4 K was first reported in the original study of β-(ET)$_2$I$_3$ by Yagubskii et al.[52] The more pronounced resistive anomaly near 8 K was first reported in another study by some of the present authors,[22] and it has been observed by others in subsequent investigations.[57] The magnitude of these anomalies is apparently quite variable from sample to sample. Although these anomalies were not understood before reports appeared on the existence of a higher superconducting state of the salt, it now seems clear that the anomaly near 8 K is due to superconducting shunts from small quantities of a higher T_c state within regions of the single-crystal specimens.

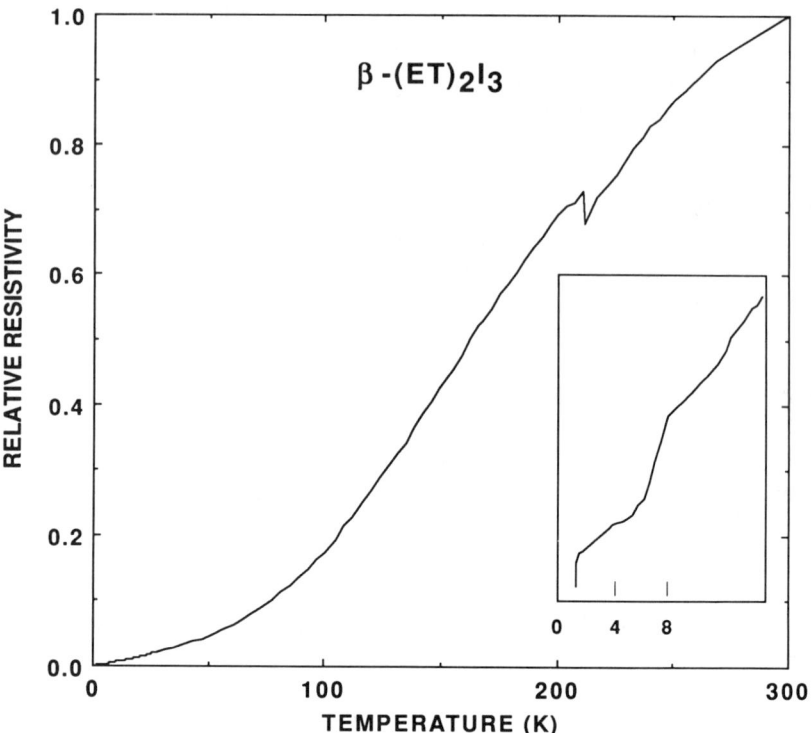

Figure 4.3 Relative electrical resistivity, $\rho(T)/\rho(300\ K)$, along the crystallographic a axis of β-$(ET)_2I_3$ at ambient pressure and zero applied magnetic field. The abrupt jump in resistivity near 200 K is likely due to the development of a crack in the crystal specimen. The inset shows the relative resistivity scaled by a factor of 50 below 15 K. This low-temperature region displays a large drop in the resistivity near 8 K, a smaller drop in the resistivity near 4 K, and the onset near 1.6 K of the superconducting transition. This figure was prepared from the original data of ref. 22.

The small amount of this state had not been detected either by diffraction studies or by measurements of volume superconductivity.[22] It is not clear, on the other hand, whether the small quantity of higher T_c material is an ambient-pressure variant of β^*-$(ET)_2I_3$ or the result of pressure induced by the stresses of the resistance-measuring probes or by some other effect. Although the anomaly near 4 K may be an artifact of the measurements represented in Figure 4.3, it has been suggested in other studies[73] that this may represent the coexistence of some other higher T_c phase (or phases). Superconducting transitions near 4 K are somewhat common in the ET-I_3^- system, as will be shown in later discussions.

A common second phase occurring in the electrocrystallization of the β-phase metals is the nonsuperconducting α modification: α-$(ET)_2I_3$,[74] α-$(ET)_2IBr_2$,[54] and α-$(ET)_2I_2Br$.[75] These α-phase salts have structures involving a herringbone arrangement of the ET donor molecule stacks, in contrast to the uniform arrangement of adjacent stacks in the β-phase structures. The α-phase salts are 2D systems with conductivities at room temperature which are comparable to those of other ET salts.

α-(ET)$_2$I$_3$ is metallic from 300 K down to 135 K, at which temperature the salt undergoes a sharp MI transition[74] to a nonmagnetic insulator.[73] α-(ET)$_2$I$_2$Br is metallic at 300 K and undergoes a similar MI transition near 265 K.[75] The α-(ET)$_2$IBr$_2$ salt is semiconductive at all temperatures below 300 K.[54,76] There is evidence for all three salts that further phase transformations occur at low temperatures. Thermoelectric power measurements of α-(ET)$_2$I$_3$ indicate that above 120 K, the electrical transport is governed by hole conduction.[74] An interesting observation, although it may be an artifact arising from differences in crystal defects, is that the conductivity is highest along the stacking axis of the organic donor molecules in α-(ET)$_2$I$_2$Br[75] but highest along the interstacking axis in α-(ET)$_2$I$_3$.[74] The nature of the metallic state of these salts is not well understood. Energy band calculations[77] suggest that the metallic state of α-(ET)$_2$I$_3$ is a semimetal or a very small gap semiconductor behaving as a semimetal at high temperatures. This result, however, is not consistent with the sharp nature of the MI transition and the negligible change in structure from the metallic to the insulating states.[6] The α-phase salts are not superconducting but may be converted to β-like superconductors by thermal treatment. The α_t-(ET)$_2$I$_3$ superconducting salt is one such thermally converted product of α-(ET)$_2$I$_3$.[20,78] We discuss these thermal conversions in a later section. There is at present no α-phase analogue for the ET-AuI$_2^-$ system. However, this system does have other phases: γ'-(ET)$_2$AuI$_2$,[79] which is a possible 1D metal, and δ-(ET)$_2$AuI$_2$,[80] which is a semiconductor.

Modified β-type structures occur for ET salts with linear anions which are shorter in length than that of the IBr$_2^-$ anion, but the physical properties of these salts are different from those with the true β-phase structure. Salts of the structural type β'-(ET)$_2$X, where X^- = BrICl$^-$, AuCl$_2^-$, and ICl$_2^-$, are semiconductors at all temperatures from room temperature on down.[81] For example, β'-(ET)$_2$ICl$_2$ is semiconducting with a room-temperature conductivity of 0.01 (Ω cm)$^{-1}$ and thermal activation energy of 100 meV.[82] The β'-phase salts are quasi-1D conductors because pairs of ET-donor molecules are more strongly dimerized than they are in the β-phase crystals, and the only effective extended S$\cdots$S interactions are those along the interstacking direction. Energy band calculations indicate that the semiconducting ground state is a magnetic insulator.[82] Salts of the structural type β''-(ET)$_2$X, where X^- = IAuBr$^-$, AuBr$_2^-$, and ICl$_2^-$, are metallic to very low temperatures but are not superconducting.[83-86] These salts are quasi-2D metals but differ in structure from the β-phase superconductors in that the layers of adjacent stacks of ET donor molecules are nearly coplanar as in (ET)$_2$ClO$_4$(TCE)$_{0.5}$. The conductivity of 1000 (Ω cm)$^{-1}$ at room temperature for β''-(ET)$_2$ICl$_2$ is among the highest values measured for metallic organic salts.[86] The conductivity is reported to be largest along the ET donor interstacking direction (c axis) in the AuBr$_2^-$ salt[85] but largest along the stacking direction (a axis) in the ICl$_2^-$ salt.[86] This difference is some artifact, however, because others have reported that the higher conductivity axis in the AuBr$_2^-$ salt switches to the stacking direction on cooling the salt below 100 K, and it remains reversed on subsequent warming and thermal cycling.[87] The Fermi surface of these salts shows an unusual coexistence of both 1D and 2D surfaces.[85,87] This suggests that electrical transport in these salts is more complicated than it is in the simpler 1D and 2D organic salts.

In addition to the abundant α and β phases, the ET/I system yields at least nine additional phases, most of which are metallic and three of which are superconducting in the region of 4 K. The electrical conductive properties of these additional phases are summarized in Table 4.3. γ-$(ET)_3(I_3)_{2.5}$ is an orthorhombic modification which is metallic near room temperature and superconducting near 1–2 K with a marked drop in its resistance beginning near 4 K.[73,88] This superconducting salt has been proposed as the origin of the resistive anomaly near 4 K sometimes observed in the resistivity of ambient-pressure β-$(ET)_2I_3$.[73] The salt δ-$(ET)(I_3)$ is a monoclinic modification with a 1:1 stoichiometry, and it is metallic to 130 K, at which temperature it undergoes an MI transition.[73,89] The room-temperature conductivities of the γ- and δ-phase salts are comparable to those of the β-phase salts.

The ζ-, ϵ-, and η-phase salts have monoclinic structures which incorporate the polyiodide anions I_5^- and I_8^-, and neutral I_2 molecules in the case of the ζ-phase salt. These salts are produced by chemical oxidation of ET in solution by use of iodine vapor. ζ-$(ET)_2(I_2)(I_8)$ is a semiconductor.[90] ϵ-$(ET)_2(I_3)(I_8)_{0.5}$[89,90] and η-$(ET)_2(I_3)(I_5)$[91,92] are metallic as illustrated in Fig. 4.4 by plots of the resistivity as a function of temperature.[91] This figure shows metallic behavior with some interesting but unexplained structure and no evidence of superconductivity down to 1.7 K. Further inductive measurements gave no indication of volume superconductivity at ambient pressure down to 0.5 K.[91] Originally, the ϵ-phase salt was reported to be superconducting.[89] Subsequent studies indicated that this salt is semiconducting and that the apparent superconductivity was caused by the presence of β-$(ET)_2I_3$.[93] Energy-band calculations show that both the ϵ- and η-phase salts should be semiconductors at room temperature, so that the metallic behavior shown in Fig. 4.4 is

TABLE 4.3 Electrical Conductivities (σ) at Room Temperature and MI Transition Temperatures (T_{MI}) of Some Additional Salts of ET with Triiodide and Other Polyiodide Anions

Salt	State[a]	$\sigma_\|^b$ (Ω cm)$^{-1}$	T_{MI} (K)	Reference[c]
γ-$(ET)_3(I_3)_{2.5}$	2D-M	20 (a)	—[d,e]	88
δ-$(ET)(I_3)$	?-M	20	130	89
ζ-$(ET)_2(I_2)(I_8)$	?-S	0.1		90
ϵ-$(ET)_2(I_3)(I_8)_{0.5}$	2D-M[f]	1 (a)	—[d]	89,91
η-$(ET)_2(I_3)(I_5)$	2D-M[f]	0.1 (c)	—[d]	91,92
θ-$(ET)_2I_3$	2D-M	30 (b)	—[d,e]	95,96
κ-$(ET)_2I_3$	2D-M	50	—[d,e]	97,98
β_d-$(ET)_2I_3$	2D-M	≥ 100	140	105
λ_d-$(ET)_2I_3$	2D-M	≥ 10	220	105

[a] State indicates dimensionality and metallic (M) or semiconducting (S) behavior at 300 K.
[b] Conductivity in the plane of the ET donor molecules, designated where possible by a crystallographic axial direction as given by the cited references. Conductivities are approximate and have a range of values.
[c] See text for additional references.
[d] Salt is metallic to low temperatures.
[e] This metallic salt is superconducting.
[f] This salt apparently is metallic if nonstoichiometric but semiconducting if stoichiometric.

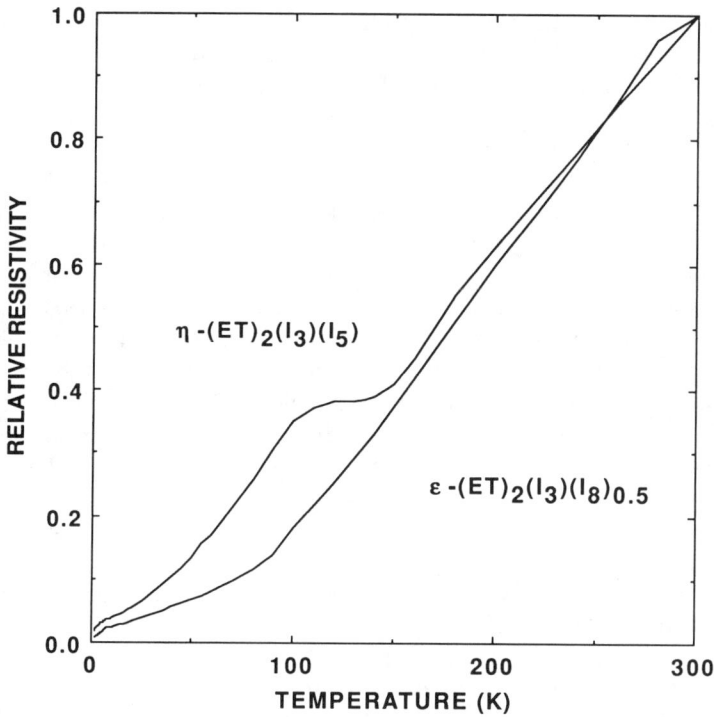

Figure 4.4 Relative electrical resistivity, $\rho(T)/\rho(300\text{ K})$, at ambient pressure and zero applied magnetic field of η-$(ET)_2(I_3)(I_5)$ along the crystallographic c axis and ϵ-$(ET)_2(I_3)(I_8)_{0.5}$ along the a axis. This figure was prepared from the original data of ref. 91 (note: η was originally denoted as ζ).

possibly caused by anion vacancies. Present evidence suggests, therefore, these two salts are semiconductors if fully stoichiometric but metallic-like if nonstoichiometric. This might account for the unusually low conductivities, 1 $(\Omega \text{ cm})^{-1}$ for ϵ-$(ET)_2(I_3)(I_8)_{0.5}$ and 0.1 $(\Omega \text{ cm})^{-1}$ for η-$(ET)_2(I_3)(_5)$. These values are comparable to those of the semiconducting rather than metallic salts of ET. It is reported that the iodine-rich ζ- and ϵ-phase salts lose iodine on heating and yield β-$(ET)_2I_3$ with a T_c near 6–7 K at ambient pressure.[90,93,94] Whether or not this product is similar to the pressurized β-phase salt, β^*-$(ET)_2I_3$, is a question we will discuss later in a broader context.

The salts θ-$(ET)_2I_3$[95,96] and κ-$(ET)_2I_3$[97,98] are quasi-2D organic metals to low temperatures, and both are reported to be superconducting in the region of 3.6 K. These salts were originally synthesized as mixed-anion salts of composition $(ET)_2(I_3)_{1-x}(AuI_2)_x$ in an attempt to alter the structural modulation in β-$(ET)_2I_3$ by the introduction of crystallographic disorder in the anions.[95] X-ray microanalysis of the θ-phase salt indicated the presence of Au ($x < 0.02$) but in concentrations too small for precise determination.[95] Similar analysis of the κ-phase salt indicated that the Au content was below the limits of detection ($x < 0.006$).[97]

The θ-phase salt, originally identified as having a pseudo-orthorhombic structure,[95] has a monoclinic lattice[99] with a molecular packing which resembles that of α-$(ET)_2I_3$, and it possesses a similar 2D network of sulfur interactions. The room-temperature conductivity and conductive anisotropy ratios are comparable to those of β-$(ET)_2I_3$.[95,100-102] Not all of the θ-phase specimen crystals have superconducting transitions, which suggests that these crystals are very sensitive to lattice defects and crystal imperfections.[100,102] It has been observed that θ-phase salts produced with higher concentrations of AuI_2^- ($x \approx 0.03$) and with the use of the alternate counter ion I_2Br^- ($x \approx 0$) yield metallic salts which undergo an appreciable drop in the resistivity near 4 K but no complete superconducting transition to temperatures as low as 1.5 K.[103] The κ-phase salt is a monoclinic phase with an unusual packing motif for organic salts. This structure has no segregated stacks of the organic donor molecules but consists of dimeric units $(ET)_2^+$ arranged nearly perpendicular to one another in neighboring sites (see Chap. 3). The proximity of dimeric units in all directions is sufficient to yield a 2D sulfur network in the plane of the ET donor molecule layer. The room-temperature conductivity and anisotropy ratios are comparable to those of the metallic β-phase salts.[97-100,102] In contrast to the θ-phase salts, all crystals of the κ-phase salt are reported to exhibit a superconducting transition.[102] In addition to anion doping, the structural modulation in β-$(ET)_2I_3$ can be altered by doping the ET-donor cation. β-$(ET)_{2-x}(MET)_xI_3$ is such a product,[104] but this salt will be discussed later in connection with the superconducting properties of organic salts.

The β_d-$(ET)_2I_3$[105,106] and λ_d-$(ET)_2I_3$[105,107] salts have been synthesized by a diffusion technique, which is the origin of the "d" subscription notation in the crystal designations, β_d and λ_d. The β_d-phase salt has a triclinic lattice and a packing motif which is different from that of the α- and β-phase salts of ET-I_3^-. This salt is metallic at 300 K and undergoes an MI transition at 140 K, which is very near the MI transition temperature of α-$(ET)_2I_3$ (135 K). The room-temperature conductivity in the plane of the ET-donor layer is reported to be 100 to 1000 (Ω cm)$^{-1}$.[105] The λ_d-phase salt has a unit cell very much like that of the θ-phase salt. This salt is metallic at 300 K and undergoes an MI transition near 220 K. The room-temperature conductivity is given in one report[105] as 10-100 (Ω cm)$^{-1}$ and in another[107] as 500-5000 (Ω cm)$^{-1}$. These unusually large conductivities, the closeness of the MI transition temperatures of the β_d- and α-phase salts, and the similarity of the λ_d- and θ-phase unit cells indicate that these salts need further investigation and confirmation.

The κ-Phase Salts and Other Organic Metals

The prototypical packing motif for the organic donor molecules of the κ-phase, and κ-like organic salts, is that possessed by the metallic and superconducting salt κ-$(ET)_2I_3$, which consists of the twisted arrangement of neighboring $(ET)_2^+$ dimers. Several of these salts are formed from donor molecules other than ET. The electrical conductive properties of these salts, including those of κ-$(ET)_2I_3$ which have been discussed in the previous section, are summarized in Table 4.4. Actually, the first known organic charge-transfer salt with the κ-like packing motif is $(MT)_2Au(CN)_2$, where MT is the radical-cation donor molecule bis-(methylenedithio)tetrathiafulvalene.[108] This κ-like salt is metallic at 300 K with a

TABLE 4.4 Electrical Conductivities (σ) at Room Temperature and MI Transition Temperatures (T_{MI}) of κ-phase and κ-like Salts

Salt	State[a]	$\sigma_{\|}$[b] (Ω cm)$^{-1}$	T_{MI} (K)	Reference[c]
κ-(ET)$_2$I$_3$[d]	2D-M	50	—[e,f]	97,98
(MT)$_2$Au(CN)$_2$	2D-M	300	80	108
(ET)$_2$Ag(CN)$_2$(H$_2$O)	2D-M		150	109
κ-(ET)$_2$Cu(NCS)$_2$	2D-M	40 (c)	—[f,g]	110,111
κ-(ET)$_2$Cu[N(CN)$_2$]Br	2D-M	48	—[f,g]	204,205
κ-(ET)$_2$Cu[N(CN)$_2$]Cl	2D-M	2	40[h]	206
(ET)$_4$Hg$_{2.89}$Br$_8$	2D-M	5	—[e,f]	114,116
(ET)$_4$Hg$_{3-x}$Cl$_8$	2D-M	30	4-20	118
(MDT-TTF)$_2$AuI$_2$	2D-M	20	—[e,f]	9
(DMET)$_2$AuBr$_2$	2D-M		—[f,g]	120

[a] State indicates dimensionality and metallic (M) or semiconducting (S) behavior at 300 K.
[b] Conductivity in the plane of the organic donor molecules. Conductivities are approximate and have a range of values.
[c] See text for additional references.
[d] Repeated from Table 4.3.
[e] Salt is metallic to low temperatures.
[f] This metallic salt is superconducting.
[g] Conductivity initially is semiconductive to temperatures near 100–150 K and then metallic to lower temperatures.
[h] Superconducting under pressure.

fairly high conductivity of ~ 300 (Ω cm)$^{-1}$ in a direction parallel to the MT layer, and it undergoes a broad MI transition near 80 K. The structural features indicate that this is a 2D salt. (ET)$_2$Ag(CN)$_2$(H$_2$O) is an ET-based salt with a κ-like structure which incorporates water molecules connected to the anions to form a polymeric layer.[109] This salt is metallic at 300 K and has an MI transition near 150 K.

κ-(ET)$_2$Cu(NCS)$_2$ has the κ-phase packing motif and a polymeric network of anions, and it possesses a superconducting transition temperature of ~ 10 K at ambient pressure,[7,8] This salt is metallic at room temperature with $\sigma_{\|1} \approx 40$ (Ω cm)$^{-1}$, and the anisotropy ratio for conductivities within the plane (bc) of the ET-donor molecules is $\sigma_{\|2}/\sigma_{\|1} \approx 0.5$.[8,110,111] In a direction (along a^*) perpendicular to this plane, the conductivity is several orders-of-magnitude smaller, $\sigma_\perp/\sigma_{\|1} \approx 5 \times 10^{-4}$. Thus, κ-(ET)$_2$Cu(NCS)$_2$ is a quasi-2D metal comparable to the β-phase superconductors. However, the temperature dependence of the resistivity of this salt is unusual for an ambient-pressure organic superconductor. Figure 4.5 illustrates the resistivity of κ-(ET)$_2$Cu(NCS)$_2$ as a function of temperature. The resistivity decreases only slightly from 300 K to 250–270 K. As the temperature is further lowered, the resistivity increases to a maximum near 100 K, after which the resistivity decreases rapidly to the onset of a broad superconducting transition that begins near 11 K and ends with zero resistance below 9 K.

The resistive behavior represented by Fig. 4.5 is typical of most measurements of κ-(ET)$_2$Cu(NCS)$_2$ carried out with different crystal specimens in different laboratories. The apparent semiconductive region, where the resistivity increases

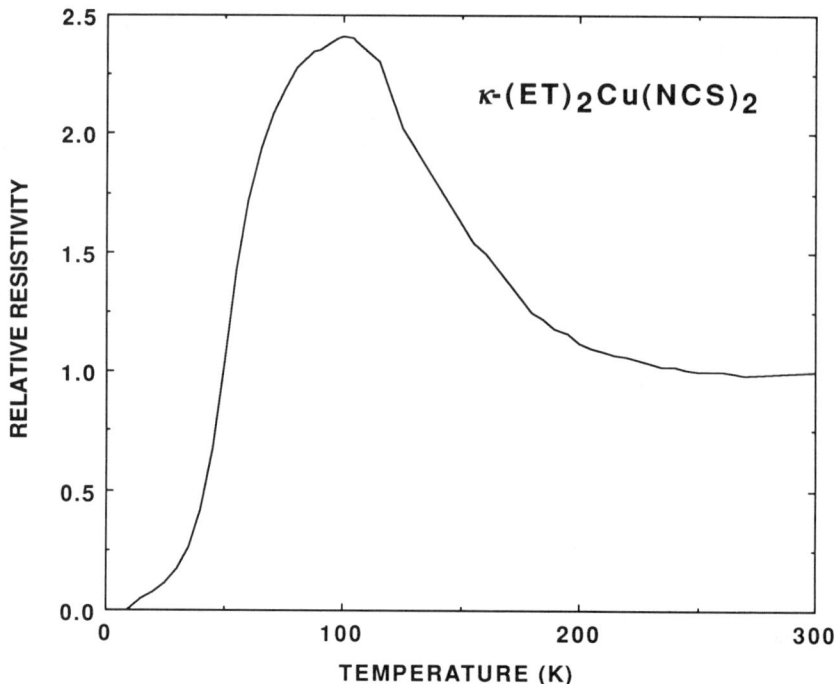

Figure 4.5 Relative electrical resistivity, $\rho(T)/\rho(300\ K)$, at ambient pressure and zero applied magnetic field of $\kappa\text{-}(ET)_2Cu(NCS)_2$ within the plane of the organic donor molecules. This figure was prepared from unpublished measurements by the authors.

with decreasing temperature, has been attributed to some disorder in the anions leading to increased scattering rather than to an opening of a band gap.[7] This semiconductive behavior is suppressed by applied pressures,[110] and it seems to be suppressed at ambient pressure in thinner crystals.[112] This suggests that defects play a role in this behavior. It has been reported that crystals of $\kappa\text{-}(ET)_2Cu(NCS)_2$ synthesized from meticulously purified starting materials show no resistive maximum and exhibit a sharp superconducting transition at 10.5 K.[113] Thermoelectric power measurements show hole conduction in one direction and electron conduction in another in the ET-donor molecule layer which correlates with the existence of both open and closed Fermi surfaces.[111]

The latest κ-phase superconducting salt is $\kappa\text{-}(ET)_2Cu[N(CN)_2]Br$ with T_c near 11.6 K.[204] This salt possesses the κ-phase packing arrangement of the organic donor molecules and has a polymeric network of the anions. The room temperature conductivity is $\sigma_\parallel \approx 48\ (\Omega\ cm)^{-1}$, which is comparable to that of the other κ-phase organic metals. The temperature dependence of the resistivity, which is illustrated in Fig. 4.6, is very similar to that of $\kappa\text{-}(ET)_2Cu(NCS)_2$. The resistivity of $\kappa\text{-}(ET)_2Cu[N(CN)_2]Br$ exhibits weak metallic behavior below 300 K, a change to semiconductive behavior below ~ 220 K, and a resistive maximum near 100 K.[205] The origin of the semiconductive behavior and resistive maximum, perhaps arising from impurities, defects, or possibly crystallographic disorder, is likely to be the same as that for $\kappa\text{-}(ET)_2Cu(NCS)_2$. Below 100 K, the resistivity of

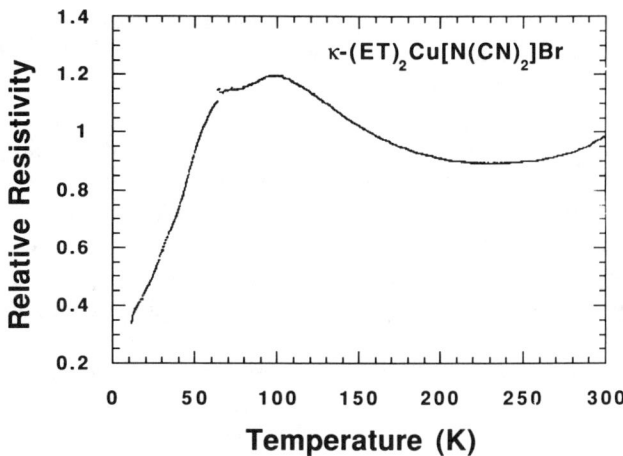

Figure 4.6 Relative electrical resistivity, $\rho(T)/\rho(300\ K)$, at ambient pressure and zero applied magnetic field of a single-crystal specimen of $\kappa\text{-(ET)}_2\text{Cu[N(CN)}_2\text{]Br}$ within the plane of the organic donor molecules. This figure was prepared from the original data cited in refs. 204 and 205.

$\kappa\text{-(ET)}_2\text{Cu[N(CN)}_2\text{]Br}$ drops rapidly with decreasing temperatures to a superconducting transition with an onset at 12.5 K in resistive measurements and 11.6 K in inductive measurements.[204,205]

The isostructural salts $\text{(ET)}_4\text{Hg}_{2.89}\text{Br}_8$ and $\text{(ET)}_4\text{Hg}_{3-x}\text{Cl}_8$ (x is undetermined) have κ-like packing motifs for the ET-donor molecules, polymeric networks of the anions, and complex structures involving incommensurate Hg sublattices.[114,115] Both salts are superconducting but under different conditions. $\text{(ET)}_4\text{Hg}_{2.89}\text{Br}_8$ is a 2D metal to low temperatures and an ambient-pressure superconductor below 4 K.[116] Recent experiments have shown that the superconductivity near 4 K is a volume property of the salt and not an artifact of entrapped Hg atoms.[117] The conductivity measured parallel to the layer (ab plane) of the ET-donor molecules is approximately isotropic with a room-temperature value of $\sim 5\ (\Omega\ \text{cm})^{-1}$, and this conductivity is metallic down to the superconducting transition. In the direction perpendicular to the plane of the organic-donor molecules, the conductivity is several orders-of-magnitude smaller, and it exhibits strong semiconductive behavior with decreases in temperature down to 20 K, below which temperature the conductivity in this direction becomes metallic. This semiconductive behavior along the direction of the insulating anion layers is an expected feature of 2D metals, but it is not consistently observed in all salts of similar dimensionality. As previously described, for example, the $\sigma_\perp$ conductivity for the β- and β^*-phase salts of $\text{(ET)}_2\text{I}_3$ is metallic in its temperature dependence, despite its much lower magnitude than that of the metallic conductivity within the plane of the ET-donor molecules.[57] This contrasting behavior probably represents the occasional existence of shunting circuits and nonuniform electrical fields in different crystal specimens. This illustrates some of the practical difficulties in determining accurate conductivities.

The salt $\text{(ET)}_4\text{Hg}_{3-x}\text{Cl}_8$ is another 2 D metal but a superconductor only under applied high pressures.[118] The conductivity of this salt is approximately isotropic in the plane of the donor molecules with a room-temperature value of $\sim 30~(\Omega~\text{cm})^{-1}$. This conductivity is metallic to low temperatures, and in some but not all crystal specimens the conductivity exhibits some kind of apparent MI transition in the region 4–20 K. This salt is superconducting near 1.8 K under an applied pressure of 12 kbar. The pressure dependence of the resistivity suggests that the salt undergoes a phase transformation at higher pressures, resulting in a phase that is superconducting near 5 K under a pressure of 29 kbar.[118] The apparent superconducting transition of this "new" phase, however, is incomplete at this higher temperature, so that it is not clear whether or not this is indeed a phase of a higher T_c superconductor.

The discussed κ-phase salts consist of organic donor molecules (MT and ET) that are symmetrical. The κ-phase packing motif, however, occurs also with the use of the unsymmetrical donor molecules MDT-TTF and DMET. The organic donor molecule MDT-TTF is the unsymmetrical fulvalene derivative methylenedithiotetrathiafulvalene, and it yields the κ-like superconducting salt (MDT-TTF)$_2$AuI$_2$, which has an ambient-pressure superconducting transition near 4 K.[9,10] The relatively high value of T_c for (MDT-TTF)$_2$AuI$_2$ is significant in establishing, along with several (DMET)$_2$X salts, that there is no intrinsic limitation to the occurrence of superconductivity in salts derived from unsymmetrical organic-donor molecules as opposed to the symmetrical donor molecules such as TMTSF and ET. Structurally, (MDT-TTF)$_2$AuI$_2$ is a 2D system, and it is a typical organic metal with a conductivity of $\sim 20~(\Omega~\text{cm})^{-1}$ parallel to the layer of the organic-donor molecules and with a normal metallic behavior down to the superconducting transition temperature.

(DMET)$_2$AuBr$_2$[119] is another κ-phase salt composed of an unsymmetical donor, dimethyl(ethylenedithio)diselenadithiafulvalene (DMET), which is a hybrid of the TMTSF and ET molecules. Apparently, three different salts of the DMET and AuBr$_2{}^-$ anion exist, but only one of these is known to have the κ-like packing motif. κ-phase (DMET)$_2$AuBr$_2$ has a resistance behavior somewhat similar to that of κ-(ET)$_2$Cu(NCS)$_2$ and κ-(ET)$_2$[N(CN)$_2$]Br: the resistance initially increases with decreasing temperature and reaches a maximum near 150 K. With further decreasing temperatures, the resistance is metallic, and the salt becomes superconducting near 2 K at ambient pressure.

The other two salts of DMET and AuBr$_2{}^-$ have different normal state and superconductive properties.[120,121] One salt is metallic to low temperatures but is not superconducting. Another salt, (DMET)$_2$AuBr$_2$, has a resistivity behavior somewhat similar to that of the κ-phase crystal and becomes superconducting near 1 K under an applied pressure of 1.5 kbar. The DMET donor molecule yields a variety of other (not κ-like) metallic and superconducting salts. (DMET)$_2$Au(CN)$_2$ and (DMET)$_2$AuI$_2$ are metallic salts, $\sigma \approx 200\text{--}300~(\Omega~\text{cm})^{-1}$, which undergo MI transitions to a spin-density-wave state at low temperatures under ambient pressure and become superconducting below 1 K under applied pressures.[122] (DMET)$_2$AuCl$_2$ appears to have a low-temperature MI transition, but it also becomes superconducting near 0.8 K at ambient pressure.[123] (DMET)$_2$I$_3$ and (DMET)$_2$IBr$_2$ are metallic to low temperatures without MI transitions and superconducting below 0.6 K at ambient pressure.[124] As with β-(ET)$_2$I$_2$Br, the salt (DMET)$_2$I$_2$Br is metallic to low temperatures but not superconducting.[124] The conductive dimensionalities of these salts are not yet well characterized, but their structures suggest some degree of 2D character. The properties of these (DMET)$_2$X salts are reviewed in several sources.[121,125,126]

The remaining salts we wish to mention here briefly are additional organic metals, derived from symmetrical organic donor or acceptor molecules, which are superconductors only under applied pressures. (ET)$_3$Cl$_2$(H$_2$O)$_2$[127] is one of several known salts of ET composed of monoatomic anions but the only one of this kind presently known to be superconducting. The anion layer is composed of planar Cl$_4$(H$_2$O)$_4$ units with Cl connected to the O atoms through the hydrogen bonds. This salt has an unusually large room-temperature conductivity for an ET salt, $\sigma \approx 500$ $(\Omega$ cm$)^{-1}$. This high value might be accounted for in part by a larger number density of conducting electrons (a two-thirds filled band). (ET)$_3$Cl$_2$(H$_2$O)$_2$ undergoes an MI transition near 100 K. This transition is suppressed by applied pressures, and the salt becomes superconducting near 2 K at a pressure of 16 kbar. (ET)$_3$Cl$_2$(H$_2$O)$_2$ and β$_m$-(BEDO-TTF)$_3$Cu$_2$(NCS)$_3$ are the only organic superconductors so far known with a 3:2 stoichiometry for the ratio of the organic donor molecule to the inorganic acceptor; all others have a 2:1 stoichiometry, or a 1:2 stoichiometry in the case of the superconductors derived from the Ni(dmit)$_2$ acceptor molecule.

The Ni(dmit)$_2$ acceptor, or its metal-replacement derivatives, yield three, and possibly four, salts which are superconducting but only under high-applied pressures. The conductive dimensionalities of these materials are not yet clearly established, but the initial intent in the use of these acceptor molecules would appear to be to enhance the dimensionality through the interactions of the terminal sulfur atoms. The salt (TTF)[Ni(dmit)$_2$]$_2$ has $\sigma \approx 300$ $(\Omega$ cm$)^{-1}$ at room temperature and is metallic to low temperatures.[128] It is not superconducting at ambient pressure but becomes superconducting near 1.6 K at 7 kbar applied pressure. Isostructural with this salt is the Pd-replacement analogue α'-(TTF)[Pd(dmit)$_2$]$_2$,[129,130] which has a room-temperature conductivity of $\sigma \approx 770$ $(\Omega$ cm$)^{-1}$. This salt undergoes a broad MI transition near 250 K. This can be suppressed by high-applied pressures, and the salt becomes superconducting near 6 K at 24 kbar. The relatively high values of σ for these TTF-Metal(dmit)$_2$ salts may be due to the presence of conducting networks for both the organic donor (TTF) and acceptor [Metal(dmit)$_2$] molecules. The TTF-Pd(dmit)$_2$ system yields several different salts. One of these, α-(TTF)[Pd(dmit)$_2$]$_2$, is very similar in structure to the α'-phase, and it is reported (without published details at this time) to be superconducting near 1.7 K at 22 kbar applied pressure.[131] [N(CH$_3$)$_4$][Ni(dmit)$_2$)]$_2$, with $\sigma \approx 50$ $(\Omega$ cm$)^{-1}$ at room temperature, is metallic to low temperatures without an MI transition, but it is superconducting only under pressure: $T_c \approx 3$ K at 3.2 kbar applied pressure, and $T_c \approx 5$ K at 7 kbar applied pressure.[132] It is suggested that this change in T_c represents a positive pressure dependence for a single-phase material rather than a change in structure with increased pressure. Most organic superconductors typically exhibit a negative pressure dependence for T_c. However, (ET)$_4$Hg$_{2.89}$Br$_8$ possesses a positive pressure dependence,[117] which will be discussed later.

SUPERCONDUCTING PROPERTIES

The Superconducting Transition Temperature T_c

Among the several measured and derived properties that characterize the superconducting state, the superconducting transition temperature T_c—the so-called superconducting "critical" temperature—is paramount but not simply defined for the organic metals by a single temperature. First of all, single crystals of the superconducting organic salts, although of apparent high quality, are generally inhomogeneous with respect to T_c, so that the transition from the normal state to the fully developed volume superconducting state frequently occurs over a temperature range of several K. Secondly, different experimental techniques measure different aspects of the superconducting state and, furthermore, are sometimes better suited to different operational definitions of T_c. Resistive measurements probe changes in the electrical conductive properties, but they are subject to filamentary shunting circuits and surface effects which may not be representative of the bulk material. On the other hand, resistive measurements can be very sensitive to the presence of trace quantities of a superconducting material. Inductive measurements probe changes in the magnetic properties or changes in the induced shielding currents in closed surfaces, which are more direct indications of the bulk state of the test sample.

For resistive measurements, T_c is commonly defined as the temperature at the midpoint of the transition, where the resistance is one-half of its normal-state value. Although not consistently reported, the temperature of the onset of the transition and the temperature at which the resistance first becomes zero below instrument resolution, or some other indicator of the width of the superconducting transition, are useful supplementary data for analyzing the superconducting properties. The resistive onset temperature, which is sometimes difficult to determine precisely, is the temperature at which the resistance first begins to drop below the value of the normal-state resistance. For inductive measurements, T_c is commonly reported as the onset temperature because the transitions, particularly for the low-T_c salts, are frequently incomplete to the lowest measured temperature and there is no absolute final value of the property equivalent to zero resistance. The onset temperature for inductive measurements is defined as the temperature at which the inductive signal first begins to change from that of the normal state. The completion of the transition for inductive measurements is the temperature at which the signal reaches a maximum (or a minimum, depending on the measurement) and remains constant with further decreases in temperature.

Table 1.1, which lists the presently known superconductors for introductory purposes, gives approximate T_c's without experimental qualifications. In this section, however, we will identify the definition of T_c and the experimental methods for its determination in examples where this seems to be important. T_c's and properties based on the onset temperatures are representative of the superconducting properties of the higher quality regions of the crystal specimen, whereas those based on the midpoint temperature are representative of the lower quality regions. It is possible that temperature-dependent properties, such as the slopes of the critical magnetic fields as a function of temperature, are different for different definitions of

T_c and for different detection methods. Such differences may account for quantitative discrepancies between several reported measurements of the same property for a given superconducting material.

Organic superconductors are extreme type II superconductors which have widely different values for the lower critical magnetic fields H_{c1}, where the magnetic flux initially penetrates the sample, and the upper critical magnetic field H_{c2}, where the magnetic field completely restores the normal metallic state. The electrical conductive anisotropy of the organic salts is reflected in the anisotropy of these critical magnetic fields. We denote these anisotropic properties by the notations $H_{c\parallel}$ and $H_{c\perp}$, in which the notations "$\parallel$" and "$\perp$" indicate, respectively, the application of magnetic fields parallel and perpendicular to the crystallographic plane of the organic-donor molecules. There is little anisotropy in $H_{c2\parallel}$ determined within the plane of the organic donor molecules. $H_{c\perp}$ corresponds to a value measured with the magnetic field applied approximately (e.g., the reciprocal lattice c^* instead of the real lattice c) along the direction of the anion layers. Similar to the conductivity, the upper critical magnetic field along the direction of the anion layers is much lower than those measured along directions parallel to the donor molecule plane. Another important property of superconductors, especially for practical applications, is the critical current density, J_c. Very few studies of this property have been reported for organic superconductors. Apparently, these are rather small quantities, of the order of 10 A cm^{-2} for the (TMTSF)$_2X$ and (ET)$_2$ReO$_4$ salts[2,39] and 1000 A cm^{-2} for κ-(ET)$_2$Cu(NCS)$_2$.[176]

The β-Phase Superconductors at Ambient Pressure

The superconducting properties of the β-phase salts, comprising β-(ET)$_2$I$_3$, β-(ET)$_2$IBr$_2$, and β-(ET)$_2$AuI$_2$, have been studied extensively by a variety of different techniques which have now well established these materials as being anisotropic volume superconductors. This class of organic superconductors includes, as a very important member, the high-T_c (8 K) variant, β*-(ET)$_2$I$_3$ or β$_H$-(ET)$_2$I$_3$, but this phase is discussed later in connection with the influence of applied pressures on the β-phase superconductors. As typically synthesized by electrocrystallization techniques, the superconducting transitions of the β-phase salts at ambient pressure are sample dependent and relatively broad, extending in temperature from onset to completion by 1 K or more. The onset temperatures and breadths of the superconducting transitions are strongly but subtly influenced by the synthetic preparation techniques. Smaller crystals often have higher onset temperatures, presumably because they are of higher quality in terms of higher chemical purity and fewer defects. The resistivity ratio $\rho(300)/\rho(T_L)$, where $\rho(300)$ refers to the resistivity at 300 K and $\rho(T_L)$ refers to the resistivity at some convenient low temperature (T_L) within the normal-state region, is a useful measure of a crystal's quality. Sharper transitions often occur for samples having larger resistivity ratios.

The superconducting transitions for β-(ET)$_2$I$_3$ as determined by resistive measurements are found to have onset temperatures of 1.6–1.7 K, midpoint temperatures of 1.4–1.5 K, and temperatures near 0.5 K for the first appearance of zero resistance.[22,52] Resistivity ratios $\rho(300)/\rho(4.2)$ are typically of the order of 500 or smaller. As previously mentioned, some anomalous resistance drops have been observed at even higher temperatures, and they are still observed in current studies. But the above cited low-temperature parameters fairly well characterize the resistive transition properties of the "low-T_c state" of this salt as presently understood. Inductive studies of different β-(ET)$_2$I$_3$ crystals by rf penetration depth measurements,[133,134] low-field ESR,[135] and by ac susceptibility and dc magnetization measurements[136] have yielded superconducting onsets of 1.2, 1.4, and 1.6 K and transition completions below 0.5 K. These measurements indicate a substantial difference in the quality of various crystals. Unlike the resistive measurements, inductive studies at ambient pressure of β-(ET)$_2$I$_3$ crystals grown by the usual electrocrystallization methods generally have shown no detectable signals for superconductivity at temperatures above 2 K (but see later discussions on this point).

An important property of a superconductor is its ability to expel magnetic flux when it is cooled in a magnetic field to a temperature below the superconducting transition temperature. A convenient measure of this ability is the Meissner effect, calculated as $M = -4\pi\chi(1 - n)$, in which χ is the (dimensionless) volume diamagnetic susceptibility of the sample and n is the demagnetization factor. The Meissner effect expresses the expelled magnetic flux of the test sample as a fraction of that expelled by a perfect superconductor with volume diamagnetic susceptibility $\chi = -1/4\pi$. As in other inductive measurements, the Meissner effect determined for β-(ET)$_2$I$_3$ typically shows an incomplete transition to the lowest temperatures,[136] but effects as high as 25 percent have been determined at temperatures near 0.8 K.[137] The relatively low value is typical of type II superconductors, and it occurs because of flux pinning at the sites of defects. The diamagnetic shielding effect, calculated from the Meissner fraction M but with χ determined by applying a magnetic field to a field-free superconductor below T_c, gives larger values, ~ 60 percent.[136] The Meissner and diamagnetic shielding effects clearly indicate that β-(ET)$_2$I$_3$ is a bulk superconductor but one with substantial defects.

Figure 4.7 illustrates the broadness of the superconducting transitions of different β-(ET)$_2$I$_3$ crystals as determined by a resistive measurement and by two different inductive techniques. The broadness indicates a very inhomogeneous distribution of T_c over different regions in the crystal sample. One wonders whether these broad transition widths, ΔT, represent some intrinsic property due to the presence of the modulated structure (see Chap. 3). However, a recent report[138] of a resistive study of high-quality crystals of β-(ET)$_2$I$_3$ shows that for a crystal with a resistivity ratio of ~ 1000, a midpoint transition of 1.4 K, and a transition width of $\Delta T \approx 0.3$ K, much sharper transitions are possible for carefully synthesized crystals. From the various published measurements of T_c, one judges that the onset of superconductivity for a crystal of β-(ET)$_2$I$_3$ as synthesized by carefully executed techniques may be above 1.6 K. In fact, very recent studies have indicated that the prolonged annealing of conventionally synthesized crystals at 110 K gives specimens with an altered incommensurate superstructure and an increase in T_c to 2 K at ambient pressure.[139] For crystals of conventional quality, the volume superconducting transition temperature can be reasonably quoted as 1.4 ± 0.1 K, which would encompass both the onset and midpoint transition temperatures.

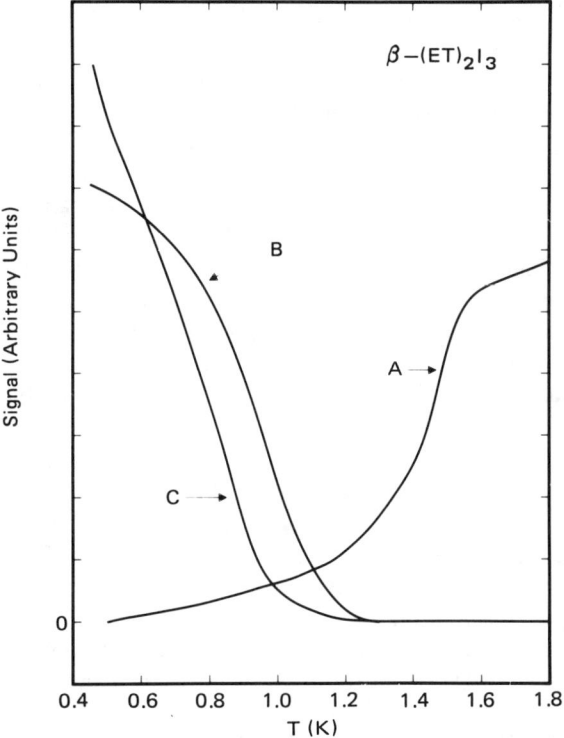

Figure 4.7 Superconducting transition curves for β-(ET)$_2$I$_3$ at ambient pressure and zero applied magnetic field as determined by three different experimental methods: A from resistivity measurements, ref. 52; B from rf penetration depth measurements, ref. 22; C from ac susceptibility measurements, ref. 136. Reproduction of Fig. 25 of ref. 6 with permission: *Progress In Inorganic Chemistry*, Vol. 35, pp. 51–218, Ed. by S. J. Lippard, an Interscience ® Publication, Copyright © **1987** by John Wiley and Sons, Inc., NY.

Resistive measurements of the superconducting transition of β-(ET)$_2$IBr$_2$ have yielded onset temperatures of 2.5–2.8 K, midpoint transitions near 2 K, and zero reistance not far below 2 K for crystals with resistivity ratios ρ(300)/ρ(4.2) within the range 200–1000.[56,140] The resistive transition widths for these crystals in general are somewhat narrower, < 1 K, than those for the I$_3$⁻ salt, and there is no problem of structural disorder in β-(ET)$_2$IBr$_2$ that could affect the transition. A recent article[138] on the resistive transition for a high-quality crystal with a resistivity ratio of 4000 reports an onset temperature of 2.8 K and a sharp transition of width ΔT = 0.1 K. Inductive studies by rf penetration depth[54,141,142] and low-field magnetization[143] measurements have yielded onset temperatures of 2.3, 2.4, 2.7, and 2.8 K and transition widths from ~ 1 K to ~ 0.2 K for different crystals of β-(ET)$_2$IBr$_2$. Magnetization measurements[143,144] have determined Meissner effects up to 40 percent and diamagnetic screening effects as high as 80 percent, thus establishing that the superconductivity is a bulk property of β-(ET)$_2$IBr$_2$. These various studies indicate that, although the crystal quality, transition temperatures, and transition widths may vary among samples, high-quality crystals of β-$(ET)_2IBr_2$ have $T_c = 2.7 \pm 0.1$ K and reasonably sharp transition widths.

Most studies of the superconducting transition of β-$(ET)_2AuI_2$ have employed inductive techniques. However, one study[143] reports a resistive onset temperature of ~ 5.5 K and a completion temperature of ~ 4.6 K, at which lower temperature the inductive signal for superconductivity in the same sample exhibits an onset. This result demonstrates that a filamentary superconducting shunt can easily occur before appreciable volume superconductivity is developed. Inductive studies by rf penetration depth[53,145] and magnetization[143,144] measurements yield a very wide distribution of superconducting onset temperatures for different crystals: 3.2, 3.4, 3.9, 4.4, and 5.0 K, which is a spread of nearly 2 K. The transition widths extend over an even larger range, 2–3 K. This unusually large distribution of temperatures suggests that a substantial problem occurs with chemical impurities. However, no resistivity ratios are available to delineate the transport quality of these crystals. Meissner and diamagnetic shielding effects[143,144] have been determined to be ~ 50 percent and ~ 80 percent, respectively, near 1.2 K, which establishes the existence of bulk superconductivity in this salt. For the purpose of listing a single transition temperature for β-$(ET)_2AuI_2$, it seems reasonable to take $T_c = 5.0 \pm 0.5$ K, which covers several experimental onset temperatures and likely indicates the superconducting transition temperature of an ideal high-quality crystal. The bulk superconducting state for typical crystals, however, is near 4 K or lower. A careful study of the synthetic aspects of β-$(ET)_2AuI_2$ in relation to T_c would be of considerable interest.

Upper[140,142,143,145,146] and lower[136,143] critical magnetic fields, $H_{c2}(T)$ and $H_{c1}(T)$, respectively, have been determined for all three ambient-pressure β-phase superconductors. These properties are summarized in Table 4.5 and identified in terms of the experimental basis for their determination, either by a resistive method (R) or by an inductive technique (I). As is usually the case, properties based on resistive methods define T_c as the temperature at the midpoint of the superconducting transition, and those based on inductive techniques (rf penetration depth measurements, ac susceptibility, or low-field dc magnetization measurements) define T_c as the temperature at the onset of the transition.

Values of $H_{c1}(T)$ are listed in Table 4.5 for the single low temperature ($T = T_{c1}$) at which they were determined, and all of these were obtained from inductive measurements. Values of $H_{c2}(T)$ are listed for the temperature $T = 0$ K, as derived from a Ginzburg-Landau (GL) analysis,[147]

$$H_{c2}(0) = -T_c \left.\frac{dH_{c2}}{dT}\right|_{T_c} \qquad (4.6)$$

This prescription for $H_{c2}(0)$ is useful because of the relationships established[148] between the normal-state properties (such as the resistivity, specific heat, and Fermi surface area) and the slope of $H_{c2}(T)$ evaluated at T_c, $(dH_{c2}/dT)_{T_c}$. Furthermore, Eq. (4.6) is a reasonable extrapolation of the experimental values, as discussed. For comparisons with $H_{c2}(0)$, this table gives the paramagnetic critical magnetic field

TABLE 4.5 Upper Critical Magnetic Fields $H_{c2}(T)$, Evaluated at $T = 0$ K, and Lower Critical Magnetic Fields $H_{c1}(T)$, Determined at $T = T_{c1}$, for the β-Phase Superconductors[a]

Property[b–d]	β-(ET)$_2$I$_3$		β-(ET)$_2$IBr$_2$		β-(ET)$_2$AuI$_2$	
	R	I	R	I	I	I
T_c (K)	1.1	1.03	2.3	2.25	4.05	4.12
H_{po} (kOe)	20.2		42.3	41.4	74.5	75.8
$H_{c2\parallel}(0)$ (kOe)	17.8		32.2	28.2	66.3	61.5
$H_{c2\perp}(0)$ (kOe)	0.8		1.3	2.9	5.1	10.5
$H_{c2\parallel}/H_{c2\perp}$	22.3		24.8	9.7	13.0	5.8
$\xi_\parallel$ (Å)	626		503	337	254	177
$\xi_\perp$ (Å)	28		20	35	20	30
Ref. (H_{c2})	146		140	142	143	145
T_{c1} (K)		0.12		0.5		1.2
$H_{c1\parallel}(T_{c1})$ (Oe)		0.09		3.9		4.0
$H_{c1\perp}(T_{c1})$ (Oe)		0.36		16.5		20.5
Ref. (H_{c1})		136		143		143

[a] R denotes a resistivity measurement based on the midpoint of the superconducting transition; I denotes an inductive measurement based on the onset of the transition.
[b] H_{po} is the Chandrasekhar-Clogston paramagnetic limit.[149]
[c] $H_\parallel$ and $H_\perp$ refer, respectively, to magnetic fields applied parallel ($\parallel$) to the plane (ab) of the ET donor molecules and perpendicular ($\perp$) to the plane, along the direction of the anion layers.
[d] $\xi_\parallel$ and $\xi_\perp$ refer, respectively, to the GL coherence lengths parallel ($\parallel$) and perpendicular ($\perp$) to the plane of the ET donor molecules.

H_{po}, which is the hypothetical critical field of the specimen due to interactions between the applied magnetic field and the spin-pairing of the superconducting electrons in the absence of orbital interactions. This critical field is evaluated in the Chandrasekhar-Clogston limit,[149]

$$H_{po} = 18.4\, T_c \text{ (kOe)} \qquad (4.7)$$

Also given in this table are the anisotropy ratios of the critical fields, $H_{c2\parallel}(0)/H_{c2\perp}(0)$, and values of the GL coherence lengths,[147] ξ, evaluated at $T = 0$, as derived from the expressions

$$H_{c2\parallel}(0) = \frac{\Phi_0}{2\pi \xi_\parallel \xi_\perp} \qquad (4.8)$$

$$H_{c2\perp}(0) = \frac{\Phi_0}{2\pi \xi_\parallel^2} \qquad (4.9)$$

in which Φ_0 is the flux quantum ($\Phi_0 = 2.07 \times 10^{-7}$ Oe cm^2). These expressions are based on the assumption of an isotropic coherence length $\xi_\parallel$ in the plane of the ET-donor molecules. The coherence length, ξ, is a useful measure of the extent of the concentration of superconducting electrons in a spatially-varying magnetic field.

One observes that, according to Eqs. (4.8) and (4.9), the ratio $H_{c2\parallel}(0)/H_{c2\perp}(0)$ is identical to the ratio of the coherence lengths, $\xi_\parallel/\xi_\perp$.

Resistive determinations reported by Ginodman et al.[150] for the upper critical magnetic fields of β-$(ET)_2I_3$ show negative curvature, $d^2H_{c2}/dT^2 < 0$, in plots of $H_{c2}(T)$ versus T over the full range of measurements from $T \approx 1.4$ K to $T \approx 0.5$ K. This curvature is qualitatively typical of that observed for conventional superconductors[151] and is expected[19] for superconductors describable by weak-coupling BCS (Bardeen, Cooper, and Schrieffer) theory.[152] However, all other determinations of H_{c2} for β-$(ET)_2I_3$,[146] β-$(ET)_2IBr_2$,[140,142] and β-$(ET)_2AuI_2$[143,145] by either resistive or inductive methods yield linear temperature dependencies for $H_{c2\parallel}(T)$ and $H_{c2\perp}(T)$, and thus GL behavior,

$$H_{c2}(T) = H_{c2}(0)\left[1 - \frac{T}{T_c}\right] \quad (4.10)$$

over a remarkably extended range of temperatures below T_c. One observes in Table 4.5 that $H_{c2}(0)$ in all cases is less than the paramagnetic limit, H_{po}, which implies that H_{c2} is essentially determined in these superconductors by the magnetic fields interacting only with the orbital motions of the electrons (orbital effects). The nonlinear critical magnetic fields reported by Ginodman et al.[150] also extrapolate in the low temperature region to values less than the paramagnetic limit.

Table 4.5 indicates discrepancies of about a factor of 2 in the slopes of $H_{c2\perp}$ for two different measurements each of β-$(ET)_2IBr_2$ and β-$(ET)_2AuI_2$. For the former salt, the discrepancy may arise from the use of different definitions of T_c (i.e., the midpoint transition for the resistive determination and the onset for the inductive determination). For the latter salt, with both determinations based on inductive measurements, the source of the discrepancy is not obvious. It may be caused by differences in purity, defects, or twinning, or some other sample-dependent effect. Nevertheless, the data are qualitatively consistent in showing that there is considerable anisotropy between $H_{c2\parallel}$ and $H_{c2\perp}$, with $H_{c2\parallel}/H_{c2\perp} \approx 6$–$25$. Only small anisotropy effects have been detected in $H_{c2\parallel}$ for different axial directions within the plane of the ET-donor molecules. This anisotropy along with the linear behavior of the critical fields is illustrated in Fig. 4.8, which shows H_{c2} determined resistively for β-$(ET)_2I_3$[146] with magnetic fields applied parallel to the crystallographic a axis ($H_{c2\parallel a}$), the b' axis ($H_{c2\parallel b'}$), and the c^* axis ($H_{c2\perp}$) (the b' axis is perpendicular to a in the ab plane). One finds from the data of Fig. 4.8 that $(H_{c2\parallel a}/H_{c2\parallel b'} = 1.05$ compared to $(H_{c2\parallel a}/H_{c2\perp} = 22$. These critical field anisotropies correlate (near T_c) with the anisotropies in the normal-state conductivities because of the relationship between the critical fields or coherence lengths and the Fermi velocities V_F (e.g., $\xi_\parallel \propto V_{F\parallel}/T_c$).[19] Thus, like the normal-state conductivities, the upper critical magnetic fields establish the fundamental 2D nature of the Fermi surface in these β-phase superconductors. The general trend in the critical field anisotropies is that the conductive dimensionality increases slightly toward 3D character with changes in the anion from I_3^- to IBr_2^- to AuI_2^-.

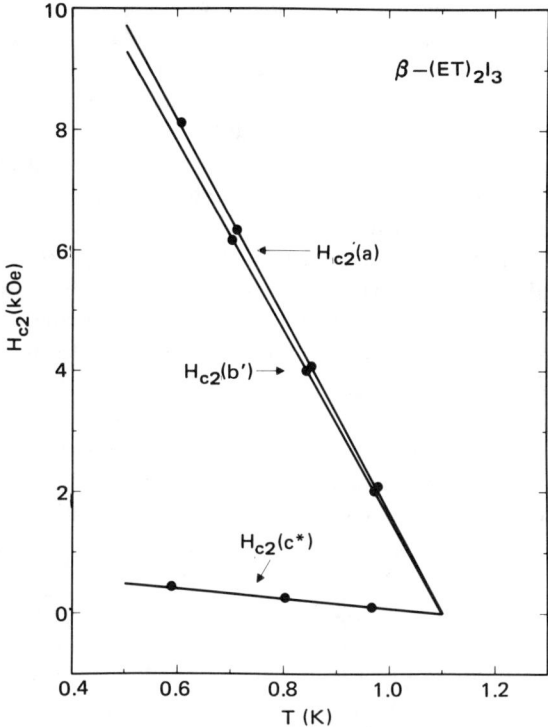

Figure 4.8 Temperature dependence of the upper critical magnetic field H_{c2} of β-$(ET)_2I_3$ as determined by Tokumoto et al.[146] from electrical resistivity measurements at ambient pressure for magnetic fields applied along the crystallographic a, b', and c^* axes (the b' axis is ⊥ to a within the ab plane). Reproduction of Fig. 32 of ref. 6 with permission: *Progress In Inorganic Chemistry*, Vol. 35, pp. 51–218, Ed. By S. J. Lippard, an Interscience® Publication, Copyright © **1987** by John Wiley and Sons, Inc., NY.

The coherence lengths ξ are small quantities, especially $\xi_\perp$ because of the high values of the upper critical magnetic fields in directions parallel to the plane of the organic donor molecules. $\xi_\perp$ is the coherence length along the c^* axis, and $c^* \approx 15$ Å for the β-phase superconductors. The fact that $\xi_\perp \approx 20\text{--}30$ Å, which is no more than double the unit-cell length, suggests that Josephson coupling or proximity effects may be important in connecting segregated superconducting layers of the organic donor molecules. This could, of course, lead to novel superconducting effects. However, there is no question that superconductivity is other than three-dimensional (a volume superconductor) at temperatures below T_c. The small values of the lower critical magnetic fields, H_{c1}, and the small values of ξ (or high values of H_{c2}) imply large values for the GL parameter,[147]

$$\kappa = \frac{\lambda_L}{\xi} \tag{4.11}$$

where λ_L is the London (magnetic field) penetration depth. $\kappa > 1/\sqrt{2}$ distinguishes type II superconductors. The data of Table 4.5 show that H_{c1} is $\sim 10^{-1}$–10^1 Oe and H_{c2} is $\sim 10^3$–10^5 Oe. From these, one can derive values of κ for the β-phase superconductors that are of the order of 10 for $\kappa_\perp$ and 100 for $\kappa_\parallel$.[143]

Important properties of a superconducting material also include the specific-heat discontinuity ΔC at T_c due to the second-order transformation of the normal state to the superconducting state and the electronic specific-heat coefficient γ, which is proportional to the density of electronic states at the Fermi level. Specific-heat measurements have been reported for β-$(ET)_2I_3$[137] and β-$(ET)_2AuI_2$,[153,154] and although the results are incomplete in defining precise properties, they reveal a novel phenomenon: β-$(ET)_2I_3$ exhibits no specific-heat discontinuity but β-$(ET)_2AuI_2$ does exhibit a discontinuity. The measurements for β-$(ET)_2I_3$ included the determination of a Meissner effect of at least 25 percent at 0.8 K for the same crystal specimens that exhibited no specific-heat discontinuity down to 0.7 K.[137] These results clearly indicate that the crystals were volume superconductors but apparently possessed no superconducting energy gap in the electronic energy levels. On the other hand, the specific heat of β-$(ET)_2AuI_2$ exhibited a broad but distinct discontinuity near T_c.[153,154] and evidence of a fully developed gap as T → 0 K.[153] The electronic specific-heat coefficient was determined to be ~ 24 mJ/mol K^2 for β-$(ET)_2I_3$ and approximately the same for β-$(ET)_2AuI_2$. Major difficulties in determining specific-heat properties of organic superconductors near T_c arise from the inhomogeneous distribution in T_c and the very large number of atoms, ~ 55 atoms per unit cell, contributing to the lattice specific heat.[154]

The width of the superconducting energy gap Δ in the β-phase superconductors has been studied by electron tunneling techniques, but the results to date are inconclusive. An investigation leading to the determination of the energy gap at 0 K for tunneling in the plane of the organic donor molecules, $\Delta_\parallel(0)$, for β-$(ET)_2AuI_2$ reports a value more than four times larger than that predicted by conventional (weak-coupling) BCS theory, $2\Delta(0)/kT_c = 3.53$ (where k is Boltzmann's constant).[155] Moreover, this investigation presents evidence that the energy gap in the perpendicular direction, $\Delta_\perp(0)$, is smaller by about 30 percent, indicating a considerable amount of anisotropy in the gap. These very large energy gaps imply that the superconducting state involves something very much beyond weak-coupling BCS theory. In consideration of the large number of atoms that contribute to the phonon spectrum and that possibly can yield an appropriate dispersion of phonon energies, the explanation of strong electron-phonon interactions (strong coupling) is attractive. However, subsequent determinations[156] of the energy gap in β-$(ET)_2AuI_2$ by tunneling measurements have yielded $2\Delta_\parallel(0)/kT_c \approx 4$, a value only ~ 13 percent larger than the BCS ratio of 3.53. On the other hand, preliminary determinations[156] of the energy gap in β-$(ET)_2I_3$ gave $2\Delta_\parallel(0)/kT_c \approx 8$. This value is more than a factor of two larger than the BCS gap, and it contradicts the specific-heat measurements,[137] which indicate the absence of a superconducting energy gap. These several results are confusing but clearly demonstrate that further experiments on the specific-heat discontinuity and the superconducting energy gap are essential to a firm understanding of the nature of superconductivity in the β-phase materials. An interesting question regarding the tunneling experiments with β-$(ET)_2I_3$ is whether or not the electrical probes might induce some kind of local high-pressure transformation, so that the sample specimen might be the β^*-$(ET)_2I_3$ phase.

The β-Phase Superconductors under Applied Pressures

The superconducting transition temperature T_c of organic superconductors typically decreases with increasing pressure P, $dT_c/dP < 0$. Some salts, notably those of the TMTSF organic donor molecule but also a variety of the newer salts listed in Table 1.1, require elevated pressures (frequently of the order of several kbar) to achieve the superconducting state, but nevertheless, they too typically exhibit decreasing T_c with further increases in the pressure. The ambient-pressure superconducting salt $(ET)_4Hg_{2.89}Br_8$ is an exception to this pressure dependence,[117] and there is evidence that the high-pressure superconducting salts $(TTF)[Ni(dmit)_2]_2$[128] and $[N(CH_3)_4][Ni(dmit)_2]_2$[132] also are exceptions, with $dT_c/dP > 0$. All three of the ambient-pressure β-phase superconducting salts, however, have conventional negative pressure dependencies and comparable slopes, $dT_c/dP \approx -1$ K/kbar.[70] This behavior is illustrated in Fig. 4.9.

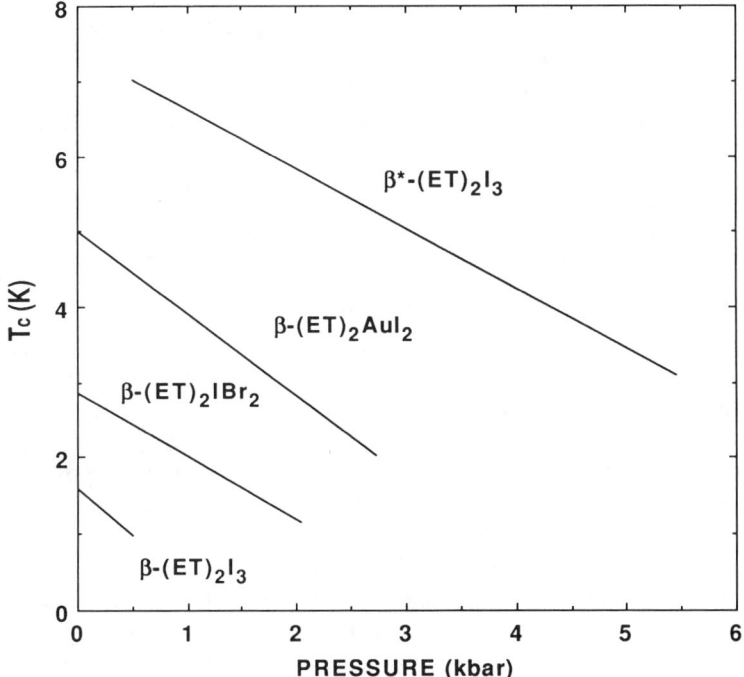

Figure 4.9 A schematic representation of the superconducting onset temperature T_c as a function of applied pressure in zero applied magnetic field for the β-phase superconductors. Note the abrupt increase at ~ 0.5 kbar in T_c near 1 K for β-$(ET)_2I_3$ to T_c near 7 K for β^*-$(ET)_2I_3$. This representation is based on the graphical data reported by Schirber et al.[70]

β-(ET)$_2$IBr$_2$ and β-(ET)$_2$AuI$_2$ exhibit regular decreases in T_c as the pressure is increased from 1 bar (atmospheric pressure) to several kbar. Although it initially decreases with increasing pressure, the T_c of β-(ET)$_2$I$_3$, undergoes a remarkable increase in T_c by about a factor of 5 near a pressure of ~ 0.5 kbar. This increase is not a manifestation of a positive pressure dependence of T_c but, as previously discussed, is the signature of a structural change from a crystallographic unit cell possessing partial crystallographic disorder to a cell that is perfectly ordered.[69,70] This ordered phase is denoted as β*-(ET)$_2$I$_3$, but it is frequently identified as β$_H$-(ET)$_2$I$_3$ and often referred to as the "high-T_c" state of β-(ET)$_2$I$_3$. There is, however, a subtle but important distinction to be made in this nomenclature: the β*-phase, with a well-established crystallographic structure, is superconducting under pressure, ~ 0.5 kbar minimum, with T_c near 8 K, but a high-T_c state of β-(ET)$_2$I$_3$ is not necessarily the β*phase in the absence of a structural confirmation. This distinction is important because elevated T_c's (8 K and lower) occur with β-(ET)$_2$I$_3$ under a variety of different conditions including ambient pressure, the origins of which are not yet well understood.

There is a very large amount of literature on the subject of elevated T_c's for β-(ET)$_2$I$_3$ published since the discovery by Laukhin et al.[67] in late 1984 and the subsequent confirmation by Murata and associates[68] of the effect of pressure on the elevation of T_c in this salt. Much of this literature is devoted to the existence of elevated T_c's at ambient pressure. It was recognized very early from the evidence of partial resistive drops that several high-T_c states of β-(ET)$_2$I$_3$, or possibly other superconducting phases, coexisted in β-(ET)$_2$I$_3$ at ambient pressure with T_c's approaching 8 K, although not necessarily in bulk superconducting quantities.[73,157] Subsequent resistive studies have well documented that high-T_c states can occur at ambient pressure in conventionally synthesized crystals of β-(ET)$_2$I$_3$[158-160] and that high-T_c states of β-(ET)$_2$I$_3$ can be obtained by thermal treatment near 400 K of other phases of the ET/I system, such as the iodine-rich phase ε-(ET)$_2$(I$_3$)(I$_8$)$_{0.5}$.[94,161] Repeated thermal cycling from 300 K to low temperatures at ambient pressure[162] and pressure cycling up to 6.5 kbar[163] seem to enhance the resistive onset of superconductivity near 8 K at ambient pressure in conventionally synthesized crystals, but the superconducting transitions tend to be very broad, extending from the onsets near 8 K to completions below 2 K. Specimens of β-(ET)$_2$I$_3$ which have been pressure-cycled at ambient temperature have been found to exhibit complete resistive superconducting transitions near 8 K at ambient pressure but possess no detectable volume superconductivity in diamagnetic shielding measurements.[143]

The production of the β*-phase structure from β-(ET)$_2$I$_3$ under applied pressures is reported[70] to require an anisotropic medium that can induce a substantial shear stress on the crystal. This means that β* cannot be accessed by applied pressure alone. This shear stress can be achieved by the presence of the attached resistive probes with the use of an isotropic pressure medium such as He or by the use of an anisotropic pressure medium such as ice. Superconductivity near 8 K in the β* phase has been confirmed by Meissner-effect measurements to be a genuine bulk property of the crystals.[164] The high T_c of the β* phase is apparently caused by the perfect structural ordering, in contrast to the disorder existing in the β phase with $T_c \approx 1.4$ K. It is sometimes stated that the low T_c of the β phase is due to the presence of the incommensurate modulated structure, but this statement is incorrect. The low T_c is caused by the structural disorder, and the modulated structure is the signature of this disorder. Several careful studies have shown that the β* phase is unstable at ambient pressure and temperature with respect to reversion to the β phase, but it can be stabilized at ambient pressure by low-temperature annealing under pressure, provided that the temperature on release of the applied pressure does not exceed ~ 125–170 K.[165–167] With the use of such preparative techniques, one can obtain in volume-superconducting quantities the β* phase at ambient pressure which exhibits a reasonably homogeneous T_c with an onset near 8 K, a midpoint near 7.5 K, and a superconducting transition width of ~ 1 K.

Returning to the question of high-T_c superconductivity coexisting in the β phase in conventionally synthesized crystals at ambient pressure, one judges, from the evidence of the very broad and often incomplete transitions and the absence of induced Meissner currents, that only small amounts of a high-T_c phase (or phases) ordinarily occur, perhaps only in trace quantities, and this can give rise to partial or complete shunting circuits in resistive measurements. However, critical-field measurements of these high-T_c resistive drops indicate the coexistence of genuine high-T_c superconducting phases.[158,163] Furthermore, a report of the thermal conversion of the ε phase to a high-T_c β-like phase gives evidence of the occurrence of diamagnetic screening currents in the samples, although they occur near 4.5 K, which is considerably below the resistive superconductive onset.[161] Therefore, the evidence is convincing that at least some amount of a high-T_c phase (or phases) exist in ambient-pressure crystals of β-$(ET)_2I_3$ that are not especially prepared from the β* phase. It is not a clear-cut issue whether or not these high-T_c phases are the same as the ordered β* phase, but one speculates that they are very similar, at least with regard to increased crystallographic order. The coexistence of two or more superconducting phases in apparent single crystals of the ET/I system is plausible because of the complexity of this system.

Recent studies by Montgomery et al.[168] demonstrate that apparent "single" crystals with α-β mixtures occur in conventionally synthesized $(ET)_2I_3$ salts and give rise to inductive superconducting transitions at ambient pressure in the range of 2–7 K. This superconductivity no doubt arises from the presence of the β phase and possibly occurs at higher T_c because of the encapsulation of the β phase within a crystal. Zvarykina and associates[169] have reported that β-$(ET)_2I_3$ crystals can experience plastic deformations by twinning that are accompanied by increases in T_c, and they suggest that the β* phase can be induced and stabilized by local strains arising from these plastic deformations. On the basis of these reports, it would seem that the occurrence of high-T_c states in conventionally prepared crystals of β-$(ET)_2I_3$, which may contain mixtures of phases and twinned or mosaic components, and the occurrence of high-T_c superconductivity in β*-$(ET)_2I_3$ under applied anisotropic pressures might be reconciled as different aspects of the requirement of a shear-stress component.

Many determinations of $H_{c2}(T)$ for β* (or $β_H$)-$(ET)_2I_3$ have been reported, and all are based on resistive measurements.[68,158,163,165,170–172] Similar to the upper critical magnetic fields of the ambient-pressure β-phase salts, H_{c2} for the β* phase shows considerable anisotropy between the field $H_{c2\parallel}$ parallel to the plane of the donor molecules and the field $H_{c2\perp}$ perpendicular, along the c^* axis, but small anisotropy between fields within the plane ($H_{c2\|b'}/H_{c2\|a} \approx 1.5$).[171] $H_{c2\perp}(T)$ has been determined from T_c ($\sim$ 7–8 K) down to 2 K or lower, and over this range all but one study show strong positive curvature, $d^2H/dT^2 > 0$. The discrepant study[68] yields a strictly linear behavior. Because of the very large values of $H_{c2\|}(T)$ at temperatures near T_c, these have been determined only over a limited range of temperature, so that it is impossible to assess the temperature behavior of this property. Quantitative differences among results occur in the initial slopes, $-(dH_{c2}/dT)_{T_c}$. These slopes are in the range 2.1–4 kOe/K for $H_{c2\perp}$ and 25–48 kOe/K for $H_{c2\|}$. Table 4.6 lists the Ginzburg-Landau parameters for H_{c2} as derived from Eqs. (4.6) to (4.9) with the use of data reported in two of the more complete and detailed studies of the critical magnetic fields.[170,172]

One observes in Table 4.6 that $H_{c2\|}(0)$ exceeds the paramagnetic limit H_{po}, thus suggesting that spin-pair breaking may be more important than orbital effects. On the other hand, the experimental $H_{c2\|}(T)$ may have negative curvature at temperatures lower than those measured and approach a value smaller than H_{po} as T → 0. The anisotropy ratio $H_{c2\|}/H_{c2\perp} \approx 12$ for β*-$(ET)_2I_3$ appears to be comparable to that of β-$(ET)_2AuI_2$. It is considerably smaller than that of ambient-pressure β-$(ET)_2I_3$, indicating a substantial increase in conductive dimensionality in the β → β* transformation. The GL coherence lengths are much smaller than those of the ambient-pressure β-phase superconductors, and with $\xi_\perp \approx 10$ Å or less, the coherence length along the c^* axis is smaller than the unit cell length, again suggesting the possibility of novel properties. Bulaevskii et al.[172] have estimated the low-temperature, normal-state electron-scattering mean-free paths $l_\|$ in the plane of the organic donor molecules to be ~ 70 Å for the β phase and ~ 200 Å for the β* phase. A comparison of these lengths to the coherence lengths indicates that the β phase is a dirty superconductor, $\xi_\|/l_\| > 1$, whereas the β* is a clean superconductor, $\xi_\|/l_\| < 1$. Bulaeviskii[19] also estimates that β-$(ET)_2IBr_2$ and β-$(ET)_2AuI_2$ are clean superconductors, which again indicates that there is something unusual about β-$(ET)_2I_3$ with respect to the other superconductors of this class. As is the case for the linear behavior in $H_{c2}(T)$ for the ambient-pressure β-phase superconductors, the positive curvature in $H_{c2\perp}$ for the β* phase indicates something beyond a conventional BCS superconductor. We will return later to this interesting point.

TABLE 4.6 Upper Critical Magnetic Fields H_{c2} and Coherence Lengths ξ Evaluated as Ginzburg-Landau Parameters for the High-Pressure Phase β*-$(ET)_2I_3$[a]

T_c (K)	7.2	7.5
H_{po} (kOe)	132	138
$H_{c2\|}(0)$ (kOe)	250	360
$H_{c2\perp}(0)$ (kOe)	20	30
$H_{c2\|}/H_{c2\perp}$	12.5	12
$\xi_\|$ (Å)	128	105
$\xi_\perp$ (Å)	10	8.8
Reference	170	172

[a] See footnotes of Table 4.5 and Eqs. (4.6)–(4.9) for definitions.

The κ-Phase Superconductors at Ambient Pressure

The κ-phase superconductors consist of κ-$(ET)_2I_3$, κ-$(ET)_2Cu(NCS)_2$, κ-$(ET)_2Cu[N(CN)_2]Br$, κ-$(ET)_2Cu[N(CN)_2]Cl$, $(ET)_4Hg_{2.89}Br_8$, $(ET)_4Hg_{3-x}Cl_8$, $(MDT-TTF)_2AuI_2$, and one of the $(DMET)_2AuBr_2$ phases. The $(ET)_4Hg_{3-x}Cl_8$ and κ-$(ET)_2Cu[N(CN)_2]Cl$ salts are the only members that are not ambient-pressure superconductors. This group does not possess the isostructural integrity of the β-phase superconductors because of the occurrence of different organic donor molecules and anions of different geometries, but each of its members possesses the unique arrangement of twisted pairs of neighboring dimers of the organic donor molecules as exemplified by the prototypical structure of the κ-$(ET)_2I_3$ salt.[97]

At present, only resistive measurements of the superconducting properties of κ-$(ET)_2I_3$ have been reported,[98,100] so that it is not yet established that this phase is a bulk superconductor. This circumstance possibly exists because this phase occurs only as a minor constituent of electrocrystallization and is therefore difficult to isolate for study. However, genuine upper critical magnetic fields have been determined for this phase, and the resistive superconducting transitions are rather sharp and complete not far below the onset. The resistive onset temperatures occur near 4 K, and completion temperature occurs near 3 K. Resistivity ratios $\rho(300)/\rho(4)$ are reported to be in the range of 30–385 with midpoint transition temperatures of 3.6–3.7 K.[100] Although these results suggest the existence of a volume superconducting phase, we have previously cited examples for which complete and narrow resistive transitions occur in the absence of detectable volume superconductivity, for example in β*-$(ET)_2I_3$ prepared by release of pressure at ambient temperature.[143]

κ-$(ET)_2Cu(NCS)_2$ is relatively easy to synthesize, and, therefore, its superconducting properties have been studied extensively in many laboratories by a variety of techniques. Resistive superconducting transitions rather consistently have onsets near 11 K, completions near 6–8 K, and midpoint transitions (generally quoted as T_c) of 10.2–10.5 K.[7,8,173] Resistive transitions, however, with onsets lower by about 1.5 K have been reported,[174] and it has been observed that onset temperatures can be influenced by synthetic procedures.[175] Resistivity ratios $\rho(300)/\rho(10)$ are of the order of 200.[8,174] Inductively determined superconducting transitions of κ-$(ET)_2Cu(NCS)_2$ typically occur with onsets at temperatures lower by 1–2 K than those determined by resistive measurements.[7,8,176,177] In general, both resistively and inductively determined superconducting transitions are sharper and more reproducible than those of the β-phase superconductors. This has permitted unambiguous measurements to be made of the isotopic shift on substitution of deuterium for hydrogen in the ET-donor molecule. The T_c of the deuterated salt exhibits a positive isotopic shift,[173,175] with T_c increasing by 0.5–1.0 K, instead of the negative shift expected for electron-phonon coupled superconductivity [$T_c \propto (mass)^{-1/2}$]. It is not clear how this isotopic effect might relate to the superconducting mechanism.

The occurrence of field-induced diamagnetic screening currents in inductive measurements of superconductivity in κ-$(ET)_2Cu(NCS)_2$ is reasonable evidence of volume superconductivity. Reports of a sharp onset in the dc diamagnetic susceptibility near 9.8 K with Meissner effects of ~ 20 percent at 6 K and 83 percent at 2 K[176], along with the observation of a sharp onset in the ac diamagnetic susceptibility near 10.3 K and diamagnetic shielding effects of ~ 100 percent near 5 K[177], clearly establish this salt to be a volume superconducting phase. For general purposes of quoting a superconducting transition temperature of κ-(ET)$_2$Cu(NCS)$_2$, $T_c = 10.4 \pm 0.6$ K is a reasonable choice that encompasses most resistive midpoint transitions and several inductively determined onset temperatures.

κ-(ET)$_2$Cu[N(CN)$_2$]Br and κ-(ET)$_2$Cu[N(CN)$_2$]Cl are currently the most interesting organic superconductors because they are the highest T_c superconducting organic materials. At this writing, few of the superconducting properties of these salts have been reported, but many studies are now in progress. The superconducting transition[204,205] of κ-(ET)$_2$Cu[N(CN)$_2$]Br as determined by rf penetration depth and resistivity measurements is illustrated in Fig. 4.10. The inductive (rf penetration depth) measurements show a weak superconducting onset at 11.60 $\pm$ 0.05 K, a strong increase in the signal beginning near 11.4 K, a transition midpoint near 10.5 K, and saturation in the signal beginning near 9.6 K. Low-field dc magnetization measurements of the same crystal specimen yield identical parameters and confirmation of volume superconductivity. The resistively determined superconducting transition has an onset at 12.5 $\pm$ 0.1 K, a midpoint of 11.2 K, and completion (zero resistance) at 10.5 K. The extrapolated resistive onset, taken as the intersection of the extrapolated resistivity of the normal state above onset and the linear portion of the drop in the resistivity below onset, is 11.6 K, in excellent agreement with the inductive onset temperature. Both the inductive and resistive transitions have 10 to

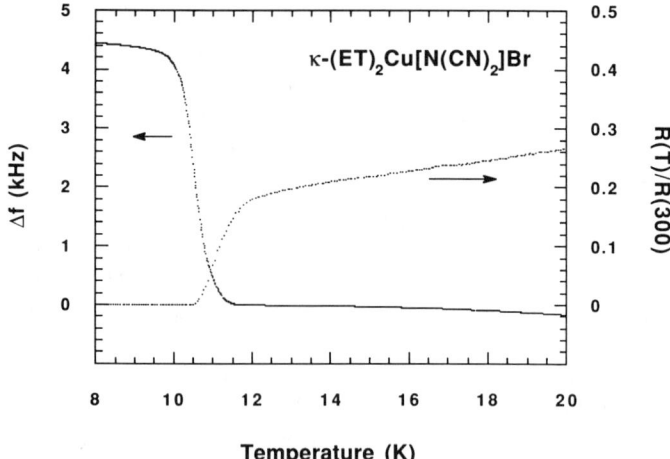

Figure 4.10 The superconducting transition of κ-(ET)$_2$Cu[N(CN)$_2$]Br in zero magnetic field as determined for a single-crystal specimen by rf penetration depth measurements (Δf is the change in resonant frequency relative to ~ 594 kHz) and for a different single-crystal specimen from the same synthetic batch by 4-probe resistivity measurements [R(T) is the resistance at temperature T]. This figure was prepared from the original data cited in refs. 204 and 205. The resistivity curve represents the low-temperature region of the curve shown in Fig. 4.6.

90 percent transition widths of ~ 1 K, indicating reasonably high-quality crystal specimens. Depending on the basis of the comparison (onset, midpoint, etc.), the T_c of κ-(ET)$_2$Cu[N(CN)$_2$]Br is 1–2 K higher than that of κ-(ET)$_2$Cu(NCS)$_2$. The κ-(ET)$_2$Cu[N(CN)$_2$]Cl salt is nonsuperconducting at ambient pressure but has a T_c of ~ 12.8 K under a very slight applied pressure of ~ 0.3 kbar.[206] This salt presently represents the highest transition temperature organic superconductor.

The κ-like salt (ET)$_4$Hg$_{2.89}$Br$_8$ is an ambient-pressure superconductor with a broad superconducting transition. Resistively determined onset and completion temperatures occur near 5 K and 3 K, respectively, with midpoint transitions near 4 K.[114,116] The resistivity ratio is abnormally low, $\rho(300)/\rho(6) \approx 5$–10, and this has been attributed to the presence of the incommensurate lattices in the crystal structure.[114] Furthermore, it has been suggested that the crystal may contain an additional superconducting phase.[116]

An inductive determination of the superconducting transition of (ET)$_4$Hg$_{2.89}$Br$_8$ by rf penetration depth measurements[178] is illustrated in Fig. 4.11. This figure shows a broad transition with an onset of 4.0 ± 0.2 K, completion near 2.4 K, and a midpoint near 3.5 K. Measurements of the dc magnetization of the same crystal specimen gave an onset of 3.7 ± 0.2 K and a completion near 2.5 K.[117] Pressure studies,[117] as described later, confirm that the superconductivity is not due to the presence of elemental Hg, which itself has T_c = 4.15 K. The

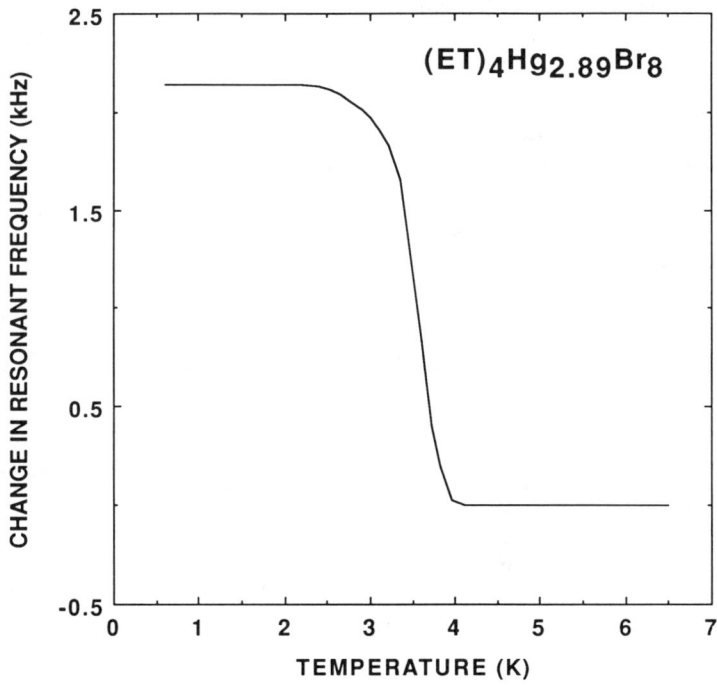

Figure 4.11 Superconducting transition of (ET)$_4$Hg$_{2.89}$Br$_8$ in zero applied magnetic field as determined by rf penetration depth measurements (~ 575 kHz). The unpublished data used for this figure were determined by the authors in connection with the study reported by Schirber et al.[117]

inductive measurements demonstrate that $(ET)_4Hg_{2.89}Br_8$ is a volume superconductor. It is not clear whether or not this salt is a pure phase, but in all presently reported studies it appears to have an inhomogeneous distribution of T_c's. A T_c of 4.0 ± 0.3 K seems to be a reasonable specification of the superconducting transition. This temperature encompasses the resistive midpoints and the onset of diamagnetic screening currents.

The occurrence of superconductivity in the κ-like salts $(MDT-TTF)_2AuI_2$[9] and $(DMET)_2AuBr_2$,[119] each having unsymmetrical organic donor molecules, is a significant demonstration that symmetrical donor molecules such as ET are not essential to the achievement of organic superconductivity. They also represent the first examples of members of an organic structural class that possess different kinds of organic-donor molecules within the same class of organic superconductors. Both resistive and inductive determinations of the superconducting transition in $(MDT-TTF)_2AuI_2$ show broad transitions with onsets near 4.5 K and completions near 3 K for the resistive measurements and below 2 K for the inductive measurements.[10] The inductive measurements indicated that this salt is a volume superconductor. Only resistive studies of κ-like $(DMET)_2AuBr_2$ have been reported, and these present few details.[119,121] The resistive onset of superconductivity is near 2 K, and the midpoint of the transition is 1.9 K, indicating a sharp transition. Applications of a magnetic field are reported[121] to suppress the superconducting transition, although upper critical magnetic fields have not yet been published.

Upper critical magnetic fields H_{c2} have been reported for κ-$(ET)_2I_3$,[100] κ-$(ET)_2Cu(NCS)_2$,[175,177,179,180] and $(ET)_4Hg_{2.89}Br_8$.[114] All but one[180] of these determinations have been obtained from resistive measurements and based on the midpoint of the superconducting transition. The Ginzburg-Landau parameters and the paramagnetic limits derived from the resistive data by the use of Eqs. (4.6) to (4.9) are summarized in Table 4.7. Similar to the β-phase superconductors, there is considerable anisotropy between the fields $H_{c2\parallel}$ and $H_{c2\perp}$. Although not generally reported in the description of each study, differences in $H_{c2\parallel}$ for magnetic fields oriented along different axial directions within the plane of the organic donor molecules are expected to be small.

TABLE 4.7 Upper Critical Magnetic Fields H_{c2} and Coherence Lengths ξ Evaluated as Ginzburg-Landau Parameters for Several κ-Phase Superconductors as Determined from Resistive Measurements

Property[a]	κ-$(ET)_2I_3$	κ-$(ET)_2Cu(NCS)_2$[b]	$(ET)_4Hg_{2.89}Br_8$
T_c (K)	3.6	10.4	4.3
H_{po} (kOe)	66	191	79
$H_{c2\parallel}(0)$ (kOe)	68	190	56
$H_{c2\perp}(0)$ (kOe)	2.6	10	2.8
$H_{c2\parallel}/H_{c2\perp}$	26	19	20
$\xi_\parallel$ (Å)	356	182	343
$\xi_\perp$ (Å)	14	9.6	17
Reference	100	179	114

[a] See footnotes of Table 4.5 and Eq. (4.6) to (4.9) for definitions.

[b] $(dH_{c2}/dT)_{T_c}$ determined from specific heat measurements for the $\parallel$ and $\perp$ components are reported to be very much larger.[180] See text for discussion.

$H_{c2\perp}(T)$ for κ-(ET)$_2$I$_3$ is linear over the measured temperature range from T_c (3.6 K) to 1.4 K,[100] the lowest temperature of the study, so that a linear extrapolation to T = 0 K is a reasonable estimate of the experimental limit for $H_{c2\perp}(0)$. $H_{c2\|}(T)$ for this salt is linear but determined over a very limited range of $\sim$ 0.6 K below T_c, which is insufficient to establish its exact temperature behavior. However, a comparison of the GL critical fields (equivalent to slopes evaluated near T_c) for κ-(ET)$_2$I$_3$ and β-(ET)$_2$I$_3$ (Table 4.5) indicates that these two phases of identical chemical composition are similar in their critical field behavior, at least near T_c. This is evident first of all in the anisotropy ratios $H_{c2\|}/H_{c2\perp}$ = 26 for κ-(ET)$_2$I$_3$ and 22 for β-(ET)$_2$I$_3$. Secondly, the similarities are evident in the slopes, $-(dH_{c2}/dT)_{T_c}$ [see Eq. (4.6)]. These slopes are 18.9 kOe/K and 16.2 kOe/K for $H_{c2\|}$ of the κ and β phases, respectively, and 0.7 kOe/K for $H_{c2\perp}$ of both salts. So the essential difference in H_{c2} for the two phases is simply a scaling by the difference in T_c. This scaling difference is reflected also in the coherence lengths. It is interesting to note that $\xi_\perp$ for the κ phase is comparable to the unit cell length along the direction of the anion layers ($a \approx$ 16 Å). The paramagnetic limit, H_{po}, for κ-(ET)$_2$I$_3$ is slightly less than $H_{c2\|}(0)$, which suggests that spin effects may be more important for this κ phase salt.

The upper critical magnetic fields of κ-(ET)$_2$Cu(NCS)$_2$ for both $\|$ and $\perp$ components as determined from resistive measurements exhibit unusual temperature dependencies, which are illustrated in Fig. 4.12. $H_{c2\perp}(T)$ over the measured temperature range from T_c to 0.5 K possesses extreme positive curvature.[179] The experimental values of $H_{c2\perp}(T)$ are $\sim$ 8 kOe near 5 K and $\sim$ 48 kOe near 1 K compared to 10 kOe for the GL limit at 0 K. This curvature is considered[181] to be associated with the same effect that yields an anomalous but unexplained enhancement in relaxation rates observed near 4 K in NMR studies of this phase.[182] $H_{c2\|}(T)$ exhibits a different kind of unusual behavior by the occurrence of two linear segments of different slope which intersect near 9 K.[175,179] From T_c to $\sim$ 9 K, $-(dH_{c2\|}/dT) \approx$ 18 kOe/K, and from $\sim$ 9 K to $\sim$ 6 K, $-(dH_{c2\|}/dT) \approx$ 34 kOe/K. This change in slope is attributed to a change in dimensionality from 3D superconducting character near T_c to Josephson-coupled 2D superconductivity below 9 K.[175,179] The plausibility of 2D behavior is indicated by the small coherence length $\xi_\perp \approx$ 10 Å compared to the unit cell dimension perpendicular to the layers of the organic donor molecules ($a \approx$ 16 Å) and by the implication of an even smaller coherence length due to the much larger values of $H_{c2\|}(T)$ below 9 K. Direct measurements of the magnetic penetration depths in κ-(ET)$_2$Cu(NCS)$_2$ indicate that superconductivity in this phase approaches the dirty limit, with $\zeta_\|/1_\| \approx 2.7$.[183]

A recent article[180] on specific heat measurements of κ-(ET)$_2$Cu(NCS)$_2$ near T_c under different applied magnetic fields has reported initial slopes of the upper critical magnetic fields, as determined from the heat-capacity data, that are enormously large: $-(dH_{c2\|}/dT)_{T_c}$ = 160 kOe/K and $-(dH_{c2\perp}/dT)_{T_c}$ = 7.5 kOe/K. These slopes are larger than those determined resistively by a factor of 8–10. The discrepancy has been suggested to arise from flux flow in the resistive measurements.[180] This is an interesting proposition, but it requires further testing by additional measurements and techniques, including determinations of the critical magnetic fields by dc magnetization measurements.

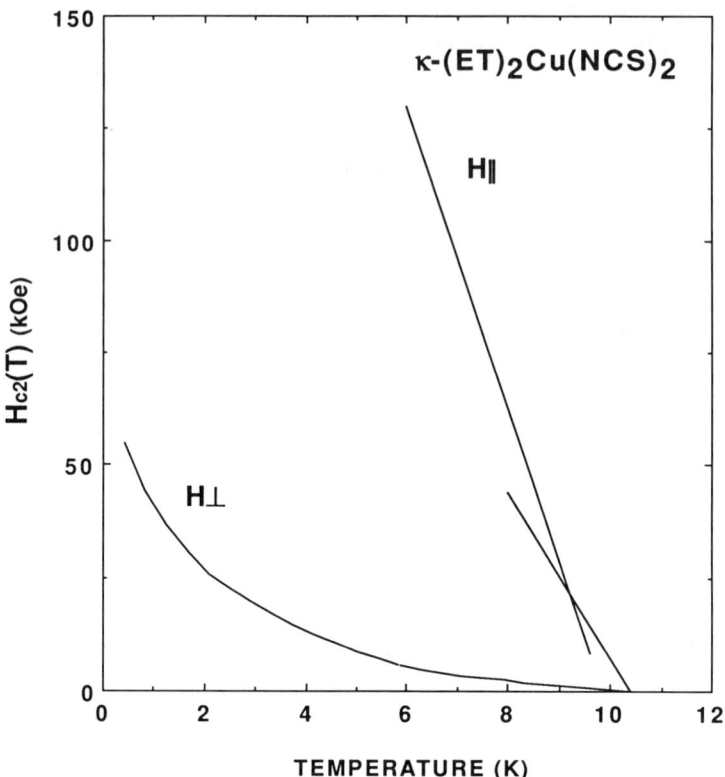

Figure 4.12 A representation of $H_{c2\|}(T)$ and $H_{c2\perp}(T)$ for κ-(ET)$_2$Cu(NCS)$_2$ based on the experimental determinations reported by Oshima et al.[179] H$\|$ refers to a magnetic field applied along the crystallographic c axis within the plane (bc) of the ET donor molecules. The straight-line extensions beyond the intersection near 9 K are for visual purposes only. H$\perp$ refers to a magnetic field applied perpendicular to the bc plane.

A comparison of the slopes of H_{c2} evaluated near T_c (the GL parameters) for the resistive determinations indicates that the initial critical magnetic fields of κ-(ET)$_2$I$_3$ and κ-(ET)$_2$Cu(NCS)$_2$ for both $\|$ and $\perp$ components differ primarily by the scaling in T_c. The slope of $H_{c2\perp}$ is somewhat larger for the κ-(ET)$_2$Cu(NCS)$_2$ salt, giving a smaller anisotropy ratio ($H_{c2\|}/H_{c2\perp}$) than that of the iodide salt (19 versus 26). There is also an interesting similarity with β^*-(ET)$_2$I$_3$ in that the slope of $H_{c2\|}$ for the low-temperature segment (38 kOe/K) is comparable to that of the parallel component for β^*-(ET)$_2$I$_3$ [from Table 4.6, $-(dH_{c2\|}/dT) = 35$–48 kOe/K]. This similarity and the comparably small coherence length of ~ 10 Å for the β^* phase suggest that 2D Josephson coupling also occurs in β^*-(ET)$_2$I$_3$, but in this case, because there is no observed change in slope, the 2D character extends up to the onset of superconductivity. Although there is small anisotropy in $H_{c2\|}$ determined along different axes in the normal phase, the deuterated phase exhibits a ratio $H_{c2\|b}/H_{c2\|c}$ of more than a factor of 2.[175] $H_{c2\|}$ for the deuterated phase also exhibits the presumed 3D-to-2D change in both axial components. Lower critical magnetic fields of κ-(ET)$_2$Cu(NCS)$_2$ have been reported[177] to be ~ 15 Oe for $H_{c1\|}$

and ~ 45 Oe for $H_{c1\perp}$ at 4.2 K. These values, combined with H_{c2}, yield $\kappa_\perp \approx 20$ and $\kappa_\parallel \approx 200$, which are about double the values estimated for β-$(ET)_2AuI_2$.

Both $H_{c2\parallel}(T)$ and $H_{c2\perp}(T)$ for $(ET)_4Hg_{2.89}Br_8$ exhibit strong positive curvature over the measured temperature range from T_c (near 4 K) down to ~ 1 K for the $\perp$ component and ~ 3 K for the $\parallel$ component.[114] To date, this salt is the first example of an organic superconductor with such a temperature behavior for both components of H_{c2}. The slopes evaluated near T_c, which yield the parameters listed in Table 4.7, are $-(dH_{c2\parallel}/dT)_{T_c} = 13$ kOe/K and $-(dH_{c2\perp}/dT)_{T_c} = 0.65$ kOe/K. However, extrapolations of the data at the low temperatures yield the remarkably large slopes of 100 kOe/K and 18 kOe/K for the $\parallel$ and $\perp$ components, respectively, as $T \rightarrow 0$ K.[114] The coherence length $\xi_\perp = 17$ Å is about half the unit cell dimension in the same direction ($c \approx 37$ Å), which suggests that some of the initial curvature in $H_{c2\parallel}$ may be due to a 3D-to-2D crossover similar to that observed in $H_{c2\parallel}$ for the κ-$(ET)_2Cu(NCS)_2$ superconductor. The anisotropy ratio $H_{c2\parallel}/H_{c2\perp}$ for $(ET)_4Hg_{2.89}Br_8$ is comparable to that for κ-$(ET)_2Cu(NCS)_2$. On the other hand, the slopes of H_{c2} evaluated near T_c are more comparable to those determined resistively[140] for β-$(ET)_2IBr_2$. This similarity may not be significant, however, because it is obvious from the available duplicate measurements of H_{c2} for a given phase such as β-$(ET)_2IBr_2$ that the reproducibility among measurements is not especially good.

It has been reported that two breaks appear in the resistivity of $(ET)_4Hg_{2.89}Br_8$ measured with applied magnetic fields.[116] One of these occurs at high-magnetic fields and corresponds to the superconducting phase with $T_c \approx 4$ K in zero applied field. The second appears at lower fields and corresponds to a superconducting phase with $T_c \approx 3$ K in zero applied field. This is the evidence previously mentioned that this salt may contain an additional superconducting phase. In the absence of chemical analyses and measurements of a Meissner effect, it is not clear whether this possible second phase is a major superconducting constituent of the test specimen, a minor constituent, or simply a chemical contaminant.

Three articles dealing with measurements of the specific heat of κ-$(ET)_2Cu(NCS)_2$ near T_c have been published. One describes the observation of a "kink" in the specific heat near T_c but presents no quantitative data.[184] The second describes a precise determination of the electronic specific-heat coefficient γ and a determination of the specific-heat discontinuity ΔC at T_c.[185] A coefficient $\gamma = 25 \pm 3$ mJ/mol K^2 was obtained from measurements with oriented single crystals and magnetic fields applied along the $H_{c2\perp}$ direction (H $\parallel$ a) to quench the superconducting state to low temperatures. This value of γ is essentially identical to those reported for β-$(ET)_2I_3$ and β-$(ET)_2AuI_2$.[153,154] The specific-heat discontinuity at T_c, relative to T_c, was found to be $\Delta C/T_c \cong 50$ mJ/mol K^2 as determined from the difference in C/T_c at zero applied magnetic field and an applied field of 10 Tesla (100 kOe). These quantities give $\Delta C/\gamma T_c \cong 2$, as opposed to a ratio of 1.43 predicted for a weak-coupling BCS superconductor. The ratio of 2 was considered to be a lower limit owing to the broadness of the superconducting transition. Extrapolation of the data to a superconductor with an ideally sharp transition was reported to give $\Delta C/\gamma T_c \approx 2.8$. Such a ratio implies strong electron-phonon coupling, as exemplified by Pb and several other conventional superconductors.[186] A third article[180] on the specific heat near T_c of κ-(ET)$_2$Cu(NCS)$_2$ reports a discontinuity $\Delta C/T_c \approx 51 \pm 5$ mJ/mol K^2, in good agreement with the previous determination. This article, however, reports no independent determination of γ.

Preliminary determinations of the superconducting energy gap $\Delta(0)$ of κ-(ET)$_2$Cu(NCS)$_2$ by tunneling measurements have shown considerable variability in the experimental gap, possibly caused by strong anisotropy in the gap, but in at least one measurement have yielded the ratio $2\Delta(0)/kT_c = 4.5$, as opposed to the weak-coupling BCS value of 3.53.[173,187] This result is in good agreement with the specific heat measurements according to the strong-coupling theory developed by Marsiglio and Carbotte.[186] In conventional BCS theory, which is a description of superconductivity with weak electron-phonon coupling (coupling parameter $\lambda_c \ll 1$), there is a unique relationship between T_c, $\Delta(0)$, and $\Delta C/\gamma$.[152] For intermediate or strong-coupling superconductors ($\lambda_c \sim 1$ and larger[188,189]), this unique relationship no longer exists,[19] so that the properties of each superconductor are separately determined by the particular phonon spectrum of the material (assuming that the formation of Cooper pairs is achieved by electron-phonon coupling). In the strong-coupling theory of Marsiglio and Carbotte,[186] reasonable adjustments to the BCS properties can be obtained by approximate analytical expressions of the type

$$\frac{\Delta C}{\gamma T_c} = 1.43 + f_1\left(\frac{T_c}{\omega_{ln}}\right) \tag{4.12}$$

$$\frac{2\Delta(0)}{kT_c} = 3.53 + f_2\left(\frac{T_c}{\omega_{ln}}\right) \tag{4.13}$$

in which f_1 and f_2 are functions of a strong-coupling variable T_c/ω_{ln}, and ω_{ln} is a measure[189] of the average phonon energy. On the basis of this theory, $\Delta C/\gamma T_c = 2.8$ gives $T_c/\omega_{ln} \approx 0.15$, $2\Delta(0)/kT_c = 4.5$ gives $T_c/\omega_{ln} \approx 0.13$, both in reasonably good agreement, and $\lambda_c \approx 1.6$,[147] which indicates that electron-phonon coupling in κ-(ET)$_2$Cu(NCS)$_2$ is similar to that in Pb, which has $T_c = 7.2$ K. These data for the specific-heat discontinuity and the superconducting energy gap of κ-(ET)$_2$Cu(NCS)$_2$ are presently the only consistent experimental evidence of strong electron-phonon coupling in the organic superconductors as derived from measurements pertaining to the superconducting energy gap.

The κ-Phase Superconductors under Applied Pressures

The effects of applied pressure for the κ-phase salts have been reported for κ-(ET)$_2$Cu(NCS)$_2$, (MDT-TTF)$_2$AuI$_2$, (ET)$_4$Hg$_{2.89}$Br$_8$, and (ET)$_4$Hg$_{3-x}$Cl$_8$. κ-(ET)$_2$Cu(NCS)$_2$[190,191] and (MDT-TTF)$_2$AuI$_2$[10] exhibit the typical decrease in T_c with increase in pressure, $dT_c/dP < 0$, as is observed for most of the organic superconductors. The pressure dependence of T_c for (MDT-TTF)$_2$AuI$_2$ is about the same as that determined for the ET-based β-phase superconducting salts, ~ -1 K/kbar. However, the pressure dependence of T_c for the κ-(ET)$_2$Cu(NCS)$_2$ phase,

$dT_c/dP \approx -3$ K/kbar, is the largest yet observed for any presently known superconductor.[190]

Studies of the superconducting transition of $(ET)_4Hg_{2.89}Br_8$ under applied pressure, with T_c occurring near 4 K at ambient pressure, are significant for two reasons: they have established that the superconductivity is not due to the presence of elemental Hg (T_c near 4.2 K) in the material, and that $dT_c/dP > 0$.[117] A pressure-temperature cycling of elemental mercury (α-Hg) in a material will transform it into another phase (β-Hg), which has substantially different superconducting properties. Because the inductive superconducting signal was large in $(ET)_4Hg_{2.89}Br_8$, and the sample possessed a Meissner effect of ~ 40 percent, such a change in properties could have been detected with ease. However, no such transformation was observed, and therefore it can be concluded that the superconductivity in the crystal specimen is an intrinsic property of the phase. With the application of hydrostatic pressures to $(ET)_4Hg_{2.89}Br_8$, the T_c increased up to a maximum of 6.7 K at ~ 3.5 kbar. The initial rate of increased dT_c/dP was ~ 0.7 K/kbar. With further application of pressure, T_c remained approximately constant and then began to decrease slowly above 4 K. It was suggested[117] that the decrease in T_c with pressure above 4 kbar indicates a change in structure of the phase, but this has not yet been confirmed by structural studies. It is clear that much more research is needed on $(ET)_4Hg_{2.89}Br_8$ to fully explain its unusual structural aspects and superconducting properties.

κ-like $(ET)_4Hg_{3-x}Cl_8$ is superconducting only under applied pressures.[118] Near ambient pressures, the resistivity of this phase undergoes some kind of an MI transition in the region of 4–20 K. This MI transition is suppressed by increasing pressures, and at ~ 12 kbar, the crystal specimen exhibits a fairly sharp resistive superconducting transition with an onset of 2.2 K, a transition midpoint of 1.8 K, and a completion near 1.5 K. Magnetic fields of the order of several kOe apparently restore the normal state, but specific data on the upper critical magnetic fields have not been reported. As the pressure is increased above 12 kbar, the T_c of the sample is suppressed, initially at a rate of about -0.05 K/kbar. At a pressure of 29 kbar, a second resistive superconducting transition appears with an onset of 5.4 K and a transition midpoint of 5.3 K. After the initial sharp drop in resistance, however, the transition tails off with a small but finite resistance extending below 1.5 K. This second superconducting transition is attributed to the occurrence of a new phase.

BCS or Non-BCS Superconductivity

The main features of the superconducting properties of the presently known organic superconductors are embodied in the previous discussions of the β-phase and κ-phase superconductors, and it is of interest at this juncture to consider briefly the origin of superconductivity in these materials. Are these conventional BCS superconductors, possibly strong-coupled BCS superconductors, or something more exotic? Certainly the material source of superconductivity—the organic salt—is exotic in relation to its recent discovery. But, what else is new? Deducing the mechanism of superconductivity in these materials is enormously complicated by the lack of consistent or definitive experimental results.

The central question of the superconducting mechanism is, of course, the pairing mechanism. The isotope effect as a signature of electron-phonon coupling of Cooper pairs is presently the only clean-cut experiment capable of probing this issue, but the results nevertheless can be inconclusive. In BCS theory,[152] it is easily demonstrated that $T_c \propto M^{-\alpha}$, where M is an elemental mass and $\alpha \approx 1/2$, because of the inverse relationship of vibrational frequencies and the square-root of a mass. However, any scientist who has carried out a normal coordinate analysis of vibrational modes knows that this simple relationship, although qualitatively correct in its general aspects, is considerably over simplified. The β-phase superconductors frequently exhibit no definite positive or negative isotope effects, owing at least in part to the sample-dependent variations in T_c observed with the natural salts. In any case, the isotopic substitutions have been confined generally to replacements of hydrogen by deuterium, which is not necessarily the most important mass substitution in the electron-phonon coupling. A normal isotope effect (a decrease in T_c) with deuterium substitution has been reported for β*-(ET)$_2$I$_3$.[154] In the case of deuterium substitution in κ-(ET)$_2$Cu(NCS)$_2$, there is a definite positive rather than negative isotopic shift in T_c.[175] However, in detailed theories of electron-phonon coupling, the isotopic shifts in T_c may be negative, zero, or even positive.[152] Thus, one concludes that the observation of a negative isotopic shift in T_c is reasonable evidence that phonons are involved, but the absence of a shift or an opposite shift does not necessarily rule out the involvement of phonon coupling.

Tests of the BCS theory, which is remarkably absent of adjustable parameters, require reliable values for the ratios $\Delta C/\gamma T_c$ and $2\Delta(0)/kT_c$ and reliable values of the critical magnetic field as a function of temperature. But consider the evidence presently available on the nature of the superconducting energy gap. There is no observed heat-capacity discontinuity at T_c in β-(ET)$_2$I$_3$,[137] suggesting the absence of a superconducting energy gap, but there is a heat-capacity discontinuity in β-(ET)$_2$AuI$_2$ that is consistent with conventional BCS theory.[154] However, direct measurements of the superconducting energy gap in β-(ET)$_2$I$_3$ indicate a gap of more than a factor of two larger than that predicted by BCS theory.[156] Furthermore, measurements of the superconducting energy gap in β-(ET)$_2$AuI$_2$ indicate that, on the one hand, the gap is more than a factor of four larger than the conventional BCS gap[155] but, on the other hand, not much larger than the weak-coupling BCS value.[156] These experimental results, therefore, prove nothing else but the fact that there are serious experimental problems involved in the measurements of superconducting properties. The situation for the κ-phase superconductors is intuitively more satisfying, but it is deceptive because of the lack of duplicate measurements and the absence of studies of several different materials. Both specific-heat measurements of the discontinuity at T_c[185] and determinations of the superconducting energy gap[187] in κ-(ET)$_2$Cu(NCS)$_2$ indicate strong-coupling superconductivity, but the gap determinations exhibit such large variations that they must be considered to be rather inconclusive. These experimental results give at least a hint that this organic superconductor might be describable within the context of strong electron-phonon coupling.

The only consistent evidence for something beyond weak-coupling BCS superconductivity in these organic superconductors is found in the temperature dependencies of H_{c2}. For the β-phase superconductors, $H_{c2\parallel}(T)$ and $H_{c2\perp}(T)$ of β-$(ET)_2I_3$, β-$(ET)_2AuI_2$, and β-$(ET)_2IBr_2$ exhibit linear behavior over the entire temperature range of their determinations, with one exception, and $H_{c2\perp}(T)$ of β*-$(ET)_2I_3$ exhibits positive curvature. $H_{c2}(T)$ of a weak-coupling BCS superconductor, however, exhibits negative curvature. These contrasting behaviors are compared in Figure 4.13. The critical fields in this figure are plotted as reduced (dimensionless) upper critical magnetic fields h*(t) expressed as a function of a reduced temperature t, where

$$h^*(t) = \frac{-H_{c2}(T)}{T_c \frac{dH_{c2}}{dT}\Big|_{T_c}} = \frac{-H_{c2}(t)}{\frac{dH_{c2}}{dt}\Big|_{t=1}} \quad (4.14)$$

$$t = \frac{T}{T_c} \quad (4.15)$$

and $H_{c2}(T)$ is the experimental or theoretical (BCS) critical magnetic field. Expressed as h*(t), the Ginzburg-Landau (GL) upper critical magnetic field is a

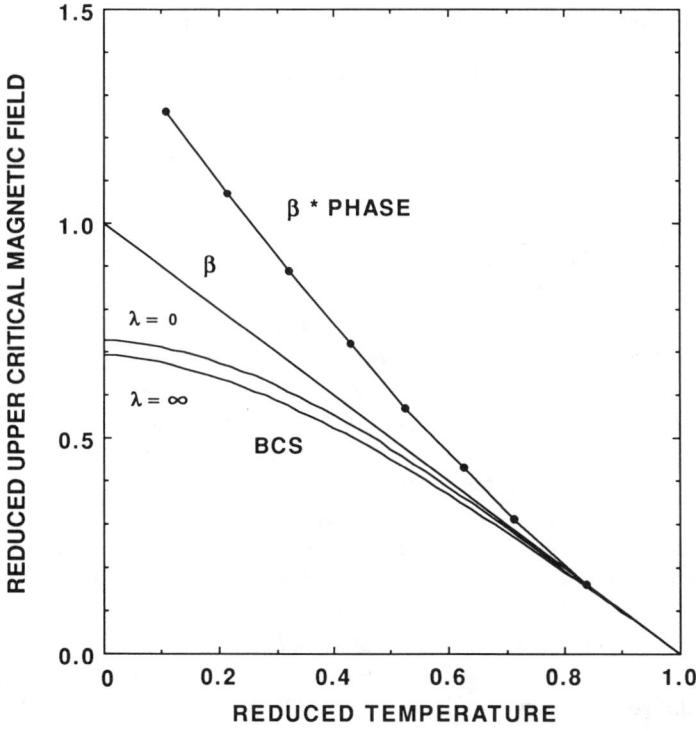

Figure 4.13 Reduced upper critical magnetic fields h*(t) as a function of reduced temperature $t = T/T_c$ for weak-coupling pure (λ = 0) and dirty (λ = ∞) BCS superconductors, the β-$(ET)_2X$ ($X^- = I_3^-$, AuI_2^-, IBr_2^-) superconductors, and the β*-$(ET)_2I_3$ superconductor for the ⊥ component. The values for the β* phase were taken from Bulaevskii et al.[172]

straight line $h^*(t) = (1 - t)$ of slope -1. Because of its linear behavior, $h^*(t)$ for the β-phase superconductors is represented in Fig. 4.13 by the GL straight line. The values of $h^*(t)$ for the ⊥ component of β-$(ET)_2I_3$ in this figure have been taken from the graphical data reported by Bulaevskii et al.[172] Figure 4.13 conveniently illustrates the differences between the experimental upper critical magnetic fields of the β-phase superconductors and those of a weak-coupling BCS superconductor.[192]

The behavior of $h^*(t)$ for a BCS superconductor is represented in Fig. 4.13 by curves for two different limits of the electron-scattering mean-free path l:

$$l \to 0, \lambda \equiv \frac{\xi}{l} = \infty \quad \text{(dirty limit),} \quad (4.16)$$

$$l \to \infty, \lambda = 0 \quad \text{(clean or pure limit).} \quad (4.17)$$

These curves are taken from one article[192] among a group of publications known collectively as the WHH (Werthamer, Helfand, and Hohenberg) theory of superconducting critical magnetic fields.[148] The BCS representations of $h^*(t)$ apply to an isotropic 3D superconductor (spherical Fermi surface) with electrons interacting according to the weak-coupling BCS potential and with magnetic fields interacting only with the orbital motions of the electrons. The limiting fields are

$$h^*(t \to 1) = (1 - t) \quad (4.18)$$

$$h^*(0) = 0.727 \ (\lambda = 0) \quad (4.19)$$

$$h^*(0) = 0.693 \ (\lambda = \infty) \quad (4.20)$$

Thus, the BCS curves near T_c are equivalent to the GL critical magnetic fields, and at $T = 0$ they are about 40 percent lower than the GL values.

Inclusion of the interactions of the magnetic field with the singlet-state electron spins of the Cooper pairs (the paramagnetic effect) decreases $H_{c2}(T)$ at $T \ll T_c$ and thus enhances the negative curvature in $h^*(t)$ of a BCS superconductor.[193] The inclusion of anisotropy in the Fermi surface leads to anisotropic scaling of $H_{c2}(T)$ for different axial orientations of the applied magnetic field, which scaling on normalization to a reduced field does not alter $h^*(t)$, but it also leads to an enhancement of $H_{c2}(T)$ at low temperatures, thus diminishing the negative curvature in $h^*(t)$.[194] The influence of such effects, however, may be negated by impurity scattering (finite l). Thus, in the absence of some combination of unusual circumstances, $h^*(t)$ of a weakly coupled anisotropic BCS superconductor is expected to exhibit a substantial negative curvature, although the curvature may be less pronounced than that of the isotropic BCS superconductor.

The linear behavior of the ambient-pressure β-phase superconductors and especially the positive curvature in $H_{c2}(T)$ of β*-$(ET)_2I_3$ indicates that these materials are *not* weak-coupling BCS superconductors. However, one must consider the experimental precision required to draw an unequivocal conclusion. For a BCS superconductor, $h^*(t)$ is closely linear to ≈ 0.8, and the divergence from linearity is only ~ 4 percent at $t \approx 0.6$. A precision of about 10 percent in the determination of $h^*(t)$ (which is established by the propagated precision in both $H_{c2}(T)$ and its slope near T_{c_3}) is a reasonable precision for a reliable comparision of the temperature dependence of h*(t), and this means that $H_{c2}(T)$ is required to be measured down to relative temperatures below t ≈ 0.4. For this value of the reduced temperature, determinations of $H_{c2}(T)$ for β-$(ET)_2I_3$ (T_c ≈ 1.4 K) should be carried out to temperatures below ~ 0.5 K, which is the lower limit of temperatures achieved by use of pumped He^3. Thus, the critical fields, determined down to ~ 0.6 K, reported by Tokumoto et al.[146] may be insufficient in establishing linearity in $H_{c2}(T)$ of β-$(ET)_2I_3$. Bulaevskii et al.,[172] in fact, have determined that the critical fields of β-$(ET)_2I_3$ reported by Ginodman et al.,[150] which fields exhibit negative curvature, are in good agreement with the predictions of weak-coupling BCS theory. Determinations of the critical fields of the other β-phase superconductors have been carried out to sufficiently low temperatures that the occurrence of linearity and positive curvature for β*-$(ET)_2I_3$ is strong evidence against weak-coupling BCS superconductivity. Similar evidence against weak coupling for the κ-phase superconductors is provided by the linear critical magnetic fields of κ-$(ET)_2I_3$ and by the critical magnetic fields of very strong positive curvature for κ-$(ET)_2Cu(NCS)_2$ and $(ET)_4Hg_{2.89}Br_8$. However, this evidence for the last two salts is complicated by the possible occurrence of a low-temperature phase transformation in κ-$(ET)_2Cu(NCS)_2$, a phase contamination in $(ET_4Hg_{2.89}Br_8$, and possibly by a change in the superconducting state in both salts from a volume superconductor to a Josephson-coupled 2D superconductor.

If not weak-coupling BCS superconductors, are the β- and κ-phase organic metals strong-coupling superconductors? Certainly some of the experiments pertaining to the energy gap support the proposition of strong-coupling BCS superconductivity (electron-phonon coupling parameter $\lambda_c \geq 1$). The question of strong coupling, however, is not easily answered in the context of a general description because with strong coupling one must consider the individual material properties and the electron-phonon spectrum of each superconductor. The basis of strong-coupling BCS superconductivity is the Eliashberg theory and the Eliashberg electron-phonon spectral function, $\alpha^2F(\omega)$, which describes the electron-phonon coupling strength and density of phonon states (ω is a phonon frequency).[195] The issue of strong-coupling superconductivity in the organic metals has been analyzed by Bulaevskii,[19] and from this analysis it seems that strong coupling might explain at least the qualitative features of the temperature behavior of $H_{c2}(T)$. For example, positive curvature in h*(t) can occur over an extended range of temperatures for rather extreme values of the electron-phonon coupling parameter, $\lambda_c \geq 4$. Nevertheless, even with very strong coupling there is negative curvature in h*(t) at very low temperatures.

At the present time, it is premature to conclude that the organic superconductors involve some kind of exotic new superconducting mechanism. On balance, the most consistent evidence to date, the critical magnetic fields $H_{c2}(T)$, suggests that they may be examples of very strong electron-phonon coupling superconductors, but the evidence is inconclusive. The authors believe that a major problem in establishing the superconducting mechanism lies in the synthesis of high-quality materials with consistent physical properties. There is ample evidence already cited that material properties can vary from one sample specimen to another. There is also the problem of appropriately defining T_c because of the intrinsic broadness of the superconducting transitions. The resistive midpoint, for example, may be an inappropriate choice in the determination of the critical magnetic fields.

Superconductivity in Other Salts of the ET/I System

As is evident from the previous discussions, the ET-donor molecule provides a remarkably large number of charge-transfer salts composed of various homonuclear anionic species of iodine. These are summarized in Table 4.8 as a focused display of the presently known phases comprising this system. Most of these salts have been discussed, or at least briefly mentioned, in the previous sections, but there are additional aspects of interest in this system pertaining to superconductivity that is induced or altered by chemical compositional effects and by thermal processing. Most of these aspects essentially relate to the production of elevated superconductivity ($T_c > 2$ K) in ambient-pressure β-$(ET)_2I_3$.[196]

We first consider some compositional effects, which we believe to have some bearing on superconductivity in the γ-phase salt. γ-$(ET)_3(I_3)_{2.5}$ is reported to be an ambient-pressure superconductor with a resistive superconducting onset of ~ 4 K, completion in the range 1–2 K, and a transition midpoint of 2.5 K.[73,88] No studies of its volume superconducting properties have been reported, but the resistive superconducting transition does exhibit upper critical magnetic field behavior typical of organic superconductors, thus confirming the presence of some quantity of a superconducting phase. The γ-phase salt was identified shortly after the discovery

TABLE 4.8 Conductive States, Metal-Insulator Transition Temperatures (T_{MI}), and Superconducting Transition Temperatures (T_c) of the Charge-Transfer Salts of the ET/I System

Salt	State[a]	T_{MI} (K)	T_c (K)[b]	Reference
α-$(ET)_2I_3$	M	135		74
α_t-$(ET)_2I_3$	M		2–8	78
β-$(ET)_2I_3$	M		1.4	52
β^*(or β_H)-$(ET)_2I_3$	M		8 (0.5 kbar)	67
β-$(ET)_{2-x}(MET)_xI_3$	M		4.6	104
β_d-$(ET)_2I_3$	M	140		105
β_t-$(ET)_2I_3$	M		4.5	198
γ-$(ET)_3(I_3)_{2.5}$	M		2.5	88
δ-$(ET)(I_3)$	M	130		89
ϵ-$(ET)_2(I_3)(I_8)_{0.5}$	M[c]			89
ζ-$(ET)_2(I_2)(I_8)$	S			90
η-$(ET)_2(I_3)(I_5)$	M[c]			91
θ-$(ET)_2(I_3)$	M		3.6	95
κ-$(ET)_2(I_3)$	M		3.6	97
λ_d-$(ET)_2I_3$	M	220		105

[a] Metallic (M) or semiconducting (S) behavior at 300 K.
[b] At ambient pressure unless specified otherwise.
[c] Either M or S depending on stoichiometry (see ref. 91).

of superconductivity in β-(ET)$_2$I$_3$, and therefore, it was considered to be the likely source of the resistive drops frequently observed near 4 K in the β phase.[73] However, a variety of evidence now indicates that the resistive drops near 4 K in β-(ET)$_2$I$_3$ are intrinsic to the β phase.[196] Furthermore, the resistivity measurements for the γ phase suggest that the source of superconductivity attributed to this phase is instead due to the presence of β-(ET)$_2$I$_3$ in an impure phase.

The resistivity behavior reported[88] for γ-(ET)$_3$(I$_3$)$_{2.5}$ from 300 K to low temperatures is shown in Fig. 4.14. This figure illustrates the resistivity of two different crystal specimens that exhibit pronounced resistive "humps" in the region of 100 K. These crystals have metallic resistance down to ~150 K, semiconductive resistance from ~150 K to ~100 K, then metallic behavior to lower temperatures, and finally superconducting transitions near 4 K. A variety of different crystals was found to exhibit more or less similar behavior. Experiments showed that different placements of the voltage probes on the crystal specimens could either enhance or diminish the resistive humps near 100 K, and these observations were taken as evidence that the humps were unrelated to the presence of the γ phase but indications of the coexistence of different phases within the apparent single crystals.[73] Certainly, these data indicate the presence of at least two phases, one being semiconducting and the other being metallic and superconducting. No structural evi-

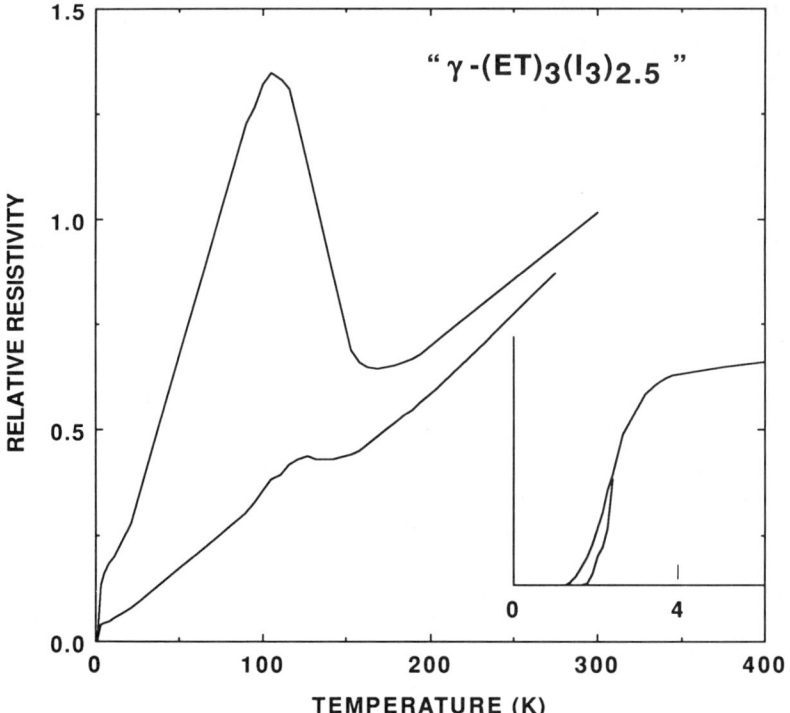

Figure 4.14 Relative electrical resistivity, ρ(T)/ρ(300 K), of two so-called λ-phase crystals as reported by Yagubskii et al.[88] The inset shows the resistive superconducting transitions with onsets near 4 K, completions near 1–2 K, and midpoints near 2.5 K.

dence was presented to support the presence of the γ phase, however. It seems likely, in fact, that the occurrence of superconductivity near 4 K was simply accepted as evidence of some superconducting phase other than β-(ET)$_2$I$_3$, with the γ phase being the only known possible alternative at the time. With our present knowledge, one could propose several other possible sources of superconductivity near 4 K, for example, κ-(ET)$_2$I$_3$ or θ-(ET)$_2$I$_3$. The most reasonable interpretation of the experiments, however, is that the semiconducting component was α-(ET)$_2$I$_3$, which has a strong metal-to-semiconductor transition near 135 K, and the superconducting component was β-(ET)$_2$I$_3$, which in mixed α-β "single" crystals is superconducting at ambient pressure with elevated T_c's in the range 2.5–6.9 K.[168] Thus, although there is structural evidence for the existence of the γ phase salt, there is no persuasive evidence that this is a superconducting phase.[196]

α-β crystals with elevated T_c's occur not only naturally in the synthesis of (ET)$_2$I$_3$ salts, but they were first detected as products of a chemical modification in which crystals were synthesized from ET in the presence of the PT-donor molecule, where PT represents bis(propylenedithio)tetrathiafulvalene.[168] No evidence was found that PT was incorporated into the crystal products, but the synthesis seemed to promote the production of mixed α-β crystals. The original intent of this compositional modification was to prevent or reduce the normal ethylene group disorder in the synthesis of β-(ET)$_2$I$_3$ by the incorporation of the PT-donor molecules, thus possibly achieving a β-type phase with a higher T_c at ambient pressure. This objective was achieved by a similar chemical modification in the synthesis of β-(ET)$_{2-x}$(MET)$_x$I$_3$, where MET is 4, 5-methylenedithio-4′, 5′-ethylenedithiotetrathiafulvalene.[104] X-ray structural studies of this salt gave no direct evidence for the presence of the MET donor, but they confirmed that the phase possessed the prototypical structure of the perfectly ordered β*-(ET)$_2$I$_3$ phase, the only difference being a slight (~ 2 %) disorder in the I$_3^-$ anions. This phase has an inductively determined superconducting onset of ~ 4.6 K, and it represents the first and only well-documented example of a β*-phase material that is stable at ambient pressure and temperature without application of any special temperature-pressure cycling procedure.

A different attempt to modify the disorder in β-(ET)$_2$I$_3$ by compositional changes was responsible for the discovery of the θ-(ET)$_2$I$_3$[95,96] and κ-(ET)$_2$I$_3$[97,98] superconducting phases, as previously discussed. Only resistive studies of the superconducting transition in the θ phase have been reported, so that it is not known whether or not this phase is a bulk superconductor. Determinations of the upper critical magnetic fields[100] and the superconducting transition curves with an onset of 4 K and transition midpoint of 3.6 K[95,100] suggest that the superconducting properties of θ-(ET)$_2$I$_3$ are very similar to those of κ-(ET)$_2$I$_3$, except that not all crystal specimens of the θ phase exhibit a complete superconducting transition.[100] Although we have no compelling reason to doubt that superconductivity in the θ and κ phases is a genuine attribute of these phases, one must wonder if a β-phase contaminant might be the source of the superconductivity.

Thermal processing of several salts of the ET/I system converts nonsuperconducting phases into superconducting phases and even converts β-(ET)$_2$I$_3$ into a higher temperature ambient-pressure superconductor. These thermal conversions require temperatures near 400 K, and in all cases the product seems to be some form of β-$(ET)_2I_3$ with an elevated T_c at ambient pressure. The iodine-rich ε-$(ET)_2(I_3)(I_8)$ phase was the first of these thermal conversions to be reported.[94] This phase when heated several hours in vacuum yields crystals that exhibit superconducting transitions with onsets near 7–8 K, completions near 2 K, and transition midpoints ranging from 2.5 K to 7 K.[93,94,161,197] These wide transitions indicate considerable inhomogeneity in T_c. X-ray analyses of the product showed only the presence of the β-phase and the existence of twinned and mosaic crystals. The ζ–$(ET)_2(I_2)(I_8)$ phase apparently undergoes a similar thermal transformation, but no details of this conversion have been reported.[90]

Nonsuperconducting α-$(ET)_2I_3$ can be thermally converted to a phase, denoted as $α_t$-$(ET)_2I_3$, which has a β-type structure and elevated superconductivity within the range of 2–8 K at ambient pressure.[20,78,198,199] It has been shown that with prolonged heating over a period of several days resistive completions of the superconducting transition can be obtained near 5–7 K.[78,198] Inductive measurements confirm that the $α_t$ phase is a volume superconductor.[198,199] Apparently no trace of the original α phase remains in the thermally tempered product, and this indicates that the elevation of the superconducting transition of the β-phase product is not directly related to the occurrence of mixed α-β crystals. The superconducting phase denoted as $β_t$-$(ET)_2I_3$ with $T_c \approx 4.5$ K is a thermally converted product of the ambient-pressure β-phase salt.[198] Inductive measurements indicate that this is a bulk superconducting phase. Finally, the salt θ-$(ET)_2I_3$, or at least a nonsuperconducting variant of this phase, has been observed to undergo a thermal conversion to a β-type bulk superconducting phase with a superconducting onset of ~ 3.6 K.[196] Thermal transformations are not necessarily restricted to salts of the polyiodides. Thermal processing of nonsuperconducting α-$(ET)_2IBr_2$ easily converts this phase into superconducting β-$(ET)_2IBr_2$.[200] In this case, there is no problem with structural order-disorder or higher T_c phases, so that the β-phase product is simply the conventionally synthesized β-phase salt. The frequency with which elevated superconductivity occurs at ambient pressure in β-$(ET)_2I_3$ is really quite a remarkable circumstance, one well worth further study. It is not clear what the structural differences are between the β*-phase superconductor and the β-type superconductors with elevated T_c's. It is likely that crystallographic order-disorder plays a role, but mechanical effects, such as plastic deformations from twinning, may be the determining factors.

The α-$(ET)_2(NH_4)Hg(SCN)_4$ and $β_m$-$(BEDO-TTF)_3Cu_2(NCS)_3$ Salts

The α-$(ET)_2(NH_4)Hg(SCN)_4$[201] and $β_m$-$(BEDO-TTF)_3Cu_2(NCS)_3$[11] salts, where BEDO-TTF represents the oxygen-containing organic donor molecule bis(ethylenedioxy)tetrathiafulvalene, are very recently discovered superconductors with superconducting transitions near 1 K. Despite their low T_c's, however, both are important discoveries, the former because it establishes a *new structural class* of organic superconductors, the α-phase class, and the latter because it is the first oxygen-containing organic superconductor.

α-(ET)$_2$(NH$_4$)Hg(SCN)$_4$ has the packing motif of nonsuperconducting α-(ET)$_2$I$_3$ for the organic donor molecules and is the second of two "α-phase" superconductors, the first being θ-(ET)$_2$I$_3$ (T_c = 3.6 K). α-(ET)$_2$(NH$_4$)Hg(SCN)$_4$ is metallic at all temperatures from 300 K to below with σ(300 K) ≈ 5 (Ω cm)$^{-1}$. The superconducting transition as determined by rf penetration depth measurements has an onset of 1.15 ± 0.05 K and a midpoint near 0.8 K.[201] Specific heat measurements have definitely established α-(ET)$_2$(NH$_4$)Hg(SCN)$_4$ as a volume superconductor.[202] The initial slope of the upper critical magnetic field for H$_{||}$, $-(dH_{c2||}/dT)_{T_c}$, as determined from specific heats measured in applied magnetic fields[202] appears to be about a factor of two or more larger than those determined by magnetoresistance measurements for any other organic superconductor except β*-(ET)$_2$I$_3$. Further measurements by other methods, however, are required to confirm this result.

β$_m$-(BEDO-TTF)$_3$Cu$_2$(NCS)$_3$ has a donor-molecule packing arrangement similar to all of the other known salts of the BEDO-TTF donor, but thus far it is the only superconducting salt of this donor. Inductive studies by rf penetration depth measurements give T_c = 1.06 ± 0.02 K for the onset of the superconducting transition.[11] Measurements of the specific heat indicate a distinct but broad anomaly, thus confirming that this salt is a volume superconductor.[203] The discovery of an oxygen-containing organic superconductor provides exciting new opportunities in the study of organic superconductivity and the potential for a unification of the organic and high-T_c ceramic oxide superconductors.

REFERENCES

1. Reviews of the (TMTSF)$_2$X salts include the following: (a) Greene, R. L.; Chaikin, P. M. *Physica* **1984**, *126B*, 431; (b) Jérome, D.; Schultz, H. J. *Adv. Phys.* **1982**, *31*, 299; (c) Williams, J. M.; Beno, M. A.; Wang, H. H.; Leung, P. C. W.; Emge, T. J.; Geiser, U.; Carlson, K. D. *Acc. Chem. Res.* **1985**, *18*, 261; (d) Williams, J. M. *Prog. Inorg. Chem.* **1985**, *33*, 183; (e) Williams, J. M.; Carneiro, K. *Adv. Inorg. Chem. Radiochem.* **1985**, *29*, 249.
2. Jérome, D.; Mazaud, A.; Ribault, M.; Bechgaard, K. *J. Phys. Lett. (Paris)* **1980**, *41*, L95.
3. Peierls, R. E. *Quantum Chemistry of Solids*, Clarendon Press: New York, **1964**.
4. Carneiro, K.; Olsen, M.; Rasmussen, F. B.; Jacobsen, C. S. *Phys. Rev. Lett.* **1981**, *46*, 852; Bechgaard, K.; Rasmussen, F. B.; Olsen, M.; Rindorf, G.; Jacobsen, C. S.; Pedersen, H. J.; Scott, J. C. *J. Am. Chem. Soc.*, **1981**, *103*, 2440.
5. Gubser, D. U.; Fuller, W. W.; Poehler, T. O.; Stokes, J.; Cowan, D. O.; Lee, M.; Block, A. N. *Mol. Cryst. Liq. Cryst.* **1982**, *79*, 225; Garoche, P.; Brusetti, R.; Bechgaard, K. *Phys. Rev. Lett.* **1982**, *49*, 1346.
6. Williams, J. M.; Wang, H. H.; Emge, T. J.; Geiser, U.; Beno, M. A.; Leung, P. C. W.; Carlson, K. D.; Thorn, R. J.; Schultz, A. J.; Whangbo, M.-H. *Prog. Inorg. Chem.* **1987**, *35*, 51.
7. Urayama, H.; Yamochi, H.; Saito, G.; Nozawa, K.; Sugano, T.; Kinoshita, M.; Sato, S.; Oshima, K.; Kawamoto, A.; Tanaka, J. *Chem. Lett.* **1988**, 55.
8. (a) Gärtner, S.; Gogu, E.; Heinen, I.; Keller, H. J.; Klutz, T.; Schweitzer, D. *Solid*

State Commun. **1988**, *65*, 1531; (b) Carlson, K. D.; Geiser, U.; Kini, A. M.; Wang, H. H.; Montgomery, L. K.; Kwok, W. K.; Beno, M. A.; Williams, J. M.; Cariss, C. S.; Crabtree, G. W.; Whangbo, M.-H.; Evain, M. *Inorg. Chem.* **1988**, *27*, 965.

9. Papavassiliou, G. C.; Mousdis, G. A.; Zambounis, J. S.; Terzis, A.; Hountas, A.; Hilti, B.; Mayer, C. W.; Pfeiffer, J. *Synth. Met.* **1988**, *27*, B379.

10. Kini, A. M.; Beno, M. A.; Son, D.; Wang, H. H.; Carlson, K. D.; Porter, L. C.; Welp, U.; Vogt, B. A.; Williams, J. M.; Jung, D.; Evain, M.; Whangbo, M.-H.; Overmyer, D. L.; Schirber, J. E. *Solid State Commun.* **1989**, *69*, 503.

11. Beno, M. A.; Wang, H. H.; Kini, A. M.; Carlson, K. D.; Geiser, U.; Kwok, W. K.; Thompson, J. E.; Williams, J. M. *Inorg. Chem.* **1990**, *29*, 1599.

12. Bryce, M. R.; Murphy, L. C. *Nature (London)* **1984**, *309*, 119.

13. Greene, R. L.; Street, G. B. *Science* **1984**, *226*, 651.

14. Cowan, D. O.; Wiygul, F. M. *Chem. Eng. News* **1986**, *64(29)*, 28.

15. Greene, R. L.; Grant, P. M.; Street, G. B. *Phys. Rev. Lett.* **1975**, *34*, 89; Greene, R. L.; Street, G. B.; Suter, L. S. *Phys. Rev. Lett.* **1975**, *34*, 577.

16. Duke, C. B.; Schein, L. B. *Physics Today* **1980**, *33(2)*, 42.

17. An excellent illustration of these two band filling mechanisms is given in ref. 14.

18. A very readable introduction to the subject of metallic conductivity and to the concepts discussed here are contained in the article "Electrical Conductivity in Metals": Bardeen, J. *J. Appl. Phys.* **1940**, *11*, 88.

19. Bulaevskii, L. N. *Adv. Phys.* **1988**, *37*, 443.

20. Weger, M.; Bender, K.; Klutz, T.; Schweitzer, D.; Gross, F.; Heidmann, C.-P.; Probst, C.; Andres, K. *Synth. Met.* **1988**, *25*, 49.

21. Bechgaard, K.; Jacobsen, C. S.; Mortensen, K.; Pedersen, H. J.; Thorup, N. *Solid State Commun.* **1980**, *33*, 1119.

22. Carlson, K. D.; Crabtree, G. W.; Choi, M.; Hall, L. N.; Copps, P. T.; Wang, H. H.; Emge, T. J.; Beno, M. A.; Williams, J. M. *Mol. Cryst. Liq. Cryst.* **1985**, *125*, 145.

23. Thompson, A. H. *Phys. Rev. Lett.* **1975**, *35*, 1786.

24. Weger, M.; Kaveh, M.; Gutfreund, H. *Solid State Commun.* **1981**, *37*, 421.

25. Hale, P. D.; Ratner, M. A. *J. Chem. Phys.* **1985**, *83*, 5277.

26. Bednorz, J. G.; Müller, K. A. *Z. Phys.* **1986**, *B64*, 189; Cava, R. J.; Batlogg, B.; van Dover, R. B.; Murphy, D. W.; Sunshine, S.; Siegrist, T.; Remeika, J. P.; Reitman, E. A.; Zahurak, S.; Espinosa, G. P. *Phys. Rev. Lett.* **1987**, *58*, 1676.

27. Schafer, D. E.; Wudl, F.; Thomas, G. A.; Ferraris, J. P.; Cowan, D. O. *Solid State Commun.* **1974**, *14*, 347.

28. Coleman, L. B.; *Rev. Sci. Instrum.* **1975**, *46*, 1125.

29. Montgomery, H. C. *J. Appl. Phys.* **1971**, *42*, 2971.

30. Saito, G.; Enoki, T.; Toriumi, K.; Inokuchi, H. *Solid State Commun.* **1982**, *42*, 557.

31. This is the *ac* plane, and σ_1 is approximately along the *a* axis.

32. Enoki, T.; Imaeda, K.; Kobayashi, M.; Inokuchi, H.; Saito, G. *Phys. Rev. B: Condens. Matter* **1986**, *33*, 1553.

33. Tajima, H.; Yakushi, K.; Kuroda, H.; Saito, G.; Inokuchi, H. *Solid State Commun.* **1984**, *49*, 769.

34. Mori, T.; Kobayashi, A.; Sasaki, Y.; Kobayashi, H.; Saito, G.; Inokuchi, H. *Chem. Lett.* **1982**, 1963.

35. Kobayashi, H.; Kato, R.; Mori, T.; Kobayashi, A.; Sasaki, Y.; Saito, G.; Enoki, T.; Inokuchi, H. *Chem. Lett.* **1984**, 179.
36. Kobayashi, H.; Kato, R.; Mori, T.; Kobayashi, A.; Sasaki, Y.; Saito, G.; Enoki, T.; Inokuchi, H. *Mol. Cryst. Liq. Cryst.* **1984**, *107*, 33.
37. Imaeda, K.; Enoki, T.; Saito, G.; Inokuchi, H. *Bull. Chem. Soc. Jpn.* **1988**, *61*, 3332.
38. Parkin, S. S. P.; Miljak, M.; Cooper, J. R. *Phys. Rev. B: Condens. Matter* **1986**, *34*, 1485.
39. Parkin, S. S. P.; Engler, E. M.; Schumaker, R. R.; Lagier, R.; Lee, V. Y.; Scott, J. C.; Greene, R. L. *Phys. Rev. Lett.* **1983**, *50*, 270.
40. The stacking and interstacking directions of the $(ET)_2ReO_4$ crystal are the a and b axes, respectively, as assigned in ref. 39.
41. Ravy, S.; Moret, R.; Pouget, J. P.; Comes, R.; Parkin, S. S. P. *Phys. Rev. B: Condens. Matter* **1986**, *33*, 2049.
42. Kwak, J. F.; Azevedo, L. J.; Schirber, J. E.; Williams, J. M.; Beno, M. A. *Mol. Cryst. Liq. Cryst.* **1985**, *125*, 365.
43. Williams, J. M.; Beno, M. A.; Wang, H. H.; Reed, P. E.; Azevedo, L. J.; Schirber, J. E. *Inorg. Chem.* **1984**, *23*, 1790.
44. Whangbo, M.-H.; Beno, M. A.; Leung, P. C. W.; Emge, T. J.; Wang, H. H.; Williams, J. M. *Solid State Commun.* **1986**, *59*, 813.
45. Kobayashi, H.; Kato, R.; Mori, T.; Kobayashi, A.; Sasaki, Y.; Saito, G.; Inokuchi, H. *Chem. Lett.* **1983**, 759.
46. Kobayashi, H.; Mori, T.; Kato, R.; Kobayashi, A.; Sasaki, Y.; Saito, G.; Inokuchi, H. *Chem. Lett.* **1983**, 581.
47. Mori, T.; Kobayashi, A.; Sasaki, Y.; Kato, R.; Kobayashi, H. *Solid State Commun.* **1985**, *53*, 627.
48. Laversanne, R.; Amiell, J.; Delhaes, P.; Chasseau, D.; Hauw, C. *Solid State Commun.* **1984**, *52*, 177.
49. Laversanne, R.; Amiell, J.; Delhaes, P.; Chasseau, D.; Hauw, C. *Mol. Cryst. Liq. Cryst.* **1985**, *119*, 405.
50. Leung, P. C. W.; Beno, M. A.; Blackman, G. S.; Coughlin, B. R.; Miderski, C. A.; Joss, W.; Crabtree, G. W.; Williams, J. M. *Acta Crystallogr.* **1984**, *C40*, 1331.
51. The interstacking direction is the c axis for β-$(ET)_2PF_6$ of ref. 47 and the b axis for $(ET)_2AsF_6$ of ref. 50 in the space group chosen for each salt. However, the two salts are isostructural.
52. Yagubskii, É. B.; Shchegolev, I. F.; Laukhin, V. N.; Kononovich, P. A.; Kartsovnik, M. V.; Zvarykina, A. V.; Buravov, L. I. *JETP Lett.* **1984**, *39*, 12.
53. Wang, H. H.; Beno, M. A.; Geiser, U.; Firestone, M. A.; Webb, K. S.; Nuñez, L.; Crabtree, G. W.; Carlson, K. D.; Williams, J. M.; Azevedo, L. J.; Kwak, J. F.; Schirber, J. E. *Inorg. Chem.* **1985**, *24*, 2465.
54. Williams, J. M.; Wang, H. H.; Beno, M. A.; Emge, T. J.; Sowa, L. M.; Copps, P. T.; Behroozi, F.; Hall, L. N.; Carlson, K. D.; Crabtree, G. W. *Inorg. Chem.* **1984**, *23*, 3839.
55. Unpublished data by the Authors.
56. Yagubskii, É. B.; Shchegolev, I. F.; Shibaeva, R. P.; Fedutin, D. N.; Rozenberg, L. P.; Sogomonyan, E. M.; Lobkovskaya, R. M.; Laukhin, V. N.; Ignat'ev, A. A.; Zvarykina, A. V.; Buravov, L. I. *Pis'ma Zh. Eksp. Teor. Fiz.*, **1985**, *42*, 167.

57. Buravov, L. I.; Kartsovnik, M. V.; Kononovich, P. A.; Laukhin, V. N.; Pesotskii, S. I.; Shchegolev, I. F. *JETP Lett.* **1986**, *64*, 1306.
58. Mortensen, K.; Williams, J. M.; Wang, H. H. *Solid State Commun.* **1985**, *56*, 105.
59. Whangbo, M.-H.; Williams, J. M.; Leung, P. C. W.; Beno, M. A.; Emge, T. J.; Wang, H. H.; Carlson, K. D.; Crabtree, G. W. *J. Am. Chem. Soc.* **1985**, *107*, 5815.
60. Talham, D. R.; Kurmoo, M.; Day, P.; Obertelli, S. D.; Parker, I. D.; Friend, R. H. *J. Phys. C: Solid State Phys.* **1986**, *19*, L383.
61. Koch, B.; Geserich, H. P.; Ruppel, W.; Schweitzer, D.; Dietz, K. H.; Keller, H. J. *Mol. Cryst. Liq. Cryst.* **1985**, *119*, 343.
62. Emge, T. J.; Wang, H. H.; Beno, M. A.; Leung, P. C. W.; Firestone, M. A.; Jenkins, H. C.; Cook, J. D.; Carlson, K. D.; Williams, J. M.; Venturini, E. L.; Azevedo, L. J.; Schirber, J. E. *Inorg. Chem.* **1985**, *24*, 1736.
63. Bando, H.; Tokumoto, M.; Murata, K.; Anzai, H.; Saito, G.; Kajimura, K.; Ishiguro, T. *J. Phys. Soc. Jpn.* **1985**, *54*, 4265.
64. Kobayashi, H.; Kato, R.; Kobayashi, A.; Saito, G.; Tokumoto, M.; Anzai, H.; Ishiguro, T. *Chem. Lett.* **1985**, 1293.
65. Emge, T. J.; Leung, P. C. W.; Beno, M. A.; Schultz, A. J.; Wang, H. H.; Sowa, L. M.; Williams, J. M. *Phys. Rev. B: Condens. Matter* **1984**, *30*, 6780; Leung, P. C. W.; Emge, T. J.; Beno, M. A.; Wang, H. H.; Williams, J. M.; Petricek, V.; Coppens, P. *J. Am. Chem. Soc.* **1985**, *107*, 6184.
66. Schultz, A. J.; Beno, M. A.; Wang, H. H.; Williams, J. M. *Phys. Rev. B: Condens. Matter* **1986**, *33*, 7823.
67. Laukhin, V. N.; Kostyuchenko, E. É.; Sushko, Yu. V.; Shchegolev, I. F.; Yagubskii, É. B. *JETP Lett.* **1985**, *41*, 81.
68. Murata, K.; Tokumoto, M.; Anzai, H.; Bando, H.; Saito, G.; Kajimura, K.; Ishiguro, T. *J. Phys. Soc. Jpn.* **1985**, *54*, 1236.
69. Schultz, A. J.; Wang, H. H.; Williams, J. M.; Filhol, A. *J. Am. Chem. Soc.* **1986**, *108*, 7853.
70. Schirber, J. E.; Azevedo, L. J.; Kwak, J. F.; Venturini, E. L.; Leung, P. C. W.; Beno, M. A.; Wang, H. H.; Williams, J. M. *Phys. Rev. B: Condens. Matter* **1986**, *33*, 1987.
71. Schirber, J. E.; Kwak, J. F.; Beno, M. A.; Wang, H. H.; Williams, J. M. *Physica B + C* **1986**, *143*, 343.
72. Merzhanov, V. A.; Kostyuchenko, E. É.; Faber, O. E.; Shchegolev, I. F.; Yagubskii, E. B. *Sov. Phys. JETP* **1985**, *62*, 165.
73. Yagubskii, É. B.; Shchegolev, I. F.; Topnikov, V. N.; Pesotskii, S. I.; Laukhin, V. N.; Kononovich, P. A.; Kartsovnik, M. V.; Zvarykina, A. V.; Dedik, S. G.; Buravov, L. I. *Sov. Phys. JETP* **1985**, *61*, 142.
74. Bender, K.; Dietz, K.; Endres, H.; Helberg, H. W.; Hennig, I.; Keller, H. J.; Schäfer, H. W.; Schweitzer, D. *Mol. Cryst. Liq. Cryst.* **1984**, *107*, 45.
75. Daoben, Z.; Ping, W.; Meixiang, W.; Zhaolou, Y.; Naijue, Z. *Solid State Commun.* **1986**, *57*, 843.
76. Nogami, Y.; Kagoshima, S.; Sugano, T.; Saito, G. *Synth. Met.* **1986**, *16*, 367.
77. (a) Mori, T.; Kobayashi, A.; Sasaki, Y.; Kobayashi, H.; Saito, G.; Inokuchi, H. *Chem. Lett.* **1984**, 957; (b) Emge, T. J.; Leung, P. C. W.; Beno, M. A.; Wang, H. H.; Williams, J. M.; Whangbo, M.-H.; Evain, M. *Mol. Cryst. Liq. Cryst.* **1986**. *138*, 393.

78. Baram, G. O.; Buravov, L. I.; Degtyarev, L. S.; Kozlov, M. É.; Laukhin, V. N.; Laukhina, E. É.; Onishchenko, V. G.; Pokhodnya, K. I.; Sheinkman, M. K.; Shibaeva, R. P.; Yagubskii, É. B. *JETP Lett.* **1986,** *44,* 293.

79. Geiser, U.; Wang, H. H.; Beno, M. A.; Firestone, M. A.; Webb, K. S.; Williams, J. M.; Whangbo, M.-H. *Solid State Commun.* **1986,** *57,* 741.

80. Kobayashi, A.; Kato, R.; Kobayashi, H.; Tokumoto, M.; Anzai, H.; Ishiguro, T. *Chem. Lett.* **1986,** 1117.

81. See Table 2.1 and references cited therein.

82. Emge, T. J.; Wang, H. H.; Bowman, M. K.; Pipan, C. M.; Carlson, K. D.; Beno, M. A.; Hall, L. N.; Anderson, B. A.; Williams, J. M.; Whangbo, M.-H. *J. Am. Chem. Soc.* **1987,** *109,* 2016.

83. Ugawa, A.; Yakushi, K.; Kuroda, H.; Kawamoto, A.; Tanaka, J. *Chem. Lett.* **1986,** 1875.

84. Mori, T.; Sakai, F.; Saito, G.; Inokuchi, H. *Chem. Lett.* **1986,** 1037.

85. Kajita, K.; Nishio. Y.; Moriyama, S.; Sasaki, W.; Kato, R.; Kobayashi, H.; Kobayashi, A. *Solid State Commun.* **1986,** *60,* 811.

86. Ugawa, A.; Okawa, Y.; Yakushi, K.; Kuroda, H.; Kawamoto, A.; Tanaka, J.; Tanaka, M.; Nogami, Y.; Kagoshima, S.; Murata, K.; Ishiguro, T. *Synth. Met.* **1988,** *27,* A407.

87. Pratt, F. L.; Fisher, A. J.; Hayes, W.; Singleton, J.; Spermon, S. J. R. M.; Kurmoo, M.; Day, P. *Phys. Rev. Lett.* **1988,** *61,* 2721.

88. Yagubskii, É. B.; Shchegolev, I. F.; Pesotskii, S. I.; Laukhin, V. N.; Kononovich, P. A.; Kartsovnik, M. V.; Zvarykina, A. V. *JETP Lett.* **1984,** *39,* 328.

89. Shibaeva, R. P.; Kaminskii, V. F.; Yagubskii, É. B. *Mol. Cryst. Liq. Cryst.* **1985,** *119,* 361.

90. Shibaeva, R. P.; Lobkovskaya, R. M.; Yagubskii, É. B.; Kostyuchenko, E. É. *Sov. Phys. Crystallogr.* **1986,** *31,* 657.

91. Beno, M. A.; Geiser, U.; Kostka, K. L.; Wang, H. H.; Webb, K. S.; Firestone, M. A.; Carlson, K. D.; Nuñez, L.; Whangbo, M.-H.; Williams, J. M. *Inorg. Chem.* **1987,** *26,* 1912.

92. This η-phase salt was originally designated by ζ in ref. 91.

93. Shibaeva, R. P.; Lobkovskaya, R. M.; Yagubskii, É. B.; Kostyuchenko, E. É. *Sov. Phys. Crystallogr.* **1986,** *31,* 267.

94. Merzhanov, V. A.; Kostyuchenko, E. É.; Laukhin, V. N.; Lobkovskaya, R. M.; Makova, M. K.; Shibaeva, R. P.; Shchegolev, I. F.; Yagubskii, É. B. *JETP Lett.* **1985,** *41,* 179.

95. Kobayashi, H.; Kato, R.; Kobayashi, A.; Nishio, Y.; Kajita, K.; Sasaki, W. *Chem. Lett.* **1986,** 789.

96. Kobayashi, H.; Kato, R.; Kobayashi, A.; Nishio, Y.; Kajita, K.; Sasaki, W. *Chem. Lett.* **1986,** 833.

97. Kobayashi, A.; Kato, R.; Kobayashi, H.; Moriyama, S.; Nishio, Y.; Kajita, K.; Sasaki, W. *Chem. Lett.* **1987,** 459.

98. Kato, R.; Kobayashi, H.; Kobayashi, A.; Moriyama, S.; Nishio, Y.; Kajita, K.; Sasaki, W. *Chem. Lett.* **1987,** 507.

99. Kobayashi, A.; Kato, R.; Kobayashi, H.; Moriyama, S.; Nishio, Y.; Kajita, K., Sasaki, W. *Chem. Lett.* **1986,** 2017.

100. Kajita, K.; Nishio, Y.; Moriyama, S.; Sasaki, W.; Kato, R.; Kobayashi, H.; Kobayashi, A. *Solid State Commun.* **1987,** *64,* 1279.

101. (a) Kajita, K.; Nishio, Y.; Takahashi, T.; Sasaki, W.; Kato, R.; Kobayashi, H.; Kobayashi, A. *Solid State Commun.* **1989,** *70,* 1181; (b) Kajita, K.; Nishio, Y.; Takahashi, T.; Sasaki, W.; Kato, R.; Kobayashi, H.; Kobayashi, A.; Iye, Y. *Solid State Commun.* **1989,** *70,* 1189.

102. Kobayashi, H.; Kato, R.; Kobayashi, A.; Moriyama, S.; Nishio, Y.; Kajita, K.; Sasaki, W. *Synth. Met.* **1988,** *27,* A283.

103. Kato, R.; Kobayashi, H.; Kobayashi, A.; Nishio, Y.; Kajita, K.; Sasaki, W. *Chem. Lett.* **1986,** 957.

104. Beno, M. A.; Kini, A. M.; Montgomery, L. K.; Whitworth, J. R.; Carlson, K. D.; Williams, J. M. *Synth. Met.* **1988,** *27,* A219.

105. Qian, M.; Wang, X.; Zhu, Y.; Zhu, D.; Li, L.; Ma, B.; Duan, H.; Zhang, D. *Synth. Met.* **1988,** *27,* A277.

106. Ma, B.; Lu, L.; Zhang, D.; Wang, X.; Zhu, D. *Solid State Commun.* **1988,** *68,* 433.

107. Lu, L.; Ma, B.; Duan, H.; Lin, S.; Zhang, D.; Wang, X.; Zhu, D. *Synth. Met.* **1988,** *27,* A311.

108. Nigrey, P. J.; Morosin, B.; Kwak, J. F.; Venturini, E. L.; Baughman, R. J. *Synth. Met.* **1986,** *16,* 1.

109. Kurmoo, M.; Talham, D. R.; Pritchard, K. L.; Day, P.; Stringer, A. M.; Howard, J. A. K. *Synth. Met.* **1988,** *27,* A177.

110. Oshima, K.; Urayama, H.; Yamochi, H.; Saito, G. *Physica C* **1988,** *153,* 1148.

111. Urayama, H.; Yamochi, H.; Saito, G.; Sato, S.; Sugano, T.; Kinoshita, M.; Kawamoto, A.; Tanaka, J.; Inabe, T.; Mori, T.; Maruyama, Y.; Inokuchi, H.; Oshima, K. *Synth. Met.* **1988,** *27,* A393.

112. Ugawa, A.; Ojima, G.; Yakushi, K.; Kuroda, H. *Synth. Met.* **1988,** *27,* A445.

113. Urayama, H. in *The Physics and Chemistry of Organic Superconductors;* Saito, G., Kagoshima, S., Eds.; Springer-Verlag: Heidelberg, FRG, **1990**; p. 276.

114. Lyubovskaya, R. N.; Zhilyaeva, E. I.; Pesotskii, S. I.; Lyubovskii, R. B.; Atovmyan, L. O.; D'yachenko, O. A.; Takhirov, T. J. *JETP Lett.* **1987,** *46,* 188.

115. Shibaeva, R. P.; Rozenberg, L. P. *Kristallografiya* **1988,** *33,* 1402.

116. Lyubovskaya, R. N.; Zhilyaeva, E. A.; Zvarykina, A. V.; Laukhin, V. N.; Lyubovskii, R. B.; Pesotskii, S. I. *JETP Lett.* **1987,** *45,* 530.

117. Schirber, J. E.; Overmyer, D. L.; Venturini, E. L.; Wang, H. H.; Carlson, K. D.; Kwok, W. K.; Kleinjan, S.; Williams, J. M. *Physica C* **1989,** *161,* 412.

118. Lyubovskii, R. B.; Lyubovskaya, R. N.; Kapustin, N. V. *Sov. Phys. JETP* **1987,** *66,* 1063.

119. Kikuchi, K.; Honda, Y.; Ishikawa, Y.; Saito, K.; Ikemoto, I.; Murata, K.; Anzai, H.; Ishiguro, T.; Kobayashi, K. *Solid State Commun.* **1988,** *66,* 405.

120. Kikuchi, K.; Murata, K.; Honda, Y.; Namiki, T.; Saito, K.; Kobayashi, K.; Ishiguro, T.; Ikemoto, I. *J. Phys. Soc. Jpn.* **1987,** *56,* 2627.

121. Ikemoto, I. Superconducting Materials (*Jpn. J. Appl. Phys., Series 1*) **1988,** 170.

122. Nogami, Y.; Tanaka, M.; Kagoshima, S.; Kikuchi, K.; Saito, K.; Ikemoto, I.; Kobayashi, K. *J. Phys. Soc. Jpn.* **1987,** *56,* 3783.

123. Kikuchi, K.; Murata, K.; Honda, Y.; Namiki, T.; Saito, K.; Anzai, H.; Kobayashi, K.; Ishiguro, T.; Ikemoto, I. *J. Phys. Soc. Jpn.* **1987,** *56,* 4241.

124. Kikuchi, K.; Murata, K.; Honda, Y.; Namiki, T.; Saito, K.; Ishiguro, T.; Kobayashi, K.; Ikemoto, I. *J. Phys. Soc. Jpn.* **1987,** *56,* 3436.

125. Murata, K.; Kikuchi, K.; Takahashi, T.; Kobayashi, K.; Honda, Y.; Saito, K.; Kanoda, K.; Tokiwa, T.; Anzai, H.; Ishiguro, T.; Ikemoto, I. *Proc. Elorma (Tashikent)* Nov. 17–20, **1987.**

126. Kikuchi, K.; Saito, K.; Ikemoto, I.; Murata, K.; Ishiguro, T.; Kobayashi, K. *Synth. Met.* **1988,** *27,* B269.

127. Mori, T.; Wang, P.; Imaeda, K.; Enoki, T.; Inokuchi, H.; Sakai, F.; Saito, G. *Synth. Met.* **1988,** *27,* A451.

128. Brossard, L.; Ribault, M.; Bousseau, M.; Valade, L.; Cassoux, P. *C. R. Acad. Sci.* **1986,** *302,* 205.

129. Brossard, L.; Ribault, M.; Valade, L.; Cassoux, P. *J. Phys. France* **1989,** *50,* 1521.

130. Brossard, L.; Hurdequint, H.; Ribault, M.; Valade, L.; Legros, J.-P.; Cassoux, P. *Synth. Met.* **1988,** *27,* B157.

131. Brossard, L. unpublished results, as reported in ref. 4 of the following paper: Legros, J.-P.; Valade, L.; Cassoux, P. *Synth. Met.* **1988,** *27,* B347.

132. Kobayashi, A.; Kim, H.; Sasaki, Y.; Kato, R.; Kobayashi, H.; Moriyama, S.; Nishio, Y.; Kajita, K.; Sasaki, W. *Chem. Lett.* **1987,** 1819; Kobayashi, A.; Kim, H.; Sasaki, Y.; Moriyama, S.; Nishio, Y.; Kajita, K.; Sasaki, W.; Kato, R.; Kobayashi, H. *Synth. Met.* **1988,** *27,* B339.

133. Williams, J. M.; Emge, T. J.; Wang, H. H.; Beno, M. A.; Copps, P. T.; Hall, L. N.; Carlson, K. D.; Crabtree, G. W. *Inorg. Chem.,* **1984,** *23,* 2558.

134. Crabtree, G. W.; Carlson, K. D.; Hall, L. N.; Copps, P. T.; Wang, H. H.; Emge, T. J.; Beno, M. A.; Williams, J. M. *Phys. Rev. B: Condens. Matter* **1984,** *30,* 2958.

135. Azevedo, L. J.; Venturini, E. L.; Schirber, J. E.; Williams, J. M.; Wang, H. H.; Emge, T. J. *Mol. Cryst. Liq. Cryst.* **1985,** *119,* 389.

136. Schwenk, H.; Heidmann, C.-P.; Gross, F.; Hess, E.; Andres, K.; Schweitzer, D.; Keller, H. J. *Phys. Rev. B: Condens. Matter* **1985,** *31,* 3138.

137. Stewart, G. R.; O'Rourke, J.; Crabtree, G. W.; Carlson, K. D.; Wang, H. H.; Williams, J. M.; Gross, F.; Andres, K. *Phys. Rev. B: Condens. Matter* **1986,** *33,* 2046.

138. Laukhina, E. É.; Laukhin, V. N.; Khomenko, A. G.; Yagubskii, É. B. *Synth. Met.* **1989,** *32,* 381.

139. Kagoshima, S.; Nogami, Y.; Hasumi, M.; Anzai, H.; Tokumoto, M.; Saito, G.; Mori, N. *Solid State Commun.* **1989,** *69,* 1177.

140. Tokumoto, M.; Anzai, H.; Bando, H.; Saito, G.; Kinoshita, N.; Kajimura, K.; Ishiguro, T. *J. Phys. Soc. Jpn.* **1985,** *54,* 1669.

141. Carlson, K. D.; Crabtree, G. W.; Hall, L. N.; Behroozi, F.; Copps, P. T.; Sowa, L. M.; Nuñez, L.; Firestone, M. A.; Wang, H. H.; Beno, M. A.; Emge, T. J.; Williams, J. M. *Mol. Cryst. Liq. Cryst.* **1985,** *125,* 159.

142. Nuñez, L.; Carlson, K. D.; Hall, L. N.; Crabtree, G. W.; Perozzo, M. T.; Wang, H. H.; Williams, J. M. *Physica B + C,* **1986,** *143,* 369.

143. Schwenk, H.; Parkin, S. S. P.; Lee, V. Y.; Greene, R. L. *Phys. Rev. B: Condens. Matter* **1986,** *34,* 3156.

144. Heidmann, C.-P.; Veith, H.; Andres, K.; Fuchs, H.; Polborn, K.; Amberger, E. *Solid State Commun.* **1986,** *57,* 161.

145. Carlson, K. D.; Crabtree, G. W.; Nuñez, L.; Wang, H. H.; Beno, M. A.; Geiser, U.; Firestone, M. A.; Webb, K. S.; Williams, J. M. *Solid State Commun.* **1986**, *57*, 89.
146. Tokumoto, M.; Bando, H.; Anzai, H.; Saito, G.; Murata, K.; Kajimura, K.; Ishiguro, T. *J. Phys. Soc. Jpn.* **1985**, *54*, 869.
147. Tinkham, M. "Introduction to Superconductivity"; Robert E. Krieger Publishing Co.: Huntington, New York, 1980; Chap. 4, p. 104.
148. Decroux, M.; Fisher, Ø. "Superconductivity in Ternary Compounds II", Maple, M. B.; Fisher, Ø., eds.; Springer-Verlag: Berlin, 1982; Chap. 3, p. 57. Especially see original references cited therein.
149. Clogston, A. M. *Phys. Rev. Lett.* **1962**, *9*, 266.
150. Ginodman, V. B.; Gudenko, A. V.; Zherikhina, L. N. *JETP Lett.* **1985**, *41*, 49.
151. See many examples given in ref. 148.
152. Bardeen, J.; Cooper, L. N.; Schrieffer, J. R. *Phys. Rev.* **1957**, 106, 162; **1957**, 108, 1175. See ref. 147, Chap. 2, p. 16, for a review of BCS theory.
153. Stewart, G. R.; Williams, J. M.; Wang, H. H.; Hall, L. N.; Perozzo, M. T.; Carlson, K. D. *Phys. Rev. B: Condens. Matter* **1986**, *34*, 6509.
154. Andres, K.; Schwenk, H.; Veith, H. *Physica B + C*, **1986**, *143*, 334.
155. Hawley, M. E.; Gray, K. E.; Terris, B. D.; Wang, H. H.; Carlson, K. D.; Williams, J. M. *Phys. Rev. Lett.*, **1986**, *57*, 629.
156. Nowack, A.; Poppe, U.; Weger, M.; Schweitzer, D.; Schwenk, H. Z. *Phys. Rev. B: Condens. Matter* **1987**, *68*, 41.
157. Buravov, L. I.; Kartsovnik, M. V.; Kaminskii, V. F.; Kononovich, P. A.; Kostuchenko, E. É.; Laukhin, V. N.; Makova, M. K.; Pesotskii, S. I.; Schegolev, I. F.; Topnikov, V. N.; Yagubskii, É. B. *Synth. Met.* **1985**, *11*, 207.
158. Murata, K.; Tokumoto, M.; Anzai, H.; Bando, H.; Saito, G.; Kajimura, K.; Ishiguro, T. *J. Phys. Soc. Jpn.* **1985**, *54*, 2084.
159. Buravov, L. I.; Kartsovnik, M. V.; Kononovich, P. A.; Laukhin, V. N.; Pesotskii, S. I.; Shchegolev, I. F. *JETP Lett.* **1986**, *64*, 1306.
160. Ginodman, V. B.; Gudenko, A. V.; Kononovich, P. A.; Laukhin, V. N.; Shchegolev, I. F. *JETP Lett.* **1986**, *44*, 673.
161. Kartsovnik, M. V.; Laukhin, V. N.; Shchegolev, I. F. *JETP Lett.* **1986**, *63*, 1273.
162. Ginodman, V. B.; Gudenko, A. V.; Zherikhina, L. N. *JETP Lett.* **1985**, *41*, 49.
163. Tokumoto, M.; Murata, K.; Bando, H.; Anzai, H.; Saito, G.; Kajimura, K.; Ishiguro, T. *Solid State Commun.* **1985**, *54*, 1031.
164. Veith, H.; Heidmann, C.-P.; Gross, F.; Lerf, A.; Andres, K.; Schweitzer, D. *Solid State Commun.* **1985**, *56*, 1015.
165. Creuzet, F.; Creuzet, G.; Jérome, D.; Schweitzer, D.; Keller, H. J. *J. Phys. Lett (Les Ulis, Fr.)* **1985**, *46*, L-1079.
166. Creuzet, F.; Jérome, D.; Schweitzer, D.; Keller, H. J. *Europhys. Lett.* **1986**, *1*, 461.
167. Ginodman, V. B.; Gudenko, A. V.; Zasavitskii, I. I.; Yagubskii, É. B. *JETP Lett.* **1985**, *42*, 472.
168. Montgomery, L. K.; Geiser, U.; Wang, H. H.; Beno, M. A.; Schultz, A. J.; Kini, A. M.; Carlson, K. D.; Williams, J. M.; Whitworth, J. R.; Gates, B. D.; Cariss, C. S.; Pipan, C. M.; Donega, K. M.; Wenz, C.; Kwok, W. K.; Crabtree, G. W. *Synth. Met.* **1988**, *27*, A195.

169. Zvarykina, A. V.; Kartsovnik, M. V.; Laukhin, V. N.; Laukhina, E. É.; Lyubovskii, R. B.; Pesotskii, S. I.; Shibaeva, R. P.; Shchegolev, I. F. *Sov. Phys. JETP*, **1988**, *67*, 1891.
170. Murata, K.; Tokumoto, M.; Anzai, H.; Bando, H.; Kajimura, K.; Ishiguro, T.; Saito, G. *Synth. Met.* **1986**, *13*, 3.
171. Tokumoto, M.; Murata, K.; Bando, H.; Anzai, H.; Kajimura, K.; Ishiguro, T. *Physica B + C*, **1986**, *143*, 338.
172. Bulaevskii, L. N.; Ginodman, V. B.; Gudenko, A. V. *JETP Lett.* **1987**, *45*, 452.
173. Saito, G.; Urayama, H.; Yamochi, H.; Oshima, K. *Synth. Met.* **1988**, *27*, A331.
174. Veith, H.; Heidmann, C.-P.; Müller, H.; Fritz, H. P.; Andres, K.; Fuchs, H. *Synth. Met.* **1988**, *27*, A361.
175. Schweitzer, D.; Polychroniadis, K.; Klutz, T.; Keller, H. J.; Hennig, I.; Heinen, I.; Haeberlen, U.; Gogu, E.; Gärtner, S. *Synth. Met.* **1988**, *27*, A465.
176. Nozawa, K.; Sugano, T.; Urayama, H.; Yamochi, H.; Saito, G.; Kinoshita, M. *Chem. Lett.* **1988**, 617.
177. Sugano, T.; Terui, K.; Mino, S.; Nozawa, K.; Urayama, H.; Yamochi, H.; Saito, G.; Kinoshita, M. *Chem. Lett.* **1988**, 1171.
178. From unpublished measurements by the authors.
179. Oshima, K.; Urayama, H.; Yamochi, H.; Saito, G. *J. Phys. Soc. Jpn.* **1988**, *57*, 730.
180. Graebner, J. E.; Haddon, R. C.; Chichester, S. V.; Glarum, S. H. *Phys. Rev. B: Condens. Matter* **1990**, *41*, 4808.
181. Oshima, K.; Urayama, H.; Yamochi, H.; Saito, G. *Synth. Met.* **1988**, *27*, A419.
182. Takahashi, T.; Tokiwa, T.; Kanoda, K.; Urayama, H.; Yamochi, H.; Saito, G. *Physica C*. **1988**, 153–155, 487.
183. Harshman, D. R.; Kleiman, R. N.; Haddon, R. C.; Chichester-Hicks, S. V.; Kaplan, M. L.; Rupp, Jr., L. W.; Pfiz, T.; Williams, D. L.; Mitzi, D. B. *Phys Rev. Lett.* **1990**, *64*, 1293.
184. Katsumoto, S.; Kobayashi, S.; Urayama, H.; Yamochi, H.; Saito, G. *J. Phys. Soc. Jpn.*, **1988**, *57*, 3672.
185. Andraka, B.; Kim, J. S.; Stewart, G. R.; Carlson, K. D.; Wang, H. H.; Williams, J. M. *Phys. Rev. B: Condens. Matter* **1989**, *40*, 11345.
186. Marsiglio, F.; Carbotte, J. P. *Phys. Rev. B. Condens. Matter* **1986**, *33*, 6141.
187. Maruyama, Y.; Inabe, T.; Urayama, H.; Yamochi, H.; Saito, G. *Solid State Commun.* **1988**, *67*, 35.
188. McMillan, W. L. *Phys. Rev.*, **1968**, *167*, 331.
189. Allen, P. B.; Dynes, R. C. *Phys. Rev. B. Condens. Matter* **1975**, *12*, 905.
190. Schirber, J. E.; Venturini, E. L.; Kini, A. M.; Wang, H. H.; Whitworth, J. R.; Williams, J. M. *Physica C*, **1988**, *152*, 157.
191. Oshima, K.; Mori, T.; Inokuchi, H.; Urayama, H.; Yamochi, H.; Saito, G. *Synth. Met.* **1988**, *27*, A165.
192. Helfand, E.; Werthamer, N. R. *Phys. Rev.* **1966**, 147, 288.
193. Werthamer, N. R., Helfand, E.; Hohenberg, P. C. *Phys. Rev.* **1966**, *147*, 295.
194. Hohenberg, P. C.; Werthamer, N. R. *Phys. Rev.* **1967**, *153*, 493.
195. Eliashberg, G. M. *Sov. Phys. JETP* **1960**, *11*, 696; **1967**, *12*, 1000.

196. Carlson, K. D.; Wang, H. H.; Beno, M. A.; Kini, A. M.; Williams, J. M. *Mol. Cryst. Liq. Cryst.* **1990,** *181,* 91.
197. Zvarykina, A. V.; Kononovich, P. A.; Laukhin, V. N.; Molchanov, V. N.; Pesotskii, S. I.; Simonov, V. I.; Shibaeva, R. P.; Shchegolev, I. F.; Yagubskii, É. B. *JETP Lett.* **1986,** *43,* 329.
198. Schweitzer, D.; Bele, P.; Brunner, H.; Gogu, E.; Haeberlen, U.; Hennig, I.; Klutz, I.; Swietlik, R.; Keller, H. J. *Z. Phys. B: Condens. Matter* **1987,** *67,* 489.
199. Wang, H. H.; Ferraro, J. R.; Carlson, K. D.; Montgomery, L. K.; Geiser, U.; Williams, J. M.; Whitworth, J. R.; Schlueter, J. A.; Hill, S.; Whangbo, M.-H.; Evain, M.; Novoa, J. J. *Inorg. Chem.* **1989,** *28,* 2267.
200. Wang, H. H.; Carlson, K. D.; Montgomery, L. K.; Schlueter, J. A.; Cariss, C. S.; Kwok, W. K.; Geiser, U.; Crabtree, G. W.; Williams, J. M. *Solid State Commun.* **1988,** *66,* 1113.
201. Wang, H. H.; Carlson, K. D.; Geiser, U.; Kwok, W. K.; Vashon, M. D.; Thompson, J. E.; Larsen N. F.; McCabe, G. D.; Hulscher, R. S.; Williams, J. M. *Physica C,* **1990,** *166,* 57.
202. Andraka, B.; Stewart, G. R.; Carlson, K. D.; Wang, H. H.; Vashon, M. D.; Williams, J. M. *Phys. Rev. B: Condens. Matter* **1990,** *42,* 9963.
203. Andraka, B.; Stewart, G. R.; Present Authors, work in progress.
204. Kini, A. M.; Geiser, U.; Wang, H. H.; Carlson, K. D.; Williams, J. M.; Kwok, W. K.; Vandervoort, K. G.; Thompson, J. E.; Stupka, D. L.; Jung, D.; Whangbo, M.-H. *Inorg. Chem.* **1990,** *29,* 2555.
205. Williams, J. M.; Kini, A. M.; Geiser, U.; Wang, H. H.; Carlson, K. D.; Kwok, W. K.; Vandervoort, K. G.; Thompson, J. E.; Stupka, D. L.; Jung, D.; Whangbo, M.-H. *International Conference on Organic Superconductors,* South Lake Tahoe, CA, May 20–24, **1990,** p.33.
206. Williams, J. M.; Kini, A. M.; Wang, H. H.; Carlson, K. D.; Geiser, U.; Montgomery, L. K.; Pyrka, G. J.; Watkins, D. M.; Kommers, J. M.; Boryschuk, S. J.; Strieby Crouch, A. V.; Kwok, W. K.; Schirber, J. E.; Overmyer, D. L.; Jung, D.; Whangbo, M.-H. *Inorg. Chem.* **1990,** *29,* 3262.

5

Single Crystal ESR Studies

INTRODUCTION

Electron Spin Resonance spectroscopy (ESR) is an extremely valuable technique for studying synthetic metals. It is highly sensitive, enabling very small crystalline samples weighing 10 to 100 μg, to be easily measured. Due to the multiphasic nature of many organic metals (see Chaps. 2 and 3), ESR spectroscopy is an ideal tool for the study of each individual phase. In this chapter, we will focus on ESR studies of single crystals of organic conductors. A typical ambient temperature ESR spectrum for a single crystal of an organic metal consists of a single absorption line. In solid state samples, hyperfine coupling to the hydrogen atoms of the organic-donor molecule is not observed, owing to the broad linewidth of the signals. Nevertheless, the peak-to-peak linewidth is a useful parameter for the identification of different phases and the investigation of various phase transitions at high or low temperatures. Other information that can be obtained from ESR measurements includes the g-values, the spin susceptibility, and the lineshape. The g-values for organic metals usually vary from 2.000 to 2.014, depending on the crystal orientation with respect to the static magnetic field. The spin susceptibility of most organic conductors is on the order of 10^{-4} to 10^{-3} emu/mole. From the temperature dependence of the spin susceptibility, the conductive nature of an organic charge-transfer salt can be inferred. The ESR lineshapes of organic conductors are usually Lorentzian, except for highly conductive samples which display Dysonian lineshapes in certain orientations. In this chapter, single crystal ESR data on BEDT-TTF (or ET) based materials and related salts will be summarized. Their variable temperature behavior will be grouped into categories depending on their structural packing motifs. In addition, a section will be devoted to thermal transformations in the solid state.

ROOM TEMPERATURE ESR STUDIES

Most of the known organic conductors and superconductors are structurally layered two-dimensional materials (see Chap. 3 for details). The organic-donor layer (often parallel to the ab plane) alternates with an inorganic anion layer. Among the different crystal morphologies observed, such as needles, plates, hexagons or cubes, the platelet morphology is the most predominant. The electrical conductivities are usually high within the crystal plane that corresponds to the organic-donor molecule layer. The conductivities are typically one order of magnitude less along the direction perpendicular to the conducting crystal plane, that is, along the c^* axis. The conductivity results derived from ESR measurements are generally consistent with those obtained from 4-probe conductivity measurements. A new layered organic superconductor, the α-$(ET)_2(NH_4)Hg(SCN)_4$ salt, will be presented here as a typical example of an organic metal.[1]

Orientational Studies

A square platelet single crystal of α-$(ET)_2(NH_4)Hg(SCN)_4$ salt is used here to demonstrate the room temperature orientational dependence of the g-value, the peak-to-peak linewidth, and the ESR lineshape. The crystal belongs to the α-type packing motif, and the unit cell parameters are as follows: $a = 9.968$Å, $b = 10.089$Å, $c = 20.668$Å, $\alpha = 93.58°$, $\beta = 104.28°$, $\gamma = 89.53°$, $V = 2010$Å^3. The order of the abc axes used here is different from that given in the literature,[2] but they are consistent with other known α-$(ET)_2X$ salts ($X^- = I_3^-$, I_2Br^-, IBr_2^-, etc.). An X-band ESR spectrometer operated at 9.5 GHz and equipped with a TE$_{102}$ rectangular microwave cavity was used in these studies. The orientational study was accomplished by use of a goniometer which enables one to rotate the sample 360° around a single axis. The single crystal specimen was mounted on a quartz rod by use of a small amount of silicone vacuum grease.

The square platelet single crystal was first oriented vertically in the microwave cavity [b axis pointed up, Fig. 5.1(a)]. The 0° and 90° orientations corresponded to that of the crystal plane normal (c^*) which were perpendicular and parallel to the static magnetic field (H), respectively. The resulting ESR spectra at these orientations revealed a symmetric Lorentzian lineshape. A typical single absorption ESR spectrum determined at 0° orientation is shown in Fig. 5.1(b). The above to below peak ratio (A/B) is 1.06 at this orientation which is consistent with the symmetric nature of a Lorentzian curve. The g-value and peak-to-peak linewidth of the α-$(ET)_2(NH_4)Hg(SCN)_4$ salt, with respect to rotational angle, are plotted in Fig. 5.2.

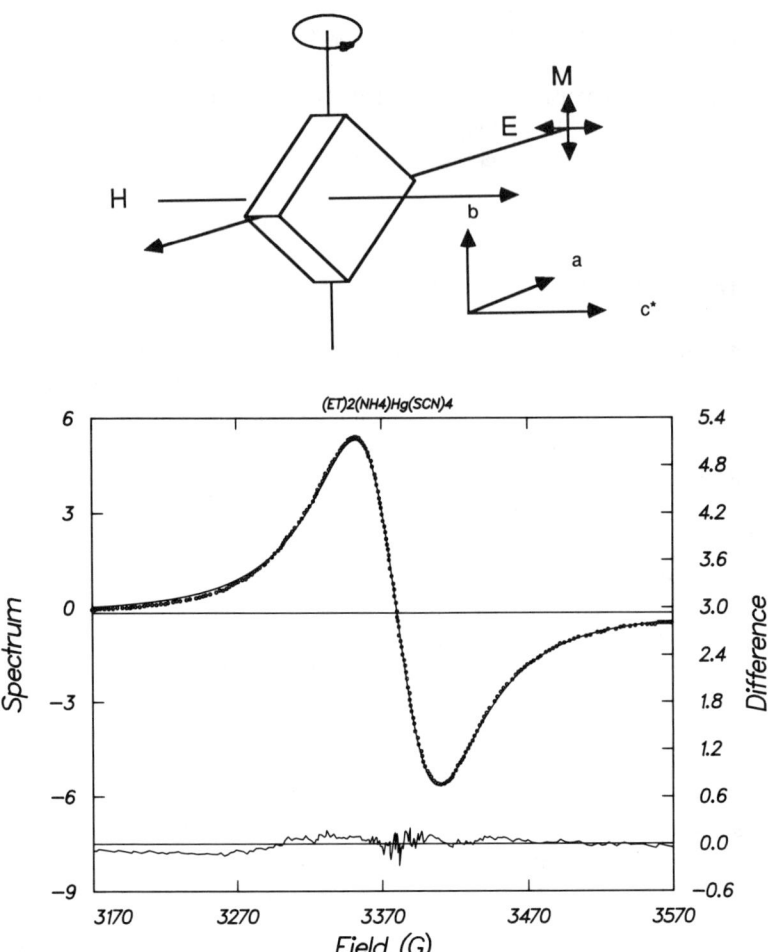

Figure 5.1 The crystal orientations (5.1a, top) and the resulting ESR spectra at 0° (5.1b).

As shown in Fig. 5.2, the open circle is the experimentally measured g-value and a least squares fit is carried out by use of the following equation:

$$g_{obs}^2 = \sum_{i,j=1}^{3} g_{ij}^2 l_i l_j \qquad (5.1)$$

where l_i and l_j are the direction cosines of the principal axes. The solid line is the calculated g-value, and it is in good agreement with the observed value. Anisotropy in the g-value arises from coupling of the spin angular momentum with the orbital angular momentum, and these details have been described elsewhere.[3] The g-values vary from a minimum of 2.002 at 0° to a maximum of 2.013 at 90°, and are quite similar to those of other ET-based synthetic metals.[4] The g-value maximum corresponds to a special orientation where the static magnetic field is approximately

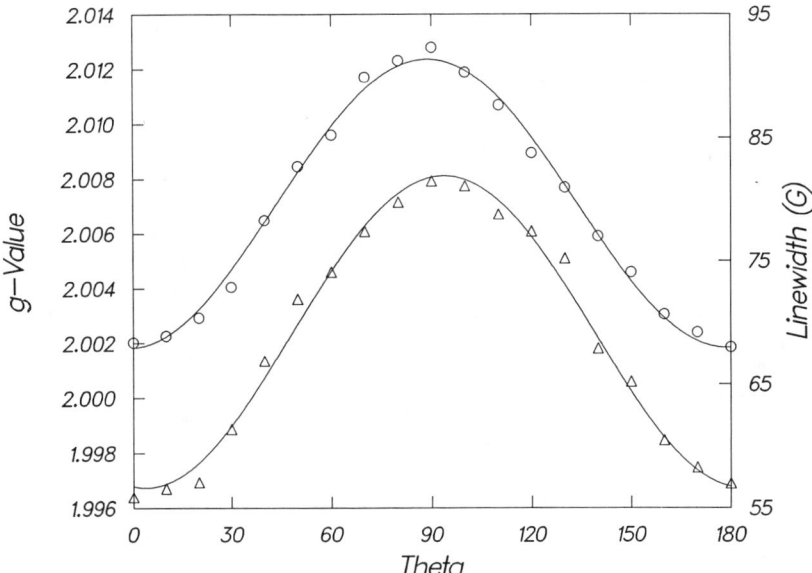

Figure 5.2 The g-values (circles) and peak-to-peak linewidths (triangles) of the α-$(ET)_2(NH_4)Hg(SCN)_4$ salt at room temperature.

parallel to the long molecular axis (or central C=C double bond) of the ET molecule. The principal g-values of many ET salts for a few major structural types, such as α, β, α', θ, and κ-phases are given in Table 5.1. The descriptions of these structural packing motifs are included in Chap. 3. The g_1, g_2, and g_3 correspond to the maximum, intermediate, and minimum g-values, respectively. A general conclusion can be drawn here, that is, the principal g-values of ET-based materials are in the range of 2.014 (maximum), 2.006 (intermediate) to 2.001 (minimum), respectively. These values correspond approximately to orientations where the static magnetic field, with respect to the ET molecule in the solid, is parallel to the long molecular axis, the short molecular axis, and the normal to the molecular plane, respectively.[6] This result suggests that the g-values of ET-based salts are essentially determined by the g tensor of the ET radical cation. Different structural phases do not appear to change the g-value distribution significantly.

The peak-to-peak linewidths, on the other hand, are usually quite characteristic of different phases (vide infra). The linewidth variations of the α-$(ET)_2(NH_4)Hg(SCN)_4$ salt[1] with different rotation angles are also shown in Fig. 5.2. The open triangles are the measured peak-to-peak linewidths, and the solid line is a calculated curve obtained by use of the following equation:

$$\Delta H = \sum_{i,j=1}^{3} \Delta H_{ij} l_i l_j \qquad (5.2)$$

The linewidths range from 55 to 80 G, approximately following the g-value variation in this case, and are consistent with that observed for the α-type packing motif.

TABLE 5.1 Principal g-Values for BEDT-TTF Salts at Room Temperature

Compounds	g_1	g_2	g_3	Reference
α-(ET)$_2$I$_3$	2.0113		2.0033	4
α-(ET)$_2$(NH$_4$)Hg(SCN)$_4$	2.013		2.002	1
α-(ET)$_2$RbHg(SCN)$_4$	2.011		2.002	79
θ-(ET)$_2$I$_3$	2.010		2.002	5
β-(ET)$_2$I$_3$	2.0111	2.0065	2.0025	4,6
β-(ET)$_2$IBr$_2$	2.0104	2.0073	2.0019	6
α'-(ET)$_2$AuBr$_2$	2.010	2.006	2.004	7
α'-(ET)$_2$CuCl$_2$	2.010	2.005	2.004	7
α'-(ET)$_2$Ag(CN)$_2$	2.009	2.006	2.003	7
(ET)$_2$AsF$_6$	2.0116	2.0067	2.0023	8
(ET)$_2$SbF$_6$	2.0115	2.0070	2.0026	8
(ET)$_2$(ClO$_4$)(TCE)$_{0.5}$	2.0125		2.0022	9
(ET)$_3$(HSO$_4$)$_2$	2.010		2.001	10
(ET)$_3$Br$_2$(H$_2$O)$_2$	2.014		2.003	11
(ET)$_5$(HgBr$_4$)$_2$(HgBr$_3$)	2.0094		2.0015	12
(ET)(Re$_6$Se$_5$Cl$_9$)(DMF)	2.0145	2.0065	2.0020	13
(ET)$_8$SiW$_{12}$O$_{40}$	2.011		2.005	14
κ-(ET)$_2$Cu(NCS)$_2$	2.0095		2.0057	15
κ-(ET)$_2$Cu[N(CN)$_2$]Br	2.007		2.002	79
κ-(ET)$_4$Hg$_{2.89}$Br$_8$	2.012		2.005	5
κ-(ET)$_4$Hg$_3$Cl$_8$	2.0135		2.0017	16

When the aforementioned square-platelet crystal of the α-(ET)$_2$(NH$_4$)Hg(SCN)$_4$ salt is oriented horizontally in the microwave cavity, that is, the microwave electric field is parallel to the crystal plane (*ab* plane), and the microwave magnetic field is parallel to the c^* axis, a Dysonian lineshape is observed. The sample orientation and the resulting ESR spectrum are shown in Fig. 5.3(a) and (b), respectively. This phenomenon is understood in terms of an eddy current set up by the microwave electric field. The Dysonian lineshape at room temperature was only observed in highly conductive specimens and only at the proper orientation. The associated g-values and peak-to-peak linewidths at this orientation showed very weak angular dependence which was consistent with the isotropic conductivity within the *ab* molecular plane.

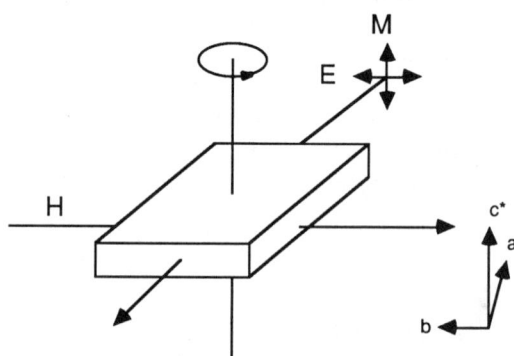

Figure 5.3(a) The crystal orientation of α-(ET)$_2$ (NH$_4$) Hg (SCN)$_4$.

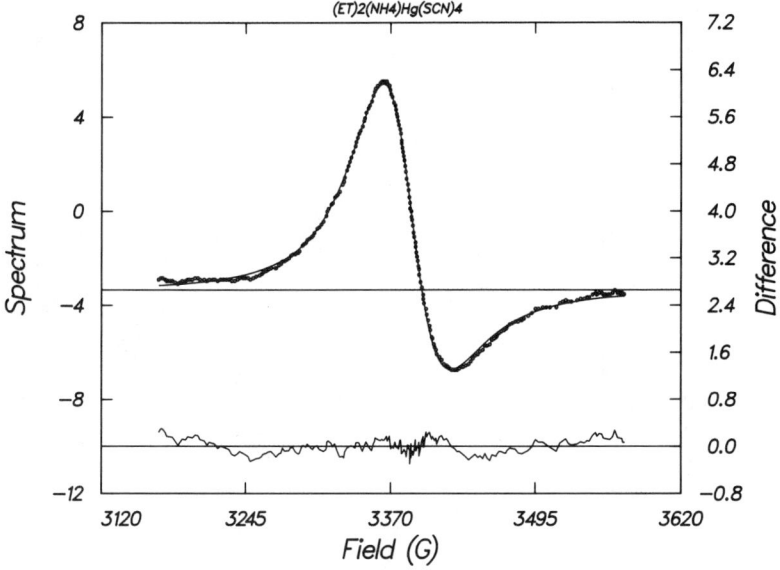

Figure 5.3(b) The resulting ESR spectra associated with the crystal orientation in Fig. 5.3a.

Phase Identification

We will now focus on the room temperature ESR peak-to-peak linewidth (ΔH) and its use in identifying different crystalline phases. In contrast to the g-values, which usually fall within a small range (2.014 to 2.001), the linewidths are quite characteristic for each structural phase. With a proper understanding of the linewidth behavior, fairly accurate phase identification can be made. The term "phase" is used here in a broad sense, and it implies both different structural phases and different stoichiometries. The room temperature ESR peak-to-peak linewidths of many ET-based materials are listed in Table 5.2. The compounds are arranged by following their anion types in approximately the same order as those that appeared in the ET crystallographic data table (Table 3.1a). The first entry next to each compound is the ESR peak-to-peak linewidth at room temperature (300 K). A single number implies that the crystal studied was at an arbitrary orientation and the value could be anywhere within the linewidth range limit of that particular compound. Whenever data from orientation studies are available, the corresponding ranges for peak-to-peak linewidths are listed. As shown in Table 5.2, the ESR linewidths range from a minimum of near 4 Gauss to a maximum near 120 Gauss, at 300 K, for a majority of the ET derivatives except when magnetic anions such as $FeX_4^- (X^- = Cl^-, Br^-)$ and $MnCl_4^{2-}$ are present.

It is important to point out that the origin of the ESR linewidths for organic metals is not clearly understood. Both theoretical equations and empirical modifications have been proposed (vide infra). The contribution from spin-orbit coupling certainly plays a major role and there are at least three considerations associated with this coupling. The structural packing motif, or the surrounding symmetry of each ET radical cation in the solid, is a predominant factor that determines the midpoint value of the peak-to-peak linewidth (ΔH_m). We use the notation ΔH_m to

TABLE 5.2 Summary of ESR Data for BEDT-TTF Based Salts

Compound	ΔH (300 K) G	ΔH (T,K) G	χ (300 K) emu/mol	Property	Reference
			Linear Anions		
α-(ET)$_2$I$_3$	70–110	2 (5 K)	6.8×10^{-4} [a]	T_{MI} = 135 K	4,17,18
β-(ET)$_2$I$_3$	17.60–23.35	10 (80 K)	4.6×10^{-4} [a]	Pauli paramagnetism T_c = 1.4 K anomaly at 140 K	4,17,19
β_D-(ET)$_2$I$_3$	22–32	28 (104 K)			20
θ-(ET)$_2$I$_3$	60–80	1.31 (10 K)		Pauli paramagnetism T_c = 3.6 K	5
β^*-(ET)$_2$I$_3$		0.5 (8 K)		Pauli paramagnetism T_c = 8 K	19,21
α_t-(ET)$_2$I$_3$	20		4.5×10^{-4} [a]	Pauli paramagnetism T_c ~ 7 K	18,22
β-(ET)$_2$I$_2$Br	19	3 (5 K)	1.5×10^{-3}	Pauli paramagnetism	19,23
α-(ET)$_2$IBr$_2$	45			Curie-Weiss	24
β-(ET)$_2$IBr$_2$	15.28–21.11	0.2 (5 K)	8×10^{-4}	Pauli paramagnetism T_c = 2.8 K	6,19
α-(ET)$_2$BrICl	50				25
β'-(ET)$_2$BrICl	10				25
β'-(ET)$_2$ICl$_2$	10		2×10^{-3}	Bonner Fisher/Spin density wave χ decreases at 17 K	25
β''-(ET)$_2$ICl$_2$	10				26
β-(ET)$_2$AuI$_2$	15–20	2.4 (5 K) 0.6 (2 K)	3.4×10^{-4}	Pauli paramagnetism T_c = 4.98 K	19,27
β''-(ET)$_2$IAuBr	45			Bonner Fisher	26
α'-(ET)$_2$AuBr$_2$	37–50	100 (4.2 K)		Bonner Fisher	28
β''-(ET)$_2$AuBr$_2$	40, 45	0.6 (4.2 K)		ΔH decreases at 24 K	29
δ-(ET)$_2$AuBr$_2$	10	7 (110 K)		Semiconductor	30
β'-(ET)$_2$AuCl$_2$	6	0.6 (33 K)	9.0×10^{-4}	Bonner Fisher/Spin density wave	31
α'-(ET)$_2$Au(CN)$_2$	34				32
α'-(ET)$_2$Ag(CN)$_2$	24–34	~500 (4.2 K)	9.3×10^{-4}	Bonner Fisher/Spin Peierls	28
(ET)$_2$Ag(CN)$_2$(II)	25				33
(ET)$_2$Ag(CN)$_2$(III)	16–28	1.2 (35 K)		T_{MI} = 100 K	33
α'-(ET)$_2$CuCl$_2$	41–52	~300 (4.2 K)	9.0×10^{-4}	Bonner Fisher	28
			Planar Anions		
(ET)$_2$C(CN)$_3$	25–35	4 (100 K)		T_{MI} = 180 K	34
(ET)$_2$[C$_5$(CN)$_5$](TCE)$_x$	18	43 (30 K)		Bonner Fisher	35
(ET)$_4$Pt(C$_2$O$_4$)$_2$	43–46	3.5 (4.2 K)		Metal-to-metal Transition at 200 K	36

[a] Determined from magnetic susceptibility measurements.

TABLE 5.2 Summary of ESR Data for BEDT-TTF Based Salts (*continued*)

Compound	ΔH (300 K) G	ΔH (T,K) G	χ (300 K) emu/mol	Property	Reference
β-(ET)$_4$[Pt(CN)$_4$]	32			T_{MI} = 200 K	37
γ-(ET)$_4$[Pt(CN)$_4$]	35–65	4 (4.2 K)		Bonner Fisher	37
δ-(ET)$_4$[Pt(CN)$_4$]	25–40			Semiconductor	37
(ET)(AuBr$_2$Cl$_2$)	19	2.1 (100 K)		Semiconductor	38
(ET)$_2$AuCl$_4$	9	2 (100 K)		Disruptive phase transition at 120 K	39
(ET)$_2$Au(CN)$_2$Cl$_2$	20	6 (200 K)		T_{MI} ~ 295 K	39
			Tetrahedral Anions		
(ET)$_2$ReO$_4$	13–19	2 (5 K)		T_{MI} 80 ~ 90 K	40
(ET)$_3$(ReO$_4$)$_2$	33–39	2 (5 K)		T_{MI} = 88 K	40
(ET)$_2$(ReO$_4$)(THF)$_{0.5}$	12–18	3 (5 K)			40
(ET)$_2$(ClO$_4$)(TCE)$_{0.5}$	22–34	5 (3 K)	4.2 × 10^{-4}		9
(ET)$_3$(ClO$_4$)$_2$	44–65	1 (3 K)	6.2 × 10^{-4}		9
(ET)$_2$BrO$_4$		5.5 (100 K) 0.3 (6 K)		T_{MI} = 180 K (from conductivity)	41
(ET)$_3$(HSO$_4$)$_2$	44–65	11 (100 K)		T_{MI} = 130 K	10
(ET)$_2$GaCl$_4$	9–10			Semiconductor	42
(ET)$_2$InCl$_4$	8–10	5.5 (100 K)		Semiconductor	42
(ET)$_2$TlCl$_4$	8–10			Semiconductor	42
(ET)$_2$InBr$_4$	8–10, 42–47[a]			Semiconductor	42
(ET)$_2$TlBr$_4$	8–11	4.5 (100 K)		Semiconductor	42
(ET)$_2$GaI$_4$	14–16, 39–51[a]			Semiconductor	42
(ET)$_2$InI$_4$	14–19			Semiconductor	42
(ET)$_2$TlI$_4$	17–21			Semiconductor	42
(ET)$_2$CuCl$_4$	50–120	30 (4.2 K)		T_{MI} = 100 K	43
(ET)$_2$FeCl$_4$	700	700 (4.2 K)		Semiconductor	43
(ET)FeBr$_4$	1400			Insulator	43
(ET)$_3$(MnCl$_4$)$_2$	220	~250 (10 K)		Curie-Weiss	44
(ET)$_5$(HgBr$_4$)$_2$(HgBr$_3$)	16	5 (10 K)	1.2 × 10^{-4}	T_{MI} ~ 120 K	12
			Octahedral Anions		
(ET)$_2$AsF$_6$	16–24	2 (100 K)		T_{MI} = 264 K	8
(ET)$_2$SbF$_6$	16.25–25	7 (100 K)		T_{MI} = 273 K	8

TABLE 5.2 Summary of ESR Data for BEDT-TTF Based Salts (*continued*)

Compound	ΔH (300 K) G	ΔH (T,K) G	χ (300 K) emu/mol	Property	Reference
		Monoatomic Anions			
(ET)$_3$Cl$_2$(H$_2$O)$_2$	37	5 (50 K)		T_{MI} = 100 K	45
(ET)$_3$Br$_2$(H$_2$O)$_2$	38–50	3.5 (12 K)	9.7×10^{-4}	T_{MI} = 185 K	11
		Other Polyatomic Discrete Anions			
(ET)$_4$(Hg$_2$Br$_6$)(TCE)	4				46
(ET)$_2$(CH$_3$C$_6$H$_4$SO$_3$)	29	140 (30 K)		Bonner Fisher	47
(ET)(Re$_6$Se$_5$Cl$_9$)(DMF)$_2$	3.7–4.3	~ 0.3 (10 K)		Bonner Fisher	13
(ET)$_8$SiW$_{12}$O$_{40}$	30–45			Semiconductor	14
		Polymeric Anions			
(ET)$_3$Ag$_{6.4}$I$_8$	30	1.95 (2 K)	2.8×10^{-3}		48
(ET)Ag$_4$(CN)$_5$	12–14			Curie-Weiss	49
(ET)Ag$_{1.6}$(SCN)$_2$	38	0.78 (4.2 K)		Curie-Weiss	50
(ET)Ag$_{2.4}$Br$_3$	45–65	20 (104 K)		T_c = 5.0 K	51
κ-(ET)$_2$Ag(CN)$_2$(H$_2$O)	70			Pauli paramagnetism T_c = 10.4 K	33,86
κ-(ET)$_2$Cu(NCS)$_2$	58–68	200 (15 K)		T_c = 11.6 K	15,52
κ-(ET)$_2$Cu[N(CN)$_2$]Br	60–80	40 (10 K)		T_c = 12.8 K (0.3 kbar)	79
κ-(ET)$_2$Cu[N(CN)$_2$]Cl	55–75			Pauli paramagnetism T_c = 4.2 K	85
κ-(ET)$_4$Hg$_{2.89}$Br$_8$	69–83	148 (10 K)		Pauli paramagnet at T > 220 K	5
κ-(ET)$_4$Hg$_3$Cl$_8$	80–110	< 20 (4 K)	8×10^{-4}	Semiconductor	53,54
(ET)Cu$_2$(NCS)$_3$	7			Bonner Fisher	50
(ET)$_2$HgBr$_3$(TCE)	16,20	0.9 (10 K)		Pauli paramagnetism	55
α-(ET)$_2$(K)Hg(SCN)$_4$	70, 83	~ 2 (10 K)		Pauli paramagnetism T_c = 1.15 K	56
α-(ET)$_2$(NH$_4$)Hg(SCN)$_4$	56–84	0.33 (4.1 K)		Pauli paramagnetism	1
α-(ET)$_2$(Rb)Hg(SCN)$_4$	55–81	24 (100 K)		Pauli paramagnetism	79
(ET)Hg$_{0.776}$(SCN)$_2$	22–34	1.1 (4.2K)		Pauli paramagnetism	66

denote an averaged value of the maximum and minimum linewidths. The linewidth can also be broadened by the nature of the anions although this is a relatively minor effect. For example, the interactions between the triiodide anion (I_3^-) and the ET radical cation causes strong spin-orbit coupling.[4,6] This argument is based on the comparisons between α- and β-$(ET)_2X$ ($X^- = I_3^-$ and IBr_2^-) salts, where data from isostructural compounds are available. For example, ΔH of β-$(ET)_2I_3$ (17.6–23.4 G) is slightly larger than that of β-$(ET)_2IBr_2$ (15.3–21.1 G). Additional information on the isostructural β"-$(ET)_2X$ salts ($X^- = IAuBr^-$, $AuBr_2^-$, and ICl_2^-) also supports it. In other words, the chloride containing anions always give rise to narrower linewidths compared to isostructural salts containing bromide or iodide anions. The linewidth anisotropy caused by different crystal orientations can be estimated with Eq. (5.3) and usually falls within the following range, $\pm(\Delta H_m \times 20\%)$. This is an empirical observation, and for instance, the linewidth range of α-$(ET)_2I_3$ is 90 ± 20 G, β-$(ET)_2I_3$ 21 ± 4 G, and $(ET)_3(HSO_4)_2$ 55 ± 11 G. The resulting linewidth range can be described as follows:

$$\Delta H = \Delta H_m \pm (0.2) \Delta H_m \qquad (5.3)$$

where ΔH_m is the average between maximum and minimum linewidths, and is mainly determined by the structural packing motif in the crystals under study.

We now investigate the correlation between the ΔH_m values of different crystalline ET phases and their corresponding structural packing motifs. The linewidth ranges associated with different structural types are given in Fig. 5.4. For

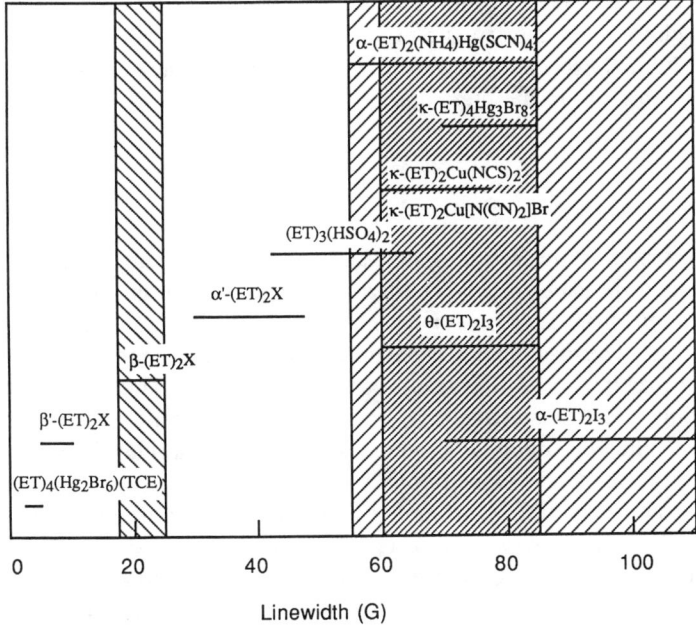

Figure 5.4 The ESR linewidth ranges associated with different structural phases. The horizontal lines indicate the linewidth ranges for each compound and the superconducting, α, β, θ, and κ-phases appear in shaded areas.

linewidths larger than 60 G, there are only three known structural types, α, θ, and κ. Four salts are known to form α-phases, that is, $(ET)_2I_3$ (70–110 G),[17] $(ET)_2(NH_4)Hg(SCN)_4$ (56–84 G),[1] $(ET)_2(K)Hg(SCN)_4$ (70, 83 G)[56] and $(ET)_2(Rb)Hg(SCN)_4$ (55–81 G).[79] Only one θ-phase is known, that is, θ-$(ET)_2I_3$ (60–80 G),[5] and its structure is very similar to α-$(ET)_2I_3$ except that the θ-phase salt has two crystallographically unique donor-molecule layers in the unit cell.[57] Within the donor-molecule layer, the inter- and intrastack packing modes are almost identical to α-$(ET)_2I_3$. There are six ET-based κ-phase salts with linewidths varying from 55 to 110 G, and they are listed in Table 5.2. Common features among these salts having large linewidths include: good metallic conductivity near room temperature and W-type (see Chap. 3) interstack packing. The conduction electron spin resonance linewidth has been correlated to the spin-orbit coupling (Δg) and electron scattering rate (τ^{-1}) by use of the Elliot formula:[9,58]

$$\Delta H \sim \frac{(\Delta g)^2}{\tau} \tag{5.4}$$

and

$$\sigma = \frac{ne^2\tau}{m} \tag{5.5}$$

where $\Delta g = g - 2.0023$, σ is the conductivity, m is effective mass and n is the number of carriers. The Elliot formula describes the temperature dependence of an ESR linewidth at a fixed orientation quite well. However, application of the Elliot formula is limited especially in anisotropic layered two-dimensional materials, as has been pointed out in the literature.[6] Equation 5.4 has been modified to:[59]

$$\Delta H \sim (\Delta g)^2 \tau_{\parallel}^{-1} \frac{\tau_{\parallel}}{\tau_{\perp}} = \frac{(\Delta g)^2}{\tau_{\perp}} \tag{5.6}$$

where $\tau_{\parallel}$ is the on-chain scattering time and $\tau_{\perp}$ is the interchain tunneling time. The ratio $\tau_{\parallel}/\tau_{\perp}$ is a measure of the 2D character of the system under study. The estimated linewidth from Eq. 5.6 for the ET-based materials is three orders of magnitude too large. The discrepancy is understood by use of a theoretical treatment that correlates the relaxation with torsional oscillations of the organic donor molecule.[60] The resulting effect is that the spin-orbit interaction cannot cause an electronic transition within a compound that consists of inversion-related molecules. The theory was originally developed for $(TMTSF)_2X$ salts and has been applied to certain ET-based salts.[59] Two schematic diagrams of views along the c* axes of the α- and β-$(ET)_2I_3$ salts, with marked unit cell and inversion centers, are shown in Fig. 5.5 (a) and (b). In the β-type structure, there is only one kind of ET stack and the inversion centers are located in-between the stacks. The α-type structure consists of two nonequivalent stacks with ET molecules in each stack inclined at about 80 ~ 90° with respect to each other, and no inversion centers are located between stacks. The effect of these two packing motifs is that the interstack spin-orbit

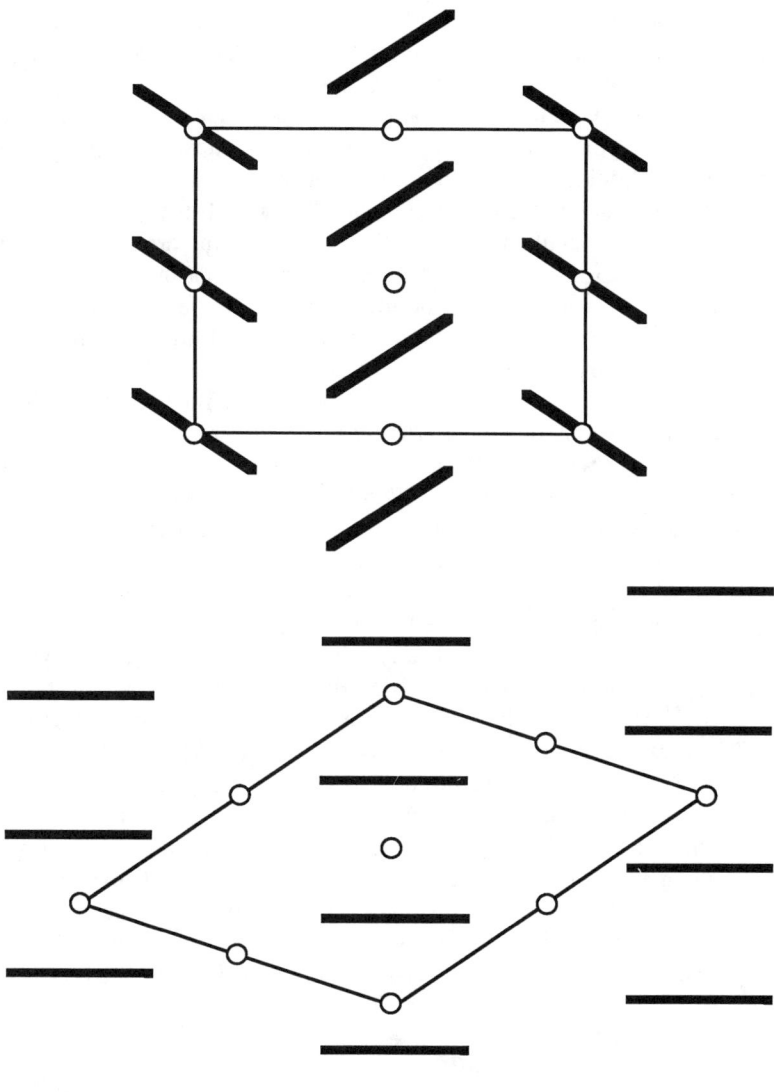

Figure 5.5 Two schematic diagrams viewed along the c* axes of α-(ET)$_2$I$_3$ (5.5a, top) and β-(ET)$_2$I$_3$ (5.5b, bottom). The thin lines indicate the unit cell and the circles are the inversion centers.

scattering is more effective in the α-phase materials. In other words, the $\tau_\perp^{-1}$ (interstack scattering rate) for α-phase systems is larger than that of the β-phase. This is the reason the structural packing motif contributes strongly to the resulting peak-to-peak linewidths (ΔH).

This rationale adequately explains the broader linewidths for α, θ, and κ-phases, and the narrower linewidths for β (17 ± 4G),[4,17] β' (~ 10 G),[25] and β"-phases (~ 40 G).[29] Additional contributions from spin-orbit coupling as a result of the anions are noted in α-(ET)$_2$X salts ($X^- $ = IBr$_2^-$ and BrICl$^-$, 45 ~ 50 G),[24,25] and α-(ET)Ag$_{2.4}$Br$_3$ (45–65 G).[51] A few ET salts contain alternating W-type and L-type interstack packing motifs (see Chap. 3), such as (ET)$_3$Ag$_{6.4}$I$_8$ and (ET)$_2$(HgBr$_3$)(TCE), with the result that the interstack packing follows an α-β-α-β pattern. The resulting linewidths are slightly broader than those of the β-phase, being 30 G for the silver-iodide salt and 16 ~ 20 G for the mercury-bromide salt.

All the aforementioned ET salts contain either a- or b-type intrastack packing motifs. A few ET salts containing c-type intrastack packing arrangements are discussed here (see Chap. 3). The c-type packing denotes that the ET molecules are crossed along the stacking axis with the rotation angle formed by two long molecular axes being near 30°. Typical examples for this class of materials are α'-(ET)$_2$X salts, with X^- = AuBr$_2^-$, Au(CN)$_2^-$, Ag(CN)$_2^-$, and CuCl$_2^-$.[28,32] A schematic diagram of the α'-(ET)$_2$X salts with the unit cell marked is shown in Fig. 5.6. In this class of materials, inversion centers are located in the anion layer and not in the donor- molecule layer with the interstack packing mode being L-type (β-like). The observed linewidths for these salts are 37–50 G and 41–52 G for the AuBr$_2^-$ and CuCl$_2^-$ salts, respectively. These values lie between that of the α- and β-phases (vide supra). The linewidths for the Au(CN)$_2^-$ and Ag(CN)$_2^-$ salts are in the range of 24 to 34 G, which overlap slightly with the β-phase, and could be caused by the weaker spin-orbit coupling due to the anions.

Another class of materials with c-type intrastack packing and L-type interstack packing is represented by δ-(ET)$_2$AuBr$_2$,[61] (ET)$_2$AsF$_6$, (ET)$_2$SbF$_6$,[8] and (ET)$_2$C(CN)$_3$.[34] The major difference between these salts and the α'-(ET)$_2$X salts is that ET dimers occur along the donor-molecule stack. The resulting intrastack packing motif is c-type within each dimer, and b-type between dimers (see Chap. 3). The first three salts contain two donor-molecule layers per unit cell, while the last salt contains only one layer per unit cell. A schematic diagram for δ-(ET)$_2$AuBr$_2$ as viewed along the c axis, with the unit cell boundaries and inversion centers marked, is shown in Fig. 5.7. The inversion centers both in the stack and between the stacks are clearly evident. The observed ESR peak-to-peak linewidths for these four compounds are: 10 G, 16–24 G, 16–25 G, and 25–35 G, respectively. Another

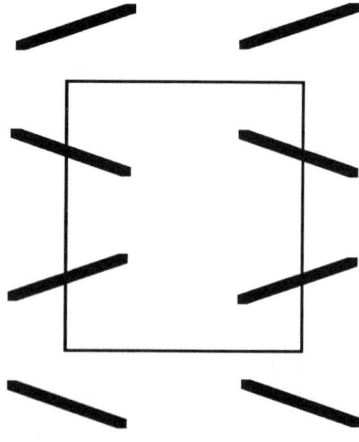

Figure 5.6 Schematic diagram of α'-(ET)$_2$X salts.

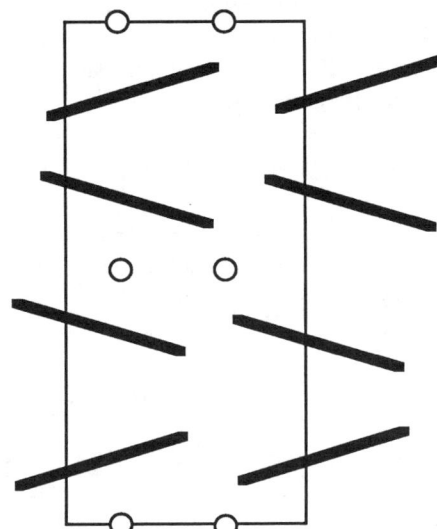

Figure 5.7 Schematic packing diagram of δ-(ET)$_2$AuBr$_2$. The thin lines indicate the unit cell and the circles are the inversion centers.

compound worth noting with c- and L-type intra- and interstack packing mode, respectively, is (ET)$_4$(Hg$_2$Br$_6$)(TCE).[46] Inversion centers are located between stacks, and the measured linewidth is very narrow, ~ 4 G.

In summary, W-type interstack packing modes are known to give rise to broad ESR peak-to-peak linewidths (> 60 G), while L-type interstack packing motifs generate narrow linewidths (~ 20 G). The rationale for this behavior depends on whether or not inversion centers are located between stacks. When inversion centers occur between stacks, the spin-orbit interaction is strongly reduced and the resulting linewidth is relatively narrow.

LOW-TEMPERATURE ESR STUDIES

The microwave conductive properties of organic metals can be obtained from the variable temperature behavior of the derived spin susceptibility and the peak-to-peak linewidth. Shown in the third column of Table 5.2 are the available peak-to-peak linewidths of numerous compounds at the lowest measured temperatures. Most of the ET salts show very narrow linewidths at low temperature except that for the α′- and κ-phase derivatives. The residual linewidths at 5 K or below vary from 3 G to 0.2 G. The large differences in ESR linewidth between α- and β-phases at 300 K seems to be totally absent at ~ 4 K. For instance, the linewidths at ~4 K for α-(ET)$_2$(NH$_4$)Hg(SCN)$_4$, β-(ET)$_2$IBr$_2$, and β-(ET)$_2$AuI$_2$ are 0.33 G, 0.2 G, and 0.6 G, respectively.[19] Caution must be exercised in comparing the values listed in column 3 because many compounds have undergone phase transitions at low temperature, and the crystal structure above and below the phase transition may be different. Listed in the fourth column of Table 5.2 is the derived spin susceptibility (χ). Typical values of χ for metallic specimens are on the order of 10^{-4} emu/mol. Some of the values presented are from magnetic susceptibility measurements. The fifth column is a summary of the electrical and magnetic properties of each salt. In the following section, we will focus on the ESR behavior observed in various ET salts between 300 and 4 K, and possible reasons for the similar ESR linewidths of the α- and β-phases at 4 K.

Organic Superconductors

Among the ET-based organic metals, four types of structural packing motifs, the α, β, θ, and κ-phases, are known to lead to superconductivity. The first three types of salts have similar ESR behavior while that of the κ-phase is different from the other three.

α, β, and θ-Phase Superconductors.

The new organic superconductor, α-$(ET)_2(NH_4)Hg(SCN)_4$, with T_c near 1.15 K is the only α-phase material that exhibits superconductivity.[1] The peak-to-peak linewidth at 300 K varies between 56 and 84 G. When a platelet crystal is mounted vertically with the b axis pointed up, and with the c^* axis parallel to the static magnetic field in the microwave cavity, the ESR linewidth is the maximum of 84 G and the lineshape is Lorentzian. Variable temperature ESR measurements have been carried out to 4.1 K with this crystal orientation. The linewidth (triangle), and relative spin susceptibility (square), from 150 K to 4 K are shown in Fig. 5.8. The linewidth (ΔH) decreases monotonically from 28.3 G at 150 K to 0.33 G at 4.1 K. The linewidth can be correlated with temperature (T) between 20 and 100 K by use of the following linear relationship:

$$\Delta H = bT + a \tag{5.7}$$

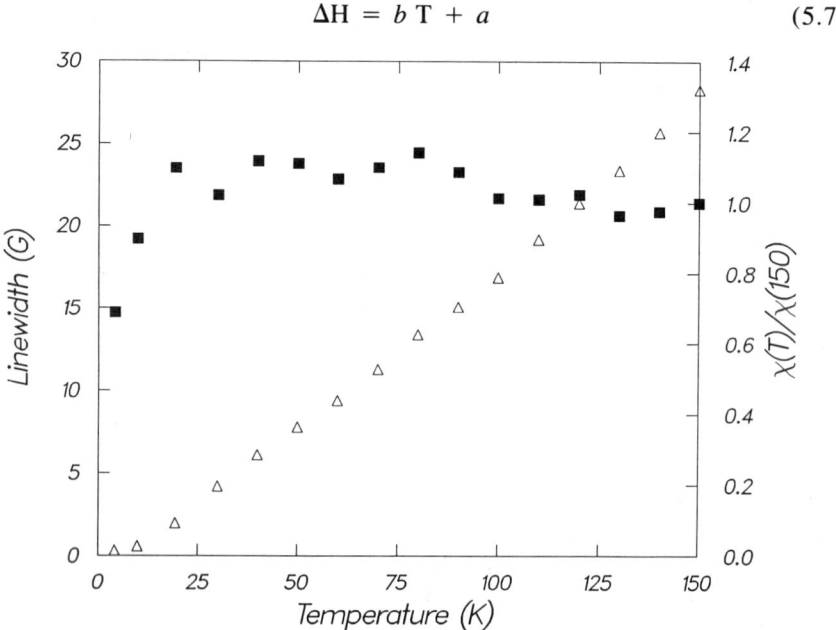

Figure 5.8 The linewidth (triangles) and relative spin susceptibility (squares) from 150 K to 4 K of α-$(ET)_2(NH_4)Hg(SCN)_4$.

where b and a are the slope (G/K) and intercept (G), respectively. Below 20 K, the linewidth decreases at a much slower rate with decreasing temperature, and reaches a residual value of 0.3 G at 4.1 K. The conduction electron spin resonance linewidth can be qualitatively understood by use of the Elliot formula [Eqs. (5.4),(5.5)]. Since Δg is a constant at a fixed crystal orientation, and the conductivity (σ) of the sample increases with decreasing temperature, the resulting linewidth is expected to decrease with decreasing temperature. The spin susceptibility is approximately constant from 150 K to 20 K. The constant susceptibility behavior is termed Pauli paramagnetism and is that observed for conduction electrons in metallic samples. Below 20 K, the spin susceptibility begins to decrease and is likely due to a skin effect. The penetration depth (δ) of a periodic field in a metal is determined by:[62]

$$\delta = \frac{c}{2\pi\sqrt{\sigma}} \quad (5.8)$$

As the conductivity (σ) increases significantly, the penetration depth begins to decrease with the result that the measured ESR signal does not correspond to that of the entire sample and consequently the signal intensity decreases.

Variable temperature ESR measurements (300 and 100 K) on θ-(ET)$_2$I$_3$ have been reported.[5] As mentioned in the previous section, the crystal structure of the θ-phase donor layer is almost identical to that of α-phase materials. The low-temperature results on θ-(ET)$_2$I$_3$ are also similar to those of α-(ET)$_2$(NH$_4$)Hg(SCN)$_4$. The linewidth drops monotonically from 69 G (300 K) to 28 G (104 K), and follows a linear temperature dependence as described in Eq. (5.7). The spin susceptibility is a constant and is consistent with the expected Pauli paramagnetism of a metal.

The low-temperature ESR properties of β-(ET)$_2$X salts ($X^- = I_3^-$, I_2Br^-, IBr_2^-, and AuI_2^-) have been reported.[17,19,23,27] Except for the I_2Br^- salt, these salts are ambient pressure superconductors. The lack of superconductivity in the I_2Br^- salt has been attributed to the disorder of the unsymmetric anion and the random periodic potential it causes.[23] The temperature dependent ESR behavior of these salts is similar to that of the aforementioned α- and θ-phase compounds, namely, a constant spin susceptibility from 300 to 4 K, and a monotonic decrease in peak-to-peak linewidth with decreasing temperature. The residual ESR linewidths of the above four β-(ET)$_2$X salts have been studied recently and the peak-to-peak linewidths follow a nearly linear decrease with decreasing temperature between 100 and 25 K.[19] The behavior can be described by use of Eq. (5.7), and the slopes, which are related to the electron scattering rate, are nearly identical for all four β-phase salts. This anion independent scattering rate is consistent with the model that correlates the relaxation rate with torsional oscillations of the organic-donor molecule (vide supra).[19,60] Further support of this interpretation is that the aforementioned α-(ET)$_2$(NH$_4$)Hg(SCN)$_4$ and θ-(ET)$_2$I$_3$ salts also have similar slopes although they are somewhat larger than those of the β-(ET)$_2$X salts. Below 15 K, the linewidths of the β-(ET)$_2$X salts reach a residual value and remain almost constant to ~ 2 K. This is probably due to the fact that the $(\Delta g)^2/\tau$ contribution to the linewidth is rather small at very low temperature, and the residual linewidth begins to dominate. This phenomenon is similar to the residual resistance of a good metal at low temperature. The origin of the residual resistance has been attributed to small amounts of impurities, on the part per million level, or strains and dislocations in the crystalline sample. One of the factors contributing to the residual ESR linewidth can be the disorder of the anion or the ethylene groups of the donor molecule.[19] For instance, the totally ordered (donor molecule and anion) β-phase IBr_2^- and AuI_2^- salts give rise to residual linewidths around 0.2 and 0.6 G, respectively, while the I_3^- salt (modulated structure) and the I_2Br^- salt (disordered anion) have residual linewidths near 2.5 G. Measurements on β*-$(ET)_2I_3$, the fully-ordered high-T_c phase of β-$(ET)_2I_3$ with the ethylene-group structural disorder suppressed by shear stress,[80] reveal a sharp residual linewidth of 0.5 at 8 K.[19,21] Recent results on α-$(ET)_2(NH_4)Hg(SCN)_4$ single crystals ($\Delta H = 0.33$ G at 4.1 K) further indicate that the residual linewidth is a measure of crystal perfection irrespective of the various ET salt structural packing motifs.

κ-phase Superconductors. There are seven ET-based κ-phase salts, viz., κ-$(ET)_2X$, $X^- = Cu(NCS)_2^-$,[15,52] I_3^-,[82] $Cu[N(CN)_2]Br^-$,[79,81] $Cu[N(CN)_2]Cl^-$,[85] $(Hg_{2.89}Br_8^{2-})_{0.5}$,[5] $(Hg_3Cl_8^{2-})_{0.5}$,[53,54] and $Ag(CN)_2^- \cdot (H_2O)$.[33,86] The $Cu[N(CN)_2]Cl^-$ and $Hg_3Cl_8^{2-}$ salts exhibit metal-insulator transitions near 42 and 100 K, respectively, while the other salts are ambient pressure superconductors with T_c's ranging from 3.6 to 11.6 K.[15,81,82,83,86] The low-temperature ESR measurements on the $(ET)_4Hg_{2.89}Br_8$ salt were carried out on a platelet sample with the *ab* crystal plane oriented horizontally in the microwave cavity.[5] The microwave electric field was parallel to the highly conductive crystal plane and a strongly Dysonian lineshape was observed at all temperatures. The peak-to-peak linewidths were extracted by use of a lineshape analysis program.[84] The relative spin susceptibilities (square) and linewidth (triangle) versus temperature are shown in Fig. 5.9. The most distinct feature of the κ-phase salts is that the peak-to-peak linewidth increases with decreasing temperature. In the $Hg_{2.89}Br_8^{2-}$ salt, the linewidth is 65 G at 300 K and 147 G at 10 K.[5] This behavior is in sharp contrast to that of the β-phase salts (vide supra) and clearly cannot be explained by use of the Elliot formula. A linewidth increase with decreasing temperature has been observed in TTF-TCNQ[63] and κ-$(MDT-TTF)_2AuI_2$.[63,70,87] A similar phenomenon has been observed in the study of the pressure dependence of β-$(ET)_2I_3$, in which the ESR linewidth increases with increasing pressure in the 0 to 5 kbar pressure range.[59] According to the model where torsional oscillations of the ET molecule can lift symmetry restrictions and enable intrachain and interchain spin relaxation to cause broader linewidths, higher pressure should stiffen the crystal lattice, which should reduce the oscillations and thereby sharpen the ESR linewidth. The results are contradictory to the experimental findings. An empirical formula for describing this situation has been proposed:

$$\Delta H = (\Delta g)^2 (a \tau_\parallel^{-1} + b \tau_\perp^{-1}) \tag{5.9}$$

where $\tau_\parallel^{-1}$ and $\tau_\perp^{-1}$ are the intrachain and interchain electron scattering rates. These two terms are related by perturbation theory through the following equation

$$\tau_\perp^{-1} = c \mid t_\perp \mid^2 \tau_\parallel \tag{5.10}$$

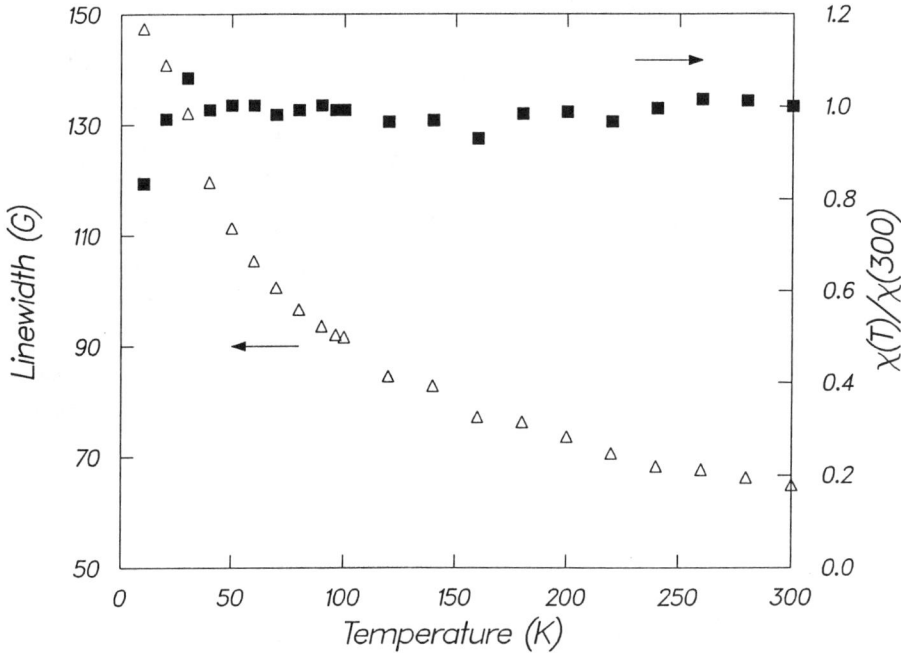

Figure 5.9 The linewidth (triangles) and relative spin susceptibility (squares) from 300 K to 4 K of κ-(ET)$_4$Hg$_{2.89}$Br$_8$.

where $t_\perp$ is the interchain transfer integral. The increase in linewidth of β-(ET)$_2$I$_3$ with pressure is brought about by the strong increase in $\tau_\parallel$, which contributes to the $b\tau_\perp^{-1}$ term in Eq. (5.9). The increase in linewidth of the κ-(ET)$_4$Hg$_{2.89}$Br$_8$ salt can be rationalized by the use of the same concept; that is, at lower temperature the *intra*dimer interactions (and $\tau_\parallel$) increase strongly, which has a secondary effect on the *inter*dimer electron scattering rate ($\tau_\perp^{-1}$). Therefore, an increase in $\tau_\parallel$ contributes to the linewidth increase at low temperature.

At 4.2 K, the broad signal observed in κ-(ET)$_4$Hg$_{2.89}$Br$_8$ disappears on the onset of superconductivity. The spin susceptibility is constant between 300 and 30 K and consistent with the expected Pauli paramagnetism. Below 30 K a significant decrease is observed, which is likely due to the skin effect (vide supra). The temperature dependence of the spin susceptibility for the κ- and β-phases is basically identical.

Variable temperature ESR measurements on the κ-(ET)$_2$Cu(NCS)$_2$ salt reveal results similar to those of the κ-(ET)$_4$Hg$_{2.89}$Br$_8$ salt, i.e., the linewidth increases from 61 G at 300 K to over 200 G at 15 K.[52] The relative spin susceptibility is approximately constant from 300 to 35 K, at which temperature a significant decrease is observed. There is one difference between these two κ-phase compounds worth noting, i.e., a distinct slope change in the linewidth versus temperature plot for κ-(ET)$_2$Cu(NCS)$_2$.[52] The log(linewidth) versus log(temperature) plots for both κ-phase compounds are shown in Fig. 5.10.[5] A linear relationship between linewidth and temperature is found for κ-(ET)$_4$Hg$_{2.89}$Br$_8$ between 300 and 20 K.

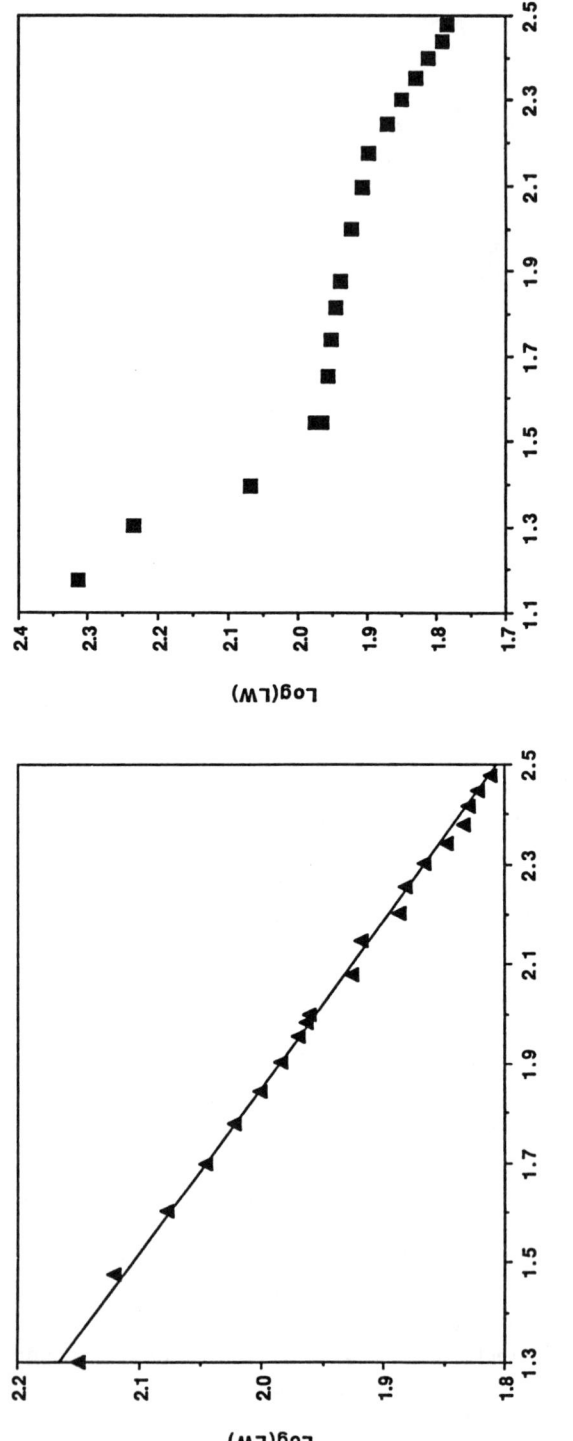

Figure 5.10 The log (linewidth) versus log (temperature) plots for κ-(ET)$_4$Hg$_{2.89}$Br$_8$ (5.10a, left) and κ-(ET)$_2$Cu(NCS)$_2$ (5.10b, right).

However, two notable slope changes are observed in κ-(ET)$_2$Cu(NCS)$_2$ at 100 and 30 K.[52] The change of slope at 100 K corresponds to an observed resistance maximum, and the change at 30 K may be associated with a structural instability. The latter point requires verification since the results from X-ray and neutron diffraction studies on different crystals of κ-(ET)$_2$Cu(NCS)$_2$ yield different results.[64,65] The differing behavior in the log (ΔH) versus log (T) plots for the two κ-phase salts also parallels the different pressure dependence of the T_c's. The T_c of the Cu(NCS)$_2$$^-$ salt decreases with increasing pressure, at a rate of −3 K/kbar, while that for the Hg$_{2.89}$Br$_8$$^{2-}$ salt shows a positive dependence of +1 K/kbar between 0 to 3.5 kbar. Additional ESR studies on the κ-phase materials are clearly needed in order to gain insight into this unusual class of compounds.

Organic Metals

For those compounds that show metallic behavior from room to very low temperature, i.e., without the occurrence of metal to insulator (MI) transitions, the ESR results are expected to be similar to those of the aforementioned organic superconductors. Typical examples include: β-(ET)$_2$I$_2$Br, β-like-(ET)Hg$_{0.776}$(SCN)$_2$, and α-(ET)$_2$(K)Hg(SCN)$_4$.[23,56,66] The ESR measurements on these compounds reveal constant spin susceptibilities and a monotonic decrease in linewidth with decreasing temperature. In the following section, organic metals are divided into two categories, those with and without sharp MI transitions.

Organic Metals Exhibiting Sharp Metal-to-Insulator Transitions.

The spin susceptibilities of metallic and insulating samples are on the order of 10^{-4} and 10^{-6} emu/mol, respectively. As a metal undergoes a MI transition, a sharp drop occurs in both the electrical conductivity and the magnetic susceptibility (one order of magnitude or greater is expected). The associated linewidths and g-values should also vary below the phase transition, and the final linewidth could be determined by the degree of crystal perfection and the associated crystal symmetry. A few typical examples, with the MI transition temperatures listed in Table 5.2, are α-(ET)$_2$I$_3$,[17,67] (ET)$_2$ReO$_4$,[40] (ET)$_3$X$_2$ (X^- = ClO$_4$$^-$ and HSO$_4$$^-$),[9,10] and (ET)$_2$C(CN)$_3$.[34] The temperature dependence of the peak-to-peak linewidth and relative spin susceptibility for (ET)$_2$C(CN)$_3$ are shown in Fig. 5.11. Above the MI phase transition, the linewidth and the derived spin susceptibility show only a very weak temperature dependence. However, a large drop in both is observed at the onset of the MI transition. In summary, the ESR properties for all five aforementioned compounds are almost identical.

The ESR behavior of (ET)$_2$AuCl$_4$ is quite different from those salts that exhibit strong MI transitions (Table 5.2). Its crystal structure is similar to the β-type systems, and the band electronic structure calculations reveal a two-dimensional Fermi surface similar to that of the β-(ET)$_2$X salts.[39] The ESR linewidth is narrower than that of the β-salts at room temperature. The derived spin susceptibility remains constant from 300 to 120 K, at which temperature a small drop in both linewidth and spin susceptibility is measured. However, the phase transition is irreversible, and the spin susceptibility for the warming cycle follows the Bonner Fisher behavior

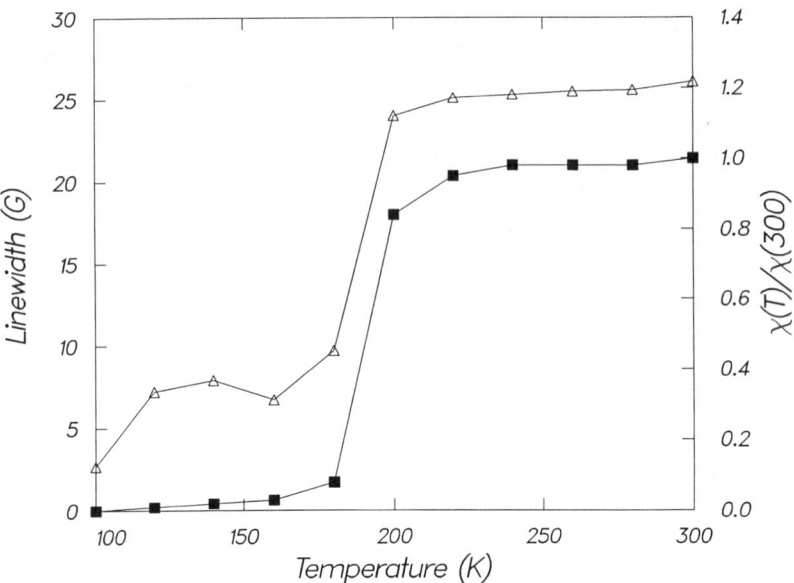

Figure 5.11 The linewidth (triangles) and relative spin susceptibility (squares) from 300 K to 100 K of $(ET)_2C(CN)_3$.

of a semiconductor.[69] On warming to room temperature, the $(ET)_2AuCl_4$ "crystal" becomes noticeably polycrystalline. Similar destructive behavior was observed during attempted low temperature X-ray studies.[39]

Organic Metals with Broad Metal-to-Insulator Transitions. For those compounds that exhibit very broad MI transitions as seen in conductivity measurements, the corresponding ESR measurements usually show the same trend. A few examples are listed below, $(ET)_2X$ ($X^- = AsF_6^-$, SbF_6^-, and $Au(CN)_2Cl_2^-$),[8,39] β-$(BEDSe\text{-}TTF)_2I_3$,[68] where BEDSe-TTF is bis(ethylenediseleno) tetrathiafulvalene. The MI transition temperatures of the first three compounds are listed in Table 5.2, and T_{MI} for the last compound is ~ 260 K. A common feature of these salts is that the phase transition temperatures are not much lower than room temperature. The linewidth and relative spin susceptibility of $(ET)_2Au(CN)_2Cl_2$ are plotted in Fig. 5.12. Both the linewidth and susceptibility from 360 K to 100 K decrease with decreasing temperature near the MI transition. No sharp discontinuities can be observed, but the rate of decrease (slope) is the largest near 295 K. Below the MI transition, both the linewidth and susceptibility continue to drop slowly with the latter becoming about 10 percent of the 300 K value at 100 K. For β-$(BEDSe\text{-}TTF)_2I_3$, the MI transition is very broad. A resistance minimum is observed at 260 K but the resistance is less than twice that of the 300 K value at 150 K.[68] The measured ESR linewidth and spin susceptibility do not indicate any anomalies between 300 and 150 K, but below 120 K the susceptibility drops sharply. Thus, the observed phase transition temperature is different in the conductivity and ESR measurements. This behavior may simply reflect differences

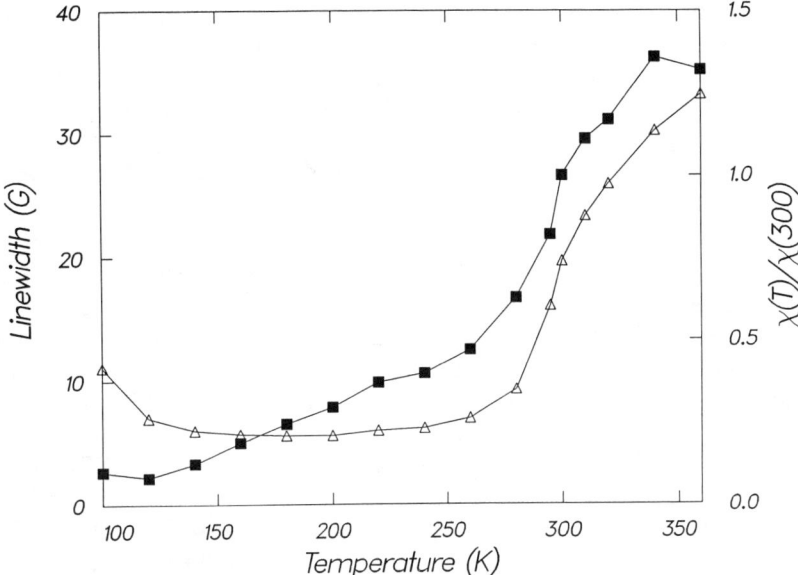

Figure 5.12 The linewidth (triangles) and relative spin susceptibility (squares) from 360 K to 100 K of $(ET)_2Au(CN)_2Cl_2$.

between a DC, or low frequency AC ($\sim$ 99 Hz) conductivity measurement, and a microwave (10^{10} Hz) conductivity measurement.

Organic Semiconductors

Many organic charge-transfer salts are electrical semiconductors, and their magnetic behavior can be described as being either paramagnetic or antiferromagnetic in nature. In this section, we will discuss their ESR properties according to these two types of behavior.

Paramagnetic Behavior. Paramagnetism from isolated spins is described by the Curie-Weiss law by use of the following equation:[3]

$$\chi = \frac{C}{T + \theta} \quad (5.11)$$

where χ is magnetic susceptibility, and C and θ are Curie and Weiss constants, respectively. To verify a Curie-Weiss behavior, the reciprocal susceptibility ($1/\chi$) is plotted against T, and a linear correlation is expected. A few ET-based salts fall into this category, for example, α-$(ET)_2IBr_2$,[24] $(ET)Ag_{1.6}(SCN)_2$,[50] and $(ET)Ag_{2.4}Br_3$,[51] and we will use the second salt as an example. When a single crystal of $(ET)Ag_{1.6}(SCN)_2$ is mounted vertically in the microwave cavity with the static magnetic field parallel to the c^* axis, the observed ESR peak-to-peak linewidth drops monotonically with temperature from 38 G (300 K) to 7 G (25 K). At 10 K, the linewidth is 0.9 G, and it becomes 0.78 G at 4.2 K. The linewidth below

10 K seems to be a residual linewidth because of an impurity, nonstoichiometric amount of Ag, or crystal defects. The reciprocal spin susceptibility, plotted against temperature between 300 and 4.2 K, is given in Fig. 5.13. A straight line is observed between 25 and 260 K, indicating Curie-Weiss behavior. The sudden increase of the spin susceptibility below 10 K, along with the apparent occurrence of a residual linewidth, suggests that the low temperature ($<$ 10 K) magnetic properties arise from different origins.

Antiferromagnetic Behavior. The magnetic properties of a uniform spin 1/2 Heisenberg antiferromagnetic chain have been studied by Bonner and Fisher.[69] The associated model consists of an array of uniformly spaced, but antiparallel, spins. The magnetic susceptibilities of such a system are expected to increase to a maximum value at low temperature then fall to a finite value at 0 K, and for a single crystal, only at a certain crystal orientation, the susceptibility could vanish. Many ET salts exhibit this Bonner Fisher behavior, viz., α'-$(ET)_2X$ (X^- = $AuBr_2^-$ and $CuCl_2^-$)[28], $(ET)_2[C_5(CN)_5](TCE)_x$,[35] $(ET)_2(CH_3\text{-}C_6H_4SO_3)$,[47] $(ET)_2HgBr_3(TCE)$,[55] and so on. The linewidths of the first four compounds are almost temperature independent from 300 to 100 K, but at lower temperatures the linewidth begins to increase rapidly. However, the linewidth of the last compound decreases with decreasing temperature. The temperature dependence of the linewidth and spin susceptibility for $(ET)_2HgBr_3(TCE)$ are shown in Fig. 5.14. The increase in linewidth and susceptibility below 20 K is likely due to a paramagnetic impurity.

Certain semiconducting ET salts follow Bonner Fisher behavior, but at a low temperature the susceptibility suddenly drops to zero irrespective of crystal orien-

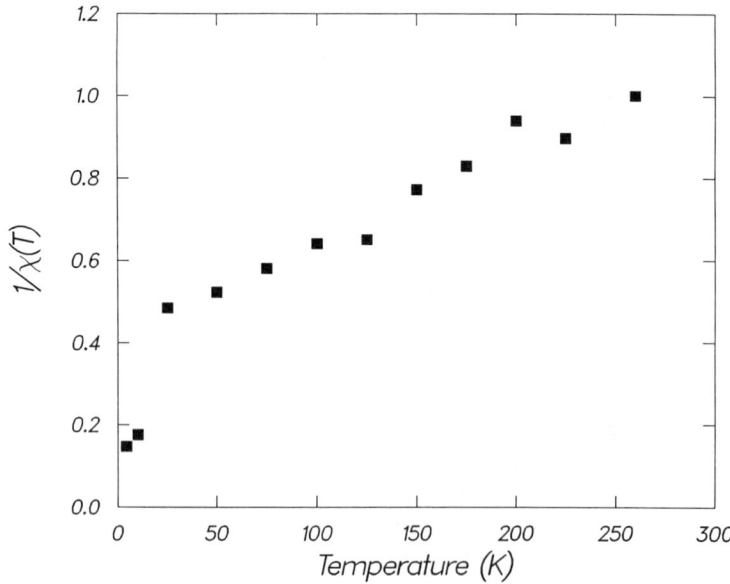

Figure 5.13 The reciprocal spin susceptibility versus temperature for $(ET)Ag_{1.6}(SCN)_2$.

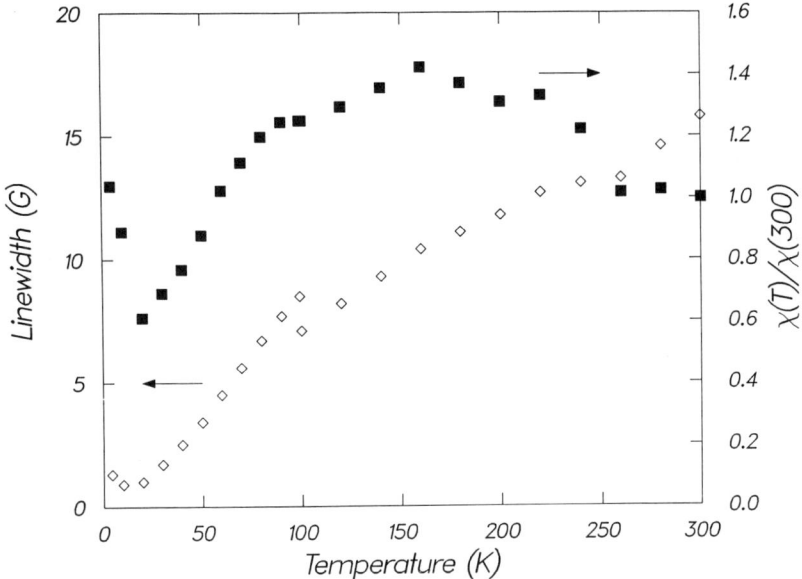

Figure 5.14 The linewidth (diamonds) and relative spin susceptibility (solid squares) versus temperature for $(ET)_2(HgBr_3)(TCE)$.

tation. This behavior is understood in terms of the Spin Peierls transition (Tsp).[71] The physical picture associated with this phenomenon is that the spin 1/2 uniform Heisenberg antiferromagnetic chain becomes dimerized. Each dimer contains two-paired spins, and the ground state is nonmagnetic. Examples are, α'-$(ET)_2Ag(CN)_2$,[28] β'-$(ET)_2ICl_2$,[25] β'-$(ET)_2AuCl_2$,[31] $(BEDSe\text{-}TTF)_2IBr_2$,[68] and $(BPDT\text{-}TTF)_2I_3$,[72] where BPDT-TTF is bis(propylenedithio)tetrathiafulvalene. The linewidth of the first compound is temperature independent from 300 to 100 K, but below 50 K it increases rapidly. The linewidths of the other four compounds decrease with decreasing temperature until the phase transition temperature where they begin to broaden rapidly. The ESR spin susceptibilities of all five compounds exhibit Bonner Fisher behavior down to the transition temperature (T_{sp}), at which point they vanish quickly. As an example, the linewidth and spin susceptibility of $(BEDSe\text{-}TTF)_2IBr_2$ are shown in Fig. 5.15.

HIGH TEMPERATURE ESR STUDIES

As pointed out in other chapters, the T_c of β-$(ET)_2I_3$ at 1.4 K can be increased to 8 K under shear stress of about 0.5 kbar[73–75] with the disorder of the ethylene group being totally suppressed, resulting in a totally ordered high-T_c phase termed β^*-$(ET)_2I_3$.[76] It has also been found that when α-$(ET)_2I_3$ (T_{MI} = 135 K) is heated above 70°C, a similar high-T_c state can be generated.[77,78] The thermally converted, or transformed, phase is also termed α_t-$(ET)_2I_3$, which shows a broad superconducting transition with onset at 7 K. The available X-ray data on converted crystals

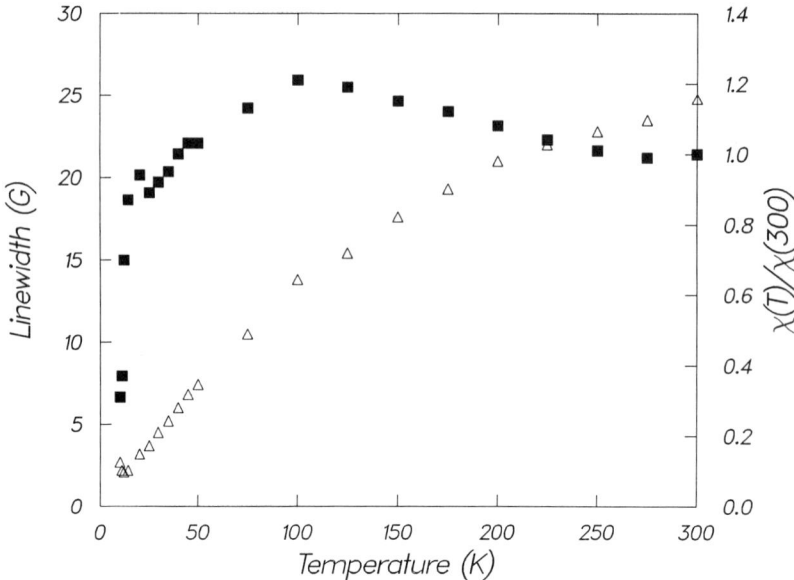

Figure 5.15 The linewidth (triangles) and relative spin susceptibility (squares) from 300 K to 10 K of (BEDSe-TTF)$_2$IBr$_2$.

suggests that they belong to the β-type packing motif.[78] However, the β*- and α$_t$-phases are not identical because the former shows a sharp T_c and is unstable at temperatures above 100 K unless under applied pressure. Because of the poor crystal quality of the α$_t$-phase material, the true nature of its structure is still not understood. We have discovered that high-temperature ESR techniques are a good method for studying thermally converted materials,[22] and two examples are given below.

The (ET)$_2$IBr$_2$ system[24] is a much simpler system than that of α$_t$-(ET)$_2$I$_3$. There are only two phases known, α and β, in contrast to more than ten different phases in the ET/I$_3^-$ system. When a single crystal of α-(ET)$_2$IBr$_2$ is maintained in a constant orientation in the microwave cavity, and the crystal is heated (warm nitrogen gas) from 300 K up at 20 K intervals, the peak-to-peak linewidth increases from 47 G (300 K) to 68 G (400 K). At 416 K, a sudden drop in linewidth to 22 G, which is characteristic of a β-phase sample, is observed. The ESR linewidth as a function of temperature for α- to β-(ET)$_2$IBr$_2$ is shown in Fig. 5.16. The converted β-(ET)$_2$IBr$_2$ crystals are indeed superconducting, with T_c at 2.66 K, which is very close to the T_c of the as-grown crystals (2.7 to 2.8 K). Another point worth noting for the thermally converted materials is that the high-temperature conversion is usually irreversible, again suggesting that β-phase materials are thermodynamically more stable than α-phase materials.[22,24]

In the ET/I$_3^-$ system two phases, α- and θ-(ET)$_2$I$_3$, are known to undergo similar thermal conversions to give α$_t$-(ET)$_2$I$_3$. A common feature of these salts is that they all appear to contain W-type interstack packing (see Chap. 3). Attempted thermal conversion of κ-(ET)$_2$Cu(NCS)$_2$ only results in eventual melting, with decomposition of the crystals.[52]

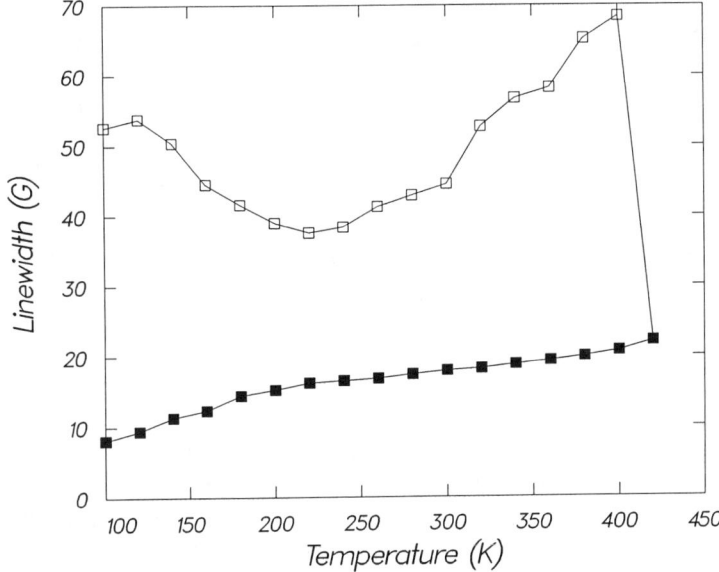

Figure 5.16 The linewidth versus temperature for α-(ET)$_2$IBr$_2$ (hollow squares, heating) and β-(ET)$_2$IBr$_2$ (solid squares, cooling).

Another type of thermal conversion has been reported recently, that is, δ-(linewidth 10 G) to α'-(ET)$_2$AuBr$_2$ (linewidth 37–50 G).[30] Both the initial and final materials are semiconductors. Finally, as described in detail in this chapter, ESR spectroscopic studies of organic conductors and superconductors have proven extremely valuable in identifying the different crystalline phases produced during electrocrystallization, in probing the low-temperature metallic properties of these systems, and in examining the nature of the phase transitions that many of these systems undergo at high and low temperatures. This technique is especially useful for investigating crystals not amenable to four-probe electrical conductivity studies in which the derived spin susceptibility serves as an alternative indicator of metallic conductivity.

REFERENCES

1. Wang, H. H.; Carlson, K. D.; Geiser, U.; Kwok, W. K.; Vashon, M. D.; Thompson, J. E.; Larsen, N. F.; McCabe, G. D.; Hulscher, R. S.; Williams, J. M. *Physica C* **1990**, *166*, 57.
2. Mori, H.; Tanaka, S.; Oshima, M.; Saito, G.; Mori, T.; Maruyama, Y.; Inokuchi, H. *Bull. Chem. Soc. Jpn.* **1990**, *63*, 2183.
3. Drago, R. S. *Physical Methods in Chemistry,* W. B. Saunders Co., Philadelphia 1977, Chap. 9.
4. Sugano, T.; Saito, G.; Kinoshita, M. *Phys. Rev. B: Condens. Matter* **1986**, *34*, 117.
5. Wang, H. H.; Vogt, B. A.; Geiser, U.; Beno, M. A.; Carlson, K. D.; Kleinjan, S.; Thorup, N.; Williams, J. M. *Mol. Cryst. Liq. Cryst.* **1990**, *181*, 135.

6. Sugano, T.; Saito, G.; Kinoshita, M. *Phys. Rev. B: Condens. Matter* **1987**, *35*, 6554.
7. Obertelli, S. D.; Friend, R. H.; Talham, D. R.; Kurmoo, M.; Day, P. *Synth. Met.* **1988**, *27*, A375.
8. Laversanne, R.; Amiell, J.; Delhaes, P.; Chasseau, D.; Hauw, C. *Solid State Commun.* **1984**, *52*, 177.
9. Enoki, T.; Imaeda, K.; Kobayashi, M.; Inokuchi, H.; Saito, G. *Phys. Rev. B: Condens. Matter* **1986**, *33*, 1553.
10. Porter, L. C.; Wang, H. H.; Beno, M. A.; Carlson, K. D.; Geiser, U.; Tomasko, M. S.; Williams, J. M.; Kang, D. B.; Whangbo, M.-H., to be submitted.
11. Urayama, H.; Saito, G.; Sugano, T.; Kinoshita, M.; Kawamoto, A.; Tanaka, J. *Synth. Met.* **1988**, *27*, A401.
12. Mori, T.; Wang, P.; Imaeda, K.; Enoki, T.; Inokuchi, H. *Solid State Commun.* **1987**, *64*, 733.
13. Penicaud, A.; Lenoir, C.; Batail, P.; Coulon, C.; Perrin, A. *Synth Met.* **1989**, *32*, 25.
14. Davidson, A.; Boubekeur, K.; Penicaud, A.; Auban, P.; Lenoit, C.; Batail, P.; Hervé, G. *J. Chem. Soc., Chem. Commun.* **1989**, 1373.
15. Urayama, H.; Yamochi, H.; Saito, G.; Sato, S.; Sugano, T.; Kinoshita, M.; Kawamoto, A.; Tanaka, J.; Inabe, T.; Mori, T.; Maruyama, Y.; Inokuchi, H.; Oshima, K. *Synth. Met.* **1988**, *27*, A393.
16. Sekretarczyk, G.; Graja, A.; *Synth. Met.* **1988**, *24*, 161.
17. Venturini, E. L.; Azevedo, L. J.; Schirber, J. E.; Williams, J. M.; Wang, H. H. *Phys. Rev. B: Condens. Matter* **1985**, *32*, 2819.
18. Klotz, S.; Schilling, J. S.; Gärtner, S.; Schweitzer, D. *Solid State Commun.* **1988**, *67*, 981.
19. Venturini, E. L.; Schirber, J. E.; Wang, H. H.; Williams, J. M. *Synth. Met.* **1988**, *27*, A243.
20. Ma, B.; Lu, L.; Zhang, D.; Wang, X.; Zhu, D. *Solid State Commun.* **1988**, *68*, 433.
21. Hurdequint, H.; Creuzet, F.; Jérome, D. *Synth. Met.* **1988**, *27*, A183.
22. Wang, H. H.; Ferraro, J. R.; Carlson, K. D.; Montgomery, L. K.; Geiser, U.; Williams, J. M.; Whitworth, J. R.; Schlueter, J. A.; Hill, S.; Whangbo, M.-H.; Evain, M.; Novoa, J. J. *Inorg. Chem.* **1989**, *28*, 2267.
23. Emge, T. J.; Wang, H. H.; Beno, M. A.; Leung, P. C. W.; Firestone, M. A.; Jenkins, H. C.; Cook, J. D.; Carlson, K. D.; Williams, J. M.; Venturini, E. L.; Azevedo, L. J.; Schirber, J. E. *Inorg. Chem.* **1985**, *24*, 1736.
24. Wang, H. H.; Carlson, K. D.; Montgomery, L. K.; Schlueter, J. A.; Cariss, C. S.; Kwok, W. K.; Geiser, U.; Crabtree, G. W.; Williams, J. M. *Solid State Commun.* **1988**, *66*, 1113.
25. Emge, T. J.; Wang, H. H.; Leung, P. C. W.; Rust, P. R.; Cook, J. D.; Jackson, P. L.; Carlson, K. D.; Williams, J. M.; Whangbo, M.-H.; Venturini, E. L.; Schirber, J. E.; Azevedo, L. J.; Ferraro, J. R. *J. Am. Chem. Soc.* **1986**, *108*, 695.
26. Ugawa, A.; Okawa, Y.; Yakushi, K.; Kuroda, H.; Kawamoto, A.; Tanaka, J.; Tanaka, M.; Nogami, Y.; Kagoshima, S.; Murata, K.; Ishiguro, T. *Synth. Met.* **1988**, *27*, A407.
27. Talham, D. R.; Kurmoo, M.; Day, P.; Obertelli, S. D.; Parker, I. D.; Friend, R. H. *J. Phys. C: Solid State Phys.* **1986**, *19*, L383.
28. Obertelli, S. D.; Friend, R. H.; Talham, D. R.; Kurmoo, M.; Day, P. *J. Phys.: Condens. Matter* **1989**, *1*, 5671.

29. Kurmoo, M.; Talham, D. R.; Day, P.; Parker, I. D.; Friend, R. H.; Stringer, A. M.; Howard, J. A. K. *Solid State Commun.* **1987**, *61*, 459.
30. Montgomery, L. K.; Wang, H. H.; Schlueter, J. A.; Geiser, U.; Carlson, K. D.; Williams, J. M.; Rubinstein, R. L.; Brennan, T. D.; Stupka, D. L.; Whitworth, J. R.; Jung, D.; Whangbo, M.-H. *Mol. Cryst. Liq. Cryst.* **1990**, *181*, 197.
31. Emge, T. J.; Wang, H. H.; Bowman, M. K.; Pipan, C. M.; Carlson, K. D.; Beno, M. A.; Hall, L. N.; Anderson, B. A.; Williams, J. M.; Whangbo, M.-H. *J. Am. Chem. Soc.* **1987**, *109*, 2016.
32. Beno, M. A.; Firestone, M. A.; Leung, P. C. W.; Sowa, L. M.; Wang, H. H.; Williams, J. M.; Whangbo, M.-H. *Solid State Commun.* **1986**, *57*, 735.
33. Kurmoo, M.; Talham, D. R.; Pritchard, K. L.; Day, P.; Stringer, A. M.; Howard, J. A. K. *Synth. Met.* **1988**, *27*, A177.
34. Beno, M. A.; Wang, H. H.; Soderholm, L.; Carlson, K. D.; Hall, L. N.; Nuñez, L.; Rummens, H.; Anderson, B.; Schlueter, J. A.; Williams, J. M.; Whangbo, M.-H.; Evain, M. *Inorg. Chem.* **1989**, *28*, 150.
35. Watson, W. H.; Kini, A. M.; Beno, M. A.; Montgomery, L. K.; Wang, H. H.; Carlson, K. D.; Gates, B. D.; Tytko, S. F.; Derose, J.; Cariss, C.; Rohl, C. A.; Williams, J. M. *Synth. Met.* **1989**, *33*, 1.
36. Gärtner, S.; Heinen, I.; Schweitzer, D.; Nuber, B.; Keller, H. J. *Synth. Met.* **1989**, *31*, 199.
37. Gärtner, S.; Heinen, I.; Schweitzer, D.; Keller, H. J.; Niebl, R. and Nuber, B. *Z. Naturforsch.*, in press.
38. Porter, L. C.; Wang, H. H.; Beno, M. A.; Carlson, K. D.; Pipan, C. M.; Proksch, R. B.; Williams, J. M. *Solid State Commun.* **1987**, *64*, 387.
39. Geiser, U.; Anderson, B. A.; Murray, A.; Pipan, C. M.; Rohl, C. A.; Vogt, B. A.; Wang, H. H.; Williams, J. M.; Kang, D. B.; Whangbo, M.-H. *Mol. Cryst. Liq. Cryst.* **1990**, *181*, 105.
40. Carneiro, K.; Scott, J. C.; Engler, E. M. *Solid State Commun.* **1984**, *50*, 477.
41. Azevedo, L. J.; Venturini, E. L.; Kwak, J. F.; Schirber, J. E.; Williams, J. M.; Wang, H. H.; Reed, P. E. *Mol. Cryst. Liq. Cryst.* **1985**, *125*, 169.
42. Wang, H. H.; Allen, T. J.; Schlueter, J. A.; Hallenbeck, S. L.; Stupka, D. L.; Chen, M. Y.; Despotes, A. M.; Kao, H.-C. I.; Carlson, K. D.; Geiser, U.; Williams, J. M. *Phosphorus and Sulfur* **1988**, *38*, 329.
43. Mallah, T.; Hollis, C.; Bott, S.; Day, P.; Kurmoo, M. *Synth. Met.* **1988**, *27*, A381.
44. Mori, T.; Wang, P.; Imaeda, K.; Enoki, T.; Inokuchi, H.; Sakai, F.; Saito, G. *Synth. Met.* **1988**, *27*, A451.
45. Rosseinsky, M. J.; Kurmoo, M.; Talham, D. R.; Day, P.; Chasseau, D.; Watkin, D. *J. Chem. Soc., Chem. Commun.* **1988**, 88.
46. Geiser, U.; Wang, H. H.; Kleinjan, S.; Williams, J. M. *Mol. Cryst. Liq. Cryst.* **1990**, *181*, 125.
47. Chasseau, D.; Watkin, D.; Rosseinsky, M. J.; Kurmoo, M.; Talham, D. R.; Day, P. *Synth. Met.* **1988**, *24*, 117.
48. Venturini, E. L.; Wang, H. H.; Geiser, U.; Williams, J. M., unpublished results.
49. Geiser, U.; Wang, H. H.; Williams, J. M.; Venturini, E. L.; Kwak, J. F.; Whangbo, M.-H. *Synth. Met.* **1987**, *19*, 599.

50. Geiser, U.; Beno, M. A.; Kini, A. M.; Wang, H. H.; Schultz, A. J.; Gates, B. D.; Cariss, C. S.; Carlson, K. D.; Williams, J. M. *Synth. Met.* **1988,** *27,* A235.
51. Geiser, U.; Wang, H. H.; Rust, P. R.; Tonge, L. M.; Williams, J. M.; *Mol. Cryst. Liq. Cryst.* **1990,** *181,* 117.
52. Wang, H. H.; Montgomery, L. K.; Kini, A. M.; Carlson, K. D.; Beno, M. A.; Geiser, U.; Cariss, C. S.; Williams, J. M.; Venturini, E. L. *Physica C* **1988,** *156,* 173.
53. Sekretarczyk, G.; Graja, A.; *Synth. Met.* **1988,** *24,* 161.
54. Sekretarczyk, G.; Graja, A.; Pichet, J.; Lyubovskaya, R. N.; Lyubovskii, R. B. *J. Phys., France* **1988,** *49,* 653.
55. Geiser, U.; Wang, H. H.; Williams, J. M., to be published.
56. Oshima, M.; Mori, H.; Saito, G.; Oshima, K. *The Physics and Chemistry of Organic Superconductors,* Eds, Saito, G.; Kagoshima, S.; Springer-Verlag, **1990,** p. 257.
57. Kobayashi, A.; Kato, R.; Kobayashi, H.; Moriyama, S.; Nishio, Y.; Kajita, K.; Sasaki, W. *Chem. Lett.* **1986** 2017.
58. Elliott, R. J. *Phys. Rev.* **1954,** *96,* 266.
59. Forró, L.; Sekretarczyk, G.; Krupski, M.; Schweitzer, D.; Keller, H. J. *Phys. Rev. B: Condens. Matter* **1987,** *35,* 2501.
60. Adrian, F. J. *Phys. Rev. B* **1986,** *33,* 1537.
61. Mori, T.; Sakai, F.; Saito, G.; Inokuchi, H. *Chem. Lett.* **1986,** 1589.
62. Al'tshuler, S. A.; Kozyrev, B. M. *Electron Paramagnetic Resonance,* Academic Press, Inc., **1964,** New York, NY, p. 260.
63. Tomkiewicz, Y. *Phys. Rev. B* **1979,** *19,* 4038.
64. Geiser, U.; Beno, M. A.; Kini, A. M.; Wang, H. H.; Montgomery, L. K.; Williams, J. M. *Annual Meeting, Amer. Cryst. Assn. Series 2,* **1988,** *16,* Poster PK11, p. 111.
65. Schultz, A. J.; Beno, M. A.; Geiser, U.; Wang, H. H.; Kini, A. M. and Williams, J. M., to be published.
66. Wang, H. H.; Beno, M. A.; Carlson, K. D.; Thorup, N.; Murray, A.; Porter, L. C.; Williams, J. M.; Maly K.; Bu, X.; Petricek, V.; Cisarova, I.; Coppens, P.; Jung, D.; Whangbo, M.-H., Schirber, J. E.; Overmyer, D. L. *Chem. of Materials,* **1991,** in press.
67. Bender, K.; Dietz, K.; Endres, H.; Helberg, H. W.; Hennig, I.; Keller, H. J.; Schäfer, H. W.; Schweitzer, D. *Mol. Cryst. Liq. Cryst.* **1984,** *107,* 45.
68. Wang, H. H.; Montgomery, L. K.; Geiser, U.; Porter, L. C.; Carlson, K. D.; Ferraro, J. R.; Williams, J. M.; Cariss, C. S.; Rubinstein, R. L.; Whitworth, J. R.; Evain, M.; Novoa, J. J.; Whangbo, M.-H. *Chem. Mater.* **1989,** *1,* 140.
69. Bonner, J. C.; Fisher, M. E. *Phys. Rev.* **1964,** 135, A640.
70. MDT-TTF is methylenedithio tetrathia fulvaline. Kini, A. M.; Beno, M. A.; Son, D.; Wang, H. H.; Carlson, K. D.; Porter, L. C.; Welp, U.; Vogt, B. A.; Williams, J. M.; Jung, D.; Evain, M.; Whangbo, M.-H.; Overmyer, D. L.; Schirber, J. E. *Solid State Commun.* **1989,** *69,* 503.
71. Bray, J. W.; Interrante, L. V.; Jacobs, I. S.; Bonner, J. C. *Extended Linear Chain Compounds,* Volume 3, Ed. J. S. Miller, Plenum Press, **1983,** p. 353.
72. Kushch, N. D.; Merzhanov, V. A.; Romanyukha, A. A. *Sov. Phys. JETP* **1989,** *69,* 205.
73. Lakhin, V. N.; Kostyuchenko, E. É.; Sushko, Yu. V.; Shchegolev, I. F.; Yagubskii, E. B. *JETP Lett (Engl. Transl.)* **1985,** *41,* 81.

74. Murata, K.; Tokumoto, M.; Anzai, H.; Bando, H.; Saito, G.; Kajimura, K.; Ishiguro, T. J. *Phys,. Soc. Jpn.* **1985,** *54,* 1236.
75. Kwak, J. F.; Venturini, E. L.; Leung, P. C. W.; Beno, M. A.; Wang, H. H.; Williams, J. M. *Phys. Rev. B: Condens. Matter* **1986,** *33,* 1987.
76. Schultz, A. J.; Wang, H. H.; Williams, J. M.; Filhol, A. *J. Am. Chem. Soc.* **1986,** *108,* 7853.
77. Baram, G. O.; Buravov, L. I.; Degtyarev, L. S.; Kozlov, M. É.; Laukhin, V. N.; Laukhina, E. É.; Onishchenko, V. G.; Pokhodnya, K. I.; Sheinkman, M. K.; Shibaeva, R. P.; Yagubskii, E. B. *JETP Lett (Engl. Transl.)* **1986,** *44,* 376.
78. Schweitzer, D.; Bele, P.; Brunner, H.; Gogu, E.; Haeberlen, U.; Hennig, I.; Klutz, I.; Swietlik, R.; Keller, H. J. *Z. Phys. B.* **1987,** *67,* 489.
79. Wang, H. H.; Beno, M. A.; Carlson, K. D.; Geiser, U.; Kini, A. M.; Montgomery, L. K.; Thompson, J. E.; Williams, J. M., in *Proceedings of International Conference on Organic Superconductors,* May 1990, South Lake Tahoe, California, *Organic Superconductivity,* Eds. V. Kresin, W. Little, Plenum Press, New York, **1990,** p. 45.
80. Schirber, J. E.; Azevedo, L. J.; Kwak, J. F.; Venturini, E. L.; Leung, P. C. W.; Beno, M. A.; Wang, H. H.; Williams, J. M. *Phys Rev. B: Condens. Matter* **1986,** *33,* 1987.
81. Kini, A. M.; Geiser, U.; Wang, H. H.; Carlson, K. D.; Williams, J. M.; Kwok, W. K.; Vandervoort, K. G.; Thompson, J. E.; Stupka, D. L.; Jung, D.; Whangbo, M.-H. *Inorg. Chem.* **1990,** *29,* 2555.
82. Kobayashi, A.; Kato, R.; Kobayashi, H.; Moriyama, S.; Nishio, Y.; Kajita, K.; Sasaki, W. *Chem. Lett.* **1987,** 459.
83. Lyubovskaya, R. N.; Zhilyaeva, E. A.; Zvarykina, A. V.; Laukhin, V. N.; Lyubovskii, R. B.; Pesotskii, S. I. *JETP Lett.* (Engl. Transl.) **1987,** *45,* 530.
84. Montgomery, L. K.; Geiser, U.; Wang, H. H.; Beno, M. A.; Schultz, A. J.; Kini, A. M.; Carlson, K. D.; Williams, J. M.; Whitworth, J. R.; Gates, B. D.; Cariss, C. S.; Pipan, C. M.; Donega, K. M.; Wenz, C.; Kwok, W. K.; Crabtree, G. W. *Synth. Met.* **1988,** *27,* A195.
85. Williams, J. M.; Kini, A. M.; Wang, H. H.; Carlson, K. D.; Geiser, U.; Montgomery, L. K.; Pyrka, G. J.; Watkins, D. M.; Kommers, J. M.; Boryschuk, S. J.; Strieby Crouch, A. V.; Kwok, W. K.; Schirber, J. E.; Overmyer, D. L.; Jung, D.; Whangbo, M.-H. *Inorg. Chem.* **1990,** *29,* 3262.
86. Mori, H.; Hirabayashi, I.; Tanaka, S.; Mori, T.; Inokuchi, H.; *Solid State Commun.* **1990,** *76,* 35.
87. Delhaes, P.; Amiell, J.; Flandrois, S.; Ducasse, L.; Fritsch, A.; Hilti, B.; Mayer, C. W.; Zambounis, J.; Papavassiliou, G. C. *J. Phys. France* **1990,** *51,* 1179.

6

Vibrational Spectroscopy

INTRODUCTION

Superconductors have generated considerable interest in recent years.[1] The high-transition temperatures (high-T_c) in copper oxide superconductors have provided immense interest, and these materials have been extensively studied. These superconductors have T_c's which exceed the boiling point of liquid nitrogen (77 K), and show great promise for important practical applications. Less well known are the organic systems involving charge-transfer (CT) molecular materials of low densities. Some of the organic materials become superconducting with pressure and at ambient pressures at T_c's of ~12K or lower. These materials also show promise for device applications purposes, and are interesting since by use of synthetic, organic chemistry techniques, they can be molecularly engineered to provide an almost infinite number of materials. Figure 1.1 in Chap. 1 illustrates structures of several donor and acceptor molecules and anions, which form the organic CT superconductors. The Bechgaard salts[2] of the $(TMTSF)_2X$ type, where TMTSF is tetramethyltetraselenafulvalene and $X = $ an inorganic anion, have been previously discussed.[1] Most of these materials become superconducting only with external pressure, with the exception of $(TMTSF)_2ClO_4$, which is superconducting at ambient pressure. Table 1.1 in Chap. 1 compiles the T_c's of the superconductors involving the donor molecules bis(ethylenedithio)tetrathiafulvalene (BEDT-TTF or

ET), and ET variants such as B(alk)DT-TTF (alk = propyl, methyl or mixed), methylenedithiotetrathiafulvalene (MDT-TTF), dimethyl(ethylenedithio)diselenadithiafulvalene (DMET), and 4,5-dimercapto-1, 3-dithiole-2-thione (dmit), and compares them with those of (TMTSF)$_2$X. At this point in time, nearly 40 organic superconductors are now known, some of which are illustrated in Table 1.1. The T_c's of the organic superconductors have doubled every few years since their initial discovery.[3] The chronology of synthetic, organic superconductors is shown in Table 1.2.

This chapter concerns itself with the role vibrational spectroscopy has or is playing in characterizing the synthetic, sulfur-based, organic superconductors.

SPECTROSCOPY OF ORGANIC CT SUPERCONDUCTORS

Vibrational spectroscopy has played an important role in the characterization of synthetic, organic, electrical superconductors.[4] The major spectroscopic interest has centered in the following areas:

1. determination of the extent of CT in CT complexes, where CT is <1,
2. assignments of vibrations in neat donor and acceptor molecules, and in their salts, for example, (ET)$_2$I$_3$.
3. determination from infrared reflective spectra of optical anisotropy, electronic structure, plasma frequencies, optical band gap, optical conductivity, electron-phonon coupling constants, dimensionality of the complex, and differentiation of various structural phases in (ET)$_2$X type of salts, where X = inorganic anion, as well as differentiation of the nature of environments in the anion cavity of these materials.

Vibrational Assignments for the Neat Donor ET

Only the donor molecule ET will be discussed in this section as it is the donor that has produced the major number of superconductors thus far. Assignments of the fundamental vibrations for ET and ET-d$_8$ have been made by Kozlov et al.[5] Table 6.1 lists the symmetry classes and selection rules for ET, and Table 6.2 tabulates the assignments made from Raman and infrared data.

TABLE 6.1 Symmetry Classes and Selection Rules for ET

$\Gamma(C_{2v})=$	$12A_1$	+	$6A_2$	+	$7B_1$	+	$11B_2$
	Raman, i.r.		Raman		Raman, i.r.		Raman, i.r.
$\Gamma(D_{2h})=$	$12A_g + 1A_u + 11B_{1w}$	+	$6B_{1g} + 6A_u$	+	$7B_{2t} + 7B_{3u}$	+	$11B_{3t} + 11B_{2w}$
	Raman ia i.r.		Raman ia		Raman i.r.		Raman i.r.

Reprinted with permission from Spectrochimica Acta, *43A*, M. E. Kozlov, K.L. Pokhadnia, and A.A. Yurchenko, "The Assignment of Fundamental Vibrations of BEDT-TTF and BEDT-TTF-d$_8$," 1987, Pergamon Press, Inc., Elmsford, N.Y.

TABLE 6.2 Infrared and Raman Spectra of BEDT-TTF and BEDT-TTF-d_8

BEDT-TTF Infrared[a] v_t	i	BEDT-TTF Raman[b] v_i	i	Assignment	BEDT-TTF-d_8 Infrared[a] v_i	i	BEDT-TTF-d_8 Raman[b] v_i	i	Assignment[c]
2958 w	66				2237 w	66			
2958 w	26			vCH$_2$	2225 w				vCD$_2$
					2169 vw	26			
2916 w	44				2141 vw				
		1552 m	2				1552 m	2	
1505 w	27	1511 m	27		1506 w	27	1511 m	27	
		1494 s	3				1494 s	3	
1420 w	28				1011 m	29			
				δCH$_2$	1002 vw	5			δCD$_2$
1406 m	45				1041 m	46	1044 w	46	
		1409 vw	56		1030 w	57	1029 w	57	
		1285 vw	5				794 w	7	vSCD$_2$ + ωCD$_2$
1282 m	29				793 w	31			
1259 w	46			ωCH$_2$	990 vw	47			ωCD$_2$
1253 vw	57	1256 vw	57		984 w	59	984 w	59	
1173 w	21	1175 vw	21		930 w	21	935 vw	21	
1132 sh	38	1132 vw	38	tCH$_2$			805 w	38	tCD$_2$
1125 w	67	1126 vw	67		806 w	67			
		1016 vw	59				1018 w	58	
996 w	30	1002 w	30	vCC	1110 m	28			vCC + ωCD$_2$
987 w	6	990 w	6						
938 vw	22			ρCH$_2$	741 w	22	741 w	22	ρCD$_2$
917 s	48	919 vw	60		827 w	50			
905 m	31	911 vw	7		905 m	30			
890 m	49	888 vw	49		879 m	49			
875 w	50	875 vw	50		879 m	48			
860 vw	61	860 vw	61						
772 s	32	765 w	32		772 s	32	784 vw	32	
687 w	68			ρCH$_2$	693 m	68			ρCD$_2$
		687 w	39		678 w	39	693 w	39	
653 w	33	653 m	8		634 w	33	635 m	8	
624 w	51	625 w	62		609 w	51	610 m	62	
499 m	34	486 m	9		500 m	34	486 m	9	
450 w	10	440 m	10		452 w	10	439 m	10	
							351 w		
390 m	35				388 m	35			
		348 w	63				339 w	63	
335 m	52	334 w	64		326 w	52	323 w	64	
		308 w	11				296 w	11	
278 m	36	272 vw	36		269 vw	36			
257 m	53	260 w	53		257 m	53	255 w	53	
		159 s	65				155 s	65	
		151 s	12				147 s	12	
96 m	72	127 vw	43		88 w	72	125 w	43	
		52 vs					52 vs		
30 w	54	31 vs	54		30 w	54	30 vs	54	

Relative intensities: vs = very strong; s = strong; m = medium; w = weak; vw = very weak; sh = shoulder
[a] KBr pellets above 400 cm^{-1} and Nujol mull below
[b] Monocrystalline sample, exciting line 676.45 nm
[c] The following symbols have been used for ethylene group vibrations: v = stretching, δ = bending, ω = wagging, t = twisting, π = rocking.
Reprinted with permission from Spectrochimica Acta, *43A*, M. E. Kozlov, K.L. Pokhadnia and A. A. Yurchenko, "The Assignment of Fundamental Vibrations of BEDT-TTF and BEDT-TTF-d_8," 1987, Pergamon Press, Inc., Elmsford, N.Y.

Vibrational Studies of Charge-Transfer Salts of ET

Resonant Raman scattering measurements in α- and β-$(ET)_2I_3$ and α- and β-$(ET)_2IBr_2$ were made by Sugai and Saito.[6] Figure 6.1 shows several low-frequency Raman spectra of various ET salts. Strong resonance effects were noted, and more than 10 overtones of the symmetric stretching modes of the I_3^- anion were observed. The halogen vibrations of the anions are clearly noticeable in the spectra of the superconductors below 200 cm^{-1}. Differences in the Raman spectra in the energy range of C=C vibrations (1400–1700 cm^{-1}) of ET and β-$(ET)_2I_3$ and β-$(ET)_2IBr_2$ are illustrated in Fig. 6.2. Sharp peaks are observed in ET, whereas shifts in the major peaks and linewidths increase in the salts. Other Raman investigations [7–14] have been directed at ascertaining the nature of the I_3^- moiety in the anion cavity of the various polymorphs of β-$(ET)_2I_3$. Table 6.3 summarizes the Raman data for the I_3^- anion and compares them with diffraction data. The nature of the I_3^- moiety in the anion cavity, and the importance of order-disorder in the I_3^- anion on T_c is noted from these studies. Table 6.3 tabulates results for the v_1 vibration in the I_3^- anion obtained for the various forms of β-$(ET)_2I_3$ and compares them with results obtained for α-$(ET)_2I_3$ and $α_t$-$(ET)_2I_3$.

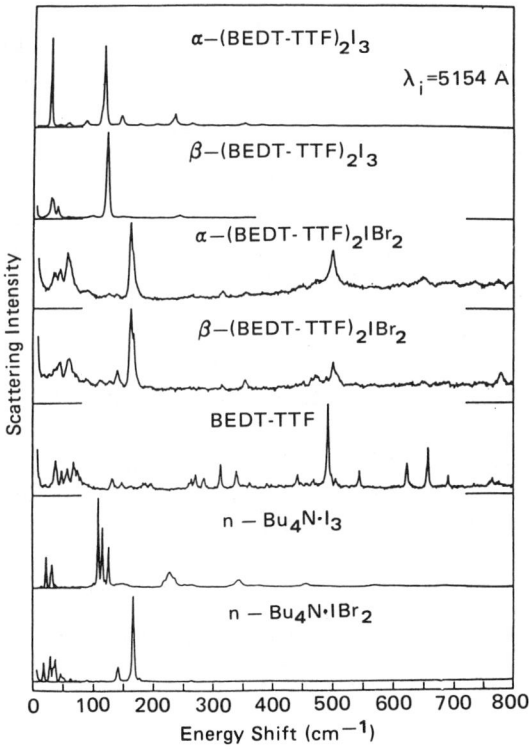

Figure 6.1 Raman spectra of $(ET)_2X$ salts at about 80 K, ET at 31 K and n-Bu_4NX at 32 K. Reprinted with permission from Solid State Commun., *58*, S. Sugai and G. Saito, "Resonant Raman Scattering in Organic Conductors α and β-$(BEDT-TTF)_2X$ ($X = I_3$ and IBr_2), **1986**, Pergamon Press, Inc., Elmsford, N.Y.

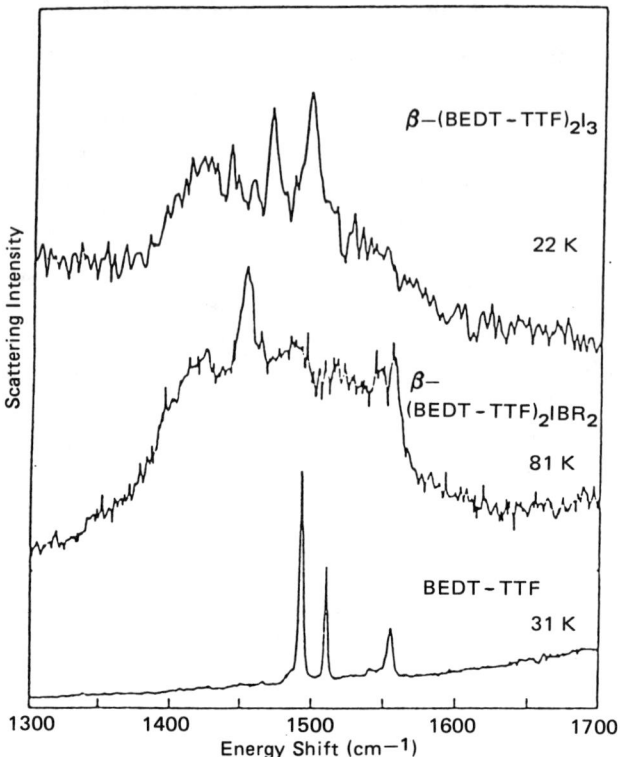

Figure 6.2 Raman spectra in the energy range of C=C vibrations. The incident laser wavelength is 4579 Å for the (ET)$_2$X spectra and 5145 Å for (ET). Reprinted with permission from Solid State Commun., *58*, S. Sugai and G. Saito, "Resonant Raman Scattering in Organic Conductors α and β-(BEDT-TTF)$_2$X(X = I$_3$ and IBr$_2$), **1986**, Pergamon Press, Inc., Elmsford, N.Y.

Polarized Reflectance Spectra of the CT Salts

The polarized reflectance spectra for several salts of the type (ET)$_2$X, where X = monovalent anions have been measured. Table 6.4 lists the various salts and the maximum range of the experimental measurements made.[15–54] Most of the

TABLE 6.3   Raman Probe of v_1 in the I$_3^-$ Anion Vibration in Various (ET)$_2$I$_3$ Salts

α-(ET)$_2$I$_3$ (T = 55K)	β-(ET)$_2$I$_3$ (T = 55K)	β*-(ET)$_2$I$_3$ (2K)	β'$_d$-(ET)$_2$I$_3$ (140K)	α$_t$-(ET)$_2$I$_3$ (2K)
115				
118				
121	122	120	121	120
disordered I$_3^-$	disordered I$_3^-$	linear ordered I$_3^-$	disordered I$_3^-$	linear ordered I$_3^-$
Ref. 7	Ref. 7	Ref. 9	Ref. 13	Refs. 9–12

| | 109 | | 111 | |

Note: α$_t$-(ET)$_2$I$_3$—obtained by thermally cycling α-(ET)$_2$I$_3$.

TABLE 6.4 Reflectance Measurements for Several CT, ET Salts

Salt	Reflectance Spectra Max. Range × 10^3(cm^{-1})	Reference
(MDT-TTF)$_2$AuI$_2$	4	37
(BO)$_2$ClO$_4$	6	46
(BO)$_3$Cu$_2$(NCS)$_3$	4+	37
(BMDT-TTF)SbF$_6$	25+	30
(BMDT-TTF)AsF$_6$	25+	30
(BMDT-TTF)I$_3$	40+	50
(PT)$_2$ICl$_2$	4	28
(PT)$_2$I$_2$Br	4	28
(PT)$_2$IBr$_2$	4	28
(PT)$_2$I$_3$	4	28
	25+	29
(PT)$_3$(PF$_6$)$_2$	25+	29
α-(ET)$_2$I$_3$	>30,15	16
	>5+	17
	>25+	18
	4	28,35
	25+	42
α'-(ET)$_2$IAuBr	25+	43
α-(ET)$_2$IBr$_2$	4	20
α'-(ET)$_2$AuBr$_2$	4	28
	25	43
α$_t$-(ET)$_2$I$_3$	4+	35
	25+	49
θ-(ET)$_2$I$_3$	25+	36
	20+	38
α-(ET)$_3$(ReO$_4$)$_2$	25+	42
β-(ET)$_2$I$_3$	>15+	15,16,20
	>25+	18
	>25+	22
	>25+	23
	4	28
	~8+	19
	20+	38
	25+	48
β'-(ET)$_2$I$_3$	20+	38
κ-(ET)$_2$I$_3$	20+	38
β-(ET)$_2$AuI$_2$	~8+	25
	4	28,40
β-(ET)$_2$IBr$_2$	>25+	21
β-(ET)$_2$I$_2$Br	~8+	25
β-(ET)$_2$PF$_6$	25+	20
	25+	44
β"-(ET)$_2$ICl$_2$	20+	38
β'-(ET)$_2$ICl$_2$	4	28
	20+	38
β'-(ET)$_2$AuCl$_2$	4	28
β"-(ET)$_2$IAuBr	~20+	26

BO is BEDO-TTF.

TABLE 6.4 Reflectance Measurements for Several CT, ET Salts *(continued)*

Salt	Reflectance Spectra Max. Range × 10^3(cm^{-1})	Reference
β''-(ET)$_2$AuBr$_2$	9+	40
(ET)$_2$Tosylate	4	28
(ET)HgBr$_3$	4+	37
(ET)$_2$HgBr$_3$(TCE)$_{\sim 1.0}$	4+	37
α-(ET)$_2$NH$_4$Hg(NCS)$_4$	4+	37
(ET)$_4$Hg$_3$Cl$_8$	25+	31
(ET)$_3$(ClO$_4$)$_2$	25+	20
(ET)$_2$ClO$_4$(C$_2$H$_3$Cl$_3$)$_{0.5}$	25+	20,24
(ET)$_2$BrO$_4$	4	28
(ET)$_2$ReO$_4$	4	28
(ET)$_2$Ag$_x$(NCS)$_2$, x = 1.6	4+	37
(ET)$_2$Au(CN)$_2$(Br)$_2$	4+	37
(ET)$_2$Au(CN)$_2$(Cl)$_2$	4+	37
(ET)$_2$Au(NCS)$_2$	4+	37
κ-(ET)$_2$Cu(NCS)$_2$	4+	33
	9+	39
	20+	27,41
	28+	32,45
	8+	51,52
κ-(ET)$_2$Cu[N(CN)$_2$]Br	4+	37
(BEDSe-TTF)$_2$I$_3$	4	34
(BEDSe-TTF)$_2$IBr$_2$	4	34
β-d$_8$-(ET)$_2$I$_3$	4	28
β-d$_8$-(ET)AuI$_2$	4	28
β-d$_8$-(ET)$_2$I$_2$Br	4	28
β-d$_8$-(ET)$_2$ICl$_2$	4	28
κ-d$_8$-(ET)$_2$Cu(NCS)$_2$	4+	33
	9+	39
	28+	45
	6+	51
α-d$_8$-(ET)$_2$I$_3$	4	28

+ Polarized spectra

materials shown have reflectance spectra that resemble one another. They all have a broad band increasing in reflectance with decreasing frequency and sharp features appearing at frequencies <1500 cm.[115-54] The nature of these sharp vibrations will be discussed in the next section.

From reflectance spectral values for R(ω), one can obtain values of $\sigma(\omega)$ (the optical conductivities) through Kramers-Kronig analysis. All of the $\sigma(\omega)$ curves resemble one another as do the R(ω) curves. They all have a broad band with a maximum and superimposed sharp bands appearing at lower frequency. The spectra and $\sigma(\omega)$ were analyzed[19] by use of the Rice model.[53] The broad band is assigned to the charge transfer between the donor and acceptor molecules. The superimposed fine structure is associated with an electron-phonon coupled transition (Williams et al.[54]). Table 6.5 shows the plasma frequency (ω_p) obtained from reflectance data for several ET salts. For a theoretical discussion on polarized reflectance infrared see Chap. 7.

TABLE 6.5 Plasma Frequencies Obtained from Reflectance Data for ET

Compound	$\omega_p(\text{cm}^{-1}) \times 10^3$
$(ET)_2PF_6$	5.1[a]
	5.3
$(ET)_2ClO_4(TCE)_{0.5}$	8.8
	4.7
$(ET)_2ClO_4$	10.5
	4.4
β-$(ET)_2I_3$	8.5
	5.3
α-$(ET)_2I_3$	4.4
β-$(ET)_2I_2Br$	9.3
β-$(ET)_2AuI_2$	10.3
	5.8

[a] The first number applies to $\sigma(\omega)$ parallel to a crystallographic axis; the second, parallel to c axis.

Micro FT-IR Reflectance Techniques in the Study of ET Superconductors

The use of the microscope FT-IR technique in the study of ET superconductors has proven to be quite useful. These materials are opaque and are usually synthesized in small quantities. The technique is suitable for such crystals as the reflection mode of the microscope can be used and only microcrystals are necessary.

The electrocrystallization method is the main source for the synthesis of CT, organic electrical conductors and superconductors.[1,54] In the instances involving ET as the donor molecule, several phases can form on the platinum anode. It is necessary to differentiate between phases, as some phases lead to superconductivity when the temperature is lowered, while others are nonsuperconductive. The basis for differentiation has been the measurement of the ESR linewidth of the material. Materials with large linewidths (e.g., >70G) are often nonsuperconductive, while other materials with low linewidths may become superconducting.[54]

The room temperature reflectance spectra of these materials are characterized by two regions of interest. The first feature extends from 1600–4000 cm^{-1} in our studies, although the high-energy side may extend well into the high-frequency region. This has been ascribed to inter- and intramolecular electronic transitions of the donor molecule and their overlap, and has been designated as the plasma spectrum.[15,17,31] The second feature occurs from 1200–1400 cm^{-1} and has been termed the vibronic region.[15,17] It has been attributed to the interaction between conduction electrons and molecular vibrations. These interactions involve the intramolecular, symmetric Raman-active vibrational modes of ET (e.g., A_g modes such as the C=C vibration).[17] Coupling of the electrons with vibrational modes causes a lower symmetry of the ET stacks, and the vibronic spectrum becomes IR-active. In some cases, superimposed vibrational structure appears on the vibronic envelope. It appears that the -C-C-H bending vibrations of the ethylene groups of ET occurring in the anion cavity are also involved in the vibronic envelope, inasmuch as deuterated ET salts show changes in the vibronic region.[28] These -C-C-H bending vibrations are considerably reinforced in intensity compared to the CH stretching vibration, which appears very weakly. This intensity reinforcement may be caused by the coupling of the -C-C-H bending vibrations with the vibron in the 1200–1400 cm^{-1} region. Two main types of spectra are obtained in this region for the ET salts, and these have been correlated with the type of structural phase of the salts, as well as with the electronegativity of the anion and to the hardness of the lattice.[28,54] The materials having the β-, β′-, κ-, α′-type structures show a split in the envelope in this region, whereas the split is absent in materials with the α-type structure and in deuterated salts. Figure 6.3 illustrates the two types of vibronic spectra obtained for κ-(ET)$_2$Cu(NCS)$_2$ and κ-d$_8$-(ET)$_2$Cu(NCS)$_2$. Table 6.6 tabulates the reflectance measurements (polarized and nonpolarized) made for several ET salts, ESR linewidths (ΔH_{pp}) and describes the type of conductor obtained.

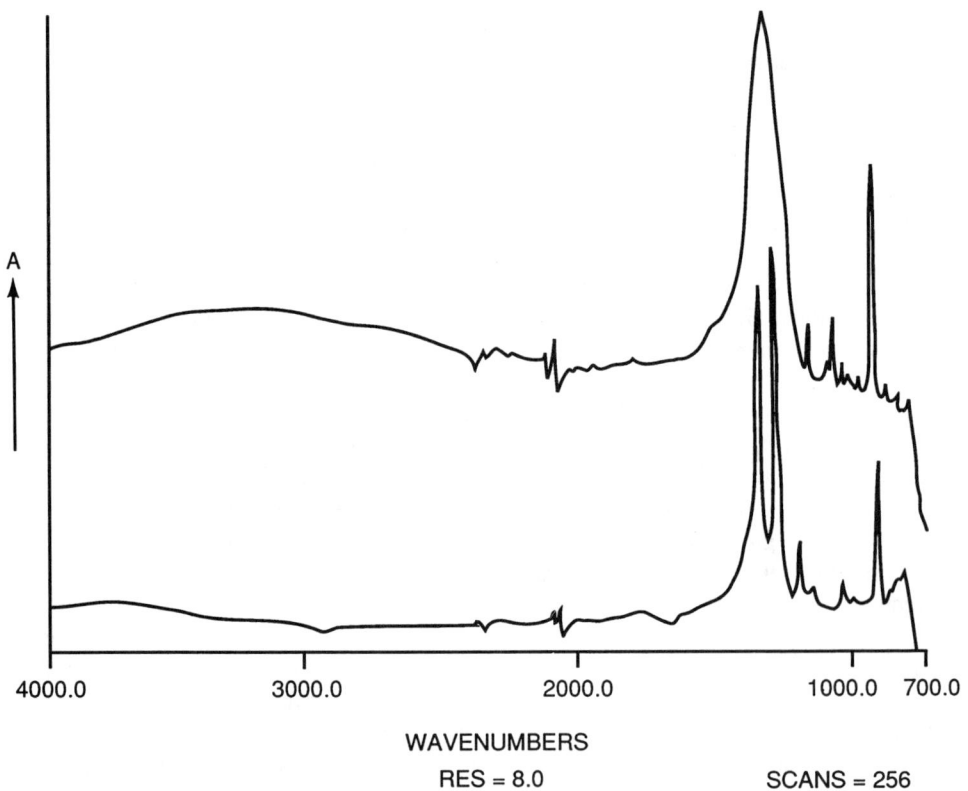

Figure 6.3. Reflectance spectrum of κ-(ET)$_2$Cu(NCS)$_2$.
Superconductor: bottom curve
Reflectance spectrum of κ-d$_8$-(ET)$_2$Cu(NCS)$_2$
Superconductor: top curve
Note: Absorbance in arbitary units. A Kramers-Kronig transformation was used (from Ferraro et al.[33]).

TABLE 6.6 ESR, Reflectance Data for ET Salts

Salt	Reflectance Spectra (1200–1400 cm^{-1})	ΔHpp	Comments
(PT)$_2$ICl$_2$	No Split	3.5	Semiconductor
(PT)$_2$I$_2$Br	Split	6–7	Semiconductor
(PT)$_2$IBr$_2$	Split	6–7	Semiconductor
(PT)$_2$I$_3$	Split	8	Semiconductor
α_t-(ET)$_2$I$_3$	Split	20*	Superconductor
β-(ET)$_2$I$_3$	Split	21–22	Superconductor
β-(ET)$_2$AuI$_2$	Split	19	Superconductor
β-(ET)$_2$IBr$_2$	Split	19	Superconductor
β-(ET)$_2$I$_2$Br	Split	20	Metallic to 0.49K
β'-(ET)$_2$ICl$_2$	Split	6–10	Semiconductor
β'-(ET)$_2$AuCl$_2$	Split	6	Semiconductor
α'-(ET)$_2$AuBr$_2$	Split	54	Semiconductor
α'-(ET)$_2$IAuBr	Split	NA	Semiconductor
κ-(ET)$_2$Cu[N(CN)$_2$]Br	Split	58–85	Superconductor
κ-(ET)$_2$Cu(NCS)$_2$	Split[a]	60–70	Superconductor
(ET)$_2$(Ag$_x$NCS)$_2$, $x=1.6$	No Split[a]	32–38	Semiconductor
(ET)$_2$Au(NCS)$_2$	NA	30–33	Semiconductor
(ET)$_2$Au(CN)$_2$Cl$_2$	No Split	15	Semiconductor
(ET)$_2$Au(CN)$_2$Br$_2$	Split[a]		Metal
(ET)$_2$Tosylate	Split	24–34	Semiconductor
(ET)$_4$Hg$_3$Cl$_8$[b]	Split[a]	NA	Superconductor with P = 12 kbar, 29 kbar
α-(ET)$_2$NH$_4$Hg(NCS)$_4$	Split[a]	56–84	Superconductor
(ET)HgBr$_3$	Split[a]	10.9	Semiconductor
(ET)$_2$HgBr$_3$(TCE)$_{\sim 1.0}$	No Split	20	Semiconductor
(ET)$_2$BrO$_4$	Split	NA	Metallic
(ET)$_2$ReO$_4$	Split	16	Superconductor with P = 4 kbar
(BEDSe-TTF)$_2$I$_3$	Split[a]	45–60	Metallic to 260° K
(BEDSe-TTF)$_2$IBr$_2$	Split[a]	22–28	Semiconductor
(MDT-TTF)$_2$AuI$_2$	No Split	100	Superconductor
α-(ET)$_2$I$_3$	No Split	95	Metallic
α-(ET)$_2$IBr$_2$	No Split	55	Semiconductor
α-(ET)$_3$(ReO$_4$)$_2$	No Split	NA	Semiconductor
β-d$_8$-(ET)$_2$I$_3$	No Split	20	Superconductor
β-d$_8$-(ET)$_2$AuI$_2$	No Split	20	Superconductor
β-d$_8$-(ET)$_2$I$_2$Br	No Split	18.5	Superconductor
β-d$_8$-(ET)$_2$ICl$_2$	No Split	10	Metallic
κ-d$_8$-(ET)$_2$Cu(NCS)$_2$	No Split	~60	Superconductor
α-d$_8$-(ET)$_2$I$_3$	No Split	75–100	Metallic
(BO)$_2$ClO$_4$	No Split	40	Semiconductor
(BO)$_3$Cu$_2$(NCS)$_3$	No Split	19–20	Superconductor

[a] Polarized reflectance spectra
[b] Kaplunov et al.[31]
PT = bis(propylenedithio)tetrathiafulvalene
Tosylate = CH$_3$-C$_6$H$_4$-SO$_3$-
(BEDSe-TTF) = bis(ethylenediseleno)tetrathiafulvalene
NA = not available
* D. Schweitzer et al.[7]
BO is BEDO-TTF

κ-(ET)$_2$Cu(NCS)$_2$. Polarized microreflectance studies were made on the (100) faces of κ-(ET)$_2$Cu(NCS)$_2$.[33] Figure 6.4 illustrates the polarized spectra of this superconductor in different orientations. Two widely different positions in the absorption maxima are observed. See Fig. 6.4(a) and (b). As previously mentioned, these features may be attributed to the two different -C-C-H bending vibrations occurring in the anion cavity of the molecule.[33] The absorption maxima (as well as the dips) occur at higher frequency for the polarization of the incident light parallel to the c-axis (||c) as opposed to the positions for the polarization parallel to the b-axis (||b). See Fig. 6.4(a) and (b). For the light polarization parallel to 45° between the b- and c-axis, the maxima in absorption is identical to that found for the unpolarized light [Fig. 6.4(c) and (d)]. The dip occurs at 1292 cm^{-1} in the 45° polarization, which is intermediate between the results for ||b and ||c polarizations (1302 cm^{-1} and 1280 cm^{-1}, respectively). Deuteration of κ-(ET)$_2$Cu(NCS)$_2$ causes the absorption maxima in this region to shift (Fig. 6.3). This occurs because the -C-C-D bending of d$_8$-(ET) absorbs at a lower frequency,[27,33,52] outside the broad

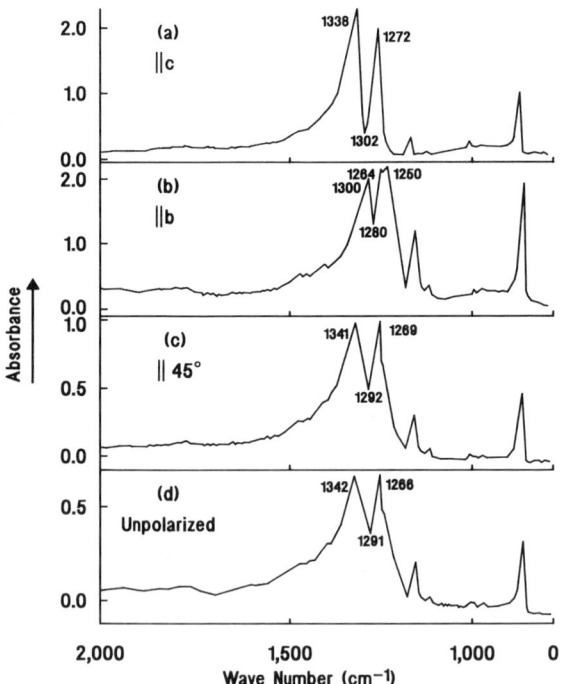

Figure 6.4 IR reflectance spectra of κ-(ET)$_2$Cu(NCS)$_2$ obtained with (a) polarized light parallel to the *c*-axis, (b) polarized light parallel to the *b*-axis, (c) polarized light parallel to 45° between the *b*- and *c*-axes, and (d) unpolarized light. Reprinted with permission from Solid State Commun., *68*, J.R. Ferraro, H. H. Wang, U. Geiser, A. M. Kini, M. A. Beno, J. M. Williams, S. Hill, M.-H. Whangbo and M. Evain, "Observation of Two Different -C-C-H Bending Modes in Ambient Pressure Organic Superconductor κ-(ET)$_2$Cu(NCS)$_2$ by Polarized Infrared Reflection Measurements and Implications on Its Structural and Physical Properties," **1988**, Pergamon Press, Inc., Elmsford, N.Y.
Note: A Kramers-Kronig transformation was used (from Ferraro et al.[33]).

vibronic reflectance region in the 1200–1400 cm^{-1} range, with much lowered intensity as compared to the -C-C-H bending vibration. It is important to examine how the polarization-dependence of the absorption dip is related to the presence of two nonequivalent ET molecules in κ-(ET)$_2$Cu(NCS)$_2$, which differ in the softness[55] of their C-H···anion interactions. As shown in Fig. 6.5, the C-H···anion contact interactions of molecule 2 (largely, C-H···S contacts) are softer (longer) than those of molecule 1 (largely, C-H···C and C-H···N contacts). Thus, the -C-C-H bending mode of molecule 2 should occur at a lower frequency, compared with that of molecule 1. It is observed that the absorption maxima (and dip) shift gradually to a lower frequency as the plane of the polarized light turns away from the *c*- to the *b*-axis. Therefore, the observed -C-C-H bending mode is a "hybrid" of the softer

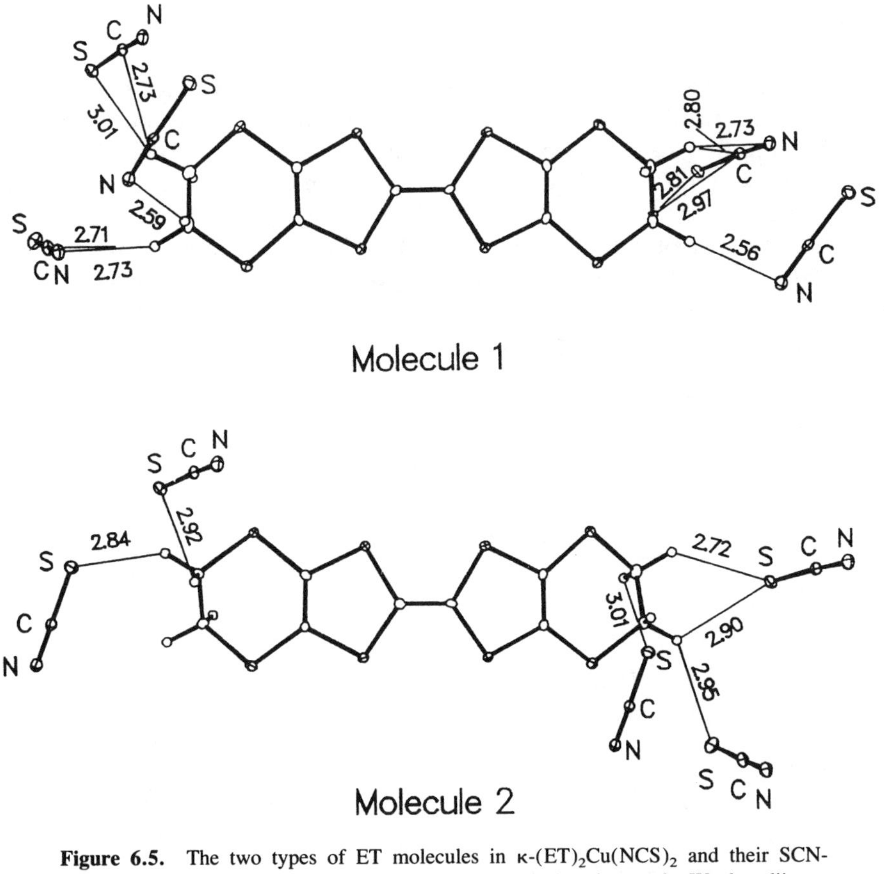

Figure 6.5. The two types of ET molecules in κ-(ET)$_2$Cu(NCS)$_2$ and their SCN-environment. Thin lines are short H···anion contacts less than the van der Waals radii sum (3.1 Å, 3.0 Å and 2.85 Å for H···S, H···C, and H···N, respectively). Reprinted with permission from Solid State Commun., *68*, J. R. Ferraro, H. H. Wang, U. Geiser, A. M. Kini, M. A. Beno, J. M. Williams, S. Hill, M. -H. Whangbo and M. Evain, "Observation of Two Different -C-C-H Bending Modes in Ambient Pressure Organic Superconductor κ-(ET)$_2$Cu(NCS)$_2$ by Polarized Infrared Reflection Measurements and Implications on Its Structural and Physical Properties," **1988**, Pergamon Press, Inc., Elmsford, N.Y.

and the harder -C-C-H bending modes expected from molecules 2 and 1, respectively, and a shift to lower frequency occurs as the weight of the softer -C-C-H bending mode increases.

As depicted in Fig. 6.6, the ethylene group hydrogen atoms of molecules 1 and 2 make short C-H···anion contacts with the $Cu(NCS)_2^-$ anion chains. Thus the -C-C-H bending vibrations of molecules 1 and 2 should interact via the anion chains. The $Cu(NCS)_2^-$ anion chains are aligned along the b-axis so that the longitudinal modes of the $Cu(NCS)_2^-$ anion chain vibrations would be more effectively excited by the $\parallel b$-polarized light than by the $\parallel c$-polarized light. Therefore, the shift of the -C-C-H bending frequency to a lower value, which occurs when the plane of the polarized light approaches the b-axis, is understandable if the longitudinal mode of the anion chain vibration promotes the -C-C-H bending of molecule 2.

(MDT-TTF)$_2$AuI$_2$. Polarized microreflectance studies on the superconductor (MDT-TTF)$_2$AuI$_2$ have been made.[37] The preparation of the superconductor (T_c = 3.5K) was first made by Papavassiliou et al.[56] The synthesis provided evidence that unsymmetrical sulfur-based organic donors can produce organic superconductors. A recent study[57] has shown that the T_c is nearly 4.5K.

The donor molecule packing in (MDT-TTF)$_2$AuI$_2$ is nearly similar to that observed for the 10 K superconductor κ-(ET)$_2$Cu(NCS)$_2$.[58–60] Both molecules have tightly linked dimers, which are orthogonally arranged in a 2-dimensional S···S network. However, MDT-TTF does not have -C-H bonds directed perpendicular to

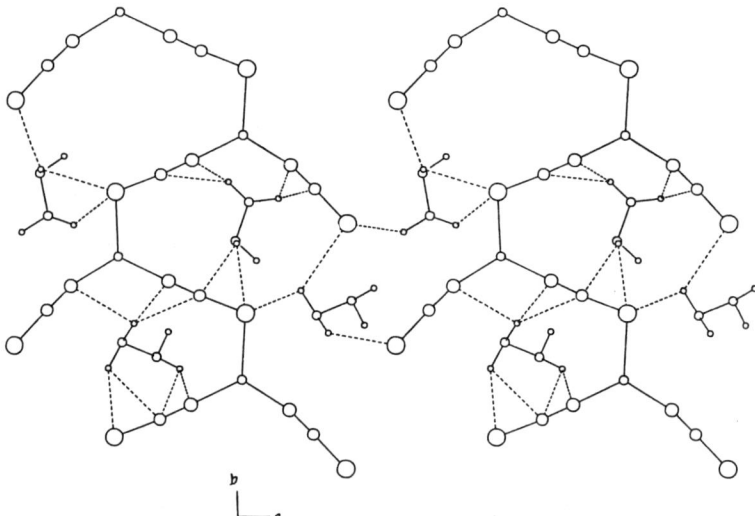

Figure 6.6 Perspective view of how the ethylene groups of ET molecules make short C-H···anion contacts (shown by dashed lines) with the $Cu(NCS)_2^-$ anion chains in κ-(ET)$_2$Cu(NCS)$_2$. For simplicity, only the ethylene groups below the plane of the anion chains are shown. In the anion chains, the smallest and the largest circles represent the Cu and S atoms, respectively. In the ethylene groups, the smaller circles represent the H atoms.

the molecular plane, which prevents the donor molecule from having short H⋯S contacts within the donor layer. The interplanar spacing within the MDT-TTF dimers is 3.35 Å, identical to that observed in κ-(ET)$_2$Cu(NCS)$_2$.[58-60] Figure 6.7 illustrates the (MDT-TTF)$_2$AuI$_2$ structure. In the (MDT-TTF)$_2$AuI$_2$ structure, the methylene and ethylene groups of the donor molecules in the dimer alternate so that a methylene group resides over an ethylene group. Each end of the (MDT-TTF)$_2$AuI$_2$ dimer has only four hydrogen atoms, as compared to eight hydrogens at each end in κ-(ET)$_2$Cu(NCS)$_2$ dimers, thus forming fewer -C-H⋯anion contacts. Expectations would be that the translational and/or librational modes of vibrations are substantially different in (MDT-TTF)$_2$AuI$_2$ than in κ-(ET)$_2$Cu(NCS)$_2$, and this would cause differences in the electron coupling in the vibronic region. The microreflectance spectrum of (MDT-TTF)$_2$AuI$_2$ bears this out. Figure 6.8 shows the microreflectance spectrum for this material. It can be observed that the splitting in the 1200–1400 cm^{-1} region is absent, whereas it occurs in κ-(ET)$_2$Cu(NCS)$_2$.[33] Since, as previously mentioned, the vibronic region involves the -C-C-H bending vibration in the anion cavity, the results point to differences in this cavity for (MDT-TTF)$_2$AuI$_2$ and κ-(ET)$_2$Cu(NCS)$_2$. The results concur with the structural data,[57] and lend further support to the deuteration studies and the notion that we are dealing with a -C-C-H bending vibration in the anion cavity, and coupled to the vibron.

The use of the micro-FT/IR reflectance technique for the study of this class of conductors and superconductors appears to be a viable, alternate method of identifying differences between the various phases obtained for PT or ET compounds, and in identifying the nature of the environment in the anion cavity of these materials. The technique is also useful in identifying microcontamination in these salts. Contamination in 10^2–10^3 parts per picogram can be analyzed for samples as small as 10μm × 10μm. The method is accomplished at room temperature, is rapid, nondestructive and requires only minute amounts of material and no sample prep-

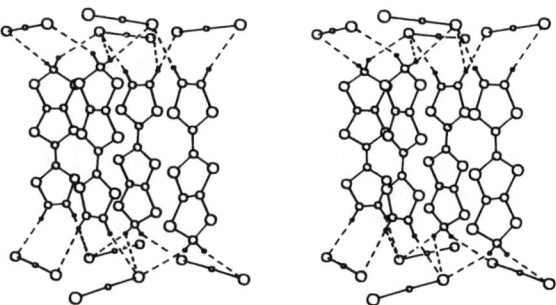

Figure 6.7 Stereodiagram showing the hydrogen⋯iodine contacts less than 3.6 Å (dashed line) present in (MDT-TTF)$_2$AuI$_2$. Reprinted with permission from Solid State Commun., 69, A. M. Kini, M. A. Beno, D. Son, H.H. Wang, K. D. Carlson, L. C. Porter, U. Welp, B. A. Vogt, J. M. Williams, D. Jung, M. Evain, M.-H. Whangbo, D. L. Overmeyer and J. E. Schirber, "(MDT-TTF)$_2$AuI$_2$: An Ambient Pressure Organic Superconductor (T_c = 4.5K) based on an Unsymmetrical Electron Donor," **1989**, Pergamon Press, Inc., Elmsford, N.Y.

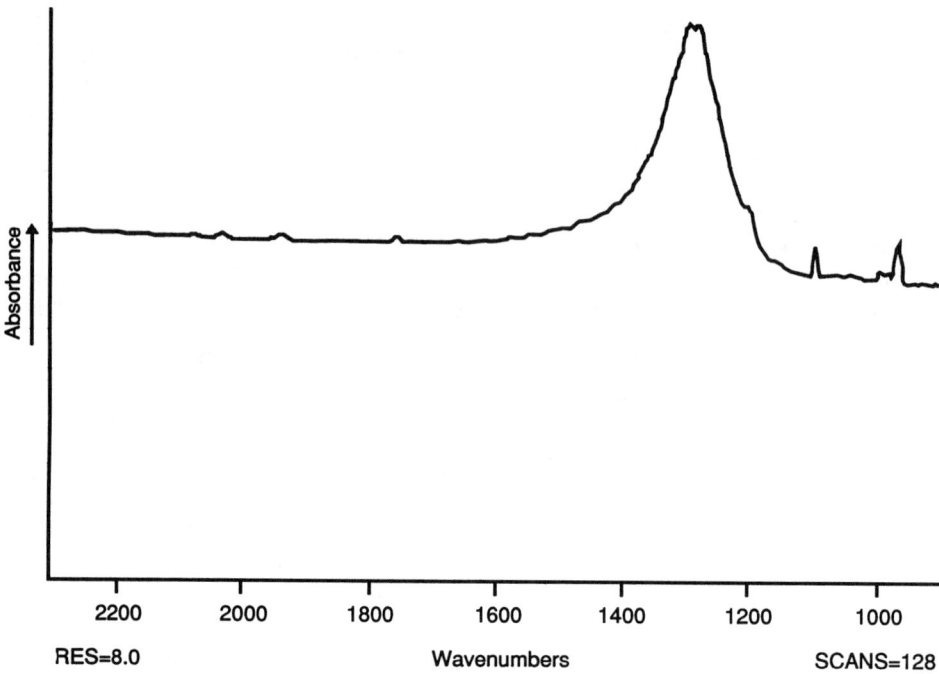

Figure 6.8 Micro-reflectance spectrum of the $(MDT\text{-}TTF)_2AuI_2$ superconductor.
Note: Absorbance in arbitrary units. A Kramers-Kronig transformation was used (from Ferraro et al.[33]).

aration. Although the room temperature reflectance measurements have proven useful, it should be understood that these materials may contain contributions from thermally generated carriers (see Chap. 7). Additionally, structural changes may ensue as the temperature lowers. Thus room temperature reflectance spectra should be used with caution, and should not attempt to be used in making interpretations concerning low-temperature properties of the charge-transfer organic superconductors.

EXPERIMENTAL

The experimental conditions used in the infrared, Raman and reflection methods reported in this chapter are listed in Tables 6.7 and 6.8. Unfortunately, as may be observed, a number of the publications failed to list the experimental conditions

TABLE 6.7 Instrumentation Used for Raman and Infrared Studies

	Instrumentation	
Reference	Raman	Infrared*
5	DPS-24	Specord 75
	(488 and 514.5 nm)	Bruker IFS-113V
6	(514.5, 457.9, 632.8 nm)	n.l.

* Instrumentation for infrared reflectance measurements is listed in Table 6.8.
n.l. = not listed in manuscripts where Raman and/or IR data are reported

TABLE 6.8 Instrumentation Used for Reflectance Measurements

Reference	Instrumentation
15	n.l.
16	n.l.
17	Bruker IFS 113V
18	Jobin-Yvon H-20F
	H-20 FIR
19	PE #98
20	Jasco MIR-300
	Olympus MMSP-RF
21	Jobin-Yvon H-20F, H-20FIR
22	Jasco MIR-300
23	Jasco MIR-300
	Olympus MMSP-RF
24	Jasco MIR-300
25	Bruker IFS 113V
	PE #98
26	n.l.
27	Jasco MIR-300
	Olympus MMSP-RF
28	IBM IR/44
29	Hitachi 340
	Jasco MIR-300
30	n.l.
31	n.l.
32	PE 1760
	Homemade spectrometer at higher frequencies
33	Digilab FTS-40*
34	Digilab FTS-40*
35	Digilab FTS-40*
36	JASCO MIR-300
37,46,47	Digilab FTS-40*
39	Bomem DA3
40	PE 1710
43	Jasco MIR-300
50	KSVU-2
	Bruker IFS 113V
51,52	Bruker IFS 113V

* UMA-300A microscope used.
n.l. = not listed in manuscripts where Raman and/or IR data are reported.

used. The materials are opaque and are synthesized in small quantities, and the method of choice to investigate them is by reflective techniques using microspectrometric instrumentation. Our microreflectance studies were made using a Digilab FTS-40 interferometer interfaced with a UMA-300A microscope and a narrow-band MCT detector. The single crystal was placed on a KBr salt plate to eliminate any reflection/absorption effects from the sample mount. A sputtered gold plate was used as instrument background. Polarized spectra were generally made with single crystals. A Kramers-Kronig transformation was applied to all microreflectance spectra. Spectra were recorded from the 4000–700 cm^{-1} region at 4 cm^{-1} resolution. Infrared data were recorded on KBr disks or on thin layers of crystals made by sublimation techniques. Raman data were obtained from powders, and in some cases, from solutions.

REFERENCES

1. Ferraro, J. R.; Williams, J. M. *Introduction to Synthetic Electrical Conductors*; Academic Press: Orlando, Florida, 1987.
2. Bechgaard, K.; Jacobsen, C. S.; Mortensen, K.; Petersen, H. J.; Thorup, N. *Solid State Commun.* **1980** *33*, 1119.
3. Carlson, K. D.; Geiser, U.; Kini, A. M.; Wang, H. H.; Montgomery, L. K.; Kwok, W. K.; Beno, M. A.: Williams, J. M.; Cariss, C. S.; Crabtree, G. W.; Whangbo M.-H.; Evain, M. *Inorg. Chem.* **1988** *27*, 965.
4. Ferraro, J. R.; Williams, J. M. In *Practical Fourier Transform Infrared Spectroscopy, Industrial and Laboratory Chemical Analysis*, Ferraro, J. R.; Krishnan, K., Eds., Academic Press; San Diego, CA, 1989.
5. Kozlov, M. E.; Pokhodnia, K. I.; Yurchenko, A. A. *Spectrochim. Acta* **1987**, *43*, 323.
6. Sugai, S.; Saito, G. *Solid State Commun.* **1986**, *58*, 759; *Synth. Met.* **1987**, *19*, 231.
7. Schweitzer, D.; Bele, P.; Brunner, H.; Gogu, E.; Haeberlen, V.; Hennig, I; Klutz, I.; Swietlik, R.; Keller, H. J. *Z. Phys. B: Condens. Matter* **1987**, *67*, 489.
8. Swietlik, R.; Schweitzer, D.; Keller, H. J. *Phys. Rev. B: Condens. Matter* **1987**, *36*, 6881.
9. Schweitzer, D.; Gogu, E.; Hennig, I.; Klutz, T.; Keller, H. J. *Ber. Bunsenges Phys. Chem.* **1987**, *91*, 890.
10. Schweitzer, D. In *Electronic Properties of Conjugated Polymers*, Kuzmany, H.; Mehring, M.; Roth, S., Eds.; Proceedings of an International Winter School, Kirchberg, Tirol, March 14–21, 1987, 354, Springer-Verlag: Berlin.
11. Zamboni, R.; Schweitzer, D; Keller, H. J., Proc. of NATO ASI *"Lower Dimensional Systems and Molecular Electronics,"* Metzer, R. M.; Day, P.; Papavassiliou, G., Eds; Spetses, Greece, 1989, Plenum Press, 1990.
12. Schweitzer, D.; Kahlich, S.; Gärtner, S.; Gogu, E.; Grimm, H.; Heinen, I.; Klutz, T.; Zamboni, R.; Keller, H. J.; Renner, G. *Ibid.*
13. Lu, L.; Liu, J.-Q.; Ma, B.-H.; Zhang, D.-L.; Wang, X.-H.; Zhu, D. *Synth. Met.* **1989**, *31*, 45.
14. Bandrauk, A. D.; Truong, K. D.; Carlone, C.; Jondl, S. *Chem. Phys. Litt.*, **1983**, *95*, 78.
15. Kaplunov, M. G.; Yagubskii, E. B.; Rosenberg L. P.; Borodko, Yu. G. *Phys. Status Solidi A* **1985**, *89*, 509.
16. Koch, B.; Geserich, H. P.; Ruppel, W.; Schweitzer, D.; Dietz, K. H.; Keller, H. J. *Mol. Cryst. Liq. Cryst.* **1985**, *119*, 343.
17. Meneghetti, M.; Bozio, R.; Pecile, C. *J. Phys. (Paris)* **1986**, *47*, 1377.
18. Sugano, T.; Yamada, K.; Saito, G.; Kinoshita, M. *Solid State Commun.* **1985**, *55*, 137.
19. Jacobsen, C. S.; Williams, J. M.; Wang, H. H. *Solid State Commun.* **1985**, *54*, 937.
20. Kuroda, H.; Yakushi, K.; Tajima, H.; Saito, G. *Mol. Cryst. Liq. Cryst.* **1985**, *125*, 135.
21. Sugano, T.; Saito, G. *J. Phys. C: Solid State Phys.* **1986**, *19*, 5471.
22. Tajima, H.; Yakushi, K.; Kuroda, H.; *Solid State Commun.* **1985**, *56*, 159.

23. Tajima, H.; Kanbara, H.; Yakushi, K.; Kuroda, H.; Saito, G. *Solid State Commun.* **1986**, *57*, 911.
24. Tajima, H.; Yakushi, K.; Kuroda, H.; Saito, G.; Inokuchi, H. *Solid State Commun.* **1984**, *49*, 769.
25. Jacobsen, C. S.; Tanner, D. B.; Williams, J. M.; Geiser, U.; Wang, H. H. *Phys. Rev. B: Condens. Matter* **1987**, *35*, 9605.
26. Ugawa, A.; Yakushi, K.; Kuroda, H.; Kawamoto, A. Tanaka, J. *Chem. Lett.* **1986**, 1875.
27. Ugawa, A.; Ojima, G.; Yakushi, K.; Kuroda, H. *Phys. Rev. B: Condems. Matter* **1988**, *38*, 5122.
28. Ferraro, J. R.; Wang, H. H.; Ryan, J.; Williams, J. M. *Appl. Spectrosc.* **1987**, *41*, 1377.
29. Yakushi, K.; Tajima, H.; Ida, T.; Tamura, M.; Hayashi, H.; Kuroda, H.; Kobayashi, A.; Kobayashi, H.; Kato, R. *Synth. Met.* **1988**, *24*, 301.
30. Yoshitake, M.; Yakushi, K.; Kuroda, H.; Kobayashi, A.; Kato, R.; Kobayashi, H. *Bull. Chem. Soc. Jpn.* **1988**, *61*, 1115.
31. Kaplunov, M. G.; Lyubovskaya, R. N.; Aldoshina, M. Z.; Borodko, Y. G. *Phys. Status Solidi Part A* **1987**, *104*, 833.
32. Sugano, T.; Hayashi, H.; Takenouchi, H.; Nishikida, K.; Urayama, H.; Yamochi, H.; Saito, G.; Kinoshita, M. *Phys. Rev. B: Condens. Matter* **1988**, *37*, 9100.
33. Ferraro, J. R.; Wang, H. H.; Geiser, U.; Kini, A. M.; Beno, M. A.; Williams, J. M.; Hill, S.; Whangbo, M.-H.; Evain, M. *Solid State Commun.* **1988**, *68*, 917.
34. Wang, H. H.; Montgomery, L. K.; Geiser, U.; Porter, L. C.; Carlson, K. D.; Ferraro, J. R.; Williams, J. M.; Cariss, C. S.; Rubinstein, R. L.; Whitworth, J. R.; Evain, M.; Novoa, J. J.; Whangbo, M.-H. *Chem. of Mater.* **1989**, *1*, 140.
35. Wang, H. H.; Ferraro, J. R.; Carlson, K. D.; Montgomery, L. K.; Geiser, U.; Williams, J. M.; Whitworth, J. R.; Schlueter, J. A.; Hill, S.; Whangbo, M.-H.; Evain, M.; Novoa, J. J. *Inorg. Chem.* **1989**, *28*, 2267.
36. Tamura, M.; Yakushi, K.; Kuroda, H.; Kobayashi, A.; Kato, R.; Kobayashi, H.; *J. Phys. Soc. Jpn.* **1988**, *57*, 3239.
37. Ferraro, J. R. unpublished data.
38. Kuroda, H.; Yakushi, K.; Tajima, H.; Ugawa, A.; Tamura, M.; Okawa, Y.; Kobayashi, A.; Kato, R.; Kobayashi, H.; Saito, G.; *Synth. Met.* **1988**, *27*, A491.
39. Tokumoto, M.; Anzai, H.; Takahashi, K.; Kinoshita, N.; Murata, K.; Ishiguro, T.; Tanaka, Y.; Hayakawa, Y.; Nagamori, H.; Nagasaka, K. *Synth. Met.* **1988**, *27*, A171.
40. Pratt, F. L.; Hayes, W.; Kurmoo, M.; Day, P. *Synth. Met.* **1988**, *27*, A439.
41. Ugawa, A.; Ojima, G.; Yakushi, K.; Kuroda, H. *Synth. Met.* **1988**, *27*, A445.
42. Yakushi, K.; Kanbara, H.; Tajima, H.; Kuroda, H.; Saito, G.; Mori, T. *Bull. Chem. Soc. Jpn.* **1987**, *60*, 4251.
43. Ugawa, A.; Yakushi, K.; Kuroda, H.; Kawamoto, A.; Tanaka, J. *Synth. Met.* **1988**, *22*, 305.
44. Tajima, H.; Yakushi, K.; Kuroda, H.; Saito, G. *Solid State Commun.* **1985**, *56*, 251.
45. Sugano, T.; Hayashi, H.; Kinoshita, M.; Nishikida, K. *Phys. Rev. B: Condens. Matter* **1989**, *39*, 11387.
46. Beno, M. A.; Wang, H. H. Carlson, K. D.; Kini, A. M.; Frankenbach, G. M.; Ferraro,

J. R.; Larsen, N.; McCabe, G. D.; Thompson, J.; Purnama, C.; Vashon, M.; Williams, J. M.; Jung, D.; Whangbo, M.-H. *Mol. Cryst. Liq. Cryst.* **1990**, *181*, 145.

47. Ferraro, J. R.; Williams, J. M. *Mol. Cryst. Liq. Cryst.* **1990**, *181*, 253.
48. Kozlov, M. E.; Pokhodnya, K. I.; Yurchenko, A. A. *Spectrochim Acta* **1989**, *45A*, 437.
49. Kozlov, M. E.; Baram, N. P.; Pokhodnya, K. I.; Yurchenko, A. A. *Soviet J. Low Temp. Phys.* **1989**, *15*, 323.
50. Kozlov, M. E.; Onishchenko, V. G.; Pokhodnya, K. I. *Soviet J. Low Temp. Phys.* **1989**, *31*, 1005.
51. Kornelsen, K.; Eldridge, J. E.; Homes, C. C.; Wang, H. H.; Williams, J. M. *Solid State Commun.* **1989**, *72*, 475.
52. Kornelsen, K.; Eldridge, J. E.; Wang, H. H.; Williams, J. M. *Solid State Commun.*, **1990**, *74*, 501.
53. Rice, M. J.; Pietronero, L.; Brüesch, L. *Solid State Commun.* **1977**, *21*, 757; Rice, M. J. *Phys. Rev. Lett.* **1976**, *37*, 36.
54. Williams, J. M.; Wang, H. H.; Emge, T. J.; Geiser, U.; Beno, M. A.; Leung, P. C. W.; Carlson, K. D.; Thorn, R. J.; Schultz, A. J.; Whangbo, M.-H. In *Prog. Inorg. Chem.*, S. Lippard, Ed., J. Wiley and Sons; N.Y., 1987, pp. 51–218.
55. Whangbo, M.-H.; Williams, J. M.; Schultz, A. J.; Emge, T. J.; Beno, M.A. *J. Am. Chem. Soc.* **1987**, *109*, 90. (Softness is defined as degree of coupling between electrons and phonons, where phonons are translational and/or rotational vibrations.)
56. Papavassiliou, G. C., Mousdis, G. A.; Zambounis, J. S.; Terzis, A.; Hountas, A.; Hilti, B.; Mayer, C. W.; Pfeiffer, J. Proceedings of International Conference on Science and Technology of Synthetic Metals, Santa Fe, New Mexico, June 26–July 2, *Synth. Met.* **1988** 27, B379.
57. Kini, A. M.; Beno, M. A.; Son, D.; Wang, H. H.; Carlson, K. D.; Porter, L. C.; Welp, U.; Vogt, B. A.; Williams, J. M.; Jung, D.; Evain, M.; Whangbo, M.-H.; Overmyer, D. L.; Schirber, J. E. *Solid State Commun.* **1989**, *69*, 503.
58. Urayama, H.; Yamochi, H.; Saito, G.; Nozawa, K.; Sugano, T.; Kinoshita, M.; Sato, S.; Oshima, K.; Kawamoto, A.; Tanaka, J. *Chem. Lett.* **1988**, 55.
59. Urayama, H.; Yamochi, H.; Saito, G.; Sato, S.; Kawamoto, A.; Tanaka, J.; Mori, T.; Maruyama, Y.; Inokuchi, H. *Chem. Lett.* **1988**, 463.
60. Gärtner, S.; Gogu, E.; Heinen, I.; Keller, H.J.; Klutz, T.; Schweitzer, D. *Solid State Commun.* **1988**, *65*, 1531.

7

Polarized Reflectance in the Infrared

INTRODUCTION

The electronic structure of the organic radical-cation compounds can be determined through band theory calculations (see Chap. 8) or by use of experimental measurements such as optical absorption, X-ray photoelectron spectroscopy (XPS), X-ray resonance absorption in the near edge (XANES), or polarized reflectance spectroscopy. The spectra are generally measured below room temperature, because such spectra contain the main contributions due to the charge carriers. At low temperatures, only energies in which ϵ/kT (k = Boltzmann constant) is small can contribute, so that energies of the order of eV's ($1/300k = 39$ eV^{-1}) are essentially those due to bound states which make little contribution. Thus, for low temperatures and for the case where there are some free carriers, the pertinent electronic structures are contained in the infrared reflectance spectra from which the optical conductivities $\sigma(\upsilon)$ can be derived and which are related to the optical absorption $\eta(\upsilon)$:

$$\eta(\upsilon) = \frac{4\pi}{c} \sigma(\upsilon) \qquad (7.1)$$

The quantity $\eta(\upsilon)$ is the petite (micro) ensemble parameter through which one can evaluate $\sigma(T)$ for a canonical ensemble. As discussed, this interrelationship is stated because $\eta(\upsilon)$ must be that for $T = 0$.

Absorption spectra of black lusterous metallic crystals of organic conductors are not readily measured, but the reflectance spectrum, from which one can derive the optical conductivity, can be measured readily on a single crystal. In order to understand polarized reflectance spectra, knowledge of the optical properties as derived from Maxwell's equations and the various models for $\sigma(\upsilon)$ is required. To recognize the nature of the optical conductivities, the $\sigma(\upsilon)$'s, by inspection, requires a knowledge of the spectra predicted by the models for free and bound electrons. Thus, discussions of the experimentally derived $\sigma(\upsilon)$'s are faciliated by an understanding of the spectral shapes predicted by use of the models for free carriers (Drude), for bound oscillators, (Lorentz), for semiconductors, and for semimetals as summarized below.

All organic charge-transfer conductors have anisotropic properties; the optical constants are tensors. By use of polarized normal reflectance and from a knowledge of the optical axes of the crystal, the extent of the anisotropy can often be determined. To derive the optical constants n, the index of refraction, and k, the extinction coefficient, from the measured reflectance, $R(\omega)$, one employs the complex reflectance,[1]

$$\tilde{R} = \sqrt{R(\omega)} \, \exp(i\theta) \qquad (7.2)$$

in which θ is the phase angle and $\omega = 2\pi\upsilon$ is the angular frequency. Through the application of the Kramers-Kronig transformation, one can interrelate $\theta(\omega)$ and $R(\omega)$:

$$\theta(\omega) = \frac{\omega}{\pi} \int_0^\infty \frac{\ln[R(\omega')/R(\omega)]}{\omega^2 - \omega'^2} \, d\omega' \qquad (7.3)$$

From $R(\omega)$ and $\theta(\omega)$, one can derive the values for n and k as follows:

$$n = \frac{1-R}{1+R-2\sqrt{R}\cos\theta} \qquad (7.4)$$

$$k = \frac{2R\sin\theta}{1+R-2\sqrt{R}\cos\theta} \qquad (7.5)$$

Through use of Maxwell's equations, the relationships between the index of refraction, the extinction coefficient, and the real and imaginary parts, ϵ_1 and ϵ_2, of the complex dielectric function, or alternately, the polarizability and the optical conductivity[2] are:

$$\epsilon_1 = n^2 - k^2 = 1 + 4\pi\alpha(\upsilon) \qquad (7.6)$$

$$\epsilon_2 = nk = \frac{\sigma(\upsilon)}{\upsilon} \qquad (7.7)$$

MODELS FOR OPTICAL CONDUCTIVITIES

Variations of the optical conductivities with frequency, $\sigma(\upsilon)$'s, are based on the following models:

1. Drude model, a free electron with damping,
2. Lorentz oscillator, an electron bound with Hooke's law and with damping,
3. Models in which the number of carriers or hopping rate depend on frequency.

The first two models are those most cited when discussing the optical properties of solids. Model 3 can be useful in effecting a more complete classification.

For the Drude model of the free electron conductor, the electric field force is equal to the sum of an inertial force and a damping force; through the evaluation of the complex dielectric function, one finds for the optical conductivity and the polarizability, respectively

$$\sigma(v) = \frac{n_o e^2}{4\pi^2 m} \frac{2\pi\gamma}{(v^2 + \gamma^2)} \qquad (7.8)$$

and

$$\alpha(v) = \frac{n_o e^2}{4\pi^2 m} \frac{1}{(v^2 + \gamma^2)} \qquad (7.9)$$

In these equations n_o is the density of carriers (with charge e and mass m), γ is a damping constant relating the damping force to the velocity, and v is the field frequency. The damping constant is a measure of the resistance. It can be related to the relaxation time for scattering through $2\pi\gamma = \tau^{-1}$ (In some cases $\Gamma = \tau^{-1}$ is used). The frequency can be related to the angular frequency ω through the relation $2\pi v = \omega$. The function has an inflection at

$$\omega_i = \frac{\tau^{-1}}{\sqrt{3}} \qquad (7.10)$$

Plots for typical values of $\sigma(\omega)/\sigma(o)$ for two organic charge-transfer conductors are shown in Fig. 7.1.

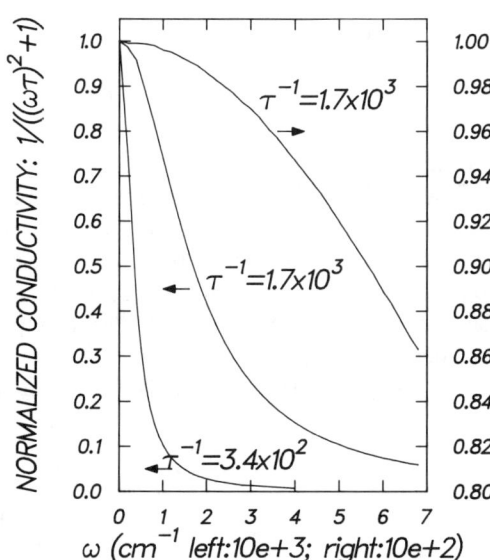

Figure 7.1 Plot of normalized optical conductivity, $\sigma(\omega)/\sigma/o$, versus ω for the Drude model. The damping constants or the relaxation times (τ), are typical of those obtained for the ET-compounds (see Table 7.1).

If the integral of the Drude $\sigma(\omega)$ is evaluated over $0 \leq \omega \leq \infty$, one obtains

$$8 \int_0^\infty \sigma(\omega) d\omega = \frac{4\pi n_o e^2}{m} = \omega_p^2 \qquad (7.11)$$

ω_p where is the plasma frequency.[3] It is a measure of the density of charge carriers or effective mass or the ratio.

For the Lorentz oscillator, the electron is bound with a Hooke's law constant, k, and it is also damped as in the Drude model. In this case

$$\sigma(v) = \frac{n_o e^2}{4\pi^2 m} \frac{2\pi \gamma v^2}{[(v_o^2 - v^2)^2 + \gamma^2 v^2]} \qquad (7.12)$$

$$\sigma(v) = \frac{n_o e^2}{4\pi^2 m} \frac{(v_o^2 - v^2)}{[(v_o^2 - v^2)^2 + \gamma^2 v^2]} \qquad (7.13)$$

$$v_o = \frac{1}{2\pi} \sqrt{\frac{k}{m}} \qquad (7.14)$$

Obviously, for $k = 0$, the Lorentz model reduces to the Drude model. The associated shape of $\sigma(v)$ is a narrow peak.[2] Transitions which can be associated with the Lorentzian shape, of course, do not contribute directly to the electrical conductivity, because they are transitions from one bound state to another, usually vibrational or vibronic. However, the vibrational modes may be coupled with other modes that are coupled directly with the conductive carriers and therein reveal a correlation with conductivity. (See Summary)

Spectral distributions for a semiconductor or a semimetal are usually not discussed in standard texts on optical properties. There are two functions that can be useful in interpreting the reported $\sigma(v)$'s for organic charge-transfer conductors. Rice[4] and Rice, Pietronero, and Brüesch[5] in a study of electron-phonon coupling in the organic conductor TEA(TCNQ)$_3$ expressed the optical conductivity for a semiconductor by use of the relation

$$\sigma(\omega) = \frac{n_o e^2}{m\omega} \left\{ \frac{\ln\left(\frac{1-S}{1+S}\right)}{2x^2 S} \right\} \qquad (7.15)$$

in which $S = (1 - x^{-2})^{1/2}$ and $x = \omega/2\Delta > 1$; 2Δ = the semiconductor energy gap in the presence of the electron phonon coupling. A plot of this function is shown in Fig. 7.2. The quantity $\ln(1 - S/1 + S)/2Sx^2$ has values extending from $(1 - \delta)$ to zero for $x = (1 + e)$ and $x = \infty$, so that the quantity measures the fraction of the oscillation period ω^{-1} (i.e., relaxation time) which is effective in the conductive process when the electrons and phonons are coupled during hopping. In the model cited in Fig. 7.2, a minimum energy of 2Δ is required to promote electrons into a conductive state.

Through the sum-rule, the effective number of carriers is given by

$$N_{\text{eff}}(\omega) = N \frac{\int_0^\omega \sigma(\omega') d\omega'}{\int_0^\infty \sigma(\omega') d\omega'} \qquad (7.16)$$

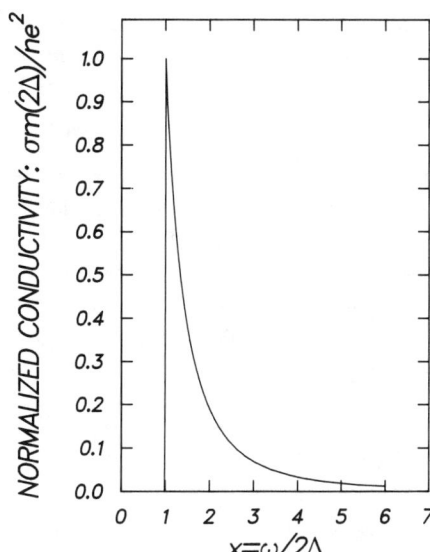

Figure 7.2 Plot of normalized optical conductivity versus the ratio of ω to the semiconductor energy gap (2Δ) for Rice's model for coupling of electrons and phonons (see Eq. 7.15).

in which N is the total density available for thermal activation. If $\sigma(\omega')$ is given by the Drude model, one obtains

$$\sigma(\omega) = \frac{e^2 N}{m} \left(\frac{2}{\pi} \tan^{-1}(\omega\tau) \frac{\tau^{-1}}{(\omega^2 + \tau^{-2})} \right) \quad (7.17)$$

The factor $(2/\pi)\tan^{-1}(\omega\tau)$ varies sigmoidally from 0 to 1 and thus measures the fraction of N which are carriers. A plot of this model, as in the case of bismuth, is shown in Fig. 7.3; it represents the behavior of a semimetal or degenerate semiconductor.

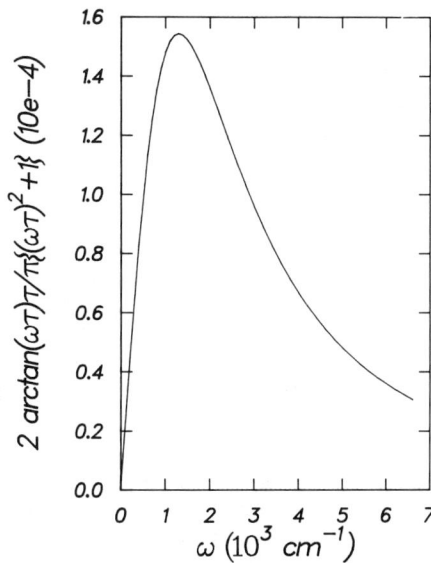

Figure 7.3 Plot of normalized optical conductivity, $\sigma m/Ne^2$, versus ω for the semiconductor model with the effective density of carriers equal to the fraction $2/\pi \tan^{-1}(\omega\tau)$ of the total density available, N (see Eq. 7.17).

In a semimetal or a degenerate semiconductor, the density of carriers at $\omega=0$ and $T=0$ may not be zero. Thus, it can have a residual conductivity. Equation (7.17) and Fig. 7.4 with $\sigma(0) \neq 0$ can then be modified so that

$$\sigma_{sm} = \sigma(0) + \sigma(\omega) \tag{7.18}$$

By use of the models discussed above, one can obtain limited information regarding the shapes of the $\sigma(\omega)$'s in relation to the characteristics of $\sigma(T)$. However, caution must be exercised in attempting to relate measured reflectances and the derived optical conductivity to the electronic structure and the electrical properties. This occurs because forms of $\sigma(v)$ or $\sigma(\omega)$ strictly apply only to measurements made at $T=0$. At any finite temperature, the situation can be quite different. If a semiconductor has a small energy gap, or if a Lorentzian oscillator has a small value of v_o, the reflectance and the optical conductivity spectra depend on the thermally generated density of carriers. If the reflectance spectra are measured at 300K, kT is 2.6×10^{-2} eV or 208 cm^{-1}. Thus reflectance spectra measured at room temperature or even at lower temperatures can contain contributions from thermally generated carriers which are difficult to take into account.

If the spectrum of $\sigma(\omega)$ for an organic conductor can be described by one of these three cases, then it can readily be classified as (a) a metal, (b) a semiconductor, or (c) a semimetal. [The spectra are usually given by σ versus wave number (i.e., $\tilde{v} = \lambda^{-1}$ cm^{-1}). The equations for the models, however, are more compact if written in terms of ω, and ω can be in units of sec^{-1} or cm^{-1} provided τ is in the corresponding units.] In some cases, $\sigma(\omega)$ is describable by Eq. (7.15) and (7.17) and thus can be classified as either a semiconductor or a semimetal. If the reflec-

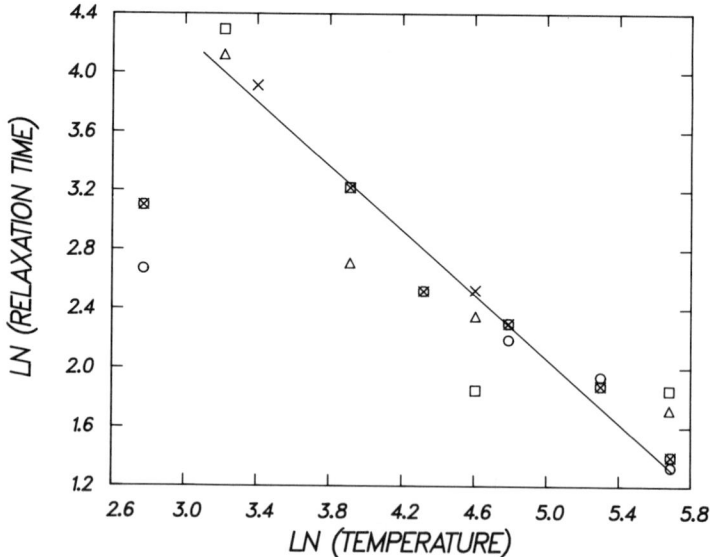

Figure 7.4 Plot of logarithm of relaxation time in the Drude model versus logarithm of temperature derived from temperature dependence of the polarized reflectance for some ET-compounds (see Table 7.1 and text) $x\beta$-$(ET)_2I_3$, ⊥100; O-Θ-$(ET)_2I_3$, ‖a; ⊗ Θ-$(ET)_2I_3$, ‖c; □ κ-$(ET)_2Cu(SCN)_2$, ‖b; and △ κ-$(ET)_2Cu(SCN)_2$, ‖c.

tance spectra have been measured at sufficiently low temperatures and low frequencies, then a distinction between these two cases can occasionally be made. If the electronic and phononic oscillations are coupled, a more comprehensive model such as presented by Rice[4] and colleagues[5] is useful. This model describes the coupling of the symmetric vibrational modes to the electronic oscillations of a charge-density wave (CDW). Thus, the model for a linear chain organic semiconductor as formulated by Rice and colleagues consists of two parts:

1. In the periodic potential of the π-bonded system, the conduction electrons "condense into charge-density waves" (CDW).
2. These conduction-electron molecular orbitals are coupled to the totally symmetric intramolecular vibrations which are infrared active.

This model is essentially the hopping part of the polaron model.[6,7]

OPTICAL CONDUCTIVITIES

Discussions of the optical conductivities reported to date are presented in two categories: (1) a general description of the common features and (2) a discussion of the individual spectra with an attempt to associate them with the models described.

General Descriptions and Features

Infrared polarized reflectance has been measured for all the BEDT-TTF or ET compounds listed herein. From measured values for $R(\omega)$, values of optical conductivity have been derived by use of the Kramers-Kronig transformation. The results are displayed in Fig. 7.5 to 7.17. One general observation is helpful in considering these materials with respect to metals. For metals, the reflectance is approximately constant at a value near unity for all values of the wave numbers corresponding to the visible region ($\sim 20 \times 10^3$ cm^{-1}). For the organic charge-transfer compounds, the reflectance is only a few tenths near 5×10^3 cm^{-1} and larger. It does not attain a value near unity until the wave number is decreased to a value below 1×10^3 cm^{-1}. However, the relaxation times listed in Table 7.1 are comparable to those for metals ($\sim 10^{-15}$ sec^{-1}).

All the curves for $\sigma(v)$ display structure, but in general they can be described as having a component resembling a broad band with a maximum and superimposed sharper peaks, or what appears like the onset of another broad band or a Drude tail. In no case has the $\sigma(v)$ curve been extended to sufficiently small wave number values, so that one can use the mentioned criteria to identify it unambiguously as a metal, semiconductor, or semimetal. Thus, most compounds appear to have metallic and/or semiconductive components in the spectra. Thus, there are significant differences between the $\sigma(v)$'s for organic charge-transfer compounds and those for metals or semiconductors. The structure in the organic compounds occurs below 5×10^3 cm^{-1} (0.62 eV). In this region (0 to 0.62 eV), the metals[8] or the usual semiconductors[9] usually display either the Drude shape or zero, respectively.

TABLE 7.1 Values for ω_p and τ^{-1} Derived from the Drude Model for $\sigma(\omega)$

Compounds	Temperature (K)	$\omega_p(\times 10^3 \text{cm}^{-1})$	$\tau^{-1}(\times 10^3 \text{cm}^{-1})$	Reference
$\alpha\text{-(ET)}_2\text{I}_3$				
E∥b	300	3.6	37.7	(11)
E⊥b		4.6	21.4	
$\beta\text{-(ET)}_2\text{I}_3$				
⊥ stack	40	9.6	3.39	(10)
∥ stack		5.7	14.3	
			6.9	(11)
⊥100	100	3.06	4.0	(15)
	50	3.71	2.0	
	30	3.87	1.0	
∥ 100	30	7.18	2.0	
$\theta\text{-(ET)}_2\text{I}_3$				
∥a	295	7.42	1.29	(18)
	200	7.82	0.70	
	120	8.39	0.55	
	16	8.47	0.34	
∥c	295	5.00	0.97	(18)
	200	5.73	0.60	
	125	5.81	0.39	
	75	5.81	0.31	
	16	5.89	0.18	
$\kappa\text{-(ET)}_2\text{Cu(SCN)}_2$				
∥b	293	2.79	19.3	(21)
	100	5.24	19.3	
	50	5.00	4.86	
	25	4.43	1.67	
∥c	293	4.84	21.8	(21)
	100	5.32	11.6	
	50	5.56	8.1	
	25	5.16	1.98	
			29.3	(22)

Conclusions concerning low-temperature properties based on $\sigma(\omega)$ derived from measurements of $R(\omega)$ at room temperature can be tenuous, because structural changes frequently occur at low temperature and such changes modify the electronic structure. For $\beta\text{-(ET)}_2\text{I}_3$ it has been demonstrated that such a change occurs in $\sigma(\omega)$. (See Figure 7.7.)[10]

The spectra, $\sigma(\tilde{v})$, have been interpreted by application of Eqs. (7.8) and (7.12) and the coupling model described by Rice. Equations (7.15) and (7.17) have not been used heretofore. In the first case, the broad band with or tending to a maximum has been associated with charge-transfer between donor and acceptor molecules. The superimposed or additional fine structure has been assigned to Lorentzian peaks or to an electron-phonon coupled transition.[11] However, even in those cases in which common bases apparently have been employed, the results derived are not consistent. (See Table 7.1).

It is meaningful to discuss the results for ET materials in two groups: (a) spectra obtained for the compounds by Kuroda et al.,[12] and (b) those obtained for β-(ET)$_2$I$_3$ by several investigations.[10-16] This makes intercomparisons and conclusions for β-(ET)$_2$I$_3$ more reliable.

The results found by Kuroda et al.,[12] for β-(ET)$_2$PF$_6$, (ET)$_2$ClO$_4$(TCE)$_{0.5}$, (ET)$_3$(ClO$_4$)$_2$, and β-(ET)$_2$I$_3$ show that the curves for R($\tilde{v}$) or σ($\tilde{v}$) are basically similar but are distinctly different in many significant features. All of the spectra show the onset of increasing reflectivity in the infrared. The onset of the increasing conductivity on the high energy side is in the region of 6000 cm^{-1} (0.7 eV). They all have a broad maximum or a shoulder indicative of the onset of a maximum. In these respects, the spectra resemble that of a semiconductor. The positions of the maxima, and the extent of the superimposed fine structure differ significantly. The spectra, σ($\tilde{v}$), for β-(ET)$_2$PF$_6$, (ET)$_2$(ClO$_4$)(TCE)$_{0.5}$, and β-(ET)$_2$I$_3$ are similar in that they all have broad maxima and well-defined sharper peaks on the low energy sides. The intensity of the sharp peak increases in the order given, so that in β-(ET)$_2$I$_3$ it is almost twice that for the broad maximum. The curve for (ET)$_3$(ClO$_4$)$_2$ displays no maximum, but only a shoulder.

The polarized reflectance spectra of β-(ET)$_2$I$_3$ have been measured by Kaplunov et al.,[11] Kuroda et al.,[12] Jacobsen et al.,[10] Koch et al.,[13] Sugano et al.,[14] and by Tajima et al.[15,16] All of the spectra derived for σ($\tilde{v}$) are almost identical. The only difference appears to be a peak at less than 1000 cm^{-1}; in Kaplunov's spectrum, it is more intense than the broad maximum; in Jacobsen's spectrum, it is approximately one-half that for the broad peak; and in Kuroda's and Sugano's spectra, it is absent because measurements were not effected to sufficiently small wave numbers. Koch's spectrum seems to be quite different. Sugano et al. reported a broad weak maximum near 20,000 cm^{-1}. This feature is also observed in Kaplunov's spectrum. The curves for σ($\tilde{v}$) obtained by Kaplunov et al., Kuroda et al., and Jacobsen et al. are shown respectively in Fig. 7.5, 7.6, and 7.7.

The spectra displayed in Fig. 7.7 illustrate the tenuous nature of conclusions regarding low-temperature properties derived from measurements at room temper-

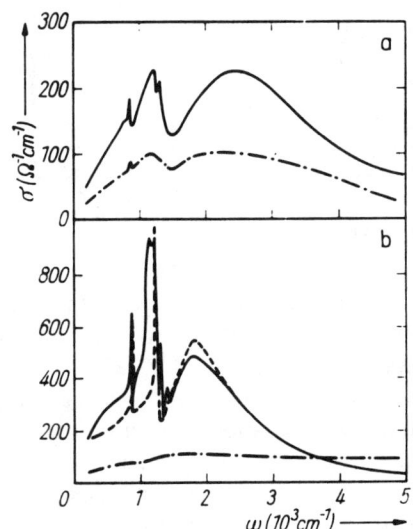

Figure 7.5 Optical conductivities of (ET)$_2$I$_3$ single crystals derived from polarized reflectance spectra. (a) α-phase with electric vector perpendicular to the *b*-axis (—) and with it parallel to *b* (–·–·–); (b) β-phase with vector parallel to *d* (—) and with it perpendicular to *d* (–·–·–); calculated for vector parallel to *d* (—) (Reproduced with permission of *Phys. Stat. Sol.* (a), Kaplunov et al., **1985**, *89*, 509).

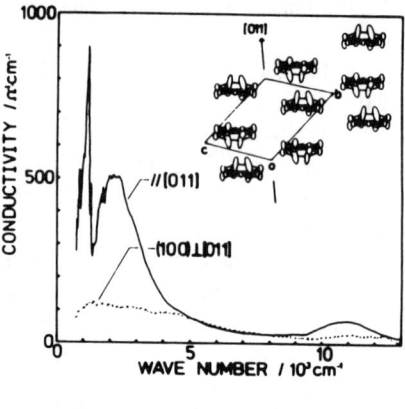

Figure 7.6 Optical conductivities of β-$(ET)_2I_3$ derived from polarized reflectance spectra. (Reproduced with permission of *Mol. Cryst. Liq. Cryst.*, Kuroda et al., **1985**, *125*, 135).

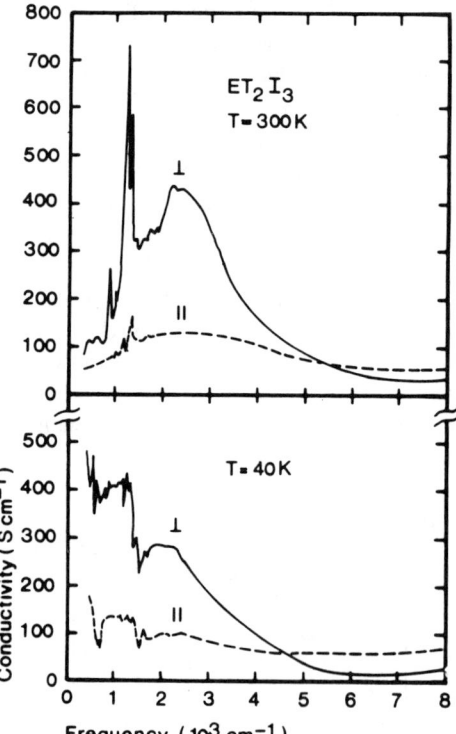

Figure 7.7 Optical conductivities of β-$(ET)_2I_3$ derived from polarized reflectance spectra at 300 K and 40 K. (Reproduced with permission of *Solid State Commun.*, Jacobsen et al., **1985**, *54*, 937).

ature. The change in the $\sigma(\tilde{\nu})$ spectrum in the region below 1500 cm^{-1}, between measurements at 300 K and 40 K, is significant with regard to the electron-phonon coupled transition. If no structural change occurs in the ET moiety between these two temperatures, then one must surmise that the change may be associated with a change in the concentration of electrons that varies with temperature and as if there exists a semiconducting component in the β-$(ET)_2I_3$ spectra.

Values for the parameters (ω_p and τ^{-1}) derived from the application of Eq. (7.8) and (7.11) to $\sigma(\tilde{\nu})$ are listed in Table 7.1. In three cases, the reflectance spectra have been measured at several temperatures, and the values of ω_p and τ^{-1} derived from the Drude model have been determined as a function of temperature. Most of the observed variation occurs in the relaxation time τ^{-1} which decreases with decreasing temperature. Because τ^{-1} is a measure of the resistance, the variation of τ^{-1} with temperature measures the variation of resistance with temperature. It is generally accepted that τ is proportional to T^{-n} with n = 1 for phononic scattering and n = 2 for electronic scattering. A plot of the ln τ versus ln T of the value listed in Table 7.1 (adjusted for the proportionality constant) is shown in Fig. 7.4. Although the precision is low, the plot does show that the variation is consistent with n = 1.

Individual Spectra

β-(ET)$_2$I$_3$. Tajima et al.,[16] have measured the reflectance spectra of β-(ET)$_2$I$_3$ at several temperatures. Using the Kramers-Kronig transformation, they derived the real part of the complex dielectric function. Although they used the Hagen-Rubens relation ($R = 1 - 2\,(\nu/\sigma)^{1/2}$) to extrapolate to zero frequency, they report that the derived parameters and the shape of $\epsilon_1(\omega)$ did not depend on the method of extrapolation. Using the Drude model, they obtained values for the relaxation time τ as a function of temperature. A plot of ln τ versus lnT in Fig. 7.4 shows that the slope is near -1. Hence, the variation of $\tau(T)$ is consistent with the expected value for electron-phonon interaction.

α − (ET)$_2$I$_3$. The polarized reflectance spectra of α-(ET)$_2$I$_3$ have been measured by Kaplunov et al.,[11] Koch et al.,[13] Sugano et al.,[14] and Yakushi et al.[17] The optical conductivities of α-(ET)$_2$I$_3$ derived from the reflectance spectra measured at several temperatures between 300 and 18 K are shown in Fig. 7.8. The spectral features resemble that for a semiconductor, given by Eq. (7.17) and shown in Fig. 7.3, with some superimposed peaks near the maximum and in the region for values less than that at the maximum. The peak near 1000 cm^{-1} resembles that for Rice's model[4,5] (see Fig. 7.2). Measured values of the resistance versus temperature indicate that the compound is a semiconductor with a metal-insulator transition at 135k.

θ-(ET)$_2$I$_3$. The polarized reflectance spectra of θ-(ET)$_2$I$_3$, with the electric field parallel to the a- and c-axes, have been measured at several temperatures from 297 to 16 K in the range from 720 cm^{-1} to 25000 cm^{-1} (see Fig. 7.9)[18]. By use of the Kramers-Kronig transformation, Tamura et al.,[18] derived the values for the optical conductivities as a function of wave numbers and have fitted the derived $\sigma(\omega)$'s to a sum of Drude and Lorentz terms. The $\sigma(\omega)$'s are essentially Drude-like edges with small, minor superimposed Lorentzian peaks. The values for ω_p vary from 0.92 eV at 295 K to 1.05 eV at 16 K for the field parallel to the a-axis and from 0.62 eV at 295 K to 0.73 eV at 16 K for the field parallel to the c-axis. Most of the variations with temperature are attributable to the variation of the relaxation time τ with temperature. (See plot in Fig. 7.4.) A comparison of the variation of the derived values of $\sigma(T)$ with the values of $\rho(T)$ measured by Kobayashi et al.[19]

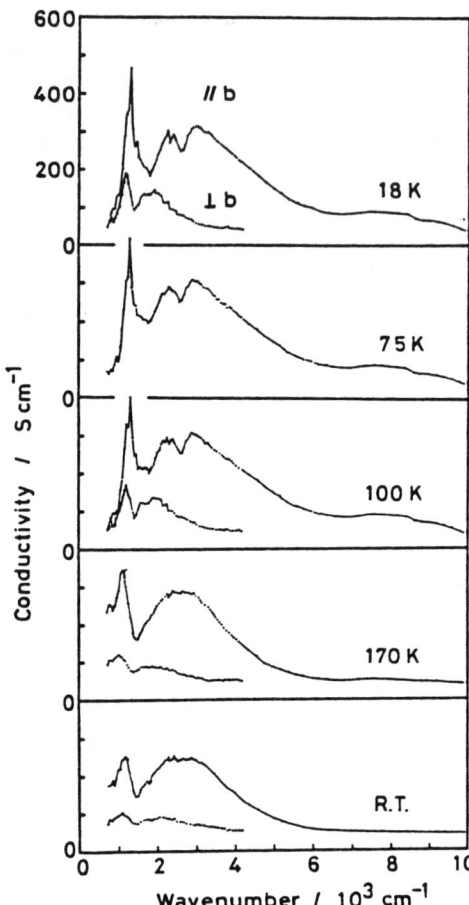

Figure 7.8 Optical conductivities of α-$(ET)_2I_3$ derived from polarized reflectance spectra at indicated temperatures. (Reproduced with permission of *Bull. Chem. Soc. Jpn.*, Yakushi, et al. **1987**, *60*, 4251).

shows that the variation of σ(T) is significantly less than that for ρ(T). Further, one finds that the variation of the relaxation time τ(T) at temperatures below 200 K is less than that at higher temperatures. A plot of ln τ versus ln T shown in Fig. 7.4, illustrates that if τ is proportional to T^{-n}, then n is smaller than one for T<200K, but may be the order of one or two for T>200 K. A possible explanation for the apparent discrepancy at T<200 K is that θ-$(ET)_2I_3$ is a semimetal, so that the density of carriers increases with temperature, and the derived values of τ measure the product of the density of carriers and the relaxation time.

β-$(ET)_2IBr_2$. Sugano and Saito[20] measured the polarized reflectance spectrum of β-$(ET)_2IBr_2$ (see Fig. 7.10) and found Drude-parameters comparable to those which Sugano[14] reported for β-$(ET)_2I_3$.

κ-$(ET)_2Cu(SCN)_2$. The polarized reflectance spectra of κ-$(ET)_2Cu(SCN)_2$ have been measured by Ugawa et al.[21] at temperatures ranging from 293 to 18 K

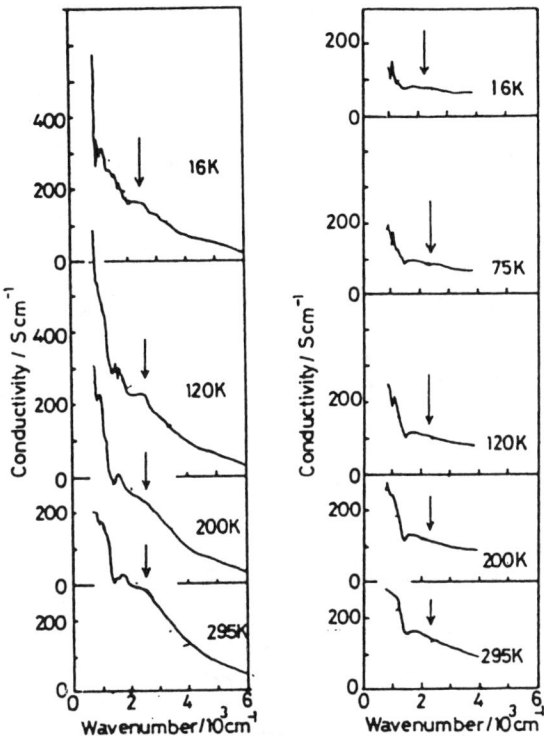

Figure 7.9 Optical conductivities of θ-(ET)$_2$I$_3$ derived from polarized reflectance spectra at indicated temperatures. The arrow indicates the wavenumber of weak absorptions attributable to interband transitions. (Reproduced with permission of *J. Phys. Soc. Jpn.*, Tamura et al., **1988**, *57*, 3239)

(see Fig. 7.11). The spectral shapes resemble the edge of the Drude model with two superimposed Lorentzian peaks. Using the derived optical conductivities, and a sum of one Drudeian and two Lorentzian forms, they have derived the parameters ω_p, τ^{-1}, and the Lorentzian parameters. The parameter which varies most with temperature is τ^{-1}. Although the precision with which τ^{-1} is determined is not high, a plot of ln τ versus ln T is approximately linear with a slope near -1, a value predicted for electron-phonon scattering. (See Fig. 7.4.) Measurements have also been made by Sugano et al.[22]

α-(ET)$_3$(ReO$_4$)$_2$. The temperature dependent polarized reflectance spectra for α-(ET)$_3$(ReO)$_2$ have been measured by Yakushi et al.,[17] from which they derived the optical conductivities (Fig. 7.12). The $\sigma(\tilde{\nu})$'s appear to be those for a semiconductor or a semimetal with some superimposed peaks having a Lorentzian shape. Theoretically derived $\sigma(\omega)$'s indicate the compound is a semimetal, but the measured resistance indicates that a semiconductor-to-metal transition occurs near 20 K. These two observations are not necessarily inconsistent, because if the change occurs at a temperature as low as 20 K, the implication is that the significant part of $\sigma(\tilde{\nu})$ is at wave number values too small to have been observable.

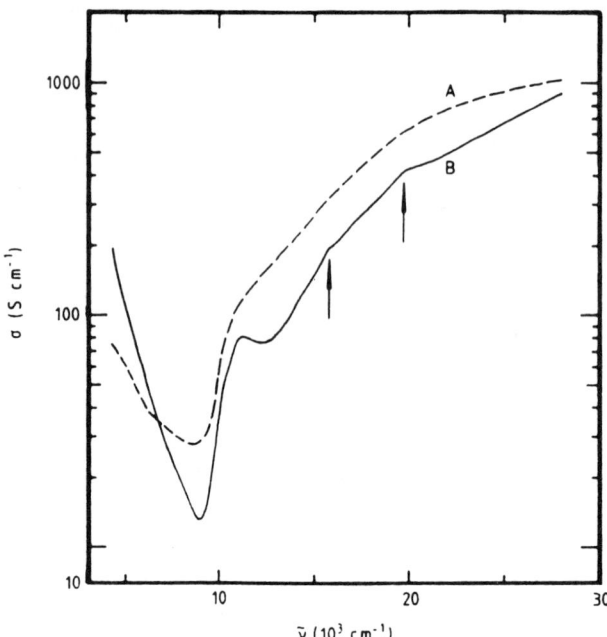

Figure 7.10 Optical conductivities of β-(ET)$_2$IBr$_2$ derived from polarized reflectance spectra. Arrows indicate the position of molecular excitations of ET. A is for perpendicular to [110]; B, for parallel to [110]. (Reproduced with permission of *J. Phys. C. Solid State Physics*, Sugano and Saito, **1986**, *19*, 5471).

(ET)$_3$(ClO$_4$)$_2$ and (ET)$_2$ClO$_4$(TCE)$_{0.5}$. Optical conductivities for (ET)$_3$(ClO$_4$)$_2$ and (ET)$_2$ClO$_4$(TCE)$_{0.5}$ have been reported by Kuroda et al.[12] Measurements for (ET)$_2$ClO$_4$(TCE)$_{0.5}$ have also been reported by Tajima et al.[23] The $\sigma(\tilde{\nu})$ for the direction parallel to the (012) plane in β-(ET)$_3$(ClO$_4$)$_2$ resembles the edge for the Drude model with some superimposed minor peaks. As shown in Fig. 7.13, the optical conductivity is anisotropic, and the conductivity in the (100) plane is small. A metal-insulator transition occurs at 170 K; hence, it is semiconductive. The optical conductivity parallel to the (102) plane in β-(ET)$_2$ClO$_4$(TCE)$_{0.5}$ is shown in Fig. 7.14; it resembles that for a composite of a semiconductor, with $\sigma(\omega)$ like that for Eq. (7.17), and a superimposed Lorentzian or Ricesian peak near 1000 cm^{-1}. It is reported to be a metal at temperatures above 1.8 K. This suggests that the extrapolation of $\sigma(\tilde{\nu})$ to zero frequency is not zero, so that the solid may be semimetallic. As in the case of β-(ET)$_3$(ClO$_4$)$_2$, the conductivity in the (010) plane is approximately one-fourth that for the (102) plane. The reported plasma frequencies are 8.8×10^3 cm^{-1} and 4.7×10^3 cm^{-1}, respectively, for the (102) and (100) planes.

β-(BMDT-TTF)AsF$_6$ and β-(BMDT-TTF)SbF$_6$. The optical conductivities of β-(BMDT-TTF)AsF$_6$ and β-(BMDT-TTF)SbF$_6$ reported by Yoshitake et al.[24] are shown in Fig. 7.15. To accomplish the Kramers-Kronig transformation, the authors extrapolated the measured reflectance curves from 4×10^3 cm^{-1} to 0 by

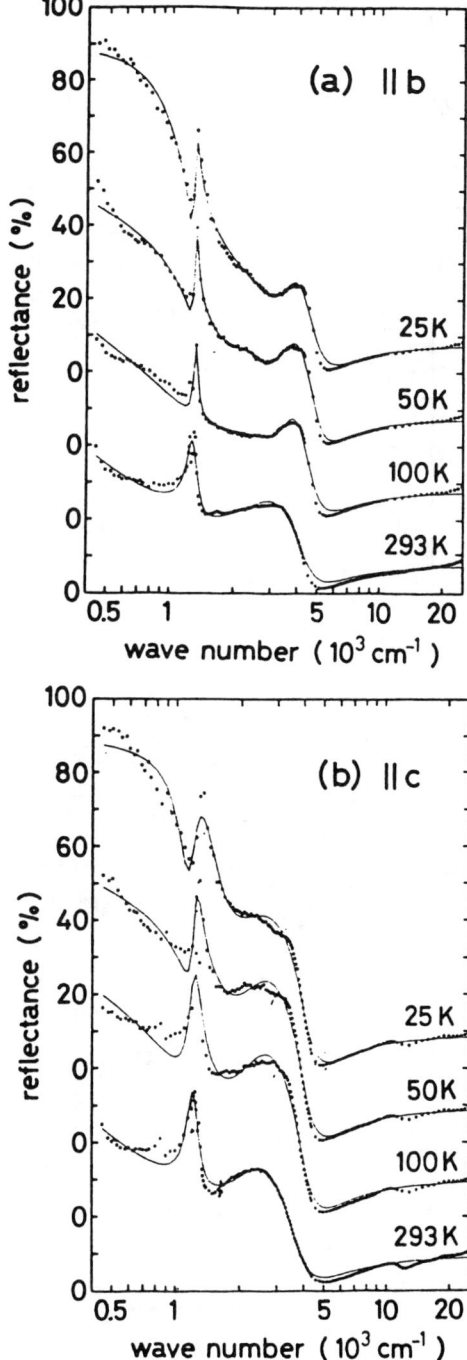

Figure 7.11 Polarized reflectance spectra of κ-(ET)$_2$Cu(SCN)$_2$ parallel to b (a) and c (b) axes at indicated temperature (···). Solid lines show simulated curves for sum of Drudeian and Lorentzian oscillators. (Reproduced with permission of *Phys. Rev.*, Ugawa et al. **1988**, *B38*, 5122).

Chap. 7   Optical Conductivities

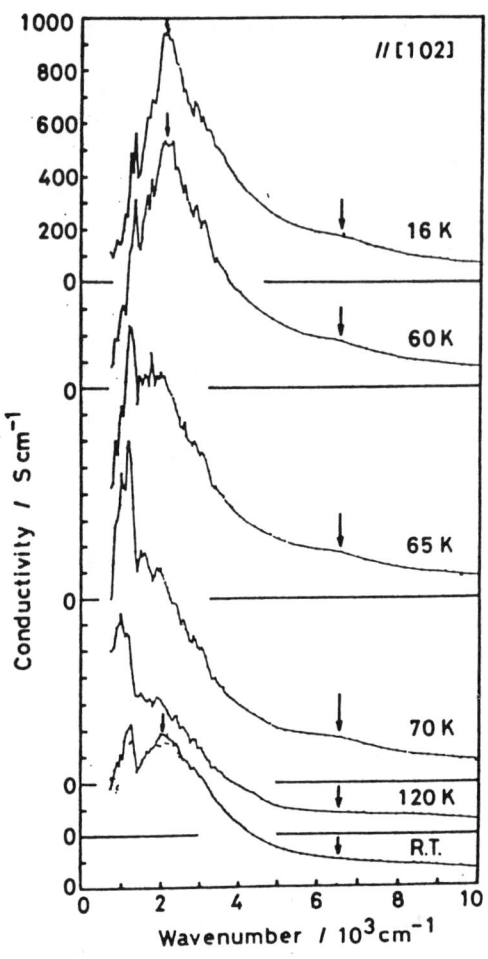

Figure 7.12 Optical conductivities of α-$(ET)_3(ReO_4)_2$ derived from polarized reflectance spectra at indicated temperatures. (Reproduced with permission of *Bull. Chem. Soc. Jpn.*, Yakushi et al., **1987**, *60*, 4251).

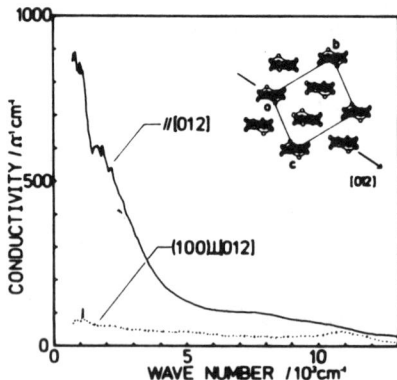

Figure 7.13 Optical conductivities of $(ET)_3(ClO_4)_2$ derived from polarized reflectance spectra. (Reproduced with permission of *Mol. Cryst. Liq. Cryst.*, Kuroda et al., **1985**, *125*, 135).

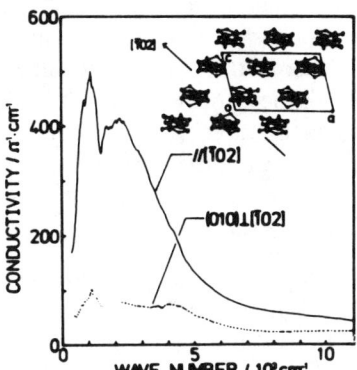

Figure 7.14 Optical conductivities of $(ET)_2ClO_4(C_2H_5Cl_3)_{0.5}$ derived from polarized reflectance spectra. (Reproduced with permission of *Mol. Cryst. Liq. Cryst.*, Kuroda et al., **1985**, *125*, 135).

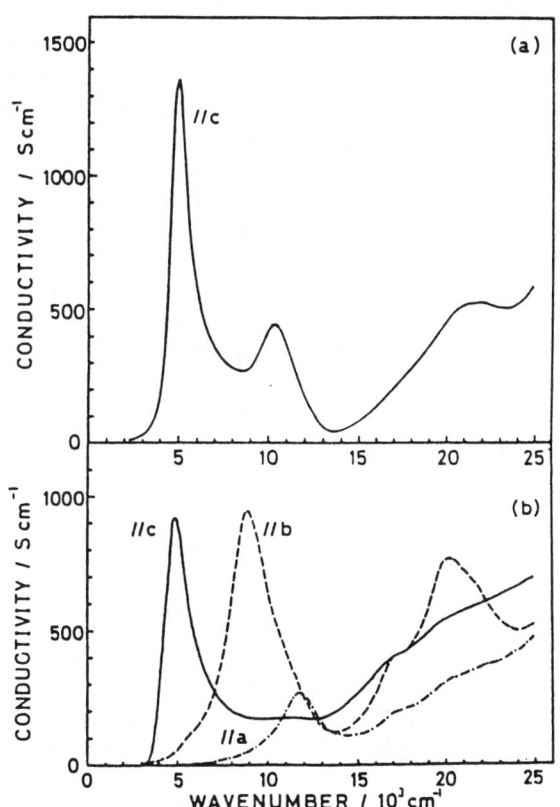

Figure 7.15 Optical conductivities of (a) (BMDT-TTF) AsF_6 and (b) (BMDT-TTF)SbF_6 derived from polarized reflectance spectra. (Reproduced with permission of *Bull. Chem. Soc. Jpn.*, Yoshitake et al., **1988**, *61*, 1115).

use of the Lorentzian model. Because the reflectance curves appear to be changing at $2 \times 10^3 \text{cm}^{-1}$ in such a way that an extrapolation to 0 yields values for the reflectances near 10 percent, it is not known how this extrapolation affects the calculated optical conductivities. If one accepts the curves for $\sigma(\tilde{v})$ as shown in Fig. 7.15, then they appear to be examples of Rice's conductivity equation for a semiconductor (see Fig. 7.3). In either case, $\sigma(\tilde{v})$ predicts that both solids are semiconductors with gaps near 4×10^{-1} eV. Kato et al.,[25] found that the logarithm of the normalized resistance versus T^{-1} is nearly linear with a slope corresponding to 1.2×10^{-1} eV.

β-(ET)$_2$PF$_6$. Optical conductivities parallel to the *a*- and to *c*-axes, at two temperatures, for β-(ET)$_2$PF$_6$ were reported by Kuroda et al.,[12] as shown in Fig. 7.16. Values have also been reported by Tajima et al.[26] The difference between the $\sigma(\tilde{v})$'s in the two directions and at the two temperatures is not large. The structure of $\sigma(\omega)$ parallel to the *a*-axis direction indicates a semiconductive part similar to that described by Eq. (7.17), shown in Fig. 7.3, and some small superimposed Lorentzian peaks near 1000 cm^{-1}. The structure of $\sigma(\tilde{v})$ for measurements parallel to the *c*-axis resembles that of the same semiconductor model with rather minor superimposed sharp peaks especially near 1000 cm^{-1}. This organic conductor has a metal-insulator transition at 297 K. Hence, below 297 K, $\rho(T)$ displays the usual semiconductor decrease. Kuroda et al.[12] interpreted the structures of $\sigma(\tilde{v})$ in terms of a Lorentzian oscillator, although they are more like that for Eq. (7.17) than that for the Lorentzian model.

(ET)$_4$Hg$_3$Cl$_8$. Kaplunov et al.[27] have measured the polarized reflectance spectra at room temperature for the two optical axes in (ET)$_4$Hg$_3$Cl$_8$ and have derived the optical conductivities (see Fig. 7.17). The structure of $\sigma(\tilde{v})$ resembles that for a semiconductor or a semimetal shown in Fig. 7.3 with superimposed Lorentzian peaks near 1000 cm^{-1}.

SUMMARY

The discussions and interpretations given previously can be summarized as follows:

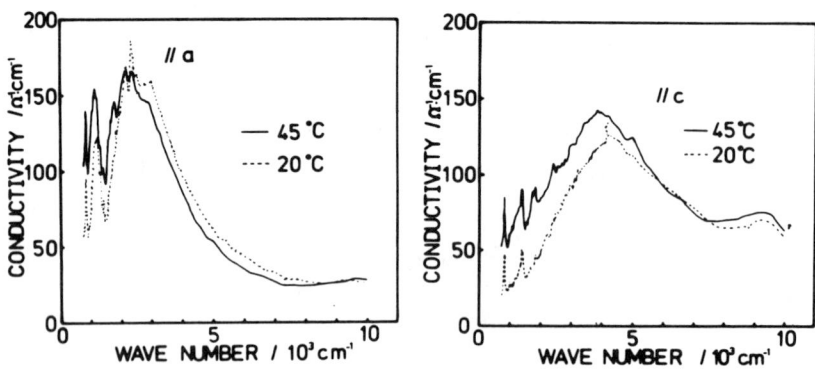

Figure 7.16 Optical conductivities of β-(ET)$_2$PF$_6$ derived from polarized reflectance spectra. (Reproduced with permission of *Mol. Cryst. Liq. Cryst.*, Kuroda et al., **1985**, *125*, 135).

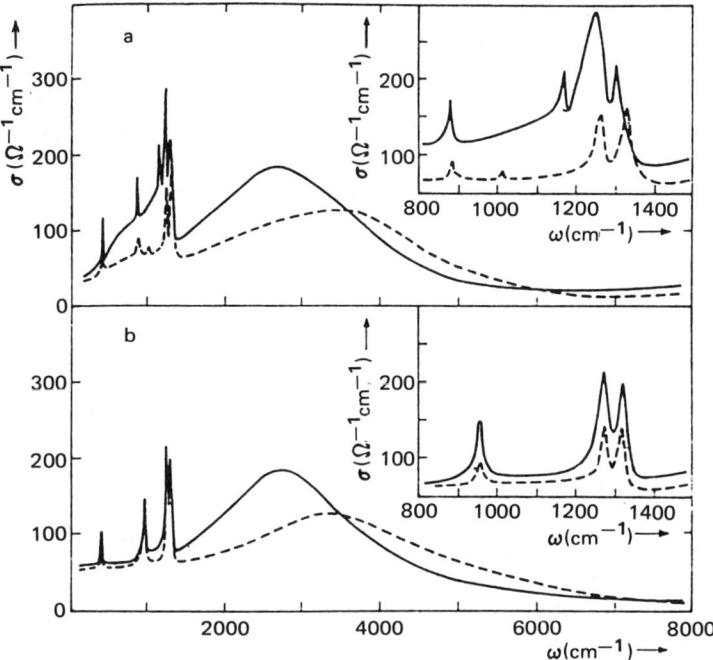

Figure 7.17 Optical conductivities of $(ET)_4Hg_3Cl_8$ derived from polarized reflectance spectra (a); theory (b); optical directions x (—) and y (----). (Reproduced with permission of *Phys. Stat. Sol.*, Kaplunov et al., (a) **1987**, *104*, 883).

1. Because the energies of the transitions and the electronic processes involved are small, the optical conductivity spectra measured at temperatures near room temperature, and even below, can reflect distortions caused by carrier densities that vary with temperature.

2. In some instances, and perhaps in all, the structures of the $\sigma(\tilde{v})$'s are composed of two of these models cited. Thus, they are composed of Drudeian and Lorentzian shapes, or of semiconductive and Lorentzian shapes. Hence, the DC conductivities as functions of temperature should reflect this composite structure. One should recognize that even the Lorentzian oscillator can contribute indirectly to the structure of $\rho(T)$.[28]

3. An examination of the temperature dependence of the relaxation time τ, derived from experimental measurements of reflectance, reveals that although the precision is not great, $\ln \tau$ versus $\ln T$ is linear with a slope near minus one. Thus, the variation $\tau(T)$ basically corresponds to that for electron-phonon scattering.

Delhaes[28] has shown for $(TMTTF)_2X$ (TMTTF is tetramethyltetrathiafulvalene) compounds that the logarithm of the conductivity at 300 K correlates linearly with the maxima in the spectra for optical conductivities near 3000–4500 cm^{-1}. These maxima are, accordingly to his surmise, to be associated with charge transfer from one molecule to another, that is, $D°D^+$. Thus he suggests that these optical gaps might scale with the electrical conductivity gap energies. Similarly, the maxima or tendency of gap formation near 1000–2000 cm^{-1} for the (ET) compounds may be associated with a scaled conductivity gap.

If the peaks near 1000–5000 cm^{-1} [in the same range which Delhaes found for the (TMTTF)$_2$X conductors] are associated with the spectral structure derived by Rice and colleagues, shown in Fig. 7.4, then these charge-transfer maxima originate from electron-phonon coupling, implying that these materials are small polaron conductors.

REFERENCES

1. Wooten, F. *Optical Properties of Solids*, Academic Press, New York, **1972**.
2. Seitz, F. *The Modern Theory of Solids*, McGraw Hill Book Co., Inc., New York, **1940**.
3. Kittel, C. *Quantum Theory of Solids*, John Wiley & Sons, Inc., New York, **1963**.
4. Rice, M. J. *Phys. Rev. Lett.* **1976**, *37*, 36.
5. Rice, M. J.; Pietronero, L.; Brüesch, L. *Solid State Commun.* **1977**, *21*, 757.
6. Allen, P. B., *Phys. Rev. B: Solid State* **1971**, *3*, 305.
7. Bottger, H. and Bryksin, V. V., *Hopping Conduction in Solids*, VCH Verlagsgesellschaft, Weinheim FRG; Deerfield Beach, FL., **1985**.
8. Landolt-Börnstein, New Series III/15b, Springer-Verlag, New York, **1985**.
9. Cohen, M. L.; Chelikowsky, J. R. *Electronic Structure and Optical Properties of Semiconductors*, Springer-Verlag, New York, **1988**.
10. Jacobsen, C. S.; Williams, J. M.; Wang, H. H. *Solid State Commun.*, **1985**, *54*, 937.
11. Kaplunov, M. G.; Yagubskii, E. B.; Rosenberg, L. P.; Borod'ko, Yu. G. *Phys. Status Solidi A* **1985**, *89*, 509.
12. Kuroda, H.; Yakushi, K.; Tajima, H.; Saito, G. *Mol. Cryst. Liq. Cryst.* **1985**, 125, 135.
13. Koch, B.; Geserich, H. P.; Ruppel, W.; Schweitzer, D.; Dietz, K. H.; Keller, H. J. *Mol. Cryst. Liq. Cryst.* **1985**, *119*, 343.
14. Sugano, T.; Yamada, K.; Saito, G.; Kinoshita, M. *Solid State Commun.* **1985**, *55*, 137.
15. Tajima, H.; Yakushi, K.; Kuroda, H. *Solid State Commun.* **1985**, *56* 159.
16. Tajima, H.; Kanbara, H.; Yakushi, K.; Kuroda, H.; Saito, G. *Solid State Commun.* **1986**, *57*, 911.
17. Yakushi, K.; Kanbara, H.; Tajima, H.; Kuroda, H.; Saito, G.; Mori, T. *Bull. Chem. Soc. Jpn.* **1987**, *60*, 4251.
18. Tamura, M.; Yakushi, K.; Kuroda, H.; Kobayashi, A.; Kato, R.; Kobayashi, H.; *J. Phys. Soc. Jpn.* **1988**, *57*, 3239.
19. Kobayashi, H.; Kato, R.; Mori, T.; Kobayashi, A.; Sasaki, Y.; Saito, G.; Enoki, T.; Inokuchi, H. *Chem. Lett.* **1984**, 179.
20. Sugano, T.; Saito, G. *J. Phys. C: Solid State Phys.* **1986**, *19*, 5471.
21. Ugawa, A.; Ojima, G.; Yakushi, K.; Kuroda, H. *Phys. Rev. B: Condens Matter* **1988**, *38*, 5122.
22. Sugano, T.; Hayashi, H.; Kinoshita, M.; Nishikida, K. *Phys. Rev. B: Condens Matter* **1989**, *39*, 11387.

23. Tajima, H.; Yakushi, K.; Kuroda, H.; Saito, G.; Inokuchi, H. *Solid State Commun.* **1984**, *49*, 769.
24. Yoshitake, M.; Yakushi, K.; Kuroda, H.; Kobayashi, A.; Kato, R.; Kobayashi, H. *Bull. Chem. Soc. Jpn.* **1988**, *61*, 1115.
25. Kato, R.; Kobayashi, H.; Kobayashi, A. *Chem. Lett.* **1986**, 2013.
26. Tajima, H.; Yakushi, K; Kuroda, H.; Saito, G. *Solid State Commun.* **1985**, *56*, 251.
27. Kaplunov, M. G.; Lyubovskaya, R. N.; Aldoshina, M. Z.; Borod'ko, Yu. G. *Phys. Status Solidi (a)* **1987**, *104*, 833.
28. Delhaes, P. Proc. of NATO ASI "Low Dimensional Systems and Molecular Electronics," Metzer, R. M.; Day, P.; Papavassiliou, G., Eds.; Spetses, Greece, 1989, Plenum Press 1990.

8

Electronic and Structural Properties of Organic Superconductors

ELECTRONIC PROPERTIES

The electronic structure of a crystalline solid is described by energy bands. A given energy band consists of **N** discrete levels, where **N** refers to the total number of unit cells in the solid. Since $N \to \infty$, all energy levels falling within a band are allowed. In a one-electron band picture, electron-electron repulsion is neglected so that each band level can be filled with two electrons. In this picture, a semiconductor (or an insulator) contains only completely filled and completely empty bands, so that an energy gap exists between the highest occupied and the lowest occupied band levels. See Fig. 8.1(a). On the other hand, a metal has at least one partially filled band [see Fig. 8.1(b)] so that no energy gap exists between the highest occupied level (i.e., the Fermi-level E_f) and the lowest unoccupied level.

The electronic instability of a metal leading to either a metal-insulator or a metal-superconductor transition may be described on the basis of orbital mixing between the occupied and unoccupied levels.[1-4] A new electronic state derived from the orbital mixing may become more stable than the metallic state when the energy gain resulting from the interactions between the occupied and unoccupied levels is greater than the inherent energy raising caused by introducing higher lying, unoccupied levels. Since the energy difference between the occupied and unoccupied levels around the Fermi level can be very small, the extent of this energy raising can be made very

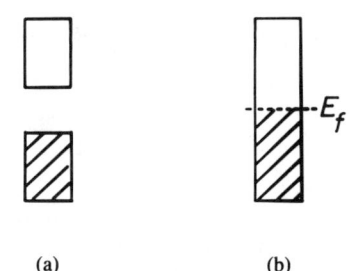

Figure 8.1 Electronic energy band representations for (a) a semiconductor and (b) a metal, where occupied band levels are shaded.

small so that orbital mixing among the band levels in the vicinity of the Fermi level is crucial for the occurrence of a metal-insulator or a metal-superconductor transition. From the viewpoint of one-electron band theory, a metal-insulator transition is most likely to occur when the Fermi surface associated with a partially filled band is well nested (vide infra).[4] A metal may become superconducting if it is free from an electronic instability toward a metal-insulator transition. In this section, we discuss how the concept of Fermi surface nesting comes about, why Fermi surface nesting leads to an electronic instability, and how the orbital mixing leading to a superconducting state differs from that leading to an insulating state.

BAND ELECTRONIC STRUCTURE

One Orbital per Site

Let us consider a three-dimensional (3D) orthorhombic lattice with repeat distances a, b, and c. The coordinate of a lattice site (i.e., the position of a unit cell) may be denoted by (m, n, p), where m, n, and p are integers, and each lattice site may be assumed to have a single orbital χ. If the orbital at the site (m, n, p) is represented by χ_{mnp}, the orbital at the coordinate origin is χ_{000} while those at its nearest neighbor sites along the a-, b-, and c-directions are χ_{100}, χ_{010}, and χ_{001}, respectively. The allowed energy levels the lattice can have are given by the Bloch orbitals[5]

$$\phi(k_a, k_b, k_c) = N^{-1/2} \sum_m \sum_n \sum_p \exp(ik_a ma) \exp(ik_b nb) \exp(ik_c pc) \chi_{mnp} \quad (8.1)$$

where k_a, k_b, and k_c are wave vectors along the a-, b-, and c-axis directions, respectively, and the terms $\exp(ik_a ma)$, $\exp(ik_b nb)$, and $\exp(ik_c pc)$ are the coefficients for the site orbital χ_{mnp}. For simplicity, the wave vectors may be limited to have the following values

$$-\pi/a \leq k_a \leq \pi/a$$
$$-\pi/b \leq k_b \leq \pi/b \quad (8.2)$$
$$-\pi/c \leq k_c \leq \pi/c$$

since they allow all possible values for the coefficients.

The band energies $E(k_a, k_b, k_c)$ associated with the Bloch orbitals $\phi(k_a,k_b,k_c)$ are expressed as

$$E(k_a,k_b,k_c) = \frac{<\phi(k_a,k_b,k_c)|H^{eff}|\phi(k_a,k_b,k_c)>}{<\phi(k_a,k_b,k_c)|\phi(k_a,k_b,k_c)>} \quad (8.3)$$

where H^{eff} is the effective Hamiltonian. Within the first nearest neighbor approximation, the denominator and the numerator of Eq. (8.3) are given as follows:

$$<\phi(k_a,k_b,k_c)|\phi(k_a,k_b,k_c)> = 1 + 2(S_a \cos k_a a + S_b \cos k_b b + S_c \cos k_c c) \quad (8.4)$$

where S_a, S_b, and S_c are the nearest neighbor overlap integrals along the a-, b-, and c-directions, respectively.

$$S_a = <\chi_{000}|\chi_{100}>$$
$$S_b = <\chi_{000}|\chi_{010}> \quad (8.5)$$
$$S_c = <\chi_{000}|\chi_{001}>$$

Likewise,

$$<\phi(k_a,k_b,k_c)|H^{eff}|\phi(k_a,k_b,k_c)> = \alpha + 2(\beta_a \cos k_a a + \beta_b \cos k_b b + \beta_c \cos k_c c) \quad (8.6)$$

where α is the Coulomb integral (i.e., the valence shell ionization potential) of the orbital χ_{mnp} while β_a, β_b, and β_c are the nearest neighbor resonance (i.e., hopping) integrals along the a-, b-, and c- directions, respectively.

$$\alpha = <\chi_{000}|H^{eff}|\chi_{000}>$$
$$\beta_a = <\chi_{000}|H^{eff}|\chi_{100}>$$
$$\beta_b = <\chi_{000}|H^{eff}|\chi_{010}> \quad (8.7)$$
$$\beta_c = <\chi_{000}|H^{eff}|\chi_{001}>$$

Consequently, $E(k_a,k_b,k_c)$ is expressed as

$$E(k_a,k_b,k_c) = \frac{\alpha + 2(\beta_a \cos k_a a + \beta_b \cos k_b b + \beta_c \cos k_c c)}{1 + 2(S_a \cos k_a a + S_b \cos k_b b + S_c \cos k_c c)} \quad (8.8)$$

When the overlap integrals S_a, S_b, and S_c are neglected, Eq. 8.8 is further simplified by

$$E(k_a,k_b,k_c) = \alpha + 2(\beta_a \cos k_a a + \beta_b \cos k_b b + \beta_c \cos k_c c) \quad (8.9)$$

According to this equation, the width of the band (i.e., W) associated with the band orbital $\phi(k_a,k_b,k_c)$ is given by

$$W = 4(|\beta_a| + |\beta_b| + |\beta_c|) \quad (8.10)$$

which differs somewhat from the corresponding value obtained from Eq. 8.3 or Eq. 8.8 due to nonzero overlap integrals between site orbitals.

Many Orbitals per Site

We now consider a general 3-dimensional (3D) lattice, with repeat vectors **a, b,** and **c**, which has more than one atom and hence a set of atomic orbitals $\{\chi_1, \chi_2, \cdots, \chi_\mu, \cdots, \chi_M\}$ per unit cell. The position of a lattice site (m, n, p) is given by the vector **R**

$$\mathbf{R} = m\mathbf{a} + n\mathbf{b} + p\mathbf{c} \tag{8.11}$$

so that an orbital χ_μ at the site (m, n, p) can be written as $\chi_\mu(\mathbf{r} - \mathbf{R})$. Thus the Bloch orbital $\phi_\mu(\mathbf{k})$ for any wave vector **k** generated by the atomic orbital $\chi_\mu(\mathbf{r} - \mathbf{R})$ is given by

$$\phi_\mu(\mathbf{k}) = N^{-1/2} \sum_\mathbf{R} \exp(i\mathbf{k} \cdot \mathbf{R}) \, \chi_\mu(\mathbf{r} - \mathbf{R}) \tag{8.12}$$

In terms of the reciprocal vectors **a*, b*,** and **c***, the wave vector **k** is expressed by

$$\mathbf{k} = k_1 \mathbf{a^*} + k_2 \mathbf{b^*} + k_3 \mathbf{c^*} \tag{8.13}$$

All possible values of **k** appropriate for $\phi_\mu(\mathbf{k})$ may be chosen from the first primitive zone (FPZ) of reciprocal space defined by

$$-1/2 \leq k_1 \leq 1/2$$

$$-1/2 \leq k_2 \leq 1/2 \tag{8.14}$$

$$-1/2 \leq k_3 \leq 1/2$$

rather than from the first Brillouin zone (FBZ) (i.e., the Wigner-Seitz cell of reciprocal space. For an orthorhombic system, the FBZ is identical with the FPZ). The allowed energy levels of the system are described by band orbitals $\Psi_i(\mathbf{k})$ (i = 1, 2, $\cdots$, M), which are given as a linear combination of the Bloch orbitals $\phi_\mu(\mathbf{k})$ (μ = 1, 2, $\cdots$, M)

$$\Psi_i(\mathbf{k}) = \sum_\mu C_{\mu i}(\mathbf{k}) \, \phi_\mu(\mathbf{k}) \tag{8.15}$$

where $C_{\mu i}(\mathbf{k})$ is the coefficient of $\phi_\mu(\mathbf{k})$. The band orbital energy $E_i(\mathbf{k})$ associated with $\Psi_i(\mathbf{k})$ is given by

$$E_i(\mathbf{k}) = \frac{\langle \Psi_i(\mathbf{k}) | H^{\text{eff}} | \Psi_i(\mathbf{k}) \rangle}{\langle \Psi_i(\mathbf{k}) | \Psi_i(\mathbf{k}) \rangle} \tag{8.16}$$

For a given **k** value, the band orbital energies $E_i(\mathbf{k})$ and the coefficients $C_{\mu i}(\mathbf{k})$ are determined by solving the eigenvalue problem associated with Eg. 8.16. The energy bands resulting from such calculations are often described by plotting the $E_i(\mathbf{k})$

values as a function **k** along certain lines of the FBZ or FPZ. Another way of describing the energy bands is to show their density of states, $n_i(E)$ (i = 1,2,···, M), as a function of energy E.

In an organic-donor molecule salt, intermolecular interactions are very weak compared with chemical bonding interactions that form the donor molecules and the anions themselves. Thus, the band electronic structure of an organic charge-transfer salt shows the molecular identity in that the major orbital character of any band is either a donor molecule or an anion, and also that the major orbital character of each donor-molecule (or anion) band reflects a particular molecular orbital of the donor molecule (or the anion).[39] The highest occupied bands of a charge-transfer salt, some of which are partially filled, originate from the HOMO of the neutral-donor molecule. If the HOMO-LUMO energy gap is large, the resulting highest occupied and lowest unoccupied bands do not overlap [see Fig. 8.2(a)] so that a partially filled band can only be found among the bands associated with the HOMO. In such a case, the highest occupied bands may be described by solving the one orbital per site problem of the previous section after substituting the HOMO for the site orbital χ. However, such an approximation can be misleading when the partially filled bands occur from the bands associated with the HOMO and the LUMO. As illustrated in Fig. 8.2(b), it is likely to occur when the HOMO-LUMO gap is small compared with the resulting bandwidths. In fact, this situation is found[6] for TTF[Ni(dmit)$_2$] and TTF[Pd(dmit)$_2$].

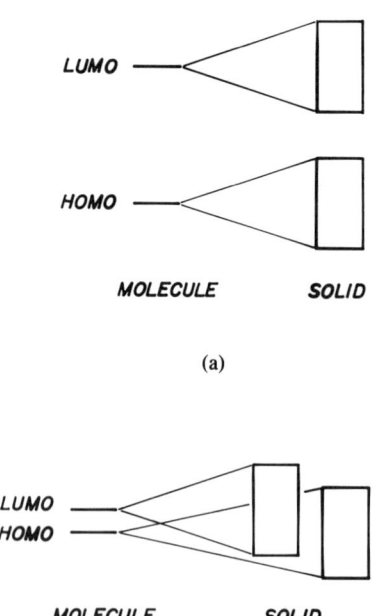

Figure 8.2 Correspondence between the energy levels of a molecule and the energy bands of its molecular solid for cases when the HOMO-LUMO difference is large in (a) and small in (b).

FERMI SURFACE NESTING AND ELECTRONIC INSTABILITY[4]

For simplicity, we consider a two-dimensional (2D) rectangular lattice with one orbital per site. From Eqs. 8.1 and 8.9, the band orbitals $\phi(k_a, k_b)$ and the band orbital energies $E(k_a, k_b)$ for a 2D lattice can be written as

$$\phi(k_a, k_b) = N^{-1/2} \sum_m \sum_n \exp(ik_a ma) \exp(ik_b nb) \chi_{mn} \quad (8.17a)$$

$$E(k_a, k_b) = \alpha + 2(\beta_a \cos k_a a + \beta_b \cos k_b b) \quad (8.17b)$$

Figure 8.3 shows two examples of band dispersions that illustrate how the ratio of the resonance integrals β_a/β_b affects the shape of band dispersion. Figure 8.3(a) and (b) represent the cases of $\beta_b/\beta_a = 0$ and $\beta_b/\beta_a < 1$, respectively, where each dashed line refers to the Fermi-level for the case when the band is half filled. Figure 8.3 shows that wave vectors in a certain region of the FBZ lead to occupied band orbitals, and those in the remaining region to unoccupied band levels. This is illustrated in Figures 8.4(a) and (b), which correspond to the band dispersions of Figures 8.3(a) and (b), respectively. In Fig. 8.4, the wave vectors of the shaded regions are occupied (i.e., they lead to occupied band levels), and those of the unshaded region are unoccupied. Since all wave vectors are equally probable, the size of the occupied wave vector region is proportional to the band filling. Thus for a half-filled band, which occurs when there is one electron per site to contribute to the band, one half of the FBZ is occupied.

The Fermi surface of a partially filled band is the boundary separating the occupied wave vector region from the unoccupied wave vector region. The Fermi surface of Fig. 8.4(a) or (b) consists of separated lines, and therefore is said to be open. For the electrons with the crystal momentum $hk/2\pi$ along a certain wave vector direction, there exists no band gap if the wave vector line crosses a Fermi

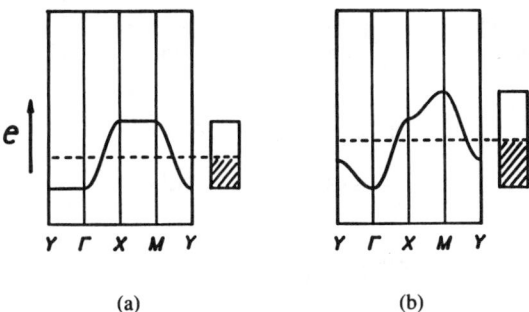

(a) (b)

Figure 8.3 Dispersion relations of a 2D band given by Eq. 8.17b for (a) $\beta_b/\beta_a = 0$ and (b) $\beta_a/\beta_b < 1$, where $\Gamma = (0,0)$, $X = (\pi/a, 0)$, $Y = (0, \pi/b)$, and $M = (\pi/a, \pi/b)$. The dashed lines refer to the Fermi level appropriate for the case when the band is half-filled. From Whangbo,[4] reproduced with permission of the American Chemical Society, Washington, D.C.

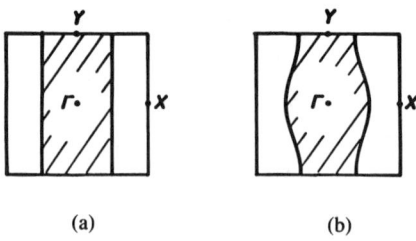

Figure 8.4 Fermi surfaces associated with the half-filled band of Fig. 8.3: (a) $\beta_b/\beta_a = 0$ and (b) $\beta_b/\beta_a < 1$. Filled and unfilled wave vectors are indicated by the presence and absence of shading, respectively. From Whangbo,[4] reproduced with permission of the American Chemical Society, Washington, D.C.

surface. In such a case, a partially filled band system under consideration exhibits a metallic character along the wave vector direction. When a straight wave vector line does not cross a Fermi surface, which is possible when the Fermi surface is open, the partially filled band system behaves as a nonmetal along that wave vector direction. The presence of an open Fermi surface such as Fig. 8.4(a) or (b) characterizes a one-dimensional (1D) metal. The wave-like Fermi surface of Fig. 8.4(b) reflects the fact that interactions along the b-direction are nonvanishing (i.e., $\beta_b \neq 0$). A piece of a Fermi surface may be translated by a single wave vector $\mathbf{q}$ and superposed on another piece of the Fermi surface. In such a case, the Fermi surface is said to be nested by the wave vector $\mathbf{q}$. For instance, the Fermi surfaces of Figure 8.4(a) and (b) have the nesting vectors shown in Fig. 8.5(a) and (b), respectively. In Fig. 8.5(b), it is not immediately obvious for one to see the effect of nesting since Fig. 8.5(b) shows only the portion of the Fermi surface belonging to the FBZ. With an extended zone representation of the Fermi surface, which is obtained by repeating the pattern of Fig. 8.5(b) in the two wave vector directions, it becomes clear that the nesting vector $\mathbf{q}$ superposes the left-hand-side piece of the Fermi surface onto the right-hand-side piece.

A metallic state predicted by one-electron band theory is not stable when its Fermi surface is nested, and becomes susceptible to a metal-insulator transition such as a charge density wave (CDW) or spin density wave (SDW) formation.[1-4] Let us now consider why a Fermi surface nesting is important in inducing a CDW or an SDW state. To simplify our notations, an occupied wave vector (k_a, k_b) may be represented by $\mathbf{k}$, and an occupied vector (k_a', k_b') by $\mathbf{k}'$. An orbital mixing between an occupied orbital $\phi(\mathbf{k})$ and an unoccupied orbital $\phi(\mathbf{k}')$ leads to new orbitals

$$\Psi(\mathbf{k}) \propto \phi(\mathbf{k}) + \gamma\phi(\mathbf{k}') \tag{8.18}$$

$$\Psi(\mathbf{k}) \propto -\gamma\phi(\mathbf{k}) + \phi(\mathbf{k}')$$

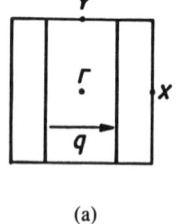
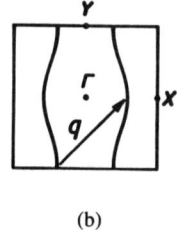

Figure 8.5 Nesting vectors associated with the Fermi surfaces of Fig. 8.2: (a) $\beta_b/\beta_a = 0$ and (b) $\beta_b/\beta_a < 1$. From Whangbo,[4] reproduced with permission of the American Chemical Society, Washington, D.C.

These new orbitals introduce electron density waves described by $\pm\cos\mathbf{q}\cdot\mathbf{R}$, respectively, where $\mathbf{q} = \mathbf{k}-\mathbf{k}'$ and $\mathbf{R} = m\mathbf{a} + n\mathbf{b}$. Thus, the density waves resulting from $\Psi(\mathbf{k})$ and $\Psi(\mathbf{k}')$ are out-of-phase in their charge distributions. When a Fermi surface has a nesting vector $\mathbf{q}$, this orbital mixing can be performed for all the wave vectors in the nested region of the FBZ thereby leading to the sets of new orbitals $\{\Psi(\mathbf{k})\}$ and $\{\Psi(\mathbf{k}')\}$ differing in their wave vectors by $\mathbf{q}=\mathbf{k}-\mathbf{k}'$. A CDW state is obtained when the orbitals $\Psi(\mathbf{k})$ are each doubly occupied, while an SDW state is obtained when the orbitals $\Psi(\mathbf{k})$ and $\Psi(\mathbf{k}')$ are each singly occupied by up-spin and down-spin electrons, respectively. With a nested Fermi surface, the sets of orbitals $\{\Psi(\mathbf{k})\}$ and $\{\Psi(\mathbf{k}')\}$ contain those derived from mixing the occupied and unoccupied levels in the vicinity of the Fermi level. For such orbitals, the energy difference between them is small, so that the orbital mixing between them is significant, and so is the interaction energy $<\phi(\mathbf{k})|H'|\phi(\mathbf{k}')>$. Here H' is a perturbation responsible for the interaction between $\phi(\mathbf{k})$ and $\phi(\mathbf{k}')$. In addition, the energy lowering associated with such an interaction matrix element can be derived from all the wave vectors $\mathbf{k}$ and $\mathbf{k}'$ related by the nesting vector $\mathbf{q}=\mathbf{k}-\mathbf{k}'$. This is why the electronic instability toward a CDW or an SDW state is strong when a metallic state possesses a Fermi surface nesting.

The perturbation H' causing a CDW state is a lattice vibration (i.e., phonon), while that causing an SDW state is an on-site repulsion U.[1-4] Both CDW and SDW states are insulating in nature since a band gap is created at the Fermi level as a result of orbital mixing. An insulating state has no Fermi surface by definition, since its highest occupied and lowest unoccupied levels are not degenerate. For a metal-insulator phase transition arising from a CDW or an SDW formation, the orbital mixing induced by a Fermi surface nesting destroys the Fermi surface. An incomplete Fermi surface nesting occurs when the pieces of the Fermi surface involved have slightly different curvatures in certain wave vector regions. For such a case, there exists as well an electronic instability resulting from the nested portions of the Fermi surface. The latter becomes destroyed by the associated orbital mixing. After a phase transition resulting from this orbital mixing, the small unnested portions of the Fermi surface give rise to small pocket-like new Fermi surfaces around them,[7] so that the resulting new electronic state is also metallic. Therefore, an incomplete Fermi surface nesting leads to an incomplete destruction of the Fermi surface, which is responsible for metal-metal transitions.

SUPERCONDUCTING STATE AND ORBITAL MIXING[4]

When the temperature is lowered, a metal may become susceptible to another electronic instability, that is, formation of a superconducting state. For a metal to become superconducting, it should avoid the electronic instability toward a metal-insulator transition associated with a good Fermi surface nesting. In general, the Fermi surface of a 1D metal is well nested, so that a 1D metal rarely undergoes a metal-superconductor transition. From the viewpoint of one-electron band theory, a superconducting state also involves orbital mixing among the band levels above and below the Fermi level, although the way this orbital mixing comes about differs from that for CDW and SDW states. Charge carriers of a superconducting state are not individual electrons but pairs of electrons (called Cooper pairs) having opposite momenta (i.e., opposite wave vectors).[8-10] Cooper pairs are described by product functions, for example, $\phi(\mathbf{k})\phi(-\mathbf{k})$ and $\phi(\mathbf{k}')\phi(-\mathbf{k}')$, where $\mathbf{k}$ and $\mathbf{k}'$ refer to occupied and unoccupied wave vectors of a normal metallic state, respectively. The energy lowering that leads to superconductivity is induced by the interaction of an occupied pair function $\phi(\mathbf{k})\phi(-\mathbf{k})$ with an unoccupied pair function $\phi(\mathbf{k}')\phi(-\mathbf{k}')$, that is, $<\phi(\mathbf{k})\phi(-\mathbf{k})|H'|\phi(\mathbf{k}')\phi(-\mathbf{k}')>$. The perturbation H' causing this mixing is the electron-phonon interaction in traditional superconductors. As a consequence of the interaction between pair functions, the character of the unoccupied pair function is mixed into that of the occupied pair function. In this indirect way, a superconducting state introduces unoccupied orbital character into occupied orbital character. Interactions between pair functions $\phi(\mathbf{k})\phi(-\mathbf{k})$ and $\phi(\mathbf{k}')\phi(-\mathbf{k}')$ introduce an energy gap at the Fermi level.

Cooper pair formation in traditional superconductors described by the BCS theory[8-10] is induced by electron-phonon interaction: A moving electron causes a slight, momentary lattice deformation around it, which affects the motion of a second electron in the wake of the first in such a way that, effectively, the two electrons move as an entity as if bound together by an attractive force. The extent of electron-phonon coupling is measured by the electron-phonon coupling constant λ. Given a lattice with Debye temperature θ_D that has a phonon spectrum (i.e., phonon band) effective for electron-phonon coupling, T_c is related to θ_D and λ as follows:[10]

$$T_c = \frac{\theta_D}{1.45} \exp\left[-\frac{1.04(1+\lambda)}{\lambda - \mu^*\left(1 + \lambda\frac{<\omega>}{\omega_o}\right)}\right] \quad (8.19)$$

where $<\omega>$ and ω_o are the average and the maximum frequencies of the phonon band, respectively. In this McMillan equation[10], the Coulomb pseudo-potential μ^* and the frequency ratio $<\omega>/\omega_o$ are on the order of 0.1 and 0.5, respectively. Among the several factors affecting the magnitude of T_c, the most important one is the electron-phonon coupling constant λ.[11] In general, T_c increases with increasing λ.

For a superconducting state to occur, the energy raising associated with the introduction of unoccupied orbital character should be smaller than the energy gain resulting from interactions among pair functions. Depending on the nature and strength of the perturbations causing orbital mixing, a normal metallic state with a nested Fermi surface may in principle lead to a superconducting state when the temperature is lowered. Fermi surface nesting gives rise to a metal-insulator transition such as CDW or SDW formation in most cases, but to a superconducting state when the interaction matrix elements $<\phi(\mathbf{k})|H|'\phi(\mathbf{k}')>$ responsible for CDW or SDW formation are small compared with the interaction matrix elements $<\phi(\mathbf{k})\phi(-\mathbf{k})|H'|\phi(\mathbf{k}')\phi(-\mathbf{k}')>$ causing a superconducting state. When the relative stabilities of CDW, SDW, and superconducting states are similar, the preference of one state over the other is delicately balanced by a change in temperature and pressure.

STRUCTURAL PROPERTIES

In attempting to understand the T_c variation in a series of closely related series of superconductors, the approximate relationship $T_c \propto \exp[-1/n(E_f)]$ is often invoked. That is, the higher T_c is expected to originate from the large $n(E_f)$ value. However, this relationship does not hold for organic salt superconductors. The T_c values of these superconductors are correlated with the softness of their lattices. Namely, a soft lattice provides a larger electron-phonon coupling constant λ and consequently a higher T_c.[11] The lattice softness is strongly influenced by the donor$\cdots$donor and donor$\cdots$anion contact interactions involving the C-H bonds of the donor molecule. Such C-H$\cdots$donor and C-H$\cdots$anion interactions are also crucial factors governing the crystal packing patterns, and therefore the electronic properties, of organic-donor molecule salts.[12] In this section, we probe important structural properties associated with the donor-molecule C-H bonds.

STRUCTURAL FACTORS AFFECTING SUPERCONDUCTIVITY

β-(BEDT-TTF)$_2$X, (X$^-$ = I$_3^-$, AuI$_2^-$, IBr$_2^-$)

Figure 8.6 shows how the T_c values of β-(BEDT-TTF)$_2$X (X$^-$ = I$_3^-$, AuI$_2^-$, IBr$_2^-$) vary as a function of applied pressure P.[13] For X$^-$ = AuI$_2^-$ and IBr$_2^-$, the T_c of β-(BEDT-TTF)$_2$X salts decreases gradually with increasing applied pressure. When P<0.5 kbar, the T_c of β-(BEDT-TTF)$_2$I$_3$ decreases as well on increasing P. At P≅0.5 kbar, however, the T_c of β-(BEDT-TTF)$_2$I$_3$ jumps to ~8 K and then decreases as P increases further. β-(BEDT-TTF)$_2$I$_3$ with the high-T_c state is designated as β*-(BEDT-TTF)$_2$I$_3$.[14] We now examine the pressure- and anion-dependence of the T_c in β-(BEDT-TTF)$_2$X on the basis of Eq. 8.19 and the crystal structures of β-(BEDT-TTF)$_2$X. By employing the observed T_c values of β-(BEDT-TTF)$_2$X salts in Eq. 8.19, along with the parameters $\mu^* = 0.1$,[15] $<\omega>/\omega_o = 0.3$[15] and $\theta_D = 200$ K,[16] we obtain the electron-phonon coupling constants λ listed in Table 8.1.[11a]

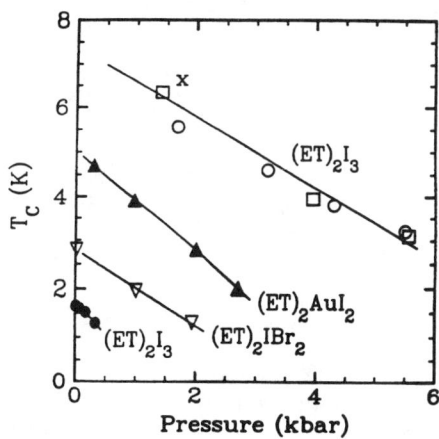

Figure 8.6 Pressure-dependence of the superconducting transition temperature T_c of β-(BEDT-TTF)$_2$X (X$^-$ = I$_3^-$, AuI$_2^-$, IBr$_2^-$). From Williams et al.[42] reproduced with permission of John Wiley & Sons, Somerset, N.Y.

TABLE 8.1 Superconducting Transition Temperature T_c, Electron-Phonon Coupling Constant λ, and Calculated Density of States at the Fermi Level $n(E_f)$ for β-(BEDT-TTF)$_2$X (X$^-$ = I$_3^-$, AuI$_2^-$, IBr$_2^-$)

Salt	T_c	λ	$n(E_f)$ electrons/eV
β-(BEDT-TTF)$_2$I$_3$	1.4 K	0.37	3.66[a]
β-(BEDT-TTF)$_2$IBr$_2$	2.8 K	0.43	3.69[a]
β-(BEDT-TTF)$_2$AuI$_2$	5.0 K	0.52	3.48[b]
β*-(BEDT-TTF)$_2$I$_3$	~8 K	0.62	3.53[c]

[a] Based on the X-ray crystal structure determined at 9 K (Ref. 17)
[b] Based on the neutron crystal structure determined at 20 K (Ref. 17)
[c] Based on the neutron crystal structure determined at 4.5 K and 1.5 kbar (Ref. 14)

Clearly, these λ values increase with T_c for all β-(BEDT-TTF)$_2$X superconductors. To better understand the variation of the T_c for all β-(BEDT-TTF)$_2$X salts, it is necessary to investigate the structural factors affecting λ because the electronic structures of β-(BEDT-TTF)$_2$X (X$^-$ = I$_3^-$, AuI$_2^-$, IBr$_2^-$) are very similar. For examples, the band dispersion relations and the Fermi surface of β-(BEDT-TTF)$_2$I$_3$, calculated on the basis of its X-ray crystal structure determined at 9K under ambient pressure,[17] are shown in Figure 8.7(a) and (b), respectively. With the formal oxidation (BEDT-TTF)$_2^+$ the highest occupied band is half filled, and its Fermi surface consists of a closed loop so that β-(BEDT-TTF)$_2$I$_3$ is a 2D metal. As shown in Figure 8.7, the band dispersion relations and the Fermi surface of β*-(BEDT-TTF)$_2$I$_3$, calculated on the basis of the crystal structure determined by neutron diffraction at 4.5K under 1.5 kbar,[14] are virtually identical with the corresponding ones of Fig. 8.7. The band dispersion relations and the Fermi surfaces of the β-(BEDT-TTF)$_2$X (X$^-$ = AuI$_2^-$, IBr$_2^-$) salts are very similar to those of Figs. 8.7 and 8.8.

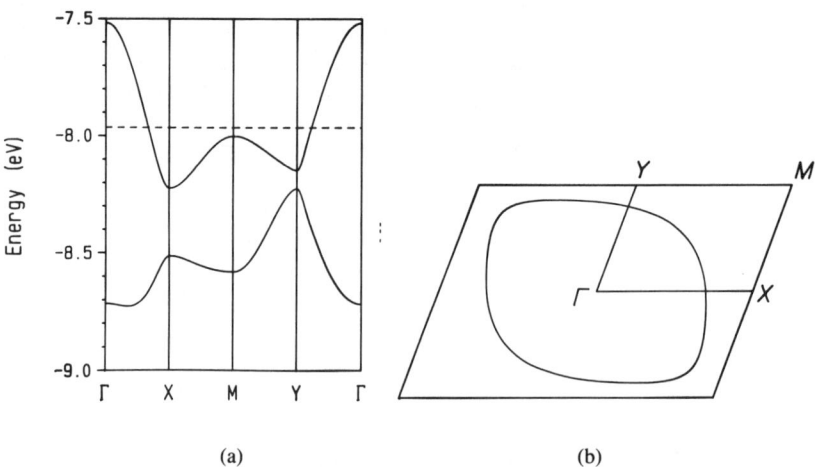

Figure 8.7 (a) Dispersion relations of the two highest-occupied bands of β-(BEDT-TTF)$_2$I$_3$, where Γ = (0, 0) = (a*/2, 0), Y = (0, b*/2), and M = (a*/2, b*/2), and (b) Fermi surface associated with the half-filled band of Fig. 8.7(a).

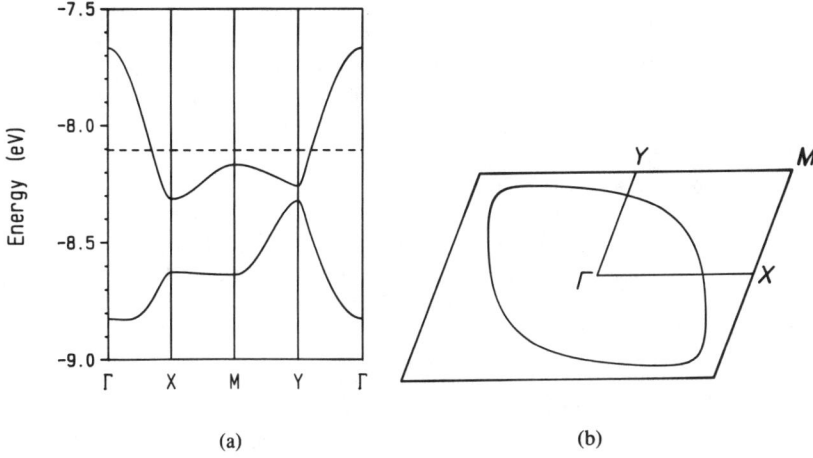

Figure 8.8 (a) Dispersion relations of the two highest occupied bands of β*-(BEDT-TTF)$_2$I$_3$, and (b) Fermi surface associated the half-filled band of Fig. 8.8(a).

For a lattice with atoms of mass M, λ is expressed by [10]

$$\lambda = \frac{N(E_f) \langle I^2 \rangle}{M \langle \omega^2 \rangle} \propto \frac{1}{M \langle \omega^2 \rangle} \qquad (8.20)$$

where n(E$_f$) is the electronic density of states at the Fermi level E$_f$, $\langle I^2 \rangle$ is the square of the electron-phonon matrix element averaged over the Fermi surface, and $\langle \omega^2 \rangle$ is the square of the phonon frequencies averaged over the phonon band. The band electronic structures of β-(BEDT-TTF)$_2$X (X$^-$ = I$_3^-$, AuI$_2^-$, IBr$_2^-$) are identical in essential characteristics, and their n(E$_f$) values vary only slightly. Thus, the n(E$_f$)$\langle I^2 \rangle$ term is expected to be constant for the β-(BEDT-TTF)$_2$X salts, so that λ depends primarily on the phonon frequency. The M$\langle \omega^2 \rangle$ term has the dimension of a force constant. Therefore, a large λ results when the lattice has soft phonons, which arise from molecular vibrations with shallow potential wells. For the β-(BEDT-TTF)$_2$X salts, phonons important for their superconductivity are likely to be translational and/or librational modes. When the lattice of β-(BEDT-TTF)$_2$X is soft toward such phonon modes, it will lead to a large value of electron-phonon coupling constant λ. The increase in the calculated λ values (from 0.37 to 0.62 in Table 8.1) of β-(BEDT-TTF)$_2$I$_3$ must be related to a corresponding increase in the lattice softness.

Each anion X$^-$ of β-(BEDT-TTF)$_2$X is enclosed in a hydrogen atom "pocket" made up of 12 BEDT-TTF molecules, as shown in Fig. 8.9, where each terminal halogen of X$^-$ is surrounded by six ethylene groups. This packing of BEDT-TTF molecules around X$^-$ is crucial for determining whether the resulting donor-molecule network becomes 1D or 2D in its electronic properties. When viewed along the central C=C bond of the BEDT-TTF molecule, the two ethylene groups of a BEDT-TTF molecule are either eclipsed or staggered with respect to each other, as shown in Fig. 8.10(a) and (b), respectively. In β-(BEDT-TTF)$_2$X (X$^-$ = AuI$_2^-$, IBr$_2^-$), all BEDT-TTF molecules have eclipsed arrangements.[17] In β-

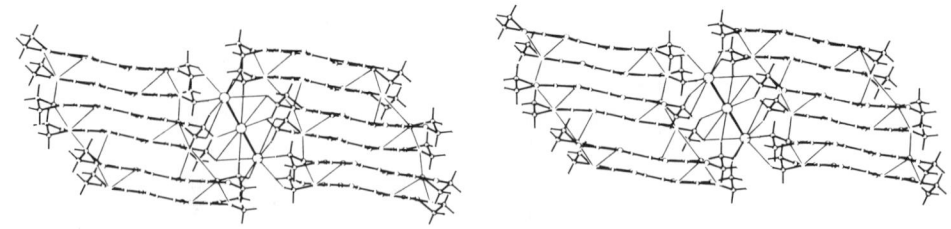

Figure 8.9 Stereoview of the β-(BEDT-TTF)$_2$AuI$_2$ structure with the 12 BEDT-TTF molecules surrounding one AuI$_2^-$ anion. The H···I and H···Au contacts less than 3.5Å and the S···S contacts less than 3.6 Å are connected by thin lines. From Whangbo et al.[11a] with permission of the American Chemical Society, Washington, D.C.

(BEDT-TTF)$_2$I$_3$, however, both eclipsed and staggered configurations of the donor molecules are found in nearly equal proportions.[17,18] As observed for β*-(BEDT-TTF)$_2$I$_3$,[14] the most striking structural change that occurs in β-(BEDT-TTF)$_2$I$_3$ under applied pressure greater than 0.5 kbar is that all BEDT-TTF molecules adopt staggered arrangements thereby removing the structural modulation that appears in β-(BEDT-TTF)$_2$I$_3$ below ~200 K under ambient pressure.

The stereoview of Fig. 8.11 shows the interactions between the terminal ethylene groups of the BEDT-TTF molecules and the I$_3^-$ anions in β*-(BEDT-TTF)$_2$I$_3$ at 4.5 K and 1.5 kbar. The ethylene groups in the A-sites have short H···I contacts with two I$_3^-$ anions while those in the B-sites have short H···I contacts with four I$_3^-$ anions. A similar arrangement of anions and terminal ethylene groups of BEDT-TTF molecules is found for the other β-(BEDT-TTF)$_2$X(X$^-$ = AuI$_2^-$, IBr$_2^-$) salts. Shown in Fig. 8.12(a) and (b) are perspective views of the ethylene groups of BEDT-TTF molecules located below a layer of I$_3^-$ anions in the ambient-

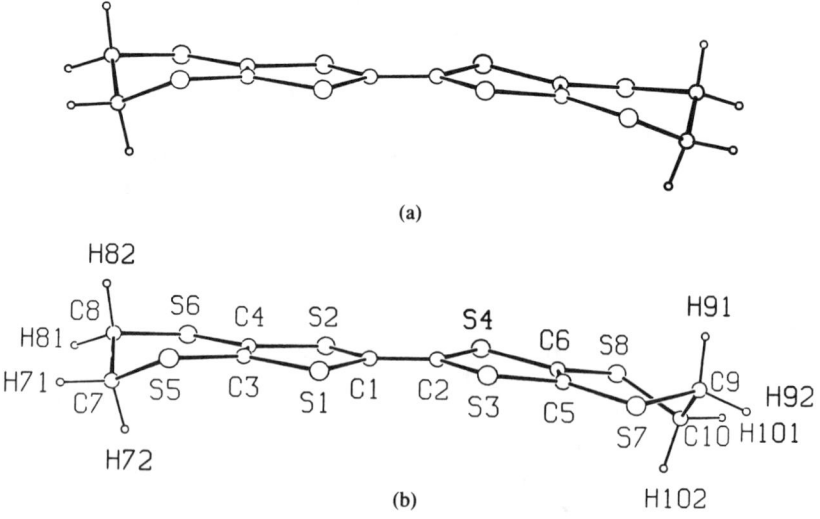

Figure 8.10 Relative arrangements of the two ethylene groups in BEDT-TTF: (a) eclipsed and (b) staggered. From Whangbo et al.[11a] with permission of the American Chemical Society, Washington, D.C.

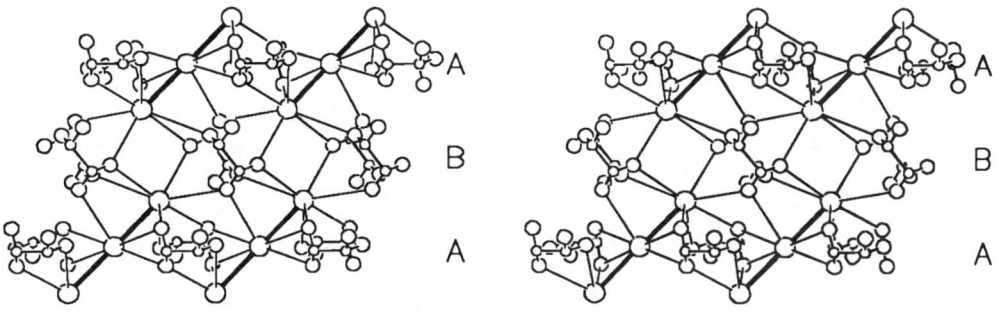

Figure 8.11 Stereodiagram of how the terminal ethylene groups of the BEDT-TTF molecules in β*-(BEDT-TTF)$_2$ make short C-H···anion contacts. From Whangbo et al.[11a] with permission of the American Chemical Society, Washington, D.C.

pressure β-(BEDT-TTF)$_2$I$_3$ structure for the cases when BEDT-TTF molecules have eclipsed and staggered arrangements, respectively. It is at the B-sites where the nature of the H···I contacts and H···H contacts varies depending upon whether BEDT-TTF has an eclipsed or staggered arrangement. As shown in Fig. 8.12(a), an eclipsed BEDT-TTF molecule leads to one extremely short H···I contact (i.e., 2.84 Å for H102···I2). This contact is removed when the ethylene group of the B-site undergoes conformational flipping as shown in Fig. 8.12(b). In the latter, however, there occurs one very short H···H contact (i.e., 2.15Å for H82···H92), which is absent in Fig. 8.12(a). Therefore, an energetically unfavorable intermolecular contact (H···I or H···H) exists in β-(BEDT-TTF)$_2$I$_3$ for either the eclipsed or the staggered arrangement. The structural modulation[18] of β-(BEDT-TTF)$_2$I$_3$ which occurs below 200 K is an effective way to reduce the extent of such unfavorable intermolecular contacts. In β*-(BEDT-TTF)$_2$I$_3$ the arrangement of ethylene groups around I$_3^-$ ions is similar to that shown in Fig. 8.12(b), except that the ethylene groups at the B-site slips slightly [see Fig. 8.12(c)] under the applied pressure so as to make the shortest H···H contact longer (i.e., 2.26Å for H82···H92). For β-(BEDT-TTF)$_2$X ($X^- $ = AuI$_2^-$, IBr$_2^-$) as well, the arrangements of the ethylene groups of eclipsed BEDT-TTF molecules around each anion X^- are similar to those shown in Fig. 8.12(a). Table 8.2 summarizes the shortest H···H and hydrogen terminal halogen (H···Y) distances found for β-(BEDT-TTF)$_2$X.[11a]

Let us now discuss how the structural features associated with the H···H and H···Y contacts of the β-(BEDT-TTF)$_2$X salts are related to the softness of their lattices and, therefore, to their electron-phonon coupling constant λ. The principal forces providing overall lattice cohesiveness are attractive Coulombic interactions of oxidized BEDT-TTF molecule layers with anion X^- layers and also the C-H···donor and C-H···anion contact interactions. Ab initio SCF-MO/MP2 calculations on model compounds show that the C-H···anion interactions are much stronger than the C-H···donor interactions. Therefore, BEDT-TTF molecules in β-(BEDT-TTF)$_2$X are anchored around the anions X^- as if effectively weak "hydrogen bonding" exists in the C-H···X^- contacts. Thus, how "soft" BEDT-TTF molecules are with respect to translational and/or librational modes of vibration depends on, and may be correlated with, the lengths of their shortest H···H and H···Y

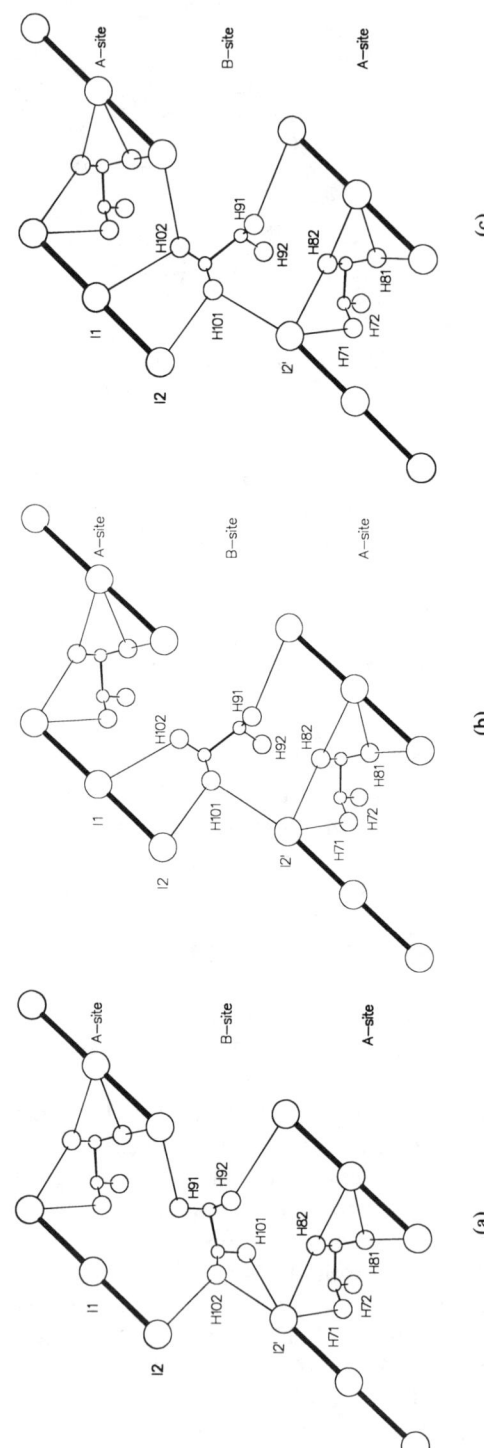

Figure 8.12 Terminal ethylene group arrangements with respect to the I_3^- anion layer in (a) β-(BEDT-TTF)$_2$I$_3$ with eclipsed BEDT-TTF molecules, (b) β-(BEDT-TTF)$_2$I$_3$ with staggered BEDT-TTF molecules, and (c) β*-(BEDT-TTF)$_2$I$_3$ with staggered BEDT-TTF molecules. From Whangbo et al.[11a] with permission of the American Chemical Society, Washington, D.C.

TABLE 8.2 Ethylene Group Arrangements and Shortest Hydrogen···Hydrogen (H···H) and Hydrogen···Anion (H···Y) Distances in β-(BEDT-TTF)$_2$X (X$^-$ = I$_3^-$, AuI$_2^-$, IBr$_2^-$)[11a]

Salt	Ethylene Group Arrangement[a]	Intermolecular Distance H···H	H···Y
β-(BEDT-TTF)$_2$I$_3$	E	2.477	2.842
	S	2.152	2.988
β-(BEDT-TTF)$_2$IBr$_2$	E	2.169	2.887
β-(BEDT-TTF)$_2$AuI$_2$	E	2.224	2.966
β*-(BEDT-TTF)$_2$I$_3$	S	2.261	3.014

[a] E and S refer to the eclipsed and the staggered ethylene group arrangements, respectively.

contacts, which indirectly represent the strengths of their interactions. Table 8.2 shows that these H···H and H···Y distances increase in the order.

$$\beta\text{-(BEDT-TTF)}_2\text{IBr}_2 < \beta\text{-(BEDT-TTF)}_2\text{AuI}_2 < \beta^*\text{-(BEDT-TTF)}_2\text{I}_3. \quad (8.21)$$

Thus the lattice softness of β-(BEDT-TTF)$_2$X would increase in the same order, which is consistent with the increase in the λ values in the same order. Under ambient pressure, β-(BEDT-TTF)$_2$I$_3$ has unfavorably short H···H and H···I contacts, the extent of which is somewhat reduced on introducing structural modulation.[18] The structural modulation in β-(BEDT-TTF)$_2$I$_3$ reflects the structural strain, that is, lattice stiffness, which would be primarily responsible for the λ value smaller than that of β-(BEDT-TTF)$_2$IBr$_2$. It is the adoption of a staggered arrangement by each BEDT-TTF molecule under a pressure greater than 0.5 kbar that makes the lattice of β*-(BEDT-TTF)$_2$I$_3$ particularly soft. A further increase in applied pressure beyond ~0.5 kbar in β*-(BEDT-TTF)$_2$I$_3$ would make its lattice stiffer, which makes its T_c lower than 8K. In general, the softness of the C-H···anion interactions, and hence that of a lattice associated with such interactions, is expected to increase in the order: H···F, H···O < H···Cl < H···Br < H···I.

κ-Phase Superconductors

Ambient-pressure κ-phase superconductors include κ-(BEDT-TTF)$_2$X (X$^-$ = Cu[N(CN)$_2$]Br,[40] Cu(NCS)$_2^-$,[19] I$_3^-$,[20] κ-(MDT-TTF)$_2$AuI$_2$,[21] and κ-(DMET)$_2$AuBr$_2$.[22] The crystal packing of a κ-phase material appears to be most favorable in achieving superconductivity,[23] since the κ-phase superconductors consist of different donor molecules and different types of anions (i.e., discrete and polymeric anions). The symmetrical donor BMDT-TTF leads to a nonsuperconducting κ-phase salt, κ-(BMDT-TTF)$_2$Au(CN)$_2$,[24] which does not become superconducting under pressure.[25] κ-(BEDT-TTF)$_4$Hg$_3$Cl$_8$[26] is not a superconductor under ambient pressure but, unlike κ-(BMDT-TTF)$_2$Au(CN)$_2$, it becomes superconducting under pressure with somewhat complicated pressure dependence of T_c. This salt first becomes a superconductor at P = 12 kbar (T_c = 1.8 K), with its T_c gradually decreasing on increasing P until it loses superconductivity at P = 23 kbar. At P = 29 kbar, this salt becomes superconducting again at a higher temperature (T_c = 5.3K).[26a,b]

As representative examples of the κ-phase salts, we show the band dispersion relations and the Fermi surfaces[23,27] of κ-(BEDT-TTF)$_2$Cu(NCS)$_2$, κ-(BEDT-TTF)$_2$I$_3$ and κ-(BMDT-TTF)$_2$Au(CN)$_2$, in Figs. 8.13, 8.14, and 8.15, respectively. The Fermi surfaces of all the κ-phase salts are essentially described as overlapping distorted circles. For κ-(BEDT-TTF)$_2$Cu(NCS)$_2$, the overlapping circles split into 1D wavy lines and small elongated elllipses due to a surface noncrossing at the intended crossing points, which reflects the fact that this salt has two nonequivalent BEDT-TTF molecules. Therefore, the κ-phase salts are all predicted to be 2D metals. Table 8.3 lists the n(E_f) values calculated for various κ-phase salts, which are greater than the n(E_f) values of the β-phase superconductors by a factor of about two (see Table 8.2). Nevertheless, as in the case of the β-phase superconductors, no apparent correlation exists between the T_c and n(E_f) values in the κ-phase salts. That the lattice softness associated with the donor-molecule C-H bonds is also important in inducing superconductivity in the κ-phase salts can be deduced from the isotope effect on T_c in κ-(BEDT-TTF)$_2$Cu(NCS)$_2$.[12b] When the hydrogen atoms of BEDT-TTF are replaced by deuterium atoms, the resulting BEDT-TTF-d$_8$ has a higher mass than BEDT-TTF. According to the BCS theory of superconductivity, therefore, the mass increase should make the T_c of κ-(BEDT-TTF)-d$_8$)$_2$Cu(NCS)$_2$ lower than that of κ-(BEDT-TTF)$_2$Cu(NCS)$_2$(10.4K). However, the T_c of κ-(BEDT-TTF-d$_8$)$_2$Cu(NCS)$_2$ is higher (T_c = 10.8K).[28] The isotope effect predicted by the BCS theory assumes that the isotope substitution does not change the electron-phonon coupling constant. This assumption is not valid for organic donor-salt superconductors in general. Since a C-D bond has a lower stretching frequency than does a C-H bond, a C-D bond is effectively shorter than a C-H bond as far as the C-H(D)···anion and C-H(D)···donor contact interactions are concerned. Thus, with respect to BEDT-TTF, BEDT-TTF-d$_8$ may provide a

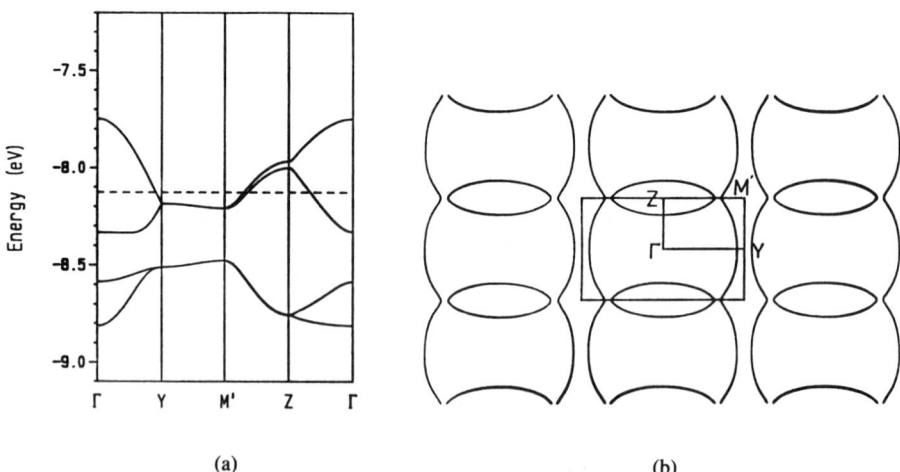

Figure 8.13 (a) Dispersion relations of the two highest occupied bands and (b) Fermi surface of κ-(BEDT-TTF)$_2$Cu(NCS)$_2$. From Jung et al.[27] with permission of the American Chemical Society, Washington, D.C.

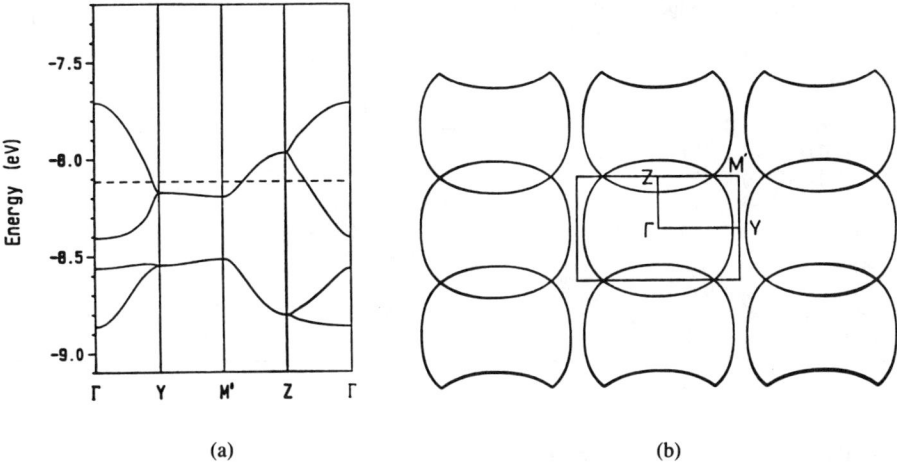

Figure 8.14 (a) Dispersion relations of the two highest occupied bands and (b) Fermi surface of κ-(BEDT-TTF)$_2$I$_3$. From Jung et al.[27] with permission of the American Chemical Society, Washington, D.C.

softer lattice and hence would lead to a larger electron-phonon coupling constant λ and a higher T_c.

For the κ-phase superconductors it is difficult to find a simple geometrical parameter that will reflect their lattice softness because they consist of different donor molecules and different types of anions. Nevertheless, the T_c trends in the three κ-phase superconductors with the linear triatomic anions (i.e., I$_3^-$, AuI$_2^-$, AuBr$_2^-$) may be rationalized on the basis of the C-H⋯anion and C-H⋯donor interactions. Since iodine is more polarizable than bromine, the C-H⋯I contact

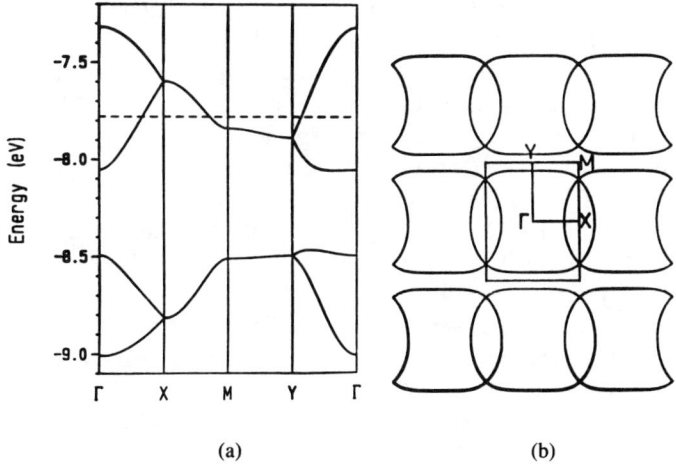

Figure 8.15 (a) Dispersion relations of the two highest-occupied bands and (b) Fermi surface of κ-(BMDT-TTF)$_2$Au(CN)$_2$. From Jung et al.[27] with permission of the American Chemical Society, Washington, D.C.

TABLE 8.3 Superconducting Transition Temperatures T_c and Densities of States at the Fermi Level $n(E_f)$ of κ-phase Salts[a]

Salt	T_c	$n(E_f)$ electrons/eV
κ-(BEDT-TTF)$_2$Cu[N(CN)$_2$]Br	11.5 K	7.71
κ-(BEDT-TTF)$_2$Cu(NCS)$_2$	10.4 K	7.57
κ-(BEDT-TTF)$_2$I$_3$	3.6 K	7.09
κ-(BEDT-TTF)$_4$Hg$_3$Cl$_8$	1.8 K (12 kbar) 5.3 K (29 kbar)	10.8
κ-(MDT-TTF)$_2$AuI$_2$	4.5 K	7.98
κ-(DMET)$_2$AuBr$_2$	1.8 K	8.92
κ-(BMDT-TTF)$_2$Au(CN)$_2$	—	7.30

[a] On the basis of the crystal structures determined at room temperature.

should provide a softer lattice than the C-H···Br contact. This is consistent with the observation that both κ-(BEDT-TTF)$_2$I$_3$ (T_c = 3.6 K) and κ-(MDT-TTF)$_2$AuI$_2$ (T_c = 4.5 K) have a higher T_c than does κ-(DMET)$_2$AuBr$_2$ (T_c = 1.8 K). The fact that κ-(MDT-TTF)$_2$AuI$_2$ has a higher T_c than does κ-(BEDT-TTF)$_2$I$_3$ may be related to the difference in their C-H···donor interactions. κ-(BEDT-TTF)$_2$I$_3$ has short C-H···S and C-H···C(sp^2) contacts between donor molecules, while such C-H···donor contacts are absent in κ-(MDT-TTF)$_2$AuI$_2$.[21b,23,27]

The Cu(NCS)$_2^-$ anions of κ-(BEDT-TTF)$_2$Cu(NCS)$_2$ and the Cu[N(CN)$_2$]Br$^-$ anions of κ-(BEDT-TTF)$_2$Cu[N(CN)$_2$]Br form ribbon-like polymeric chains, and are therefore quite different in structure from the discrete anions I$_3^-$, AuI$_2^-$ and AuBr$_2^-$. The κ-(BEDT-TTF)$_2$Cu(NCS)$_2$ salt has two nonequivalent BEDT-TTF molecules: With the Cu(NCS)$_2^-$ anions, one donor molecule makes C-H···S contacts, but the other makes C-H···C and C-H···N contacts.[29] Thus one donor is in a softer environment than the other. In κ-(BEDT-TTF)$_2$Cu[N(CN)$_2$]Br all BEDT-TTF molecules are equivalent.[40,41] One ethylene group of BEDT-TTF makes C-H···C and C-H···N contacts with the Cu[N(CN)$_2$]Br$^-$ anions, and the other ethylene group C-H···Br contacts. However, all of these C-H···anion contact interactions in κ-(BEDT-TTF)$_2$X (X^- = Cu[N(CN)$_2$]Br$^-$, Cu(NCS)$_2^-$) are not expected to be softer than the C-H···I contact interactions in κ-(BEDT-TTF)$_2$I$_3$ and κ-(MDT-TTF)$_2$AuI$_2$. Since κ-(BEDT-TTF)$_2$X (X^- = Cu[N(CN)$_2$]Br$^-$, Cu(NCS)$_2^-$) have a higher T_c than do κ-(BEDT-TTF)$_2$I$_3$ and κ-(MDT-TTF)$_2$AuI$_2$, it seems likely that the ribbon-like chains of the Cu[N(CN)$_2$]Br$^-$ and Cu(NCS)$_2^-$ anions provide a phonon spectrum conducive for superconductivity.[12b]

Finally, we examine several structural properties of the κ-phase salts (see Table 8.4) that might be relevant for understanding the complex pressure-dependence of T_c in κ-(BEDT-TTF)$_4$Hg$_3$Cl$_8$.[26] Each dimeric unit of the donor molecules has a bond-over-ring (BOR) arrangement in κ-(BEDT-TTF)$_2$X (X^- = Cu(NCS)$_2^-$, I$_3^-$), κ-(BEDT-TTF)$_4$Hg$_3$Cl$_8$, κ-(MDT-TTF)$_2$AuI$_2$, and κ-(DMET)$_2$AuBr$_2$, but a bond-over-bond (BOB) arrangement in κ-(BMDT-TTF)$_2$Au(CN)$_2$ (see Fig. 8.16 for representative examples).[23,27] It is worthwhile to note that all superconducting κ-phase salts have a BOR arrangement. Among the

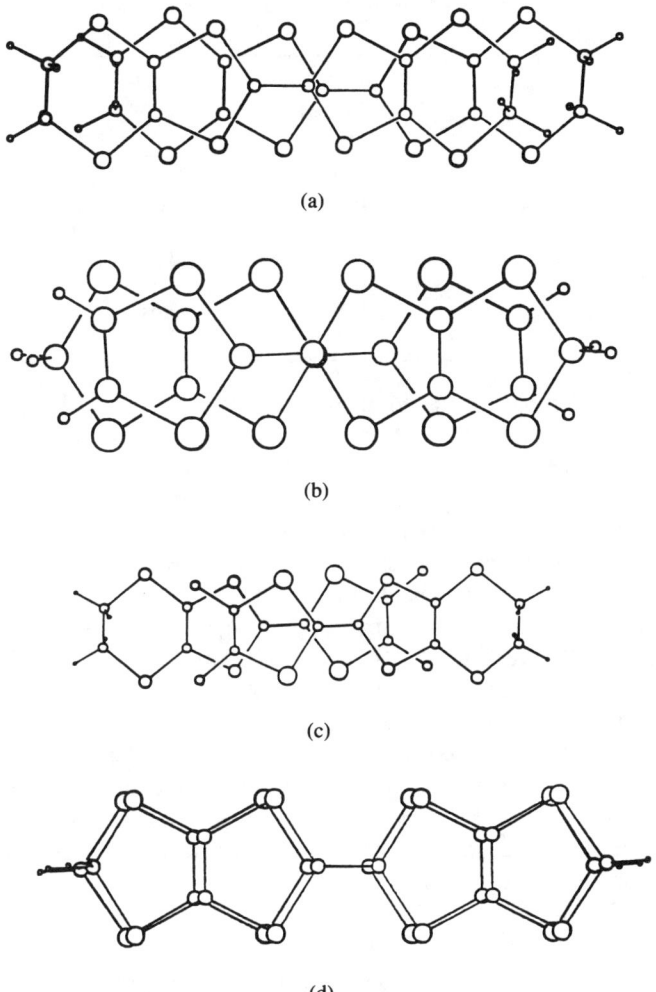

Figure 8.16 Projection views of donor molecule dimers in (a) κ-(BEDT-TTF)$_2$Cu(NCS)$_2$ and (b) κ-(MDT-TTF)$_2$AuI$_2$, (c) κ-(DMET)$_2$AuBr$_2$ and κ-(BMDT-TTF)$_2$Au(CN)$_2$. From Jung et al.[27] with permission of the American Chemical Society, Washington, D.C.

κ-phase salts of TTF-based donor molecules (i.e., BEDT-TTF and MDT-TTF), the intradimer spacing in the ambient-pressure superconducting salts is 3.35Å (Table 8.4), which is considerably shorter than the 3.59Å spacing in the nonsuperconducting salt. The large intradimer spacing for the superconducting κ-(DMET)$_2$AuBr$_2$ derives from the fact that each donor dimer has Se···Se contacts instead of S···S contacts.[22,23] The large intradimer spacing of κ-(BMDT-TTF)$_2$Au(CN)$_2$ is related to the BOB arrangement of the donor dimers.[25,27] In κ-(BEDT-TTF)$_4$Hg$_3$Cl$_8$ the intradimer spacing is large despite its BOR arrangement due probably to the way BEDT-TTF molecules make C–H···Cl contacts with the Hg$_3$Cl$_8^{2-}$ anions,[23] in

TABLE 8.4 Dimer Overlap Patterns, Intradimer Separations, and Ethylene Group Arrangement of κ-phase Salts

Salt	Dimer Overlap Pattern	Intradimer Separation (Å)	Ethylene Group Arrangement
κ-(BEDT-TTF)$_2$Cu[N(CN)$_2$]Br	BOR	3.37	E
κ-(BEDT-TTF)$_2$Cu(NCS)$_2$	BOR	3.35	S
κ-(BEDT-TTF)$_2$I$_3$	BOR	3.35	E
κ-(BEDT-TTF)$_4$Hg$_3$Cl$_8$	BOR	3.59	S
κ-(MDT-TTF)$_2$AuI$_2$	BOR	3.35	
κ-(DMET)$_2$AuBr$_2$	BOR	3.54	
κ-(BMDT-TTF)$_2$Au(CN)$_2$	BOB	3.64	

which each Hg^{2+} ion has a tetrahedral coordination. κ-(BEDT-TTF)$_4$Hg$_3$Cl$_8$ has a large $n(E_f)$ value compared with other κ-phase salts of TTF-based donor molecules (Table 8.3). This means that the electronic interaction between the donor molecules is comparatively weak in κ-(BEDT-TTF)$_4$Hg$_3$Cl$_8$. Under applied pressure, its donor molecule packing may be tightened to become similar to that found for other superconducting κ-phase salts. As summarized in Table 8.4, the BEDT-TTF molecules have a staggered arrangement in κ-(BEDT-TTF)$_2$Cu(NCS)$_2$ and κ-(BEDT-TTF)$_4$Hg$_3$Cl$_8$. By analogy with the structural change β-(BEDT-TTF)$_2$I$_3$ (T_c = 1.4 K) experiences under pressure of ~0.5 kbar to become β*-(BEDT-TTF)$_2$I$_3$ (T_c = ~8 K), it may be speculated that the donor molecules of κ-(BEDT-TTF)$_4$Hg$_3$Cl$_8$ have half staggered and half eclipsed arrangements at P = 12 kbar as in β-(BEDT-TTF)$_2$I$_3$ but only the staggered arrangement at P = 29 kbar as in β*-(BEDT-TTF)$_2$I$_3$. To test this hypothesis, it would be important to determine the crystal structures of κ-(BEDT-TTF)$_4$Hg$_3$Cl$_8$ under pressures of 12 and 29 kbar. Such studies could provide vital insight into the structural factors that govern the superconductivity of the κ-phase salts.

C-H···DONOR AND C-H···ANION INTERACTIONS AND CRYSTAL PACKING PATTERNS[12]

In determining the lattice softness of the β- and κ-phase superconductors, the donor···donor and donor···anion interactions associated with the C-H bonds are crucial, as discussed in the previous section. In the present section we show that the packing patterns of organic donor salts are also strongly influenced by the C-H···donor and C-H···anion contacts interactions.

Structural Characteristics of Donor-molecule Packing

We first examine several important structural features of neutral donor-molecule solids and charge-transfer salts which are crucial for understanding the crystal packing patterns of organic donor-molecule salts. A perspective view of how the donor molecules are packed in neutral BEDT-TTF solid[12,30] is depicted in Fig. 8.17. The C-H bonds of a donor molecule make short C-H···S and C-H···C(sp^2) contacts (shown by dashed lines) with adjacent donor molecules, the π-framework of BEDT-TTF is bent as shown in Fig. 8.18 ($\theta \cong 15°$), and one six-membered ring of BEDT-TTF has a staggered conformation [Fig. 8.19(a)] while the other ring has a conformation similar to the eclipsed one [Fig. 8.19(d)]. The six-membered ring conformation Fig. 8.19(a) is converted to Fig. 8.19(d) via the intermediate conformations Fig. 8.19(b) and (c). In Fig 8.19(b), one methylene carbon atom is on the π-plane of the ring, while in Fig. 8.19(c) both methylene carbon atoms are above the π-plane with the height of one carbon atom half that of the other carbon atom. When a methylene group is above the π-plane, one C-H bond of the methylene group is approximately perpendicular to the π-plane (C-H$_{ax}$ bond) while the

Figure 8.17 Perspective view of donor-molecule packing in neutral BEDT-TTF solid, where the dashed lines refer to short intermolecular contacts associated with the C-H bonds. From Whangbo et al.[12a] with permission of Springer-Verlag, Heidelberg, West Germany.

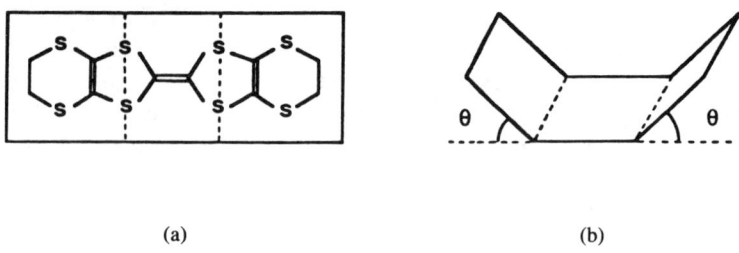

(a) (b)

Figure 8.18 π-framework bending in a BEDT-TTF molecule: (a) a top view and (b) a schematic perspective view. From Whangbo et al.[12a] with permission of Springer-Verlag, Heidelberg, West Germany.

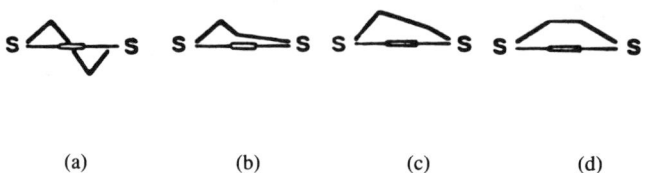

(a) (b) (c) (d)

Figure 8.19 Conformations that a six-membered ring of BEDT-TTF may adopt. From Whangbo et al.[12a] with permission of Springer-Verlag, Heidelberg, West Germany.

other C-H bond is approximately parallel to the π-plane (C-H_{eq} bond) (see Fig. 8.20).

Figure 8.21 shows a stereodiagram of how the BEDT-TTF molecules of β*-(BEDT-TTF)$_2$I$_3$ make short C-H···donor and C-H···anion contacts. The π-framework of BEDT-TTF is nearly flat, and each donor molecule has the six-membered ring conformations shown in Fig. 8.19(a) and (b). The C-H bonds of BEDT-TTF make short C-H···S, C-H···C(sp^2) and C-H···H contacts with the adjacent donor molecules, and also make short C-H···I contacts with the I$_3^-$ anions. Two C-H_{ax} bonds by each donor molecule are positioned on top of the five- and six-membered rings of the adjacent donor molecules. In a donor-molecule dimer of β*-(BEDT-TTF)$_2$I$_3$ the molecules are therefore slipped along the direction parallel to the donor-molecule central C=C bond (Fig. 8.16). In contrast, the molecules in a donor stack of (BEDO-TTF)$_2$AuBr$_2$[31] are slipped along the direction perpendicular to the donor-molecule central C=C bond, as shown by a projection view Fig. 8.22. The latter slipping allows the donor molecules of (BEDO-TTF)$_2$AuBr$_2$ to

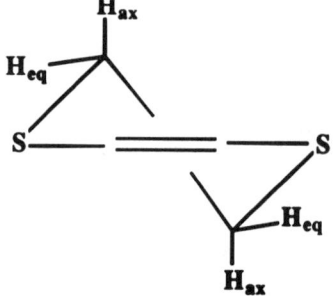

Figure 8.20 Orientations of the ethylene group C-H bonds in a six-membered ring of BEDT-TTF. From Novoa et al.[12c], with permission of Gordon and Breach Science Publishers, Inc. New York, N.Y.

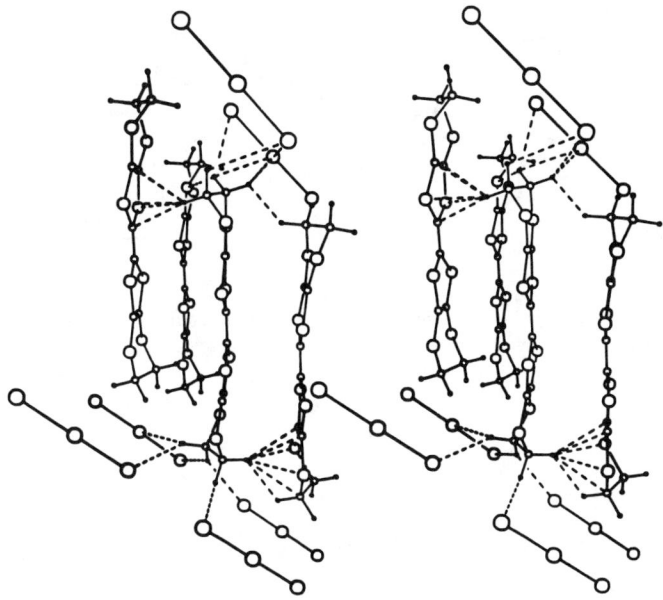

Figure 8.21 Stereodiagram of how the BEDT-TTF molecules of β*-(BEDT-TTF)$_2$I$_3$ make short C-H···donor and C-H···anion contacts. From Wang et al.[35e] with permission of the American Chemical Society, Washington, D.C.

make short C-H···O contacts both within each donor stack [Fig. 8.23(a)] and between adjacent donor stacks [Fig. 8.23(b)]. The difference between the donor molecules BEDT-TTF and BEDO-TTF is also apparent in the structural properties of their neutral solids. A layer of BEDO-TTF molecules in neutral BEDO-TTF solid is depicted in Fig. 8.24, which shows that the donor molecules make optimum use of the C-H···O contacts. Neutral BEDO-TTF molecules have a bent π-framework, which leads to short C-H···O, C-H···S, and C-H···C(sp^2) contacts between layers of donor molecules.

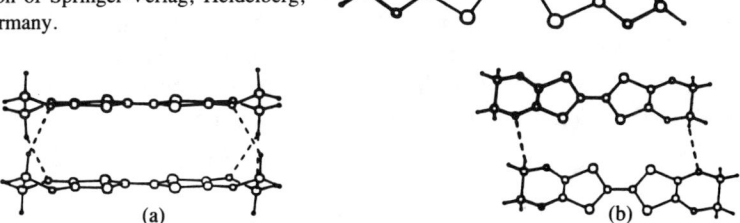

Figure 8.22 Projection view of two BEDO-TTF molecules in (BEDO-TTF)$_2$AuBr$_2$. From Whangbo et al.[12a] with permission of Springer-Verlag, Heidelberg, West Germany.

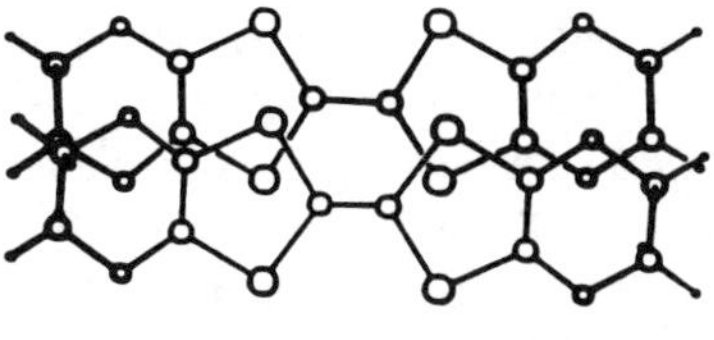

Figure 8.23 C-H···O contact interactions in (BEDO-TTF)$_2$AuBr$_2$: (a) within a donor stack and (b) between donor stacks. From Whangbo et al.[12a] with permission of Springer-Verlag, Heidelberg, West Germany.

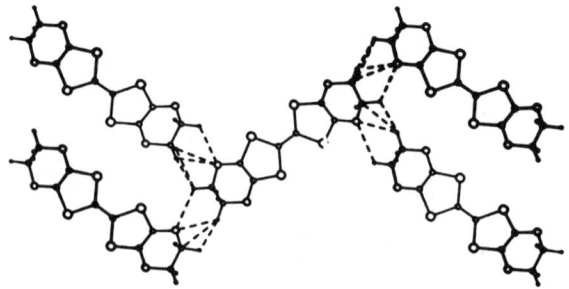

Figure 8.24 C-H⋯O contacts present within a layer of neutral BEDO-TTF solid. From Whangbo et al.[12a] with permission of Springer-Verlag, Heidelberg, West Germany.

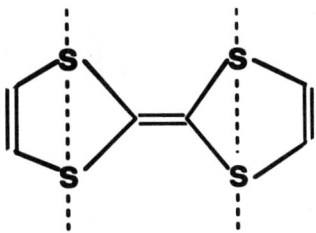

Figure 8.25 π-framework bending in TTF. From Novoa et al.[12c], with permission of Gordon and Breach Science Publishers, Inc. New York, N.Y.

Energetics of Conformational Change and Intermolecular Contact Interactions

Ab initio SCF-MO calculations on TTF (Fig. 8.25) and semiempirical MNDO SCF-MO calculations on BEDT-TTF and BEDT-TTF$^+$ (Fig. 8.18) as a function of the π-framework bending angle θ show[12] that BEDT-TTF is most stable when its π-framework is flat (θ = 0°), but the potential surface for the π-framework bending is very soft. The bending potential of BEDT-TTF$^+$ is calculated to be somewhat stiffer than that of BEDT-TTF. This finding is consistent with the observation that the π-frameworks of BEDT-TTF and BEDO-TTF are bent in their neutral solids but flat in their charge-transfer salts. Ab initio SCF-MO/MP2 calculations[12] on the model six-membered ring (Fig. 8.26), show that the staggered conformation Fig. 8.19(a) is the most stable one among Fig. 8.19(a)–(d), and a conformational change away from Fig. 8.19(a) does not require much energy as long as it avoids the fully eclipsed conformation Fig. 8.19(d). Consequently, the six-membered ring conformations other than Fig. 8.19(a) as well as the bent π-framework of the donor molecules must arise from external forces associated with short intermolecular contacts.

Figures 8.17 and 8.21 show that the short intermolecular contacts of importance are the C-H⋯donor and C-H⋯anion contacts. Though not shown in these figures, there exist short S⋯S contacts shorter than the van der Waals radii sum

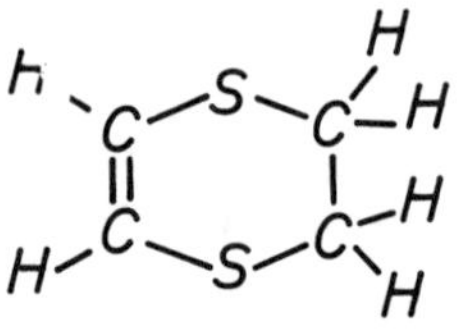

Figure 8.26 Model for a six-membered ring of BEDT-TTF. From Novoa et al.[12c], with permission of Gordon and Breach Science Publishers, Inc. New York, N.Y.

(3.60Å) between donor molecules. Ab initio SCF-MO/MP2 calculations[12] on $H_2X \cdots XH_2$ (X = S, Se) strongly suggest that $X \cdots X$ interactions are essentially nonbonding in the region of the van der Waals $X \cdots X$ contact distances. Short C-H$\cdots$donor contacts are grouped into three classes: the C-H$\cdots$heteroatom contacts (e.g., C-H$\cdots$O, C-H$\cdots$S), the C-H$\cdots$C(sp^2) contact, and the C-H$\cdots$H contact. Ab initio SCF-MO/MP2 calculations[12] on H_3C-H$\cdots XH_2$ (X = O, S, Se, Te, H_3C-H$\cdots CH_2$ = CH_2 and H_3C-H$\cdots$H-CH_3 show that all these C-H$\cdots$donor interactions are attractive. These C-H$\cdots$donor interactions are comparable in magnitude to the energies required for the π-framework bending on the six-membered ring conformational change. Since the donor molecules have numerous C-H$\cdots$donor contacts per molecule in the neutral solids and charge-transfer salts of BEDT-TTF and BEDO-TTF, the stabilization energy resulting from these C-H$\cdots$donor contacts is expected to be greater than the destabilization energy associated with the donor π-framework bending and the adoption of higher energy six-membered ring conformations. The C-H$\cdots X$ interactions energies are calculated[12] to decrease in the order C-H$\cdots$O > C-H$\cdots$S > C-H$\cdots$Se > C-H$\cdots$Te, which follows the electronegativity ordering of the X atoms. This trend in the C-H$\cdots X$ interaction energies explains why in a donor layer of neutral BEDO-TTF solid the molecules make optimum use of C-H$\cdots$O contacts and also why the BEDO-TTF salts (BEDO-TTF)$_2X$ (X^- = $AuBr_2^-$, ClO_4^-) have a donor packing arrangement in which the donor molecules have optimum use of C-H$\cdots$O contacts within a donor stack as well as between adjacent donor stacks. Ab initio SCF-MO/MP2 calculations[12] on H_3C-H$\cdots$Y-I-Y$^-$ (Y = I, Br, Cl) show that the nature of the C-H$\cdots$anion contact interaction is attractive, and, in general, the C-H$\cdots$anion contacts are energetically more favorable than the C-H$\cdots$donor contacts. This finding, along with the fact that the π-framework bending of the donor molecules becomes stiffer upon oxidation, accounts for the nearly planar π-frameworks of the donor molecules in the charge-transfer salts of BEDT-TTF and BEDO-TTF.

Structural Differences in TXF·TCNQ (X = S, Se, Te)

Projection views of the donor and the acceptor stacks of TTeF·TCNQ along the a- and b-axis directions are shown in Fig. 8.27(a) and (b), respectively.[32] TTeF·TCNQ has layers of donor molecules which have intermolecular chalcogen$\cdots$chalcogen contacts not only along the intrastack but also along the interstack directions. These donor layers alternate with layers of acceptor molecules. This mode of packing, typically found for most (TMTSF)$_2X$ and (BEDT-TTF)$_2X$ salts, is quite different from that of TXF·TCNQ (X = S, Se).[33] As shown in Fig. 8.28(a) and (b), the TXF·TCNQ (X = S, Se) salts have donor stacks separated by acceptor stacks so that there exists no short chalcogen$\cdots$chalcogen contact between donor stacks. In addition, the adjacent acceptor and donor molecules are locked in an "X"-type arrangement [Fig. 8.28 (b)]. Figure 8.29 shows a perspective view of the short intermolecular contacts between the donor and the acceptor molecules in TXF·TCNQ (X = S, Se). These contacts consist of the C-H(donor)$\cdots$N(sp) contacts, the C-H(acceptor)$\cdots X$ contacts, and the N(sp)$\cdots X$ contacts. Figure 8.30

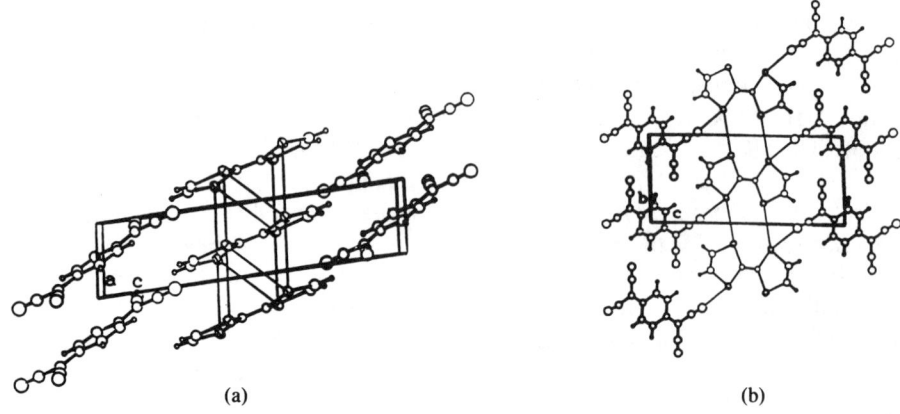

Figure 8.27 Projection views of the donor and acceptor stacks of TTeF·TCNQ along the *a*- and *b*-axis directions in (a) and (b), respectively. From Cowan et al.[32] with permission of Gordon and Breach Science Publishers, Inc. New York, N.Y.

shows a perspective view of the short intermolecular contacts between the donor and the acceptor molecules of TTeF·TCNQ. These contacts consist of the C-H(donor)···N(sp) contacts and the N(sp)···Te contacts.[32] Therefore, as in the case of the $(BEDT-TTF)_2X$ salts and their analogs, the intermolecular contacts shorter than the van der Waals radii sums in TXF·TCNQ (X = S, Se, Te) are dominated by those involving the C-H bonds. In TXF·TCNQ (X = S, Se), both the donor and the acceptor C-H bonds participate in short intermolecular contact interactions, while in TTeF·TCNQ only the donor C-H bonds do. Ab initio SCF-MO/MP2 calculations[32] on $H_3C-H \cdots N \equiv C-H$ and $H-C \equiv N \cdots XH_2$ (X = S, Se, Te) show that the C-H···N(sp) and N(sp)···Te contact interactions are attractive, but the N(sp)···S and N(sp)···Se

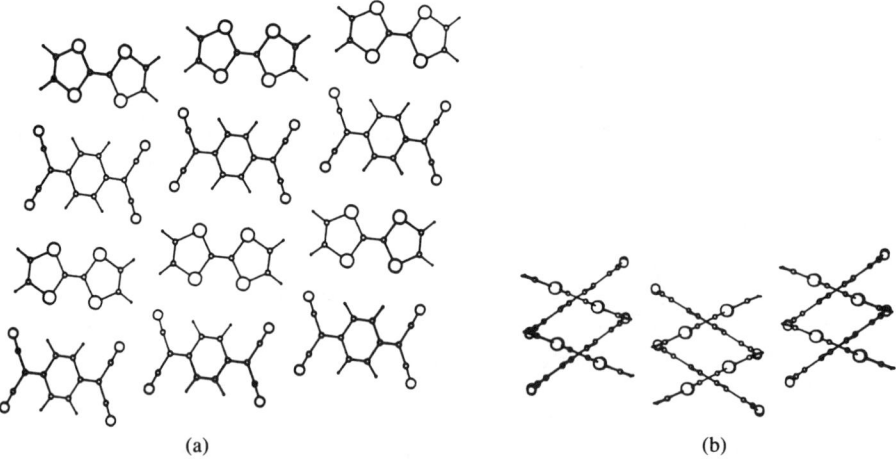

Figure 8.28 Projection views of the donor and acceptor stacks of TXF·TCNQ (X = S, Se) along the *b*- and *a*-axis directions in (a) and (b), respectively. From Cowan et al.[32] with permission of Gordon and Breach Science Publishers, Inc. New York, N.Y.

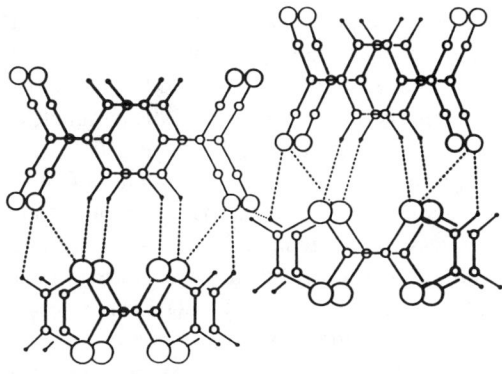

Figure 8.29 Perspective view of the short intermolecular C-H···X and N(sp)···X contacts in TXF·TCNQ (X = S, Se) (shown by dashed lines). From Cowan et al.[32] with permission of Gordon and Breach Science Publishers, Inc. New York, N.Y.

interactions are repulsive. The C-H···N(sp) interaction is calculated[32] to be more attractive than the N(sp)···Te interaction, which, in turn, is more attractive than the C-H···X (X = S, Se, Te) interaction. Therefore, the packing pattern of TTeF·TCNQ is dominated by the attractive C-H(donor)···N(sp) and N(sp)···Te interactions. Absence of such a packing pattern in TXF·TCNQ (X = S, Se) may be due in part to the nonattractive nature of the N(sp)···X (X = S, Se) interaction. The "X"-type locking between the adjacent donor and acceptor molecules of TXF·TCNQ (X = S, Se) [Fig. 8.28(b)] allows both the donor and acceptor molecule C-H bonds to engage in attractive intermolecular contact interactions.

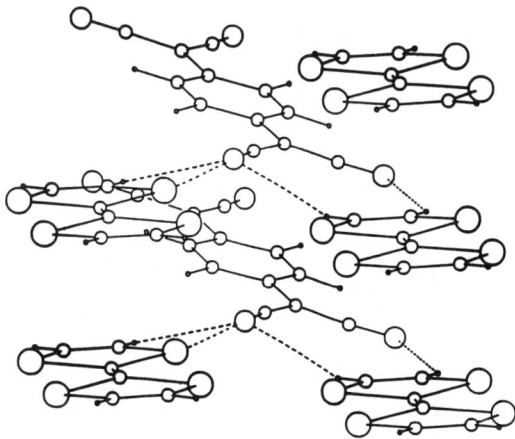

Figure 8.30 Perspective view of the short intermolecular C-H···Te and N(sp)···Te contacts in TTeF·TCNQ (shown by dashed lines). From Cowan et al.[32] with permission of Gordon & Breach Sci. Publ., Inc., New York, N.Y.

Thermal Phase Transitions

Provided that the temperature is not raised above 125 K, β^*-(BEDT-TTF)$_2$I$_3$ is stable even after the conversion pressure is released. Thermal tempering of α-(BEDT-TTF)$_2$I$_3$[34] leads to a new isomeric form α$_t$-(BEDT-TTF)$_2$I$_3$,[35] which is stable at room temperature and exhibits a superconducting transition near 8 K. Although no detailed crystal structure of α$_t$-(BEDT-TTF)$_2$I$_3$ is available, the similarities of the T_c's, the unit cell dimensions, and the symmetrical stretching modes of the I$_3^-$ ions in the Raman spectra strongly suggest that β*-(BEDT-TTF)$_2$I$_3$ and α$_t$-(BEDT-TTF)$_2$I$_3$ are probably very similar in structure. Conversion of α-(BEDT-TTF)$_2$I$_3$ to a β-like structure by thermal tempering implies that α-(BEDT-TTF)$_2$I$_3$ is thermodynamically less stable than either β-(BEDT-TTF)$_2$I$_3$ or β*-(BEDT-TTF)$_2$I$_3$. The stereodiagrams of how the ethylene group C-H bonds of one BEDT-TTF molecule in β*-(BEDT-TTF)$_2$I$_3$ and α-(BEDT-TTF)$_2$I$_3$ interact with the surrounding donor molecules and anions are shown in Figs. 8.21 and 8.31, respectively.[35d] In β*-(BEDT-TTF)$_2$I$_3$, three hydrogen atoms of each ethylene group make short C-H···anion contacts, while the remaining hydrogen atom makes short C-H···donor contacts. In α-(BEDT-TTF)$_2$I$_3$, two hydrogen atoms of each ethylene group have short C-H···anion contacts, and the other two hydrogen atoms C-H···donor contacts. For every ethylene group, therefore, β*-(BEDT-TTF)$_2$I$_3$ has more C-H···anion interactions than does α-(BEDT-TTF)$_2$I$_3$. Since the C-H···I interaction is more attractive than the C-H···S or the C-H···C(sp^2) interaction, as discussed in the previous section, it is expected[35e] that β*-(BEDT-TTF)$_2$I$_3$ is thermodynamically more stable than α-(BEDT-TTF)$_2$I$_3$. This is consistent with the conversion of α-(BEDT-TTF)$_2$I$_3$ to a β-like structure by thermal tempering.

ESR lineshape studies[36] show that δ-(BEDT-TTF)$_2$AuBr$_2$[37] is transformed into α'-(BEDT-TTF)$_2$AuBr$_2$[38] at 420 K. Thus, the latter is thermodynamically more stable. Figure 8.32(a) and (b) show projection views of two consecutive donor molecule pairs in a stack of δ-(BEDT-TTF)$_2$AuBr$_2$.[36] Both pairs have a BOB arrangement, but one pair has a mode-b donor slipping [Fig. 32(a)] while the other has a mode-c donor slipping [Fig. 32(b)]. Figure 8.33(a) and (b) show projection views of two consecutive donor molecule pairs in a donor stack of α'-(BEDT-

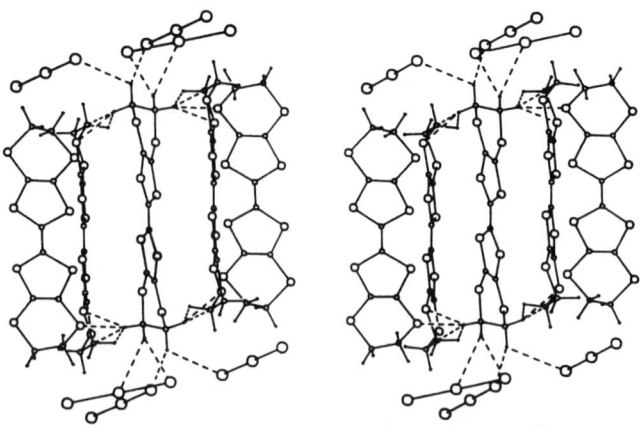

Figure 8.31 Stereodiagram that shows the short C-H···donor and C-H···anion contacts in α-(BEDT-TTF)$_2$I$_3$. From Wang et al.[35] with permission of the American Chemical Society, Washington, D.C.

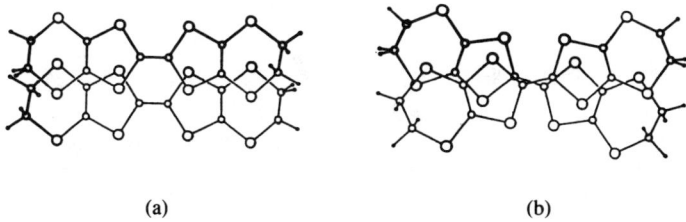

Figure 8.32 Projection views of donor molecule pairs in a stack of δ-(BEDT-TTF)$_2$AuBr$_2$: (a) the BOB/mode-b arrangement and (b) the BOB/mode-c arrangement. From Montgomery et al.[36] with the permission of Gordon and Breach. Sc. Publ. Inc., N.Y., N.Y.

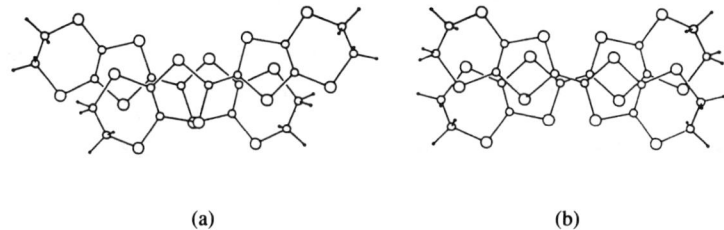

Figure 8.33 Projection views of donor molecule pairs in a stack of α'-(BEDT-TTF)$_2$AuBr$_2$: (a) the BOR/mode-c arrangement and (b) the BOB/mode-c arrangement. From Montgomery et al.[36] with the permission of Gordon and Breach. Sc. Publ. Inc., N.Y., N.Y.

TTF)$_2$AuBr$_2$. Both pairs have a mode-c donor slipping, but one pair has a BOR arrangement [Fig. 8.33(a)] while the other has a BOB arrangement [Fig. 8.33(b)]. Namely, the donor stack of δ-(BEDT-TTF)$_2$AuBr$_2$ has the BOB/mode-b and BOB/mode-c arrangements, and that of α'-(BEDT-TTF)$_2$AuBr$_2$ has the BOR/mode-c and BOB/mode-c arrangements. Thus an essential structural change associated with the δ- to α'-phase transition is the stacking rearrangement form BOB/mode-b to BOR/mode-c. The short C-H···Br contacts in δ-(BEDT-TTF)$_2$AuBr$_2$ and α'-(BEDT-TTF)$_2$AuBr$_2$ are shown in Figs. 8.34 and 8.35, respectively, where the H···Br contact distances less than the van der Waals radii sum (3.35Å) are shown by dashed lines. Clearly, the donor molecule has more short C-H···anion contacts in

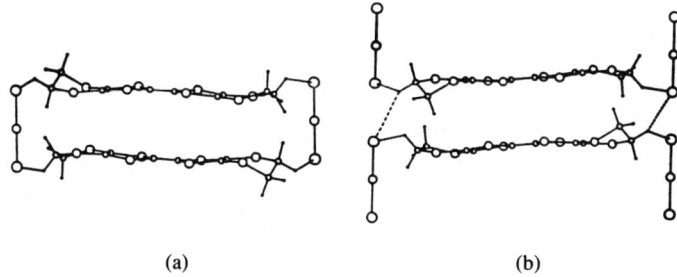

Figure 8.34 Short C-H···anion contacts in δ-(BEDT-TTF)$_2$AuBr$_2$ associated with (a) the BOB/mode-b and (b) the BOB/mode-c arrangements.

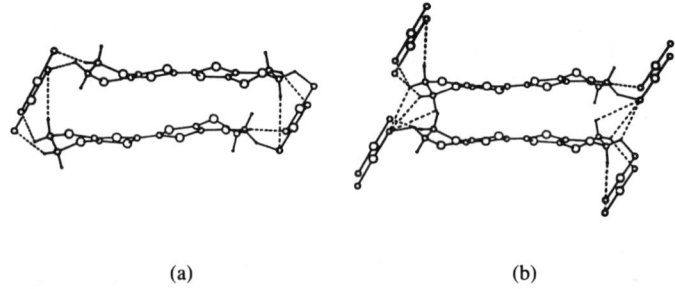

(a)   (b)

Figure 8.35 Short C-H···anion contacts in α'-(BEDT-TTF)$_2$AuBr$_2$ associated with (a) the BOR/mode-c and (b) the BOB/mode-c arrangements. From Montgomery et al.[36] with the permission of Gordon & Breach, Sc. Publ. Inc., N.Y., N.Y.

α'-(BEDT-TTF)$_2$AuBr$_2$ than in δ-(BEDT-TTF)$_2$AuBr$_2$. [Though not shown, the donor molecule has more short C-H···donor contacts in δ-(BEDT-TTF)$_2$AuBr$_2$.] Consequently, α'-(BEDT-TTF$_2$AuBr$_2$ is expected to be more stable than δ-(BEDT-TTF)$_2$AuBr$_2$, which is consistent with the thermal phase transition from the δ- to the α'-phase.

REFERENCES

1. Whangbo, M.-H. *J. Chem. Phys.* **1979**, *70*, 4963.
2. Whangbo, M.-H. *J. Chem. Phys.* **1980**, *73*, 3854.
3. Whangbo, M.-H. *J. Chem. Phys.* **1981**, *75*, 4983.
4. Whangbo, M.-H. *Electron Transfer in Biology and the Solid State: Inorganic Compounds with Unusual Properties*; Johnson, M. K.; King, R. B.; Kurtz, Jr. D. M.; Kutal, C.; Norton, M. L.; Scott, R. A., Eds.; American Chemical Society, Washington, DC, 1990; p. 269.
5. Whangbo, M.-H. *Crystal Structures and Properties of Materials with Quasi One-Dimensional Structures*, Rouxel, J. Ed., Reidel: Dordrecht, The Netherlands, 1986; p. 27.
6. Canadell, E.; Rachidi, I. E.-I; Ravy, S.; Pouget, J. P.; Brossard, L.; Legros, J. P. *J. Phys. France* **1989**, *50*, 2967.
7. Whangbo, M.-H.; Canadell, E. *Acc. Chem. Res.* **1989**, *22*, 375.
8. Bardeen, J.; Cooper, L. N.; Schrieffer, J. R. *Phys. Rev.* **1957**, *108*, 1175.
9. Solymar, L.; Walsh, D. *Lectures on the Electrical Properties of Materials*, Oxford University Press: Oxford, England. 1988; 4th Ed., Chap. 14.
10. McMillan, W. L. *Phys. Rev.* **1968**, *167*, 331.
11. (a) Whangbo, M.-H.; Williams, J. M.; Schultz, A. J.; Emge, T. J.; Beno, M. A. *J. Am. Chem. Soc.* **1987**, *109*, 90.
    (b) Whangbo, M.-H.; Williams, J. M.; Schultz, A. J.; Beno, M. A. *Organic and Inorganic Low-Dimensional Crystalline Materials*, Delhaes, P.; Drillon, M. Eds., Plenum, New York. **1987**; p. 333.
12. (a) Whangbo, M.-H.; Jung, D.; Ren, J.; Evain, M.; Novoa, J. J.; Mota, F.; Alvarez, S.; Williams, J. M.; Beno, M. A.; Kini, A. M.; Wang, H. H.; Ferraro, J. R. *The Physics*

and Chemistry of Organic Superconductors, Saito, G.; Kagoshima, S., Eds., Springer-Verlag: Heidelberg, Germany, 1990; p. 262.

(b) Whangbo, M.-H.; Novoa, J. J.; Jung, D.; Williams, J. M.; Kini, A. M.; Wang, H. H.; Geiser, U.; Beno, M. A.; Carlson, K. D. *Organic Superconductivity*, Kresin, V. Z.; Little, W. A., Eds., Plenum: New York, 1990; p. 243.

(c) Novoa, J. J.; Whangbo, M.-H.; Williams, J. M. *Mol. Cryst. Liq. Cryst.* **1990**, *181*, 25.

13. (a) Schirber, J. E.; Azevedo, L. J.; Kwak, J. F.; Venturini, E. L.; Leung, P. C. W.; Beno, M. A.; Wang, H. H.; Williams, J. M. *Phys. Rev. B: Condens. Matter* **1986**, *33*, 1987.

    (b) Schirber, J. E.; Azevedo, L. J.; Kwak, J. E.; Venturini, E. L.; Beno, M. A.; Wang, H. H.; Williams, J. M. *Solid State Commun.* **1986**, *59*, 525.

14. (a) Schultz, A. J.; Wang, H. H.; Williams, J. M.; Filhol, A. *J. Am. Chem. Soc.* **1986**, *108*, 7853.

    (b) Schultz, A. J.; Beno, M. A.; Wang, H. H.; Williams, J. M. *Phys. Rev. B: Condens. Matter* **1986**, *33*, 7823.

15. Nowack, A.; Weger, M.; Schweitzer, D.; Keller, H. J. *Solid State Commun.* **1986**, *60*, 199.

16. (a) Stewart, G. R.; O'Rourke, J.; Crabtree, G. W.; Carlson, K. D.; Wang, H. H.; Williams, J. M. *Phys. Rev. B.: Condens. Matter*, **1986**, *33*, 2046.

    (b) Stewart, G. R.; Williams, J. M.; Wang, H. H.; Hall, L. N.; Perozzo, M. T.; Carlson, K. D. *Phys. Rev. B: Condens. Matter*, **1986**, *34*, 6509.

17. Emge, T. J.; Wang, H. H.; Geiser, U.; Beno, M. A.; Webb, K. S.; Williams, J. M. *J. Am. Chem. Soc.* **1986**, *108*, 3849.

18. (a) Leung, P. C. W.; Emge, T. J.; Beno, M. A.; Wang, H. H.; Williams, J. M.; Petricek, V.; Coppens, P. *J. Am. Chem. Soc.* **1984**, *106*, 7644.

    (b) Leung, P. C. W.; Emge, T. J.; Beno, M. A.; Wang, H. H.; Williams, J. M.; Petricek, V.; Coppens, P. *J. Am. Chem. Soc.* **1985**, *107*, 6184.

19. (a) Urayama, H.; Yamochi, H.; Saito, G.; Nozawa, K.; Sugano, T.; Kinoshita, M.; Sato, S.; Oshima, K.; Kawamoto, A.; Tanaka, J. *Chem. Lett.* **1988**, 55.

    (b) Urayama, H.; Yamochi, H.; Saito, G.; Sato, S.; Kawamoto, A.; Tanaka, J.; Mori, T.; Maruyama, Y.; Inokuchi, H. *Chem. Lett.* **1988**, 463.

    (c) Gärtner, S.; Gogu, E.; Heinen, I.; Keller, H. J.; Klutz, T.; Schweitzer, D. *Solid State Commun.* **1988**, *65*, 1531.

    (d) Carlson, K. D.; Geiser, U.; Kini, A. M.; Wang, H. H.; Montgomery, L. K.; Kwok, W. K.; Beno, M. A.; Williams, J. M.; Cariss, C. S.; Crabtree, G. W.; Whangbo, M.-H.; Evain, M. *Inorg. Chem.* **1988**, *27*, 965.

20. Kobayashi, A.; Kato, R.; Kobayashi, H.; Moriyama, S.; Nishio, Y.; Kajita, K.; Sasaki, W. *Chem. Lett.* **1987**, 459.

21. (a) Papavassiliou, G. C.; Mousdis, G. A.; Zambounis, J. S.; Terzis, A.; Hountas, A.; Hilti, B.; Mayer, C. W.; Pfeiffer, J. *Synth. Met.* **1988**, *27*, B379.

    (b) Kini, A. M.; Beno, M. A.; Son, D.; Wang, H. H.; Carlson, K. D.; Porter, L. C.; Welp, U.; Vogt, B. A.; Williams, J. M.; Jung, D.; Evain, M.; Whangbo, M.-H.; Overmyer, D. L.; Schirber, J. E. *Solid State Commun.* **1989**, *69*, 503.

22. (a) Kikuchi, K.; Honda, Y.; Ishikawa, Y.; Saito, K.; Ikemoto, I.; Murata, K.; Anzai, H.; Ishiguro, T.; Kobayashi, K. *Solid State Commun.* **1988**, *66*, 405.

    (b) Kikuchi, K.; Ishikawa, Y.; Saito, K.; Ikemoto, I.; Kobayashi, K. *Synth. Met.* **1988**, *27*, B391.

23. Whangbo, M.-H.; Jung, D.; Wang, H. H.; Beno, M. A.; Williams, J. M.; Kikuchi, K. *Mol. Cryst. Liq. Cryst.* **1990**, *181*, 1.
24. Nigrey, P. J.; Morosin, B.; Kwak, J. F.; Venturini, E. L.; Baughman, R. J. *Synth. Met.* **1986**, *16*, 1.
25. Nigrey, P. J.; Morosin, B.; Kwak, J. F. *Novel Superconductivity*, Wolf, S. A.; Kresin, V. Z., Eds., Plenum: New York, 1985; p. 171.
26. (a) Lyubovskaya, R. N.; Lyubovskii, R. B.; Shibaeva, R. P.; Aldoshina, M. Z.; Goldenberg, L. M.; Rozenberg, L. P.; Khidekel, M. L.; Shulpyakov, Yu. F. *JETP Lett.* **1985**, *42*, 468.
    (b) Lyubovskii, R. B.; Lyubovskaya, R. N.; Kapustin, N. V. *Sov. Phys. JETP* **1987**, *66*, 1063.
    (c) Shibaeva, R. P.; Rozenberg, L. P. *Sov. Phys. Crystallogr.* **1988**, *33*, 834.
27. Jung, D.; Evain, M.; Novoa, J. J.; Whangbo, M.-H.; Beno, M. A.; Kini, A. M.; Schultz, A. J.; Williams, J. M.; Nigrey, P. J. *Inorg. Chem.* **1989**, *28*, 4516.
28. (a) Sugano, T.; Hayashi, H.; Kinoshita, M.; Nishikida, K. *Phys. Rev. B: Condens. Matter* **1989**, *B39*, 11387.
    (b) Tokumoto, M.; Anzai, H.; Takahashi, K.; Kinoshita, N.; Murata, K.; Ishiguro, T.; Tanaka, Y.; Hayakawa, Y.; Nagamori, H.; Nagasaka, K. *Synth. Met.* **1988**, *27*, A171.
    (c) Oshima, K.; Urayama, H.; Yamochi, H.; Saito, G. *J. Phys. Soc. Jpn.* **1988**, *57*, 730.
29. Ferraro, J. R.; Wang, H. H.; Geiser, U.; Kini, A. M.; Beno, M. A.; Williams, J. M.; Hill, S.; Whangbo, M.-H.; Evain, M. *Solid State Commun.* **1988**, *68*, 917.
30. Kobayashi, H.; Kobayashi, A.; Sasaki, Y.; Saito, G.; Inokuchi, H. *Bull Chem. Soc. Jpn.* **1986**, *59*, 301.
31. (a) Beno, M. A.; Wang, H. H.; Carlson, K. D.; Kini, A. M.; Frankenbach, G. M.; Ferraro, J. R.; Larsen, N.; McCabe, G. D.; Thompson, J.; Purnama, C.; Vashon, M.; Williams, J. M.; Jung, D.; Whangbo, M.-H. *Mol. Cryst. Liq. Cryst.* **1990**, *181*, 145.
    (b) Beno, M. A.; Kini, A. M.; Geiser, U.; Wang, H. H.; Carlson, K. D.; Williams, J. M. *The Physics and Chemistry of Organic Superconductors*, Saito, G.; Kagoshima, S., Eds., Springer-Verlag, Berlin (**1990**), *51*, 369.
32. Cowan, D. O.; Mays, M. D.; Kistenmacher, T. J.; Poehler, T. O.; Beno, M. A.; Kini, A. M.; Williams, J. M.; Kwok, Y. K.; Carlson, K. D.; Xiao, L.; Novoa, J. J.; Whangbo, M.-H. *Mol. Cryst. Liq. Cryst.* **1990**, *181*, 43.
33. (a) Kistenmacher, T. J.; Phillips, T. E.; Cowan, D. O. *Acta Crystallogr.* **1974**, *B30*, 763.
    (b) The crystal structure of TSF·TCNQ is isostructural with that of TTF·TCNQ (Grant, P. M., private communication).
34. (a) Bender, K.; Dietz, K.; Endres, H.; Helberg, H. W.; Hennig, I.; Keller, H. J.; Schäefer, H. W.; Schweitzer, D. *Mol. Cryst. Liq. Cryst.* **1984**, *107*, 45.
    (b) Emge, T. J.; Leung, P. C. W.; Beno, M. A.; Wang, H. H.; Williams, J. M. *Mol. Cryst. Liq. Cryst.* **1986**, *138*, 393.
35. (a) Baram, G. O.; Buravov, L. I.; Degtyarev, L. S.; Kozlov, M. É.; Laukhin, V. N.; Laukhina, E. É.; Onishchenko, V. G.; Pokhodnya, K. I.; Sheinkman, M. K.; Shibaeva, R. P.; Yagubskii, É. B. *JETP Lett.* **1986**, *44*, 376.
    (b) Schweitzer, D.; Bele, P.; Brunner, H.; Gogu, E.; Haeberlen, U.; Hennig, I.; Klutz, I.; Swietlik, R.; Keller, H. J. *Z. Phys. B: Condens. Matter* **1987**, *67*, 489.
    (c) Schweitzer, D.; Keller, H. J. *Organic and Inorganic Low-Dimensional Crystalline Materials*, Delhaes, P.; Drillon, M., Eds., Plenum: New York, 1987, p. 219.
    (d) Swietlik, R.; Schweitzer, D.; Keller, H. J. *ibid.*, p. 325.
    (e) Wang, H. H.; Ferraro, J. R.; Carlson, K. D.; Montgomery, L. K.; Geiser, U.;

Williams, J. M.; Whitworth, J. R.; Schlueter, J. A.; Hill, S.; Whangbo, M.-H.; Evain, M.; Novoa, J. J. *Inorg. Chem.* **1989**, *28*, 2267.

36. Montgomery, L. K.; Wang, H. H.; Schlueter, J. A.; Geiser, U.; Carlson, K. D.; Williams, J. M.; Rubinstein, R. L.; Brennan, T. D.; Stupka, D. L.; Whitworth, J. R.; Jung, D.; Whangbo, M.-H. *Mol. Cryst. Liq. Cryst.* **1990**, *181*, 197.

37. Mori, T.; Sakai, F.; Saito, G.; Inokuchi, H. *Chem. Lett.* **1986**, 1589.

38. Beno, M. A.; Firestone, M. A.; Leung, P. C. W.; Sowa, L. M.; Wang, H. H.; Williams, J. M.; Whangbo, M.-H. *Solid State Commun.* **1986**, *57*, 735.

39. Kasowski, R. V.; Whangbo, M.-H. *Inorg. Chem.* **1990**, *29*, 360.

40. Kini, A. M.; Geiser, U.; Wang, H. H.; Carlson, K. D.; Williams, J. M.; Kwok, W. K.; Vandervoort, K. G.; Thompson, J. E.; Stupka, D.; Jung, D.; Whangbo, M.-H. *Inorg. Chem.*, **1990**, *29*, 2555.

41. Williams, J. M.; Kini, A. M.; Geiser, U.; Wang, H. H.; Carlson K. D.; Kwok, W. K.; Vandervoort, K. G.; Thompson, J. E.; Stupka, D. L.; Jung, D.; Whangbo, M.-H. *Organic Superconductivity*, Kresin, V. Z.; Little, W. A., Eds., Plenum: New York, **1990**; pp. 33–44.

42. Williams, J. M.; Wang, H. H.; Emge, T. J.; Geiser, U.; Beno, M. A., Leung, P. C. W.; Carlson, K. D.; Thorn, R. J.; Schultz, A. J.; and Whangbo, M.-H. *Prog. Inorg. Chem.*, S. J. Lippard, Ed. John Wiley & Sons, N.Y., N.Y., **1987**, *35*, 51–218.

Appendix A
Bibliography of Charge-Transfer Salts, Organic Superconductors, and Conductors

In an effort to aid the interested reader in the subject matter of the book, the authors have compiled a Bibliography of publications on Organic Superconductors. The list comprises four categories. The first category relates to publications involving the donor molecule ET and its salts, as well as ET variant salts and selenium analogues of ET. Category II contains publications relative to complexes involving DMIT, dithiolates and other conductive materials containing elemental metals. Publications in Category III include salts of new and novel donor molecules, as well as unsymmetrical donor molecules. Category IV pertains to publications of BO salts, where BO is an oxygen analogue of ET.

Papers from several proceedings of International Conferences also appear in this index. These are listed as follows:

1. International Conference on Science and Technology of Synthetic Metals (ICSM), Kyoto, Japan, June 1-6, 1986 and Proceedings appearing in *Synth. Met.*, 1987, *19*.
2. ICSM Meeting held in Santa Fe, NM, June 26-July 2, 1988, and Proceedings appearing in *Synth. Met.*, 1988, 27-29.

3. International Chemical Congress of Pacific Basin Societies held in Honolulu, Hawaii, December 17-22, 1989, and Proceedings appearing in *Mol. Cryst. Liq. Cryst.*, 1990, *181*.

The reader is also directed to the references in the book entitled, "Introduction to Synthetic Electrical Conductors," by J. R. Ferraro and J. M. Williams, Academic Press, 1987.

Appendix A

ET*

CONDUCTIVITY STUDIES

Normal-pressure Superconductivity in an Organic Metal (BEDT-TTF)$_2$I$_3$ [bis(ethylenedithiolo)tetrathiofulvalene Triiodide]
Yagubskii, E. B.; Shchegolev, I. F.; Laukhin, V. N.; Kononovich, P. A.; Kartsovnik, M. V.; Zvarykina, A. V.; Buravov, L I
Pis'ma Zh. Eksp. Teor. Fiz. **1984**, *39*, 12; JETP Lett. **1984**, *39*, 12

Superconductivity at Ambient Pressure in Di[bis(ethylenedithio)tetrathiafulvalene]triiodide, (BEDT-TTF)$_2$I$_3$
Crabtree, G. W.; Carlson, K. Douglas; Hall, L. N.; Copps, P. Thomas; Wang, H. H.; Emge, T. J.; Beno, M. A.; Williams, J. M.
Phys. Rev. B **1984**, *30*, 2958

Superconducting Properties of β-(BEDT-TTF)$_2$I$_3$ Crystals
Ginodman, V. B.; Gudenko, A. V.; Zherikhina, L. N.
Pis'ma Zh. Eksp. Teor. Fiz. **1985**, *41*, 41

*Includes variants of ET such as MT, PT as well as selenium analogues.

High T_c Superconducting State in (BEDT-TTF)$_2$ Trihalides
Murata, K.; Tokumoto, M.; Bando, H.; Tanino, H.; Anzai, H.; Kinoshita, N.; Kajimura, K.; Saito, G.; Ishiguro, T.
Physica **1985**, *135B*, 515

An Increase in the Superconducting-Transition Temperature of β-(BEDT-TTF)$_2$I$_3$ to 6–7 at a Normal Pressure
Merzhanov, V. A.; Kostyuchenko, E. E.; Laukhin, V. N.; Lobkovskaya, R. M.; Makova, M. K.; Shibaeva, R. P.; Shchegolev, I. F.; Yagubskii, E. B.
Pis'ma Zh. Eksp. Teor. Fiz. **1985**, *41*, 146; *JETP Lett.* **1985**, *41*, 179

Superconductivity at Normal Pressures of Some Organic Metals of (BEDT-TTF)-I System
Shchegolev, I. F.; Yagubskii, E. B.; Laukhin, V. N.
Mol. Cryst. Liq. Cryst. **1985**, *126*, 365

Superconductivity in a (BEDT-TTF) Organic Conductor with Chloromercurate Anion
Lyubovskaya, R. N.; Lyubovskii, R. B.; Shibaeva, R. P.; Aldoshina, M. Z.; Gol'denberg, L. M.; Rozenberg, L. P.; Khidekel', M. L.; Shul'pyakov, Y. F.
Pis'ma Zh. Eksp. Teor. Fiz. **1985**, *42*, 380; *JETP Lett.* **1985**, *42*, 468

Characteristics of Conducting Properties of Highly Defective β-(BEDT-TTF)$_2$I$_3$ Crystals
Pokhodnya, K. I.; Sushko, Yu. V.; Tanatar, M. A.; Bykov, V. N.; Sheinkman, M. K.
Sov. Phys. Sol. State **1987**, *29*, 505

A Stable Superconducting State at 8K and Ambient Pressure in α_t-(BEDT-TTF)$_2$I$_3$
Schweitzer, D.; Bele, P.; Brunner, H.; Gogu, E.; Haeberlen, U.; Hennig, I.; Klutz, I.; Swietlik, R.; Keller, H. J.
Z. Phys. B - Condens. Matt. **1987**, *B67*, 489

Superconductivity in Polycrystalline Pressed Samples of Organic Metals
Schweitzer, D.; Gärtner, S.; Grimm, H.; Gogu, E.; Keller, H. J.
Solid State Commun. **1989**, *69*, 843

2K-Superconducting State with Less Disorder in the Organic Superconductor β-(BEDT-TTF)$_2$I$_3$
Kagoshima, S.; Hasumi, M.; Nogami, Y.; Kinoshita, N.; Anzai, H.; Tokumoto, M.; Saito, G.
Solid State Commun. **1989**, *71*, 843

Ambient-Pressure Superconductivity at the Highest Temperature (5K) Observed in an Organic System: β-(BEDT-TTF)$_2$AuI$_2$
Wang, H. H.; Beno, M. A.; Geiser, U.; Firestone, M. A.; Webb, K. S.; Nuñez, L.; Crabtree, G. W.; Carlson, K. D.; Williams, J. M.; Azevedo, L. J.; Kwak, J. F.; Schirber, J. E.
Inorg. Chem. **1985**, *24,* 2465

Normal- and Superconducting Properties of (BEDT-TTF)$_2$Cu(SCN)$_2$
Veith, H.; Heidmann, C.-P.; Müller, H.; Fritz, H. P.; Andres, K.; Fuchs, H.
Synth. Met. **1988**, *27,* A361

Superconductivity and Deuteration Effect in (BEDT-TTF)$_2$Cu(NCS)$_2$
Oshima, K.; Urayama, H.; Yamochi, H.; Saito, G.
Synth. Met. **1988**, *27,* A473

Effect of Alloying on the Superconductivity in Organic Metals β-(BEDT-TTF)$_2$-Trihalides
Tokumoto, M.; Anzai, H.; Murata, K.; Kajimura, K.; Ishiguro, T.
Synth. Met. **1988**, *27,* A251

Resistivity Anomaly of a New Phase of the Organic Crystal (BEDT-TTF)$_2$I$_3$
Lu, L.; Ma, B.; Duan, H.; Lin, S.; Zhang, D.; Wang, X.; Zhu, D.
Synth. Met. **1988**, *27,* A311

Competition Between Localization and Superconductivity in (BEDT-TTF)$_3$Cl$_2$·2H$_2$O
Kurmoo, M.; Rosseinsky, M. J.; Day, P.; Auban, P.; Kang, W.; Jérome, D.; Batail, P.
Synth. Met. **1988**, *27,* A425

Electrical Resistivity of DBTTF, DBTSF and TTF Salts with Dialkyltin(IV) Chloride Anions
Matsubayashi, G.; Shimizu, R.; Tanaka, T.
Synth. Met. **1987**, *19,* 715

Transport Studies of Several Novel Organic Conductors
Kwak, J. F.; Azevedo, L. J.; Schirber, J. E.; Williams, J. M.; Beno, M. A.
Mol. Cryst. Liq. Cryst. **1985**, *125,* 365

A Novel Type of Organic Semiconductors. Molecular Fastener
Inokuchi, H.; Saito, G.; Wu, P.; Seki, K.; Tang, T.; Mori, T.; Imaeda, K.; Enoki, T.; Higuchi, Y.; Inaka, K.; Yasuoka, N.
Chem. Lett. **1986**, 1265

New Organic Superconductors κ- and θ-(BEDT-TTF)$_2$I$_3$: Transport Property
Kajita, K.; Nishio, Y.; Moriyama, S.; Sasaki, W.; Kato, R.; Kobayashi, H.; Kobayashi, A.
Solid State Commun. **1987**, *64*, 1279

High T_c Superconducting State of β-(BEDT-TTF)$_2$I$_3$
Murata, K.; Tokumoto, M.; Anzai, H.; Bando, H.; Kajimura, K.; Ishiguro, T.
Synth. Met. **1987**, *19*, 151

Two Superconducting Phases in the Organic Conductor: β-(BEDT-TTF)$_2$I$_3$
Creuzet, F.; Bourbonnais, C.; Creuzet, G.; Jérome, D.; Schweitzer, D.; Keller, H. J.
Synth. Met. **1987**, *19*, 157

Study of Volume Superconductivity in β-(ET)$_2$X Superconductors
Schwenk, H.; Parkin, S. S. P.; Lee, V. Y.; Greene, R. L.
Synth. Met. **1987**, *19*, 163

The Electrical Anisotropy of the Organic Metals (BEDT-TTF)$_3$(ClO$_4$)$_2$ and α-(BEDT-TTF)$_3$(NO$_3$)$_2$
Koch, B.; Geserich, H. P.; Ruppel, W.; Schweitzer, D.; Keller, H. J.
Synth. Met. **1987**, *19*, 179

Frequency-dependent Conductivity in Pure and Iodine-doped α-(BEDT-TTF)$_2$I$_3$
Przybylski, M.; Helberg, H. W.; Schweitzer, D.; Keller, H. J.
Synth. Met. **1987**, *19*, 191

Measurements of the Microwave Conductivity of the Organic Superconductor ET$_2$(IAuI)
Tanner, D. B.; Jacobsen, C. S.; Williams, J. M.; Wang, H. H.
Synth. Met. **1987**, *19*, 197

Superconducting Transition at T_c = 3.5 K in the High-pressure Phase of the Organic Metal (BEDT-TTF)$_4$Hg$_3$Cl$_2$
Lyubovskii, R. B.; Lyubovskaya, R. N.; Kapustin, N. V.
Sov. Phys. JETP **1987**, *66*, 1063

Simple Recipe for Formation or Recovery at Ambient Pressure of the 8K Superconducting State of β-(BEDT-TTF)$_2$I$_3$
Schirber, J. E.; Azevedo, L. J.; Kwak, J. E.; Venturini, E. L.; Beno, M. A.; Wang, H. H.; Williams, J. M.
Solid State Commun. **1986**, *59*, 525

CRITICAL FIELDS/CRITICAL CURRENTS

Critical Field Anisotropy in an Organic Superconductor β-(BEDT-TTF)$_2$I$_3$
Tokumoto, M.; Bando, H.; Anzai, H.; Saito, G.; Murata, K.; Kajimura, K.; Ishiguro, T.
J. Phys. Soc. Jpn. **1985**, *54*, 869

Critical Field Anisotropy in an Organic Superconductor β-(BEDT-TTF)$_2$IBr$_2$
Tokumoto, M.; Anzai, H.; Bando, H.; Saito, G.; Kinoshita, N.; Kajimura, K.; Ishiguro, T.
J. Phys. Soc. Jpn. **1985**, *54*, 1669

Nature of the High-Temperature Superconducting State with T_c = 7-8 K in β-(BEDT-TTF)$_2$I$_3$
Zvarykina, A. V.; Kononovich, P. A.; Laukhin, V. N.; Molchanov, V. N.; Pesotskii, S. I.; Simonov, V. I.; Shibaeva, R. P.; Shchegolev, I. F.; Yagubskii, E. B.
Pis'ma Zh. Eksp. Teor. Fiz. **1986**, *43*, 257; *JETP Lett.* **1986**, *43*, 329

Peculiar Critical Field Behaviour in the Recently Discovered Ambient Pressure Organic Superconductor (BEDT-TTF)$_2$Cu(NCS)$_2$ (T_c = 10.4 K)
Oshima, K.; Urayama, H.; Yamochi, H.; Saito, G.
J. Phys. Soc. Japan **1988**, *57*, 730

Critical Field Anisotropy in "2K-Superconducting State" of Organic Superconductor β-(BEDT-TTF)$_2$I$_3$
Sasaki, T.; Toyota, N.; Hasumi, M.; Osada, T.; Kagoshima, S.; Anzai, H.; Tokumoto, M.; Kinoshita, N.
J. Phys. Soc. Jpn. **1989**, *58*, 3477

Superconducting Critical Field in (BEDT-TTF)$_2$Cu(NCS)$_2$
Oshima, K.; Urayama, H.; Yamochi, H.; Saito, G.
Synth. Met. **1988**, *27*, A419

ELECTROCHEMISTRY

Voltammetric Study of Organic Metals. I. The Determination of the Electrochemical Conditions for Crystal Growth
Sakura, S.; Imai, H.; Anzai, H.; Moriya, T.
Bull. Chem. Soc. Jpn. **1988**, *61*, 3181

Elektrochemische Untersuchungen von BEDT-TTF (Electrochemical Investigations of BEDT-TTF)
Müller, H.; Fritz, H. P.; Lerf, A.; Besenhard, J. O.
Z. Naturforsch **1989**, *44b*, 1199

ESR/EPR

Comparative ESR Study of Three (BEDT-TTF):ReO$_4$ Salts: An Organic Superconductor, A Peierls Metal and a Semiconductor
Carneiro, K.; Scott, J. C.; Engler, E. M.
Solid State Commun. **1984**, *50*, 477

ESR Study of Two Phases of Di[bis(ethylenedithio)tetrathiafulvalene]-Triiodide [(BEDT-TTF)$_2$I$_3$]
Venturini, E. L.; Azevedo, L. J.; Schirber, J. E.; Williams, J. M.; Wang, H. H.
Phys. Rev. B. **1985**, *32,* 2819

Magnetic Order in Organic Superconductors
Azevedo, L. J.; Venturini, E. L.; Schirber, J. E.; Williams, J. M.; Wang, H. H.; Emge, T. J.
Mol. Cryst. Liq. Cryst. **1985**, *119,* 389

Anisotropy in ESR g Factors and Linewidths for α- and β-(BEDT-TTF)$_2$I$_3$
Kinoshita, N.; Tokumoto, M.; Anzai, H.; Saito, G.
J. Phys. Soc. Jpn. **1985**, *54,* 4498

15. An ESR Study of the Radical Cations of Tetrathiafulvalene (TTF) and Electron Donors Containing the TTF Moiety
Cavara, L.; Gerson, F.; Cowan, D. O.; Lerstrup, K.
Helv. Chim. Acta. **1986**, *69,* 141

ESR Studies of Organic Conductors with Bis(ethylenedithio)tetrathiafulvalene (BEDT-TTF), (BEDT-TTF)$_2$ClO$_4$(C$_2$H$_3$Cl$_3$)$_{0.5}$, and (BEDT-TTF)$_3$(ClO$_4$)$_2$, and their Two-dimensionality
Enoki, T.; Imaeda, K.; Kobayashi, M.; Inokuchi, H.; Saito, G.
Phys. Rev. B. **1986**, *33,* 1553

Conduction-electron-spin Resonance in Organic Conductors: α and β Phases of di[bis(ethylenedithiolo)tetrathiafulvalene]triiodide [(BEDT-TTF)$_2$I$_3$]
Sugano, T.; Saito, G.; Kinoshita, M.
Phys. Review B **1986**, *34,* 117

Spin Relaxation and Diffusion in Quasi-two-dimensional Organic Metals: The Bis-(ethylenedithiolo)tetrathiafulvalene Compounds β-(BEDT-TTF)$_2$X (X = I$_3$ and IBr$_2$)
Sugano, T.; Saito, G.; Kinoshita, M.
Phys. Rev. B **1987**, *35,* 6554

ESR and Electrical Properties of a New Organic Metal: α-(BEDT-TTF)$_2$Cu(NCS)$_2$
Kinoshita, N.; Takahashi, K.; Murata, K.; Tokumoto, M.; Anzai, H.
Solid State Commun. **1988**, *67,* 465

ESR Study of a New Phase of Organic Conductor (BEDT-TTF)$_2$I$_3$
Ma, B.; Lu, L.; Zhang, D.; Wang, X.; Zhu, D.
Solid State Commun. **1988**, *68,* 433

Conduction Electron Spin Resonance in Organic Conductors α- and β-(ET)$_2$IBr$_2$
Romanyukha, A. A.; Shvachko, Yu. N.; Skripov, A. V.; Stepanov, A. P.
Phys. Stat. Sol. (b) **1988**, *147,* K185

Static and ESR Susceptibilities of the Ambient Pressure Superconductors β- and $α_t$-(BEDT-TTF)$_2$I$_3$ and (BEDT-TTF)$_2$Cu(SCN)$_2$
Klotz, S.; Schilling, J. S.; Gärtner, A.; Schweitzer, D.
Solid State Commun. **1988**, *67*, 981

Conduction Electron Spin Resonance in the New Organic Superconductor (ET)$_2$Cu(SCN)$_2$
Romanyukha, A. A.; Shvachko, Yu. N.; Skripov, A. V.; Ustinov, V. V.
Phys. Stat. Sol. (b) **1989**, *151*, K59

Effects of Electron Irradiation on ESR Properties of β-(ET)$_2$I$_3$
Shvachko, Yu. N.; Romanyukha, A. A.; Skripov, A. V.; Ustinov, V. V.
Phys. Stat. Sol. (b) **1989**, *151*, K65

New Organic Superconductors
Inokuchi, H.
Angew. Chem. Int. Ed. Engl. **1988**, *27*, 1747

ESR Studies of Single Crystals of Bis(ethylenedithio)tetrathiafulvalene-Based Conductor with Octaiodinemercurate Anion (BEDT-TTF)$_4$Hg$_3$I$_8$
Firlej, L.; Graja, A.; Wolak, J.; Lyubovskaya, R. N.; Goldenberg, L. M.
Phys. Stat. Sol. **1989**, *154*, 333

Spin-Peierls Transition in the Organic Complex (BPDT-TTF)$_2$I$_3$
Kushch, N. D.; Merzhanov, V. A.; Romanyukha, A. A.
JETP Lett. **1989**, *69*, 205

ESR Studies of Two BEDT-TTF Based Organic Superconductors, θ-(BEDT-TTF)$_2$I$_3$ and κ-(BEDT-TTF)$_4$Hg$_{2.89}$Br$_8$
Wang, H. H.; Vogt, B. A.; Geiser, U.; Beno, M. A.; Carlson, K. D.; Kleinjan, S.; Thorup, N.; Williams, J. M.
Mol. Cryst. Liq. Cryst. **1990**, *181*, 135

ESR Study of Optically Enhanced Phase Transition in (BEDT-TTF)$_2$Ta$_2$F$_{11}$
Acrivos, J. V.; Hughes, H. P.; Parkin, S. S. P.
J. Chem. Phys. **1987**, *86*, 1780

ESR in the β$_H$ and β$_L$ Phases of a β-(BEDT-TTF)$_2$I$_3$ Crystal
Hurdequint, H.; Creuzet, F.; Jérome, D.
Synth. Met. **1988**, *27*, A183

Effects of Anion Size and Disorder on Spin Resonance in the BEDT-TTF Salts
Venturini, E. L.; Schirber, J. E.; Wang, H. H.; Williams, J. M.
Synth. Met. **1988**, *27*, A243

ESR Studies of the 10.4 K Ambient-Pressure Organic Superconductor, κ-(BEDT-TTF)$_2$Cu(NCS)$_2$
Wang, H. H.; Montgomery, L. K.; Kini, A. M.; Carlson, K. D.; Beno, M. A.; Geiser, U.; Cariss, C. S.; Williams, J. M.; Venturini, E. L.
Physica C **1988**, *156*, 173

Conduction-electron-spin Resonance in the Organic Conductors (BEDT-TTF)$_2$X (X = I$_3$, IBr$_2$)
Sugano, T.; Saito, G.; Kinoshita, M.
Synth. Met. **1987**, *19*, 209

MAGNETIC STUDIES

Superconducting Properties of β-(BEDT-TTF)$_2$I$_3$ Crystals
Ginodman, V. B.; Gudenko, A. V.; Zherikhina, L. N.
Pis'ma Zh. Eksp. Teor. Fiz. **1985**, *41*, 41; *JETP Lett.* **1985**, *41*, 49

Observation of the Meissner Effect in the High-T_c (Pressure-) Phase of the Organic Superconductor β-(BEDT-TTF)$_2$I$_3$
Veith, H.; Heidmann, C.-P.; Gross, F.; Lerf, A.; Andres, K.; Schweitzer, D.
Solid State Commun. **1985**, *56*, 1015

Diamagnetic Evidence of Superconductivity in the Organic Conductors β-(BEDT-TTF)$_2$IBr$_2$ and β-(BEDT-TTF)$_2$AuI$_2$
Heidmann, C.P.; Veith, H.; Andres, K.; Fuchs, H.; Polborn, K.; Amberger, E.
Solid State Commun. **1986**, *57*, 161

Ordinary and Anomalous Magnetoresistances in β-(BEDT-TTF)$_2$X (X = IBr$_2$, I$_2$Br, I$_3$)
Bando, H.; Tokumoto, M.; Murata, K.; Anzai, H.; Saito, G.; Kajimura, K.; Ishiguro, T.
J. Phys. Soc. Jpn. **1985**, *54*, 4265

Magnetic Properties of the Organic Superconductor β-(BEDT-TTF)$_2$AuI$_2$
Talham, D. R.; Kurmoo, M.; Day, P.; Obertelli, D. S.; Parker, I. D.; Friend, R. H.
J. Phys. C: Solid State Phys. **1986**, *19*, L383

The High-T_c Superconducting State of β-(BEDT-TTF)$_2$I$_3$ at Atmospheric Pressure: Bulk Superconductivity and Metastability
Creuzet, F.; Jerome, D.; Schweitzer, D.; Keller, H. J.
Europhys. Lett. **1986**, *1*, 461

Antiferromagnetic Resonance in the BEDT-TTF Series
Coulon, C.; Laversanne, R.; Amiell, J.; Delhaes, P.
J. Phys. C: Solid State Phys. **1986**, *19*, L753

Study of the Superconducting Transitions in Two Modifications of β-(BEDT-TTF)$_2$I$_3$ on the Basis of the Diamagnetic-Screening Signal
Kartsovnik, M. V.; Laukhin, V. N.; Shchegolev, I. F.
Pis'ma Zh. Eksp. Teor. Fiz. **1986**, *90*, 2172; *JETP Lett.* **1986**, *63*, 1273

Critical Magnetic Fields of β-(BEDT-TTF)$_2$I$_3$ in the High-temperature Superconducting Phase
Bulaevskii, L. N.; Ginodman, V. B.; Gudenko, A. V.
Pis'ma Zh. Eksp. Teor. Fiz. **1987**, *45*, 355; *JETP Lett.* **1987**, *45*, 452

Exceeding the Paramagnetic Limit of H$_{c2}$ in an Organic Superconductor β-(ET)$_2$I$_3$ with $T_c = 7.1$ K
Laukhin, V. K.; Pesotskii, S. I.; Yagubskii, E. B.
Pis'ma Zh. Eksp. Teor. Fiz. **1987**, *45*, 394; *JETP Lett.* **1987**, *45*, 502

Meissner Effect in an Organic Superconductor (BEDT-TTF)$_2$[Cu(NCS)$_2$]
Nozawa, K.; Sugano, T.; Urayama, H.; Yamochi, H.; Saito, G.; Kinoshita, M.
Chem. Lett. **1988**, 617

A BEDT-TTF Complex Including a Magnetic Anion, (BEDT-TTF)$_3$(MnCl$_4$)$_2$
Mori, T.; Inokuchi, H.
Bull. Chem. Soc. Jpn. **1988**, *61*, 591

Time-effects in the Low-field Magnetization of a β-(BEDT-TTF)$_2$I$_3$ Single Crystal
Mota, A. C.; Visani, P.; Pollini, A.; Juri, G.; Jerome, D.
Physica C **1988**, *153*, 1153

Bulk Superconductivity in (BEDT-TTF)$_2$[Cu(NCS)$_2$]
Sugano, T.; Terui, K.; Mino, S.; Nozawa, K.; Urayama, H.; Yamochi, H.; Saito, G.; Kinoshita, M.
Chem. Lett. **1988**, 1171

Anisotropy of Magnetoresistance and the Shubnikov-de Haas Oscillations in the Organic Metal β-(ET)$_2$IBr$_2$
Kartsovnik, M. V.; Kononovich, P. A.; Laukhin, V. N.; Shchegolev, I. F.
Pis'ma Zh. Eksp. Teor. Fiz. **1988**, *48*, 498; *JETP Lett.* **1988**, *39*, 541

Observation of Giant Magnetoresistance Oscillations in the High-T_c Phase of the Two-Dimensional Organic Conductor β-(BEDT-TTF)$_2$I$_3$
Kang, W.; Montambaux, G.; Cooper, J. R.; Jérome, D.; Batail, P.; Lenoir, C.
Phys. Rev. Lett. **1989**, *62*, 2559

Anomalous Field Dependence of the Nuclear-spin Relaxation Time in the Organic Conductor (ET)$_4$Hg$_{2.89}$Br$_8$
Skripov, A. V.; Stepanov, A. P.; Merzhanov, V. A.; Lyubovskaya, R. N.; Lyubovskii, R. B.
Pis'ma Zh. Eksp. Teor. Fiz. **1989**, *49*, 229; *JETP Lett.* **1989**, *49*, 265

A New Type Oscillatory Phenomenon in the Magnetotransport of θ-(BEDT-TTF)$_2$I$_3$
Kajita, K.; Nishio, Y.; Takahashi, T.; Sasaki, W.; Kato, R.; Kobayashi, H.; Kobayashi, A.; Iye, Y.
Solid State Commun. **1989**, *70*, 1189

Anomalous Magnetotransport Phenomena in θ-(BEDT-TTF)$_2$I$_3$
Kajita, K.; Nishio, Y.; Takahashi, T.; Sasaki, W.; Kato, R.; Kobayashi, H.; Kobayashi, A.
Solid State Commun. **1989**, *70*, 1181

Magnetoresistance of the Organic Metal β_L-(ET)$_2$I$_3$: Angular Dependence and Shubnikov-de Haas Oscillations
Kartsovnik, M. V.; Kononovich, P. A.; Laukhin, V. N.; Pesotskii, S. I.; Shchegolev, I. F.
Pis'ma Zh. Eksp. Teor. Fiz. **1989**, *49*, 453; *JETP Lett.* **1989**, *49*, 519

DC Hall Effect of the Low-T_c State of β-(BEDT-TTF)$_2$I$_3$
Murata, K.; Ishibashi, M.; Honda, Y.; Tokumoto, M.; Kinoshita, N.; Anzai, H.
J. Phys. Soc. Jpn. **1989**, *58*, 3469

Flux Jumps, Critical Fields, and de Haas-van Alphen Effect in κ-(BEDT-TTF)$_2$Cu(NCS)$_2$
Swanson, A. G.; Brooks, J. S.; Anzai, H.; Konoshita, N.; Tokumoto, M.; Murata, K.
Solid State Commun. **1990**, *73*, 353

Anisotropy of Magnetization and Meissner Effect in Organic Superconductor κ-(BEDT-TTF)$_2$Cu(NCS)$_2$
Tokumoto, M.; Anzai, H.; Takahashi, K.; Murata, K.; Kinoshita, N.; Ishiguro, T.
Synth. Met. **1988**, *27*, A305

Magnetic and Optical Properties of an Ambient-pressure Organic Superconductor (BEDT-TTF)$_2$[Cu(NCS)$_2$]
Sugano, T.; Nozawa, K.; Hayashi, H.; Nishikida, K.; Terui, K.; Fukasawa, T.; Takenouchi, H.; Mino, S.; Urayama, H.; Saito, G.; Kinoshita, M.
Synth. Met. **1988**, *27*, A325

Low Dimensional Magnetic Behaviour of the Organic Conductors, α'-(BEDT-TTF)$_2$X, X = AuBr$_2$, CuCl$_2$, and Ag(CN)$_2$
Obertelli, S. D.; Friend, R. H.; Talham, D. R.; Kurmoo, M.; Day, P.
Synth. Met. **1988**, *27*, A375

Observation of de Haas-Shubnikov and de Haas-van Alphen Oscillations in β-(BEDT-TTF)$_2$AuI$_2$
Parker, I. D.; Pigram, D. D.; Friend, R. H.; Kurmoo, M.; Day, P.
Synth. Met. **1988**, *27*, A387

Magnetic Penetration Depth in the Organic Superconductor κ-(BEDT-TTF)$_2$-Cu(NCS)$_2$
Harshman, D. R.; Kleiman, R. N.; Haddon, R. C.; Chichester-Hicks, S. V.; Kaplan, M. L.; Rupp, Jr., L. W.; Pfiz, T.; Williams, D. L.; Mitzi, D. B.
Phys. Rev. Lett. **1990**, *64*, 1293

MULTIPLE DISCIPLINE STUDIES

Transverse Conduction and Metal-Insulator Transition in β-(BEDT-TTF)$_2$PF$_6$
Kobayashi, H.; Mori, T.; Kato, R.; Kobayashi, A.; Sasaki, Y.; Saito, G.; Inokuchi, H.
Chem. Lett. **1983**, 581

Crystal Structures and Electrical Properties of BEDT-TTF Compounds
Kobayashi, H.; Kato, R.; Mori, T.; Kobayashi, A.; Sasaki, Y.; Saito, G.; Enoki, T.; Inokuchi, H.
Mol. Cryst. Liq. Cryst. **1984**, *107*, 33

The Crystal Structures and Electrical Resistivities of (BEDT-TTF)$_3$(ClO$_4$)$_2$ and (BEDT-TTF)$_2$ClO$_4$(C$_4$H$_8$O$_2$)
Kobayashi, H.; Kato, R.; Mori, T.; Kobayashi, A.; Sasaki, Y.; Saito, G.; Enoki, T.; Inokuchi, H.
Chem. Lett. **1984**, 179

Crystal Structure and Electrical Conductivity of (BPDT-TTF)$_3$(PF$_6$)$_2$
Kato, R.; Mori, T.; Kobayashi, A.; Sasaki, Y.; Kobayashi, H.
Chem. Lett. **1984**, 781

Superconducting Properties of the Orthorhombic Phase of Bis-(ethylenedithiol)tetrathiafulvalene Triiodide
Yagubskii, E. B.; Shchegolev, I. F.; Pesotskii, S. I.; Laukhin, V. N.; Kononovich, P. A.; Kartsovnik, M. V.; Zvarykina, A. V.
Pis'ma Zh. Eksp. Teor. Fiz. **1984**, *39*, 275; *JETP Lett.* **1984**, *39*, 328

Synthetic Metals Based on Bis(ethylenedithio)tetrathiafulvalene (BEDT-TTF): Synthesis, Structure and Ambient-Pressure Superconductivity in (BEDT-TTF)$_2$I$_3$
Williams, J. M.; Emge, T. J.; Wang, H. H.; Beno, M. A.; Copps, P. Thomas; Hall, L. N.; Carlson, K. Douglas; Crabtree, G. W.
Inorg. Chem. **1984**, *23*, 2558

The Crystal Structure and Electrical Resistivity of (BPDT-TTF)$_2$I$_3$
Kobayashi, H.; Takahashi, M.; Kato, R.; Kobayashi, A.; Sasaki, Y.
Chem. Lett. **1984**, 1331

Neutron and X-Ray Diffraction Evidence for a Structural Phase Transition in the Sulfur-based Ambient-pressure Organic Superconductor Bis(ethylenedithio)-tetrathiafulvalene Triiodide
Emge, T. J.; Leung, P. C. W.; Beno, M. A.; Schultz, A. J.; Wang, H. H.; Sowa, L. M.; Williams, J. M.
Phys. Rev. B **1984**, *30*, 6780

$(BEDT-TTF)_2^+ J_3^-$: A Two-Dimensional Organic Metal
Bender, K.; Dietz, K.; Endres, H.; Helberg, H. W.; Hennig, I.; Keller, H. J.; Schafer, H. W.; Schweitzer, D.
Mol. Cryst. Liq. Cryst. **1984**, *107*, 45

New, Organic, Volume Superconductor at Ambient Pressure
Schwenk, H.; Heidmann, C. P.; Gross, F.; Hess, E.; Andres, K.; Schweitzer, D.; Keller, H. J.
Phys. Rev. B **1985**, *31*, 3138

Crystal Structure and Reflection Spectra of $(DBTTF)_2$-(Cu_2Br_6) Complex
Tanaka, M.; Honda, M.; Katayama, C.; Fujimoto, H.; Tanaka, J.
Chem. Lett. **1985**, 219

Chemical and Physical Properties of Cation Radical Salts of BEDT-TTF
Saito, G.; Enoki, T.; Kobayashi, M.; Imaeda, K.; Sato, N.; Inokuchi, H.
Mol. Cryst. Liq. Cryst. **1985**, *119*, 393

Crystal Structures of Organic Metals and Superconductors (BEDT-TTF)-I System
Shibaeva, R. P.; Kaminskii, V. F.; Yagubskii, E. B.
Mol. Cryst. Liq. Cryst. **1985**, *119*, 361

Superconducting and Electrical Properties of $(BEDT-TTF)_2I_3$ at Ambient Pressure
Carlson, K. Douglas; Crabtree, G. W.; Hall, L. N.; Copps, P. Thomas; Wang, H. H.; Emge, T. J.; Beno, M. A.; Williams, J. M.
Mol. Cryst. Liq. Cryst. **1985**, *119*, 357

Crystal Structures and Electrical Properties of BEDT-TTF Compounds
Kobayashi, H.; Kato, R.; Mori, T.; Kobayashi, A.; Sasaki, Y.; Saito, G.; Enoki, T.; Inokuchi, H.
Mol. Cryst. Liq. Cryst. **1984**, *107*, 33

Thermoelectric and Magnetic Properties of the α- and β-modifications of $(BEDT-TTF)_2I_3$
Merzhanov, V. A.; Kostyuchenko, E. E.; Faber, O. E.; Shchegolev, I. F.; Yagubskii, E. B.
Zh. Eksp. Teor. Fiz. **1985**, *89*, 292; *JETP Lett.* **1985**, *62*, 165

Crystal Structures and Electrochemical Properties of Organic Donors, BMDT-TTF and BEDSe-TSeF. Two Modifications of BEDT-TTF
Kato, R.; Kobayashi, H.; Kobayashi, A.; Sasaki, Y.
Chem. Lett. **1985**, 1231

X-Ray Study of Compressibilities of β-(BEDT-TTF)$_2$I$_3$ under Hydrostatic Pressure
Tanino, H.; Kato, K.; Tokumoto, M.; Anzai, H.; Saito, G.
J. Phys. Soc. Jpn. **1985**, *54*, 2390

The Crystal and Electronic Structures of (BEDT-TTF)$_2$I$_2$Br
Kobayashi, H.; Kato, R.; Kobayashi, A.; Saito, G.; Tokumoto, M.; Anzai, H.; Ishiguro, T.
Chem. Lett. **1985**, 1293

Ambient Presure Superconductivity at 4–5 K in β-(BEDT-TTF)$_2$AuI$_2$
Carlson, K. D.; Crabtree, G. W.; Nuñez, L.; Wang, H. H.; Beno, M. A.; Geiser, U.; Firestone, M. A.; Webb, K. S.; Williams, J. M.
Solid State Commun. **1986**, *57*, 89

Synthesis, Electrical Properties, and Crystal Structure of the First Organic Metal-Solid Electrolyte Hybrid: (BEDT-TTF)$_3$Ag$_x$I$_8$ ($x \sim 6.4$)
Geiser, U.; Wang, H. H.; Donega, K. M.; Anderson, B. A.; Williams, J. M.; Kwak, J. F.
Inorg. Chem. **1986**, *25*, 401

Crystal and Band and Electronic Structures of Orthorhombic γ″-(BEDT-TTF)$_2$AuI$_2$
Geiser, U.; Wang, H. H.; Beno, M. A.; Firestone, M. A.; Webb, K. S.; Williams, J. M.
Solid State Commun. **1986**, *57*, 741

New Cation-Anion Interaction Motifs, Electronic Band Structure, and Electrical Behavior in β-(ET)$_2$X Salts ($X = $ ICl$_2^-$ and BrICl$^-$)
Emge, T. J.; Wang, H. H.; Leung, P. C. W.; Rust, P. R.; Cook, J. D.; Jackson, P. L.; Carlson, K. D.; Williams, J. M.; Whangbo, M.-H.; Venturini, E.L.; Schirber, J. E.; Azevedo, L. J.; Ferraro, J. R.
J. Am. Chem. Soc. **1986**, *108*, 695

A New Molecular Superconductor, (BEDT-TTF)$_2$(I$_3$)$_{1-x}$(AuI$_2$)$_x$ ($x < 0.02$)
Kobayashi, H.; Kato, R.; Kobayashi, A.; Nishio, Y.; Kajita, K.; Sasaki, W.
Chem. Lett. **1986**, 789

Crystal and Electronic Structures of Layered Molecular Superconductor, θ-(BEDT-TTF)$_2$(I$_3$)$_{1-x}$(AuI$_2$)$_x$
Kobayashi, H.; Kato, R.; Kobayashi, A.; Nishio, Y.; Kajita, K.; Sasaki, W.
Chem. Lett. **1986**, 833

Structural Characterization and Band Electronic Structure of α-(BEDT-TTF)$_2$I$_3$ Below its 135 K Phase Transition
Emge, T. J.; Leung, P. C. W.; Beno, M. A.; Wang, H. H.; Williams, J. M.
Mol. Cryst. Liq. Cryst. **1986**, *138*, 393

Far-Infrared and Raman Spectroscopic Studies of Polyiodides
Nour, E. M.; Chen, L. H.; Laane, J.
J. Phys. Chem. **1986**, *90*, 2841

Magnetic Susceptibility and ESR of the Organic Conductor Bis(ethylenedithiolo)tetrathiafulvalene Perchlorate [(BEDT-TTF)$_3$(ClO$_4$)$_2$]: Evidence For A Peierls Transition
Parkin, S. S. P.; Miljak, M.; Cooper, J. R.
Phys. Rev. B **1986**, *34*, 1485

Superconductivity In Sulfur-Based Organic Superconductors: A Volume Property
Schwenk, H.; Parkin, S. S. P.; Lee V. Y.; Greene, R. L.
Phys. Rev. B **1986**, *34*, 3156

Crystal And Electronic Structures of δ-(BEDT-TTF)$_2$AuI$_2$
Kobayashi, A.; Kato, R.; Kobayashi, H.; Tokumoto, M.; Anzai, H.; Ishiguro, T.
Chem. Lett. **1986**, 1117

Crystal Structure and Electrical Properties of an Organic Conductor δ-(BEDT-TTF)$_2$AuBr$_2$
Mori, T.; Sakai, F.; Saito, G.; Inokuchi, H.
Chem. Lett. **1986**, 1589

Transport Property of a Newly Synthesized Organic Conductor, β''-(BEDT-TTF)$_2$AuBr$_2$
Kajita, K.; Nishio, Y.; Moriyama, S.; Sasaki, W.; Kato, R.; Kobayashi, H.; Kobayashi, A.
Solid State Commun. **1986**, *60*, 811

Critical Magnetic Fields and Related Properties of the Ambient Pressure Organic Superconductor β-(BEDT-TTF)$_2$IBr$_2$
Nuñez, L.; Carlson, K. D.; Hall, L. N.; Crabtree, G. W.; Perozzo, M. T.; Wang, H. H.; Williams, J. M.
Physica **1986**, *143B*, 369

β-(ET)$_2$X Organic Superconductors: Linear-Anion Control of Structure and Ambient-Pressure Superconductivity, and the First Point-Contact Tunnelling Studies of β-(ET)$_2$AuI$_2$ Implying Extremely Strong Coupling
Williams, J. M.; Schultz, A. J.; Wang, H. H.; Carlson, K. D.; Beno, M. A.; Emge, T. J.; Geiser, U.; Hawley, M. E.; Gray, K. E.; Venturini, E. L.; Kwak, J. F.; Azevedo, L. J.; Schirber, J. E.; Whangbo, M.-H.
Physica **1986**, *143B*, 346

Study of the Superconducting Transitions in Two Modifications of β-(BEDT-TTF)$_2$I$_3$
Kartsovnik, M. V.; Laukhin, V. N.; Shchegolev, I. F.
Pis'ma Zh. Eksp. Teor. Fiz. **1986**, *90*, 2172; JETP Lett. **1986**, *63*, 1273

X-Ray Investigations of the Low-Temperature Phases of the Organic Metals α- and β-(BEDT-TTF)$_2$I$_3$
Endres, H.; Keller, H. J.; Swietlik, R.; Schweitzer, D.; Angermund, K.; Krüger, C.
Z. Naturforsch **1986**, *41a*, 1319

Structure and Properties of a New Conducting Organic Charge-Transfer-Salt β-(BEDT-TTF)$_2$AuBr$_2$
Kurmoo, M.; Talham, D. R.; Day, P.; Parker, I. D.; Friend, R. H.; Stringer, A. M.; Howard, J. A. K.
Solid State Commun. **1987**, *61*, 459

Crystal Structure and Physical Properties of (BMDT-TTF)SbF$_6$
Kato, R.; Kobayashi, H.; Kobayashi, A.
Chem. Lett. **1986,** 2013

The Assignment of Fundamental Vibrations of BEDT-TTF and BEDT-TTF-d$_8$
Kozlov, M. E.; Pokhodnia, K. I.; Yurchenko, A. A.
Spectrochimica Acta. **1987**, *43A*, 323

The Conductivity Anisotropy of the Quasi-two-dimensional Organic Metal β-(BEDT-TTF)$_2$I$_3$
Buravov, L. I.; Kartsovnik, M. V.; Kononovich, P. A.; Laukhin, V. N.; Pesotskii, S. I.; Shchegolev, I. F.
Pis'ma Zh. Eksp. Teor. Fiz. **1986**, *91*, 2198; JETP Lett. **1986**, *64*, 1306

Is the Organic Metal (ET)$_4$Hg$_3$Br$_8$ A Quasi-2D Superconductor?
Lyubovskaya, R. N.; Zhilyaeva, E. A.; Zvarykina, A. V.; Laukhin, V. N.; Lyubovskii, R. B.; Pesotskii, S. I.
Pis'ma Zh. Eksp. Teor. Fiz. **1987**, *45*, 416; JETP Lett. **1987**, *45*, 530

Iodocuprate Bis(ethylenedithio)tetrathiafulvalene (BEDT-TTF)$_2$Cu$_5$I$_6$—The First Quasi-two-dimensional Organic Metal with a Polymer Anion Layer
Buravov, L. I.; Zvarykina, A. V.; Kartsovnik, M. V.; Kushch, N. D.; Laukhin, V. N.; Lobkovskaya, R. M.; Merzhanov, V. A.; Fedutin, L. N.; Shibaeva, R. P.; Yagubskii, E. B.
Pis'ma Zh. Eksp. Teor. Fiz. **1987**, *92*, 594; JETP Lett. **1987**, *65*, 336

A Novel Conducting Charge-Transfer Salt: (BEDT-TTF)$_3$Cl$_2$·2H$_2$O
Rosseinsky, M. J.; Kurmoo, M.; Talham, D. R.; Day, P.; Chasseau, D.; Watkin, D.
J. Chem. Soc. Chem. Commun. **1988,** 88

(BEDT-TTF)$_2$CuCl$_2$, A New Conducting Charge Transfer Salt
Kurmoo, M.; Talham, D. R.; Day, P.; Howard, J. A. K.; Stringer, A. M.; Obertelli, D. S.; Friend, R. H.
Synth. Met. **1988**, *22*, 415

Structural and Electrical Properties of (BEDT-TTF)$_3$Cl$_2$(H$_2$O)$_2$
Mori, T.; Inokuchi, H.
Chem. Lett. **1987**, 1657

Crystal Structure and Electrical Properties of (BEDT-TTF)$_3$Br$_2$(H$_2$O)$_2$
Urayama, H.; Saito, G.; Kawamoto, A.; Tanaka, J.
Chem. Lett. **1987**, 1753

Singularities of the Phase States and of the Metal-insulator Phase Transition in the α-(BEDT-TTF)$_2$I$_3$ System
Pokhodnya, K. I.; Sushko., Y. V.; Tanatar, M. A.
Pis'ma Zh. Eksp. Teor. Fiz. **1987**, *92*, 1414; *JETP Lett.* **1987**, *65*, 795

Superconductivity of (ET)$_4$Hg$_{2.89}$Br$_8$ at Atmospheric Pressure and $T_c = 4.3$ K and the Critical Field Anisotropy
Lyubovskaya, R. N.; Zhilyaeva, E. I.; Pesotskii, S. I.; Lyubovskii, R. B.; Atovmyan, L. O.; D'yachenko, O. A.; Takhirov, T. J.
Pis'ma Zh. Eksp. Teor. Fiz. **1987**, *46*, 149; *JETP Lett.* **1987**, *46*, 188

E.S.R. and Electrical Properties of the Organic Metal (BEDT-TTF)$_4$Hg$_3$Cl$_8$
Sekretarczyk, G.; Graja, A.
Synth. Met. **1988**, *24*, 161

Origin of the Anomaly in the D.C. Conductivity in (BEDT-TTF)$_x$ TCNQ$_y$ Salt
Firlej, L.; Graja, A.; Wolak, J.; Eremenko, O. N.
Synth. Met. **1988**, *24*, 157

Syntheses, Crystal Structures and Physical Properties of Conducting Salts (BEDT-TTF)$_2$X (X = CF$_3$SO$_3$ or ρ-CH$_3$C$_6$H$_4$SO$_3$)
Chasseau, D.; Watkin, D.; Rosseinsky, M. J.; Kurmoo, M.; Talham, D. R.; Day, P.
Synth. Met. **1988**, *24*, 117

A New Ambient Pressure Organic Superconductor Based on BEDT-TTF with T_c Higher than 10 K ($T_c = 10.4$ K)
Urayama, H.; Yamochi, H.; Saito, G.; Nozawa, K.; Sugano, T.; Kinoshita, M.; Sato, S.; Oshima, K.; Kawamoto, A.; Tanaka, J.
Chem. Lett. **1988**, 55

On Physical Properties of the Organic Metal BEDT-TTF$_4$(Hg$_{3-\delta}$Cl$_8$)
Sekretarczyk, G.; Graja, A.; Pichet, J.; Lyubovskaya, R. N.; Lyubovskii, R. B.
J. Phys. France **1988**, *49*, 653

Superconductivity at 10 K and Ambient Pressure in the Organic Metal (BEDT-TTF)$_2$Cu(SCN)$_2$
Gärtner, S.; Gogu, E.; Heinen, I.; Keller, H. J.; Klutz, T.; Schweitzer, D.
Solid State Commun. **1988**, *65*, 1531

Temperature and Pressure Dependence of the Resistivity of β-(BEDT-TTF)$_2$X (X = I$_3$, I$_2$Au) and α$_t$-(BEDT-TTF)$_2$I$_3$
Weger, M.; Bender, K.; Klutz, T.; Schweitzer, D.; Gross, F.; Heidmann, C. P.; Probst, Ch.; Andres, K.
Synth. Met. **1988**, *25*, 49

NMR and EPR Studies of the Organic Conductor [dimethyl(ethylenedithio)-diselenadithiafulvalene]$_2$Au(CN)$_2$: Evidence of a Spin-density-wave Transition
Kanoda, K.; Takahashi, T.; Tokiwa, T.; Kikuchi, K.; Saito, K.; Ikemoto, I.; Kobayashi, K.
Phys. Rev. B **1988**, *38*, 39

Shubnikov-de Haas Effect and the Fermi Surface in an Ambient-pressure Organic Superconductor [bis(ethylenedithiolo)tetrathiafulvalene]$_2$Cu(NCS)$_2$
Oshima, K.; Mori, T.; Inokuchi, H.; Urayama, H.; Yamochi, H.; Saito, G.
Phys. Rev. B **1988**, *38*, 938

Magnetoresistance in β-(BEDT-TTF)$_2$I$_3$ and β-(BEDT-TTF)$_2$IBr$_2$: Shubnikov-de Haas Effect
Murata, K.; Toyota, N.; Honda, Y.; Sasaki, T.; Tokumoto, M.; Bando, H.; Anzai, H.; Muto, Y.; Ishiguro, T.
J. Phys. Soc. Jpn. **1988**, *57*, 1540

A New Ambient Pressure Organic Superconductor (BEDT-TTF)$_2$Cu(NCS)$_2$ with T_c Above 10K
Oshima, K.; Urayama, H.; Yamochi, H.; Saito, G.
Physica C **1988**, *153*, 1148

Valence State of Copper Atoms and Transport Property of an Organic Superconductor, (BEDT-TTF)$_2$Cu(NCS)$_2$, Measured by ESCA, ESR, and Thermoelectric Power
Urayama, H.; Yamochi, H.; Saito, G.; Sugano, T.; Kinoshita, M.; Inabe, T.; Mori, T.; Maruyama, Y.; Inokuchi, H.
Chem. Lett. **1988**, 1057

ESR and Electrical Properties of a New Organic Metal α-(BEDT-TTF)$_2$Cu(NCS)$_2$
Kinoshita, N.; Takahashi, K.; Murata, K.; Tokumoto, M.; Anzai, H.
Solid State Commun. **1988**, *67*, 465

High T_c Superconducting States in Organic Metals, β-(BEDT-TTF)$_2$X
Tokumoto, M.; Anzai, H.; Murata, K.; Kajimura, K.; Ishiguro, T.
Jpn. J. Appl. Phys. **1987**, *26*, 1977

Transverse Magnetoresistance and Shubnikov-de Haas Oscillations in the Organic Superconductor β-$(ET)_2IBr_2$
Kartsovnik, M. V.; Laukhin, V. N.; Nizhankovskii, V. I.; Ignat'ev, A. A.
Pis'ma Zh. Eksp. Teor. Fiz. **1988**, *47*, 302; *JETP Lett.* **1988**, *47*, 363

Quantum Oscillations and Negative Magnetoresistance in the Organic Metal β''-$(BEDT-TTF)_2AuBr_2$
Pratt, F. L.; Fisher, A. J.; Hayes, W.; Singleton, J.; Spermon, S. J. R. M.; Kurmoo, M.; Day, P.
Phys. Rev. Lett. **1988**, *61*, 2721

The Non-equivalence of the Iodine Atoms in the I_3^- Anion
Gutsev, G. L.; Shul'ga, Yu. M.
Russian J. Inorg. Chem. **1988**, *33*, 597

Anomalous Behavior of ^1H Spin-Lattice Relaxation Time in β-$(BEDT-TTF)_2I_3$- Possibility of Superconducting Glass
Maniwa, Y.; Takahashi, T.; Takigawa, M.; Yasuoka, H.; Saito, G.; Murata, K.; Tokumoto, M.; Anzai, H.
J. Phys. Soc. Jpn. **1989**, *58*, 1048

Effect of Twinning on the Superconducting Transition Temperature in β-$(ET)_2X$ Organic Metals
Zvarykina, A. V.; Kartsovnik, M. V.; Laukhin, V. N.; Laukhina, E. E.; Lyubovskii, R. B.; Pesotskii, S. I.; Shibaeva, R. P.; Shchegolev, I. F.
Pis'ma Zh. Eksp. Teor. Fiz. **1988**, *94*, 277; *JETP Lett.* **1988**, *67*, 1891

Charge Transfer Salts Obtained with Organic Donors (TTF, TMTTF, TMTSF and BEDT-TTF) and Tetracyanometallate Planar Dianions $[M(CN)_4^{2-}, M = Pt^{11}, Ni^{11}]$
Ouahab, L.; Padiou, J.; Grandjean, D.; Garrigou-Lagrange, C.; Delhaes, P.; Bencharif, M.
J. Chem. Soc. Chem. Commun. **1989**, 1038

The Magnetic Susceptibility and EPR of the Organic Conductors α-$(BEDT-TTF)_2X$, $X = AuBr_2$, $CuCl_2$ and $Ag(CN)_2$
Obertelli, S. D.; Friend, R. H.; Talham, D. R.; Kurmoo, M.; Day, P.
J. Phys. Condens. Matter, **1989**, *1*, 5671

Paramagnetic Insulator-Metal and Metal-Superconductor Transitions in a Quasi-two-dimensional Organic Complex $(BEDT-TTF)_2Cu(NCS)_2$
Buravov, L. I.; Zvarykina, A. V.; Kushch, N. D.; Laukhin, V. N.; Merzhanov, V. A.; Khomenko, A. G.; Yagubskii, E. B.
Zh. Eksp. Teor. Fiz. **1989**, *95*, 322; *JETP Lett.* **1989**, *68*, 182

BEDT-TTF Salts with Square Platinates(II) as Counterions: [BEDT-TTF]$_4$[Pt(C$_2$O$_4$)$_2$], A New Organic Metal
Gärtner, S.; Heinen, I.; Schweitzer, D.; Nuber, B.; Keller, H. J.
Synth. Met. **1989**, *31*, 199

New Molecular Conductors, α- and β-(EDT-TTF)[Ni(dmit)$_2$] Metal with Anomalous Resistivity Maximum vs. Semiconductor with Mixed Stacks
Kato, R.; Kobayashi, H.; Kobayashi, A.; Naito, T.; Tamura, M.; Tajima, H.; Kuroda, H.
Chem. Lett. **1989**, 1839

Superconducting Phase α$_t$-(BEDT-TTF)$_2$I$_3$ and Its Formation
Kozlov, M. E.; Baran, N. P.; Pokhodnya, K. I.
Sov. J. Low Temp. Phys. **1989**, *15*, 323

Electrical Conductivity and Polarized Reflection Spectra of Organic Complex (BMDT-TTF)$_2$I$_3$
Kozlov, M. E.; Onishchenko, V. G.; Pokhodnya, K. I.; Yurchenko, A. A.
Sov. Phys. Solid State **1989**, *31*, 1005

Structural and Electronic Properties of κ-Phase Organic Donor Salts: κ-(DMET)$_2$-AuBr$_2$ and κ-(BEDT-TTF)$_4$Hg$_3$Cl$_8$
Whangbo, M.-H.; Jung, D.; Wang, H. H.; Beno, M. A.; Williams, J. M.; Kikuchi, K.
Mol. Cryst. Liq. Cryst. **1990**, *181*, 1

Organic Superconductor κ-(BEDT-TTF)$_2$[Cu(NCS)$_2$] and Its Related Materials
Saito, G.
Mol. Cryst. Liq. Cryst. **1990**, *181*, 65

Ubiquitous Superconductivity Near 4K in Salts of the BEDT-TTF/I System: Is There a Common Source?
Carlson, K. D.; Wang, H. H.; Beno, M. A.; Kini, A. M.; Williams, J. M.
Mol. Cryst. Liq. Cryst. **1990**, *181*, 91

Characterization of a Structural Phase Transition in δ-(ET)$_2$AuBr$_2$ at 420K by ESR and Crystal Packing Studies
Montgomery, L. K.; Wang, H. H.; Schlueter, J. A.; Geiser, U.; Carlson, K. D.; Williams, J. M.; Rubinstein, R. L.; Brennan, T. D.; Stupka, D. L.; Whitworth, J. R.; Jung, D.; Whangbo, M.-H.
Mol. Cryst. Liq. Cryst. **1990**, *181*, 197

Relation Between the Dimensionality of Electronic Structure and the Correlation Effect in (BEDT-TTF)$_2$X System
Ugawa, A.; Yakushi, K.; Kuroda, H.
Mol. Cryst. Liq. Cryst. **1990**, *181*, 269

Bulk Superconductivity in Polycrystalline Pressed Samples of α_t-(BEDT-TTF)$_2$I$_3$ and β_p-(BEDT-TTF)$_2$I$_3$
Schweitzer, D.; Kahlich, S.; Gärtner, S.; Gogu, E.; Grimm, H.; Zamboni, R.; Keller, H. J.
Mol. Cryst. Liq. Cryst. **1990**, *181*, 279

Superconductivity in BEDT-TTF Based Organic Metals: Role of Uniaxial Pressure and Inverse Isotope Effect
Tokumoto, M.; Murata, K.; Kinoshita, N.; Yamaji, K.; Anzai, H.; Tanaka, Y.; Hayakawa, Y.; Nagasaka, K.; Sugawara, Y.
Mol. Cryst. Liq. Cryst. **1990**, *181*, 295

Crystal Structure and Electrical Properties of (BEDT-TTF)$_2$Cp(CN)$_5$•(Solvent)$_x$
Watson, W. H.; Kini, A. M.; Beno, M. A.; Montgomery, L. K.; Wang, H. H.; Carlson, K. D.; Gates, B. D.; Tytko, S. F.; Derose, J.; Cariss, C.; Rohl, C. A.; Williams, J. M.
Synth. Met. **1989**, *33*, 1

ET Cation-Radical Salts with Metal Complex Anions
Shibaeva, R. P.; Lobkovskaya, R. M.; Korotkov, V. E.; Kushch, N. D.; Yagubskii, E. B.; Makova, M. K.
Synth. Met. **1988**, *27*, A457

Structural and Electrical Properties of (BEDT-TTF)$_5$Hg$_3$Br$_{11}$
Mori, T.; Wang, P.; Imaeda, K.; Enoki, T.; Inokuchi, H.
Solid State Commun. **1987**, *64*, 733

Structural and Electrical Properties of (BEDT-TTF)$_3$CuBr$_3$
Mori, T.; Sakai, F.; Saito, G.
Chem. Lett. **1987**, 927

How Well Do We Understand the Synthesis of (ET)$_2$I$_3$ by Electocrystallization? ESR and X-ray Identification of (ET)$_2$I$_3$ Crystals which Are Made of Phases and Observation of High-T_c States of (ET)$_2$I$_3$, Ranging from 2.5–6.9 K
Montgomery, L. K.; Geiser, U.; Wang, H. H.; Beno, M. A.; Schultz, A. J.; Kini, A. M.; Carlson, K. D.; Williams, J. M.; Whitworth, J. R.; Gates, B. D.; Cariss, C. S.; Pipan, C. M.; Donega, K. M.; Wenz, C.; Kwok, K. W.; Crabtree, G. W.
Synth. Met. **1988**, *27*, A195

Crystal and Electronic Structures of Molecular Conductors Based on the Multi-Sulfur π-Donor Molecules
Kobayashi, H.; Kato, R.; Kobayashi, A.
Synth. Met. **1987**, *19*, 623

Crystal and Band Electronic Structures of an Organic Salt with the First Three-Dimensional Radical-Cation Donor Network, (BEDT-TTF)Ag$_4$(CN)$_5$
Geiser, U.; Wang, H. H.; Gerdom, L. E.; Firestone, M. A.; Sowa, L. M.; Williams, J. M.
J. Am. Chem. Soc. **1985**, *107*, 8305

Fermi Surface and Pressure Effect in (BEDT-TTF)$_2$Cu(NCS)$_2$
Oshima, K.; Mori, T.; Inokuchi, H.; Urayama, H.; Yamochi, H.; Saito, G.
Synth. Met. **1988**, *27*, A165

Superconductivity of BEDT-TTF Salts: (1) Effect of Pressure and Alloying and (2) Shubnikov-de Haas Effect
Murata, K.; Tokumoto, M.; Anzai, H.; Honda, Y.; Kinoshita, N.; Ishiguro, T.; Toyota, N.; Sasaki, T.; Muto, Y.
Synth. Met. **1988**, *27*, A263

Transport Properties of κ-(BEDT-TTF)$_2$Cu(NCS)$_2$; H$_{c2}$, Its Anisotropy and Their Pressure Dependence
Murata, K.; Honda, Y.; Anzai, H.; Tokumoto, M.; Takahashi, K.; Kinoshita, N.; Ishiguro, T.; Toyota, N.; Sasaki, T.; Muto, Y.
Synth. Met. **1988**, *27*, A341

An Organic Superconductor, (BEDT-TTF)$_3$Cl$_2$(H$_2$O)$_2$, and Some Other BEDT-TTF Conductors
Mori, T.; Wang, P.; Imaeda, K.; Enoki, T.; Inokuchi, H.; Sakaj, F.; Saito, G.
Synth. Met. **1988**, *27*, A451

Superconductivity at Ambient Pressure in BEDT-TTF Radical Salts
Schweitzer, D.; Polychroniadis, K.; Klutz, T.; Keller, H. J.; Hennig, I.; Heinen, I.; Haeberlen, U.; Gogu, E.; Gärtner, S.
Synth. Met. **1988**, *27*, A465

Crystal and Electronic Structures of Cation Radical Salts Based on BEDSe-TSeF, Se-Analogue of BEDT-TTF
Kato, R.; Kobayashi, H.; Kobayashi, A.
Synth. Met. **1987**, *19*, 629

Crystal Structures, Electrical and Optical Properties of (BPDTF)$_m$X$_n$, (PEDTT-TF)$_m$X$_n$ (X = BF$_4$, I$_3$) and Similar Salts
Papavassiliou, G. C.; Terzis, A.; Underhill, A. E.; Geserich, H. P.; Kaye, B.
Synth. Met. **1987**, *19*, 703

Electronic Spectra and Crystal Structure of Organic Radical Salt: Dimethyldibenzotetrathiafulvalenium Tetrafluoroborate
Tanaka, C.; Tanaka, J.; Dietz, K.; Katayama, C.; Tanaka, M.
Bull. Chem. Soc. Jpn. **1983**, *56*, 405

Crystal Structure and Electrical Conductivity of $(TTF)_5Hg_6(SCN)_{16}$
Thorup, N.; Beno, M. A.; Cariss, C. S.; Carlson, K. D.; Geiser, U.; Kleinjan, S.; Porter, L. C.; Wang, H. H.; Williams, J. M.
Synth. Met. **1988**, *27*, B15

Electron Spin Resonance, Infrared Spectroscopic, and Molecular Packing Studies of the Thermally Induced Conversion of Semiconducting α- to Superconducting $α_t$-$(BEDT-TTF)_2I_3$
Wang, H. H.; Ferraro, J. R.; Carlson, K. D.; Montgomery, L. K.; Geiser, U.; Williams, J. M.; Whitworth, J. R.; Hill, S.; Whangbo, M.-H.; Evain, M.; Novoa, J. J.
Inorg. Chem. **1989**, *28*, 2267

A New Ambient-pressure Organic Superconductor: $(BEDT-TTF)_2(NH_4)Hg(SCN)_4$
Wang, H. H.; Carlson, K. D.; Geiser, U.; Kwok, W. K.; Vashon, M. D.; Thompson, J.E.; Larsen, N. F.; McCabe, G. D.; Hulscher, R. S.; Williams, J. M.
Physica C **1990**, *166*, 57

Determination of the Electron Phonon Coupling and the Superconducting Gap in β-$(BEDT-TTF)_2X$ Crystals ($X = I_3$, IAuI)
Nowack, A.; Poppe, U.; Weger, M.; Schweitzer, D.; Schwenk, H.
Z. Phys. B - Condens. Matt. **1987**, *68*, 41

Peculiarities of Organic Superconductors of the $(BEDT-TTF)_2X$-Family
Andres, K.; Schwenk, H.; Veith, H.
Physica **1986**, *143B*, 334

Inductive and Resistive Studies of the Ambient Pressure Organic Superconductor β-$(BEDT-TTF)_2I_3$
Carlson, K. D.; Crabtree, G. W.; Choi, M.; Hall, L. N.; Copps, P. T.; Wang, H. H.; Emge, T. J.; Beno, M. A.; Williams, J. M.
Mol. Cryst. Liq. Cryst. **1985**, *125*, 145

Recent Progress in BEDT-TTF Based Synthetic Metals
Wang, H. H.; Allen, T. J.; Schlueter, J. A.; Hallenbeck, S. L.; Stupka, D. L.; Chen, M. Y.; Despotes, A. M.; Kao, H.-C. I.; Carlson, K. D.; Geiser, U.; Williams, J. M.
Phosphorus and Sulfur **1988**, *38*, 329

Crystal and Band Electronic Structures of an Organic Salt with the First Three-dimensional Radical-cation Donor Network, $(BEDT-TTF)Ag_4(CN)_5$
Geiser, U.; Wang, H. H.; Gerdom, L. E.; Firestone, M. A.; Sowa, L. M.; Williams, J. M.
J. Am. Chem. Soc. **1985**, *107*, 8305

New Organic Synthetic Metals Derived from BEDT-TTF, Ni(dsit)$_2$ and BEDO-TTF

Beno, M. A.; Kini, A. M.; Geiser, U.; Wang, H. H.; Carlson, K. D.; Williams, J. M.

The Physics and Chemistry of Organic Superconductors, Saito, G.; Kagoshima, S., Eds., Springer-Verlag, Berlin, **1990**, vol. 51, pp. 369–372.

Similarities and Differences in the Structural and Electronic Properties of κ-Phase Organic Conducting and Superconducting Salts

Jung, D.; Evain, M.; Novoa, J. J.; Whangbo, M.-H.; Beno, M. A.; Kini, A. M.; Schultz, A. J.; Williams, J. M.; Nigrey, P. J.

Inorg. Chem. **1989**, *28,* 4516

NEUTRON DIFFRACTION

Neutron-diffraction Evidence for Ordering in the High-T_c Phase of β-di[bis-(ethylenedithio)tetrathiafulvalene] triiodide [β*-(ET)$_2$I$_3$]

Schultz, A. J.; Beno, M. A.; Wang, H. H.; Williams, J. M.

Phys. Review B **1986**, *33,* 7823

Neutron Diffraction Evidence for Unusual Cohesive H-Bonding Interactions in β-(BEDT-TTF)$_2$X Organic Superconductors

Emge, T. J.; Wang, H. H.; Geiser, U.; Beno, M. A.; Webb, K. S.; Williams, J. M.

J. Am. Chem. Soc. **1986**, *108,* 3849

Effect of Structural Disorder On Organic Superconductors: A Neutron Diffraction Study of "High-T_c" β*-(BEDT-TTF)$_2$I$_3$ at 4.5 K and 1.5 Kbar

Schultz, A. J.; Wang, H. H.; Williams, J. M.

J. Am. Chem. Soc. **1986**, *108,* 7853

The Structure of β*-(BEDT-TTF)$_2$I$_3$ at 4.5 K and 1.5 Kbar

Schultz, A. J.; Wang, H. H.; Williams, J. M.; Filhol, A.

Physica **1986**, *143B,* 354

A Neutron Diffraction Study of the Extent of Disorder in the Low Temperature (20 K) Structure of β-(BEDT-TTF)$_2$I$_2$Br

Schultz, A. J.; Emge, T. J.; Leung, P. C. W.; Beno, M. A.; Wang, H. H.; Williams, J. M.

Physica **1986**, *143B,* 351

NMR

^{1}H NMR in Organic Superconductor β-(BEDT-TTF)$_2$I$_3$

Maniwa, Y.; Takahashi, T.; Saito, G.

J. Phys. Soc. Jpn. **1986**, *55,* 47

Proton NMR Relaxation in the High T_c Organic Superconductor β-(BEDT-TTF)$_2$I$_3$
Creuzet, F.; Bourbonnais, C.; Jérome, D.; Schweitzer, D.; Keller, H.
Europhys. Lett. **1986**, *1*, 467

Anisotropic Knight Shifts and Spin Density Distribution in β-(BEDT-TTF)$_2$I$_3$
Vainrub, A.; Kheinmaa, I.; Yagubskii, E.
Pis' ma Zh. Eksp. Teor. Fiz. **1986**, *44*, 247; *JETP Lett.* **1986**, *44*, 317

Nuclear Magnetic Relation in the Organic Superconductor β-(BEDT-TTF)$_2$I$_3$
Skripov, A. V.; Stepanov, A. P.
Sov. Phys. Solid State **1986**, *28*, 1309

Proton Spin-lattice Relaxation in β-(BEDT-TTF)$_2$I$_3$
Heinmaa, I. A.; Alla, M. A.; Vainrub, A. M.; Lippmaa, E. T.
Pis' ma Zh. Eksp. Teor. Fiz. **1986**, *90*, 1748; *JETP Lett.* **1986**, *63*, 1025

Anomalous NMR Relaxation in Organic Superconductor, (BEDT-TTF)$_2$Cu(NCS)$_2$
Takahashi, T.; Tokiwa, T.; Kanoda, K.; Urayama, H.; Yamochi, H.; Saito, G.
Tech. Report of I.S.S.P., Ser. A **1988**, 1925, *Physica C* **1988**, *153*, 487

NMR Relaxation in the Organic Superconductor (BEDT-TTF)$_2$Cu(NCS)$_2$
Takahashi, T.; Tokiwa, T.; Kanoda, K.; Urayama, H.; Yamochi, H.; Saito, G.
Synth. Met. **1988**, *27*, A319

Electronic Properties of New Organic Conductors
Azevedo, L. J.; Venturini, E. L.; Kwak, J. F.; Schirber, J. E.; Williams, J. M.; Wang, H. H.; Reed, P. E.
Mol. Cryst. Liq. Cryst. **1985**, *125*, 169

PHASE DIAGRAMS

Pressure Phase Diagram of the Organic Superconductor β-(BEDT-TTF)$_2$I$_3$
Murata, K.; Tokumoto, M.; Anzai, H.; Bando, H.; Saito, G.; Kajimura, K.; Ishiguro, T.
J. Phys. Soc. Jpn. **1985**, *54*, 2084

Pressure Phase Diagram of the Organic Superconductor β-(BEDT-TTF)$_2$I$_3$
Murata, K.; Tokumoto, M.; Anzai, H.; Bando, H.; Saito, G.; Kajimura, K.; Ishiguro, T.
J. Phys. Soc. Jpn. **1985**, *54*, 2081

Investigation of the T-P Phase Diagram for α-(BEDT-TTF)$_2$I$_3$
Kartsovnik, M. V.; Kononovich, P. A.; Laukhin, V. N.; Khomenko, A. G.; Shchegolev, I. F.
Zh. Eksp. Teor. Fiz. **1985**, *88*, 1447; *JETP Lett.* **1985**, *61*, 866

T-P Phase Diagram for β-(ET)$_2$I$_3$
Ginodman, V. B.; Gudenko, A. V.; Kononovich, P. A.; Laukhin, V. N.; Shchegolev, I. F.
Pis' ma Zh. Eksp. Teor. Fiz. **1988**, *94*, 333; *JETP Lett.* **1988**, *67*, 1055

Transport Properties and Phase Transition of Organic Crystal β$_d$'-(BEDT-TTF)$_2$I$_3$
Lu, L.; Ma, B.; Duan, H.; Lin, S.; Zhang, D.; Yao, Y.; Qian, M.; Zhu, Y.; Wang, X.; Zhu, D.
Synth. Met. **1989**, *31*, 37

Study of the High Temperature Phase Diagram of β-(BEDT-TTF)$_2$I$_3$
Kang, W.; Jérome, D.; Lenoir, C.; Batail, P.
Synth. Met. **1989**, *27*, A353

Unusual Features of the Structural Phase Diagram of β-(BEDT-TTF)$_2$I$_3$: An Experimental and Theoretical Study
Ravy, S.; Moret, R.; Pouget, J. P.
Synth. Met. **1989**, *27*, A367

POINT CONTACT SPECTRA

Point-Contact Spectra of the Organic Metal β-(BEDT-TTF)$_2$I$_3$
Nowack, A.; Weger, M.; Schweitzer, D.; Keller, H. J.
Solid State Commun. **1986**, *60*, 199

POLARIZED REFLECTANCE/ABSORPTION

The Polarized Reflectance Spectrum of a Novel Organic Conductor (BEDT-TTF)$_2$ClO$_4$(C$_2$H$_3$Cl$_3$)$_{0.5}$
Tajima, H.; Yakushi, K.; Kuroda, H.; Saito, G.; Inokuchi, H.
Solid State Commun. **1984**, *49*, 769

Polarized-light Reflection Spectra of a New Organic Metal Bis(ethylenedithiolo)tetrathiafulvalene Triiodide (BEDT-TTF)$_2$I$_3$
Vlasova, R. M.; Ivanova, E. A.; Semkin, V. N.
Sov. Phys. Solid State **1985**, *27*, 326

Infrared Properties of the Ambient Pressure Organic Superconductor (BEDT-TTF)$_2$I$_3$
Jacobsen, C. S.; Williams, J. M.; Wang, H. H.
Solid State Commun. **1985**, *54*, 937

Polarized Reflectance Spectra of the Organic Conductors: α- and β-Modifications of Di[bis(ethylenedithiolo)tetrathiafulvalene] Triiodide, (BEDT-TTF)$_2$I$_3$
Sugano, T.; Yamada, K.; Saito, G.; Kinoshita, M.
Solid State Commun. **1985**, *55*, 137

Optical Investigations of the Electrical Anisotropy of α- and β-(BEDT-TTF)$_2$I$_3$
Koch, B.; Geserich, H. P.; Ruppel, W.; Schweitzer, D.; Dietz, K. H.; Keller, H. J.
Mol. Cryst. Liq. Cryst. **1985**, *119*, 343

Polarized Reflectance Spectrum of β-(BEDT-TTF)$_2$PF$_6$
Tajima, H.; Yakushi, K.; Kuroda, H.; Saito, G.
Solid State Commun. **1985**, *56*, 251

Polarized Reflectance Spectrum of β-(BEDT-TTF)$_2$I$_3$ Single Crystal
Tajima, H.; Yakushi, K.; Kuroda, H.
Solid State Commun. **1985**, *56*, 159

Optical Properties of the Two Crystal Modifications of the Organic Conductor (BEDT-TTF)$_2$I$_3$
Kaplunov, M. G.; Yagubskii, E. B.; Rozenberg, L. P.; Borodko, Yu. G.;
Phys. Stat. Sol. (a) **1985**, *89*, 509

Temperature Dependence of the Reflectance Spectrum of β-(BEDT-TTF)$_2$I$_3$
Tajima, H.; Kanbara, H.; Yakushi, K.; Kuroda, H.; Saito, G.
Solid State Commun. **1986**, *57*, 911

Electron-Molecular Vibration Coupling in 2-D Organic Conductors: High and Low Temperature Phases of α-(BEDT-TTF)$_2$I$_3$
Meneghetti, M.; Bozio, R.; Pecile, C.
J. Physique **1986**, *47*, 1377

Polarised Reflectance Spectra of Di[bis(ethylenedithiolo)tetrathiafulvalene]-dibromoiodate(I), β-(BEDT-TTF)$_2$IBr$_2$
Sugano, T.; Saito, G.
J. Phys. C: Solid State Phys. **1986**, *19*, 5471

Trends in the Infrared and Near Infrared Properties of Organic Conductors
Jacobsen, C. S.
J. Phys. C: Solid State Phys. **1986**, *19*, 5643

Crystal Structure and Reflectance Spectrum of β''-(BEDT-TTF)$_2$IAuBr
Ugawa, A.; Yakushi, K.; Kuroda, H.; Kawamoto, A.; Tanaka, J.
Chem. Lett. **1986**, 1875

Temperature Dependence of the Reflectance Spectra of the Single Crystals of Bis(ethylenedithio)tetrathiafulvalenium Salts. α-(BEDT-TTF)$_3$(ReO$_4$)$_2$ and α-(BEDT-TTF)$_2$I$_3$
Yakushi, K.; Kanbara, H.; Tajima, H.; Kuroda, H.; Saito, G.; Mori, T.
Bull. Chem. Soc. Jpn. **1987**, *60*, 4251

Crystal Structure and Polarized Reflectance Spectra of α'-Bis(ethylenedithio)-tetrathiafulvalenium)$_2$ Bromoiodoaurate, α'-(BEDT-TTF)$_2$IAuBr$_2$
Ugawa, A.; Yakushi, K.; Kuroda, H.; Kawamoto, A.; Tanaka, J.
Synth. Met. **1988**, *22*, 305

In-plane Quasi-isotropic Organic Superconductor Di[bis(ethylenedithiolo)tetrathiafulvalene] bis(isothiocyanato)cuprate(I), (BEDT-TTF)$_2$[Cu(NCS)$_2$]: Polarized Reflectance Spectra
Sugano, T.; Hayashi, H.; Takenouchi, H.; Nishikida, K.; Urayama, H.; Yamochi, H.; Saito, G.; Kinoshita, M.
Phys. Rev. B **1988**, *37*, 9100

Optical Study on Bis(propylenedithio)tetrathiafulvalenium (BPDT-TTF) Salts
Yakushi, K.; Tajima, H.; Ida, T.; Tamura, M.; Hayashi, H.; Kuroda, H.; Kobayashi, A.; Kobayashi, H.; Kato, R.
Synth. Met. **1988**, *24*, 301

Reflectance Spectra of the 1:1 Salts of Bis(methylenedithio)tetrathiafulvalene (BMDT-TTF): Estimation of the On-site Coulomb Energy
Yoshitake, M.; Yakushi, K.; Kuroda, H.; Kobayashi, A.; Kato, R.; Kobayashi, H.
Bull. Chem. Soc. Jpn. **1988**, *61*, 1115

The Reflection Spectra of the New Organic Conductor (BEDT-TTF)$_4$Hg$_3$Cl$_8$
Kaplunov, M. G.; Lyubovskaya, R. N.; Aldoshina, M. Z.; Borodko, Yu. G.
Phys. Stat. Sol. **1987**, *104*, 833

Optical and Electrical Properties of an Organic Superconductor Di[bis(ethylenedithio)tetrathiafulvalenium] Dithiocyanocuprate(I), (BEDT-TTF)$_2$[Cu(SCN)$_2$]
Ugawa, A.; Ojima, G.; Yakushi, K.; Kuroda, H.
Phys. Rev. B **1988**, *38*, 5122

Temperature Dependence of the Polarized Reflectance Spectra of the θ-Type of Bis(ethylenedithio)tetrathiafulvalenium Triiodide θ-(BEDT-TTF)$_2$I$_3$: Estimation of Band Parameters
Tamura, M.; Yakushi, K.; Kuroda, H.; Kobayashi, A.; Kato, R.; Kobayashi, H.
J. Phys. Soc. of Jpn. **1988**, *57*, 3239

Optical Properties of the New Organic Superconductor (BEDT-TTF)$_2$Cu(SCN)$_2$
Kaplunov, M. G.; Kushch, N. D.; Yagubskii, E. B.
Phys. Stat. Sol. **1988**, *110*, K111

Infrared Conductivity and Electron-molecular-vibration Coupling in the Organic Superconductor Di[bis(ethylenedithio)tetrathiafulvalene] bis(isothiocyanato)cuprate(I), κ-(BEDT-TTF)$_2$[Cu(NCS)$_2$]: Protonated and Deuterated Salts
Sugano, T.; Hayashi, H.; Kinoshita, M.; Nishikida, K.
Phys. Rev. B **1989**, *39*, 11387

FT-IR Micro-Spectroscopic Studies of Several Charge-Transfer Organic Electrical Conductors
Ferraro, J. R.; Wang, H. H.; Ryan, J.; Williams, J. M.
Applied Spectros. **1977**, *41*, 1377

Observation of Two Different -C-C-H Bending Modes in the Ambient-Pressure Organic Superconductor κ-(ET)$_2$Cu(NCS)$_2$ by Polarized Infrared Reflectance Measurements and Implications on Its Structural and Physical Properties
Ferraro, J. R.,; Wang, H. H.; Geiser, U.; Kini, A. M.; Beno, M. A.; Williams, J. M.; Hill, S.; Whangbo, M.-H.; Evain, M.
Solid State Commun. **1988**, *68*, 917

Reflectance Spectra of β-, α-, and κ-(BEDT-TTF)$_2$I$_3$ and β"- and β'-(BEDT-TTF)$_2$ICl$_2$: Relation Between the Inter-Band Transition and the Dimeric Structure
Kuroda, H.; Yakushi, K.; Tajima, H.; Ugawa, A.; Tamura, M.; Okawa, Y.; Kobayashi, A.; Kato, R.; Kobayashi, H.; Saito, G.
Synth. Met. **1988**, *27*, A491

Reflectance Spectra of BEDT-TTF Salts
Kuroda, H.; Yakushi, K.; Tajima, H.; Saito, G.
Mol. Cryst. Liq. Cryst. **1985**, *125*, 135

Electronic Structure of Some β-(C$_{10}$H$_8$S$_8$)$_2$X Compounds as Studied by Infrared Spectroscopy
Jacobsen, C. S.; Tanner, D. B.; Williams, J. M.; Geiser, U.; Wang, H. H.
Phy. Rev. B **1987**, *35*, 9605

IR Reflectance Studies of Some Conducting BEDT-TTF Salts
Pratt, F. L.; Hayes, W.; Kurmoo, M.; Day, P.
Synth. Met. **1988**, *27*, A439

Infrared Properties of a Novel Organic Superconductor, (BEDT-TTF)$_2$[Cu(SCN)$_2$]
Ugawa, A.; Ojima, G.; Yakushi, K.; Kuroda, H.
Synth. Met. **1988**, *27*, A445

Polarized Reflectance Spectra of Organic Metals α- and κ-(BEDT-TTF)$_2$Cu(NCS)$_2$ and κ-(BEDT-TTFd$_8$)$_2$Cu(NCS)$_2$
Tokumoto, M.; Anzai, H.; Takahashi, K.; Kinoshita, N.; Murata, K.; Ishiguro, T.; Tanaka, Y.; Hayakawa, Y.; Nagamori, H.; Nagasaka, K.
Synth. Met. **1988**, *27*, A171

Organic Conductors: Evidence for Correlation Effects in Infrared Properties
Jacobsen, C. S.; Johannsen, Ib.; Bechgaard, K.
Phy. Rev. Lett. **1984**, *53*, 194

Optical Properties of the 10 K Organic Superconductors $(BEDT-TTF)_2[Cu(SCN)_2]$
Kornelsen, K.; Eldridge, J. E.; Homes, C. C.; Wang, H. H.; Williams, J. M.
Solid State Commun. **1989**, *72*, 475

FT-IR Absorption Spectra of Polycrystalline Pressed Samples of the Organic Metals and Superconductors α-, β-$(BEDT-TTF)_2I_3$ and $(BEDT-TTF)_2Cu(NCS)_2$
Zamboni, R.; Schweitzer, D.; Keller, H. J.; Taliani, C.
Z. Naturforsch **1989**, *44a*, 295

Analysis and Interpretation of the Reflectance Spectra of BEDT-TTF Based Charge-Transfer Salts (Organic Superconductors and Conductors)
Ferraro, J. R.; Williams, J. M.
Mol. Cryst. Liq. Cryst. **1990**, *181*, 253

The Characterization of Organic-Charge Transfer Superconductors by Microreflectance Spectroscopy
Ferraro, J. R.; Hill, S. L.; Krishnan, K.
SPIE (Proceedings of the 7th International Conference on Fourier Transform Spectroscopy, June 19–23) **1989**, *1145*

Optical Study of the Transformation of the α-phase of $(BEDT-TTF)_2I_3$ Into the Superconducting $α_t$-phase ($T_c = 8$ K)
Helberg, H. W.; Schweitzer, D.; Keller, H. J.
Synth. Met. **1988**, *27*, A347

PREPARATION, COMPOSITION, PROPERTIES

"Organic Metals": Alkylthio Substitution Effects in Tetrathiafulvalene-Tetracyanoquinodimethane Charge-transfer Complexes
Mizuno, M.; Garito, A. F.; Cava, M. P.
J. C. S., Chem. Comm. **1978**, *687*, 18

Synthetic Metals Based on Bis(ethylenedithio)tetrathiafulvalene (BEDT-TTF): Synthesis, Structure (T = 298 and 125 K), and NMR of $(BEDT-TTF)_2X$, $X = ReO_4^-$ and BrO_4^-
Williams, J. M.; Beno, M. A.; Wang, H. H.; Reed, P. E.; Azevedo, L. J.; Schirber, J. E.
Inorg. Chem. **1984**, *23*, 1790

Synthetic Metals Based on Bis(ethylenedithio)tetrathiafulvalene (BEDT-TTF): Synthesis, Structure and Ambient-Pressure Superconductivity in $(BEDT-TTF)_2I_3$
Williams, J. M.; Emge, T. J.; Wang, H. H.; Beno, M. A.; Copps, P. Thomas; Hall, L. N.; Carlson, K. Douglas; Crabtree, G. W.
Inorg. Chem. **1984**, *23*, 2558

Synthesis, Structure and Physical Properties of a Two-Dimensional Organic Metal, Di[bis(ethylenedithiolo)tetrathiafulvalene]triiodide, $(BEDT-TTF)_2I_3$
Bender, K.; Hennig, I.; Schweitzer, D.; Dietz, K.; Endres, H.; Keller, H. J.
Mol. Cryst. Liq. Cryst. **1984**, *108*, 359

Elevation of the Temperature of Superconducting Transition in β-$(BEDT-TTF)_2I_3$ to 6-7 K at Normal Pressures
Merkhanov, V. A.; Kostyuchenko, E. E.; Laukhin, V. N.; Lobkovskaya, R. M.; Makova, M. K.; Shibaeva, R. P.; Shchegolev, I. F.; Yugubskii, E. B.
Pis'ma Eksp. Teor. Fiz. **1985**, *41*, 146

New Synthesis of the Organic Superconductor $(BEDT-TTF)_2I_3$
Kostyuchenko, E. E.; Yagubskii, E. B.; Neiland, O. Ya.; Khodorkovskii, V. Yu.
Bull. Acad. Sci. U.S.S.R. **1985**, *33*, 2598

New Results on Two Synthetic Conductors $(TMTSF)_2BrO_4$ and $(BEDT-TTF)_2I_3$
Mortensen, K.; Jacobsen, C. S.; Bechgaard, K.; Carneiro, K.; Williams, J. M.
Mol. Cryst. Liq. Cryst. **1985**, *119*, 401

Chemical and Physical Properties of Cation Radical Salts of BEDT-TTF
Saito, G.; Enoki, T.; Kobayashi, M.; Imaeda, K.; Sato, N.; Inokuchi, H.
Mol. Cryst. Liq. Cryst. **1985**, *119*, 393

Organic Superconductor β-$(BEDT-TTF)_2IBr_2$ Obtained by Diffusion Method
Saito, G.; Sugano, T.; Yamochi, H.; Kinoshita, M.; Oshima, K.; Suzuki, M.; Katayama, C.; Tanaka, J.
Chem. Lett. **1985**, 1037

Synthesis of Biethylenedithiolylenetetrathiafulvalene Donors (BEDT-TTF) and Electrochemical Preparation of their Charge Transfer Complexes
Engler, E. M.; Lee, V. Y.; Schumaker, R. R.; Parkin, S. S. P.; Green, R. L.; Scott, J. C.
Mol. Cryst. Liq. Cryst. **1984**, *107*, 19

Correlations of Anion Size and Symmetry with the Structure and Electronic Properties of β-$(BEDT-TTF)_2X$ Conducting Salts with Trihalide Anions $X = I_3^-$, I_2Br^-, IBr_2^-
Emge, T. J.; Leung, P. C. W.; Beno, M. A.; Wang, H. H.; Firestone, M. A.; Webb, K. S.; Carlson, K. D.; Williams, J. M.; Venturini, E. L.; Azevedo, L. J.; Schirber, J. E.
Mol. Cryst. Liq. Cryst. **1986**, *132*, 363

Preparation, Structure and Investigations of BEDT-TTF Trihalides
Endres, H.; Hiller, M.; Keller, H. J.; Bender, K.; Gogu, E.; Heinen, I.; Schweitzer, D.
Z. Naturforsch **1985**, *40b*, 1664

Preparation, Structure and Physical Properties of BEDT-TTF Nitrates
Weber, A.; Endres, H.; Keller, H. J.; Gogu, E.; Heinen, I.; Bender, K.; Schweitzer, D.
Z. Naturforsch **1985**, *40b,* 1658

Organic Conductors and Superconductors: Mixed (IBr_2^-) Polyhalides of BEDT-TTF
Yagubskii, E. V.; Shchegolev, I. F.; Shibaeva, R. P.; Fedutin, D. N.; Rozenberg, L. P.; Sogomonyan, E. M.; Lobkovskaya, R. M.; Laukhin, V. N.; Ignat'ev, A. A.; Zvarykina, A. V.; Buravov, L. I.
Pis'ma Eksp. Teor. Fiz. **1985**, *42,* 167

The Role of Anions on the Crystal Structures and Electrical Properties of the Organic Metals and Superconductors, $(BEDT-TTF)_2X$ (X = Trihalide Anions)
Leung, P. C. W.; Emge, T. J.; Schultz, A. J.; Beno, M. A.; Carlson, K. D.; Wang, H. H.; Firestone, M. A.; Williams, J. M.
Solid State Commun. **1986**, *57,* 93

β-$(ET)_2AuI_2$: An Organic Volume Superconductor with T_c = 3.2 K at Ambient Pressure
Amberger, E.; Fuchs, H.; Polborn, K.
Angew. Chem. Int. Ed. Engl. **1985**, *24,* 968

8. Chiral Metals? A Chiral Substrate for Organic Conductors and Superconductors
Wallis, J. D.; Karrer, A.; Dunitz, J. D.
Helvetica Chim. Acta. **1986**, *69,* 69

"Partially" Superconducting Behavior of θ-$(BEDT-TTF)_2(I_3)_{1-x}$ (A = AuI_2, I_2Br) System
Kato, R.; Kobayashi, H.; Kobayashi, A.; Nishio, Y.; Kajita, K.; Sasaki, W.
Chem. Lett. **1986,** 789

Synthesis, Structure and Electrical Properties of a Two-Dimensional Organic Conductor α-$(BEDT-TTF)_2BrI_2$
Daoben, Z.; Ping, W.; Meixiang, W.; Zhaolou, Y.; Naijue, Z.
Solid State Commun. **1986**, *57,* 843

Synthesis of Bis(Dimethylvinylenedithio)tetrathiafulvalene, BDMVDT-TTF
Inoue, K.; Tasaka, Y.; Yamazaki, O.; Nogami, T.; Mikawa, H.
Chem. Lett. **1986,** 781

t-$(ET)_2Au(CN)_2$ and t-$(ET)_2AuBr_2$, Organic Radical Cation Salts with Twist Structure
Amberger, E.; Fuchs, H.; Polborn, K.
Angew. Chem. Int. Ed. Engl. **1986**, *25,* 729

ζ-(ET)$_2$AuBr$_2$, a New Type of Structure for the Organic Radical Cation Salts (ET)$_2$AuI$_2$
Amberger, E.; Polborn, K.; Fuchs, H.
Angew. Chem. Int. Ed. Engl. **1986**, *25*, 727

"Partially" Superconducting Behavior of θ-(BEDT-TTF)$_2$(I$_3$)$_{1-x}$(A)$_x$ (A = AuI$_2$, I$_2$Br) System
Kato, R.; Kobayashi, H.; Kobayashi, A.; Nishio, Y.; Kajita, K.; Sasaki, W.
Chem. Lett. **1986,** 957

Synthesis, Structure, and Properties of (BPDT-TTF)$_2$IBr$_2$
Nigrey, P. J.; Morosin, B.; Venturini, E. L.; Azevedo, L. J.; Schirber, J. E.; Perschke, S. E.; Williams, J. M.
Physica **1986**, *143B,* 290

A New Ambient-pressure Superconductor, κ-(BEDT-TTF)$_2$I$_3$
Kato, R.; Kobayashi, H.; Kobayashi, A.; Moriyama, S.; Nishio, Y.; Kajita, K.; Sasaki, W.
Chem. Lett. **1987,** 507

A Photochemical Synthesis of Bis(ethylenedithiolo)tetrathiafulvalene
Sorm, M.; Nespurek, S.; Ryba, O.; Kubanek, V.
J. Chem. Soc., Chem. Commun. **1987,** 696

Electrochemically Prepared Radical Salts of BEDT-TTF: Molecular Metals and Superconductors
Schweitzer, D.; Gogu, E.; Hennig, I.; Klutz, T.; Keller, H. J.
Ber. Bunsenges, Phys. Chem. **1987**, *91,* 890

A New Series of Organomineral Conductors Prepared from BEDT-TTF and Dianions of Transition Metal Chlorides
Lequan, M.; Lequan, R. M.; Maceno, G.; Delhaes, P.
J. Chem. Soc., Chem. Commun. **1988**, 174

Growth of Mixed Crystals of (BEDT-TTF)-Trihalides
Anzai, H.; Tokumoto, M.; Takahashi, K.; Ishiguro, T.
J. Cryst. Growth **1988**, *91,* 225

Electrical Properties of the Organic Conductor (BEDT-TTF)$_4$Hg$_3$I$_8$ at dc and Microwave Frequencies
Vendik, I. B.; Ermolenko, A. N.; Esipov, V. V.; Pchelkin, V. M.; Sitnikova, M. F.; Gol'denberg, L. M.; Lyubovskaya, R. N.
Sov. Phys. Solid State **1988**, *30,* 765

First Salt of Bis(ethylenedithio)tetrathiafulvalene Containing Mono- and Dications
Korotkov, V. E.; Kushch, N. D.; Makova, M. K.; Shibaeva, R. P.; Yagubskii, E. B.
Bull. Academy of Sciences of the USSR **1988**, *37*, 1397

Synthesis of Bis(ethylenedithio)tetrathiafulvalene (BEDT-TTF)
Larsen, J.; Lenoir, C.
Synthesis **1989**, 134

Two Polymorphic Modifications of $(BEDT-TTF)_2I_2$ Crystals with Metallic and Semiconductor Properties
Buravov, L. I.; Zvarykina, A. V.; Ignat'ev, A. A.; Kotov, A. I.; Laukhin, V. N.; Makova, M. K.; Merzhanov, V. A.; Rozenberg, L. P.; Shibaeva, R. P.; Yagubskii, E. B.
Translated from *Izvestiya Akademii Nauk USSR, Seriya Khimicheskaya* **1988**, *9*, 2027

Mixed-valence Bis(ethylenedithio)tetrathiafulvalenium (BEDT-TTF) Monolayers Sandwiched Between Extended Close-packed Keggin-type Molecular Metal Oxide Cluster Arrays: Synthesis, Unprecedented Acentric Structure, and Preliminary Conducting and E.S.R. Properties of $(BEDT-TTF)_8SiW_{12}O_{40}$
Davidson, A.; Boubekeur, K.; Penicaud, A.; Auban, P.; Lenoit, C.; Batail, P.; Herve, G.
J. Chem. Soc., Chem. Commun. **1989**, 1373

Organic Conductors and Superconductors Based on Bis(ethylenedithio)tetrathiafulvalene and Its Derivatives
Yagubskii, E. B.; Shibaeva, R. P.
J. of Molecular Electronics **1989**, *5*, 25

Growth of High-Quality Single Crystals of the Organic Superconductors $\beta\text{-}(ET)_2I_3$ and $\beta\text{-}(ET)_2IBr_2$
Laukhina, E. E.; Laukhin, V. N.; Khomenko, A. G.; Yagubskii, E. B.
Synth. Met. **1989**, *32*, 381

Transition-Metal 1,2-Diheterolenes and Polyheterotetraheterafulvalenes: Precursors of Conducting Solids
Papavassiliou, G. C.; Kakoussis, V. C.; Lagouvardos, D. J.; Mousdis, G. A.
Mol. Cryst. Liq. Cryst. **1990**, *181*, 171

The Organic Bis(ethylenedithio)tetrathiafulvalene-based Conductors with Halidemercurate Anions
Aldoshina, M. Z.; Goldenberg, L. M.; Zhilyaeva, E. I.; Lyubovskaya, R. N.; Takhirov, T. G.; Dyachenko, O. A.; Atovmyan, L. O.; Lyubovskii, R. B.
Materials Science **1988**, *XIV*, 45

A New Organic Bis(ethylenedithio)tetrathiafulvalene-based Conductor with Octaiodinemercurate Anion, (BEDT-TTF)$_4$Hg$_3$I$_8$
Aldoshina, M. Z.; Goldenberg, L. M.; Lyubovskaya, R. N.; Takihirov, T. G.; Dyachenko, O. A.; Atovmyan, L. O.; Merzhanov, V. A.; Lyubovskii, R. B.
Materials Science **1988**, *XIV*, 53

Syntheses, Structures and Properties of (ET)$_4$(Mo$_6$Cl$_8$)Cl$_6$(THF) and (ET)$_2$X (X = AuI$_2$, (AuI$_2$)$_{0.26}$(I$_3$)$_{0.74}$, AuBr$_2$, Au(CN)$_2$)
Amberger, E.; Fuchs, H.; Polborn, K.
Synth. Met. **1987**, *19*, 605

Conductive Molecular Crystals from Three-Dimensional Building Blocks: Synthesis and Characterization of Tris(bis(ethylenedithio)tetrathiafulvalenium) Bis(tris(2-thioxo-1,3-dithiole-4,5-dithiolato) vanadate), [ET]$_3$[V(dmit)$_3$]$_2$, and the Corresponding Tetramethyltetraselenafulvalenium Salt
Broderick, W. E.; McGhee, E. M.; Godfrey, E. M.; Hoffmann, M. R.; Ibers, J. A.
Inorg. Chem. **1989**, *28*, 2902

Characterization of the Electrocrystallization Products of the Mixed-Donor System ET:MET (1:1)/I$_3$/TCE: How to Get Crystals with the Ordered β*-(ET)$_2$I$_3$ Structure and a T_c of 4.6 K Without Applied Pressure
Beno, M. A.; Kini, A. M.; Montgomery, L. K.; Whitworth, J. R.; Carlson, K. D.; Williams, J. M.
Synth. Met. **1988**, *27*, A219

Synthesis of the New Organic Metal (ET)$_2$C(CN)$_3$ and Characterization of Its Metal-Insulator Phase Transition at ~180 K
Beno, M. A.; Wang, H. H.; Soderholm, L.; Carlson, K. D.; Hall, L. N.; Nuñez, L.; Rummens, H.; Anderson, B.; Schlueter, J. A.; Williams, J. M.; Whangbo, M.-H.; Evain, M.
Inorg. Chem. **1989**, *28*, 150

Synthesis, Structure and Electrical Conductivity of (BEDT-TTF)$_x$(BrO$_4$)$_y$ Organic Metals
Beno, M. A.; Blackman, G. S.; Leung, P. C. W.; Carlson, K. D.; Copps, P. T.; Williams, J. M.
Mol. Cryst. Liq. Cryst. **1985**, *119*, 409

Synthesis and Structure of ξ-(BEDT-TTF)$_2$(I$_3$)(I$_5$) and (BEDT-TTF)$_2$(I$_3$)(TlI$_4$): Comparison of the Electrical Properties of Organic Conductors Derived from Chemical Oxidation vs. Electrocrystallization
Beno, M. A.; Geiser, U.; Kostka, K. L.; Wang, H. H.; Webb, K. S.; Firestone, M. A.; Carlson, K. D.; Nuñez, L.; Whangbo, M.-H.; Williams, J. M.
Inorg. Chem. **1987**, *26*, 1912

Superconductivity Above 2K at Ambient Pressure in Iododibromide (IBr$_2^-$) Charge-transfer Salts of Bis(ethylenedithio)tetrathiafulvalene, BEDT-TTF
Carlson, K. D.; Crabtree, G. W.; Hall, L. N.; Behroozi, F.; Copps, P. T.; Sowa, L. M.; Nuñez, L.; Firestone, M. A.; Wang, H. H.; Beno, M. A.; Emge, T. J.; Williams, J. M.
Mol. Cryst. Liq. Cryst. **1985**, *125*, 159

The Anisotropy of Intermolecular Interactions, Band Electronic Structure, and Electical Properties of β-(ET)$_2$AuCl$_2$
Emge, T. J.; Wang, H. H.; Bowman, M. K.; Pipan, C. M.; Carlson, K. D.; Beno, M. A.; Hall, L. N.; Anderson, B. A.; Williams, J. M.; Whangbo, M.-H.
J. Am. Chem. Soc. **1987**, *109*, 2016

A Test of Superconductivity vs. Molecular Disorder in (BEDT-TTF)$_2$X Synthetic Metals: Synthesis, Structure (298, 120 K), and Microwave/ESR Conductivity of (BEDT-TTF)$_2$I$_2$Br
Emge, T. J.; Wang, H. H.; Beno, M. A.; Leung, P. C. W.; Firestone, M. A.; Jenkins, H. C.; Carlson, K. D.; Williams, J. M.; Venturini, E. L.; Azevedo, L. J.; Schirber, J. E.
Inorg. Chem. **1985**, *24*, 1736

New Cation-radical Salts of BEDT-TTF with the Octahedral Metal Complex Anions PtCl$_6^{2-}$, TeCl$_6^{2-}$ and SnCl$_6^{2-}$. Synthesis, Structure and Properties
Galimzyanov, A. A.; Ignat'ev, A. A.; Kushch, N. D.; Laukhin, V. N.; Makova, M. K.; Merzhanov, V. A.; Rozenberg, L. P.; Shibaeva, R. P.; Yagubskii, E. B.
Synth. Met. **1989**, *33*, 81

BEDT-TTF Salts with Square Platinates (II) as Counterions: [BEDT-TTF]$_4$-[Pt(C$_2$O$_4$)$_2$], A New Organic Metal
Gärtner, S.; Heinen, I.; Schweitzer, D.; Nuber, B.; Keller, H. J.
Synth. Met. **1989**, *31*, 199

The Crystal Structures and Physical Properties of Polymeric (BEDT-TTF)-Metallothiocyanates
Geiser, U.; Beno, M. A.; Kini, A. M.; Wang, H. H.; Schultz, A. J.; Gates, B. D.; Cariss, C. S.; Carlson, K. D.; Williams, J. M.
Synth. Met. **1988**, *27*, A235

Synthesis, Crystal Structure, Electrical Properties, and Band Electronic Structure of Bis(1,3-propanediyldithio)tetrathiafulvalenium Tetraiodoindate(III), (BPDT-TTF)$_3$(InI$_4$)$_2$
Geiser, U.; Wang, H. H.; Schlueter, J.; Chen, M. Y.; Kini, A. M.; Kao, I. H.-C.; Williams, J. M.; Whangbo, M.-H.; Evain, M.
Inorg. Chem. **1988**, *27*, 4284

The θ-Type Cation Radical Salts of BMDT-TTF, Bis(methylenedithio)tetrathiafulvalene
Kato, R.; Kobayashi, H.; Kobayashi, A.
Chem. Lett. **1987**, 567

Crystal Structures and Electrochemical Properties of Organic Donors, BMDT-TTF and BEDSe-TSeF. Two Modifications of BEDT-TTF
Kato, R.; Kobayashi, H.; Kobayashi, A.; Sasaki, Y.
Chem. Lett. **1985**, 1231

Dimensionality Examination of Cation Radical Salts Based on EDT-TTF (EDT-TTF = Ethylenedithiotetrathiafulvalene)
Kato, R.; Kobayashi, H.; Kobayashi, A.
Chem. Lett. **1989**, 781

4,5-Methylenedithio-4′,5′-Propylenedithiotetrathiafulvalene (MPT) and 4,5-Ethylenedithio-4′, 5′-Propylenedithiotetrathiafulvalene (EPT)
Kini, A. M.; Tytko, S. F.; Hunt, J. E.; Williams, J. M.
Tetrahedron Lett. **1987**, *28*, 4153

Bis(ethylenediseleno)tetrathiafulvalene: Convenient One-pot Synthesis and X-ray Crystal Structure
Kini, A. M.; Gates, B. D.; Beno, M. A.; Williams, J. M.
J. Chem. Soc., Chem. Commun. **1989**, 169

Structures of Three New Ag(CN)$_2$ Salts of BEDT-TTF
Kurmoo, M.; Talham, D. R.; Pritchard, K. L.; Day, P.; Stringer, A. M.; Howard, J. A. K.
Synth. Met. **1988**, *27*, A177

Uniform, Non-interacting Antiferromagnetic Chains of Spins in the 1:1 Bis-ethylenedithiotetrathiafulvaleneium Salt of a Monovalent Hexarhenium Chalcohalide Cluster Anion: (BEDT-TTF)$^+$•(Re$_6$Se$_5$Vl$_9$)$^-$•(C$_3$H$_7$ON)$_2$
Penicaud, A.; Lenoir, C.; Batail, P.; Coulon, C.; Perrin, A.
Synth. Met. **1989**, *32*, 25

The X-Ray Crystal Structure and Physical Properties of (BEDT-TTF)AuBr$_2$Cl$_2$
Porter, L. C.; Wang, H. H.; Beno, M. A.; Carlson, K. D.; Pipan, C. M.; Proksch, R. B.; Williams, J. M.
Solid State Commun. **1987**, *64*, 387

Electrical Conductivity, Crystal Structure, and Variable Temperature ESR Investigations of (BEDSe-TTF)$_2$IBr$_2$
Porter, L. C.; Cariss, C. S.; Carlson, K. D.; Geiser, U.; Kini, A. M.; Montgomery, L. K.; Rubenstein, R. L.; Wang, H. H.; Whitworth, J. R.; Williams, J. M.
Synth. Met. **1988**, *27*, A223

The Synthesis, Structures and Physical Properties of New Forms of $(BEDT-TTF)_2I_3$
Qian, M.; Wang, X.; Zhu, Y.; Zhu, D.; Lu, L.; Ma, B.; Duan, H.; Zhang, D.
Synth. Met. **1988**, *27*, A277

Chemical and Physical Properties of a New Ambient Pressure Organic Superconductor with T_c Higher than 10 K
Saito, G.; Urayama, H.; Yamochi, H.; Oshima, K.
Synth. Met. **1988**, *27*, A331

Crystal Structures and Electrical Properties of $ET_3Cl_2 \cdot 2H_2O$ and $ET_4Cl_2 \cdot 4H_2O$ Organic Metals
Shibaeva, R. P.; Lobkovskaya, R. M.; Rozenberg, L. P.; Buravov, L. I.; Ignatiev, A. A.; Kushch, N. D.; Laukhina, E. E.; Makova, M. K.; Yagubskii, E. B.; Zvarykina, A. V.
Synth. Met. **1988**, *27*, A189

Crystal Structures and Electrical Properties of BEDT-TTF Salts of Mercury(II) Thiocyanate With and Without K Ion
Oshima, M.; Mori, H.; Saito, G.; Oshima, K.
Chem. Lett. **1989**, 1159

Organic Metals: Synthesis, Structure and Properties of $(BMDT-TTF)_2Au(CN)_2$
Nigrey, P. J.; Morosin, B.; Kwak, J. F.; Venturini, E. L.; Baughman, R. J.
Synth. Met. **1986**, *16*, 1

Synthesis and Properties of BEDSe-TTF
Nigrey, P. J.; Morosin, B.; Duesler, E.
Synth. Met. **1988**, *27*, B481

New Radical Cation Salts of BEDT-TTF with Complex Mercury Halide Anions
Müller, H.; Fritz, H. P.; Heidmann, C.-P.; Gross, F.; Veith, H.; Lerf, A.; Andres, K.; Fuchs, H.; Polborn, K.; Abriel, W.
Synth. Met. **1988**, *27*, A257

Physical Properties of $\beta''-(BEDT-TTF)_2ICl_2$
Ugawa, A.; Okawa, Y.; Yakushi, K.; Kuroda, H.; Kawamoto, A.; Tanaka, J.; Tanaka, M.; Nogami, Y.; Kagoshima, S.; Murata, K.; Ishiguro, T.
Synth. Met. **1988**, *27*, A407

Crystal and Electronic Structures and Physical Properties of $T_c = 10.4K$ Superconductor, $(BEDT-TTF)_2Cu(NCS)_2$
Urayama, H.; Yamochi, H.; Saito, G.; Sato, S.; Sugano, T.; Kinoshita, M.; Kawamoto, A.; Tanaka, J.; Inabe, T.; Mori, T.; Maruyama, Y.; Inokuchi, H.; Oshima, K.
Synth. Met. **1988**, *27*, A393

Bis(ethylenediseleno)tetrathiafulvalene Salts, (BEDSe-TTF)$_2$X, X$^-$ = I$_3$$^-$, AuI$_2$$^-$, and IBr$_2$$^-$
Wang, H. H.; Montgomery, L. K.; Geiser, U.; Porter, L. C.; Carlson, K. D.; Ferraro, J. R.; Williams, J. M.; Cariss, C. S.; Rubinstein, R. L.; Whitworth, J. R.; Evain, M.; Novoa, J. J.; Whangbo, M.-H.
Chem. Mat. **1989**, *1*, 140

The Synthesis, Crystal Structure, Electrical Conductivity and Band Electronic Structure of (BPDT-TTF)$_2$ICl$_2$
Williams, J. M.; Emge, T. J.; Firestone, M. A.; Wang, H. H.; Beno, M. A.; Geiser, U.; Nuñez, L.; Carlson, K. D.; Nigrey, P. J.; Whangbo, M.-H.
Mol. Cryst. Liq. Cryst. **1987**, *148*, 223

Ambient-Pressure Superconductivity at 2.7K and Higher Temperatures in Derivatives of (BEDT-TTF)$_2$IBr$_2$: Synthesis, Structure, and Detection of Superconductivity
Williams, J. M.; Wang, H. H.; Beno, M. A.; Emge, T. J.; Sowa, L. M.; Copps, P. T.; Behroozi, F.; Hall, L. N.; Carlson, K. D.; Crabtree, G. W.
Inorg. Chem. **1984**, *23*, 3839

BEDT-TTF Complexes with Percyano Substituted Organic Anions
Yamochi, H.; Tsuji, T.; Saito, G.; Suzuki, T.; Miyashi, T.; Kabuto, C.
Synth. Met. **1988**, *27*, A479

Electrical and Magnetic Properties of (BEDT-TTF)$_4$Ni(CN)$_4$ Complex
Tanaka, M.; Takeuchi, H.; Sano, M.; Enoki, T.; Suzuki, K.; Imaeda, K.
Bull. Chem. Soc. Jpn. **1989**, *62*, 1432

Charge-transfer Salts Derived from MET: Synthesis, Structure and Properties of (MET)ClO$_4$, (MET)PF$_6$ and (MET)$_3$(ReO$_4$)$_2$
Beno, M. A.; Geiser, U.; Kini, A. M.; Wang, H. H.; Carlson, K. D.; Miller, M. M.; Allen, T. J.; Schlueter, J. A.; Proksch, R. B.; Williams, J. M.
Synth. Met. **1988**, *27*, A209

Preparation, Structure and Physical Properties of an Organic Conductor, (BEDT-TTF)$_3$Br$_2$(H$_2$O)$_2$
Urayama, H.; Saito, G.; Sugano, T.; Kinoshita, M.; Kawamoto, A.; Tanaka, J.
Synth. Met. **1988**, *27*, A401

Synthesis, ESR Studies, Band Electronic Structure and Superconductivity in the (BEDT-TTF)$_2$M(NCS)$_2$ System (M = Cu, Ag, Au)
Carlson, K. D.; Geiser, U.; Kini, A. M.; Wang, H. H.; Montgomery, L. K.; Kwok, W. K.; Beno, M. A.; Williams, J. M.; Cariss, C. S.; Crabtree, G. W.; Whangbo, M.-H.; Evain, M.
Inorg. Chem. **1988**, *27*, 965

A Novel Macrocycle Containing the 4,5-dithio-1,3-dithiole-2-thione Unit and a Related Macrocycle Incorporation, The Tetrathiafulvalene Moiety
Girmay, B.; Kilburn, J. D.; Underhill, A. E.; Varma, K. S.; Hursthouse, M. B.; Harman, M. E.; Becher, J.; Bojesen, G.
J. Chem. Soc. Chem. Comm. **1989**, 1406

Conducting and Superconducting BEDT-TTF Based Synmetals
Wang, H. H.; Beno, M. A.; Whangbo, M.-H.; Emge, T. J.; Geiser, U.; Carlson, K. D.; Venturini, E. L.; Williams, J. M.
Israel J. Chem. **1986**, *27*, 309

Tetrakis(methyltelluro)tetrafulvalene (TTeC$_1$-TTF), A High-Mobility Organic Semiconductor
Inokuchi, H.; Imaeda, K.; Enoki, T.; Mori, T.; Maruyama, Y.; Saito, G.; Okada, N.; Yamochi, H.; Seki, K.; Higuchi, Y.; Yasuoka, N.
Nature **1987**, *329*, 39

Superconductivity Above 2K at Ambient Pressure in Iododibromide (IBr$_2$−) Charge-transfer Salts of Bis(ethylenedithio)tetrathiafulvalene, BEDT-TTF
Carlson, K. D.; Crabtree, G. W.; Hall, L. N.; Behroozi, F.; Copps, P. T.; Sowa, L. M.; Nuñez, L.; Firestone, M. A.; Wang, H. H.; Beno, M. A.; Emge, T. J.; Williams, J. M.
Mol. Cryst. Liq. Cryst. **1985**, *125*, 159

Tetrachalcogenafulvalenes with Outer Chalcogeno Substituents. Precursors of Organic Metals, Superconductors, LB Films, etc.
Saito, G.
Pure & Appl. Chem. **1987**, *59*, 999

Ambient Pressure Superconductivity at 4-5 K in β-(BEDT-TTF)$_2$AuI$_2$
Carlson, K. D.; Crabtree, G. W.; Nuñez, L.; Wang, H. H.; Beno, M. A.; Geiser, U.; Firestone, M. A.; Webb, K. S.; Williams, J. M.
Solid State Commun. **1986**, *57*, 89

Synthesis, Structure and Properties of BEDT-TTF Derivatives
Nigrey, P. J.; Morosin, B.; Kwak, J. F.
Novel Superconductivity, Wolf, S. A.; Kresin, V. Z., Eds., Plenum Press: New York, 1985, p. 171.

PRESSURE STUDIES

Superconductivity in a New Family of Organic Conductors
Parkin, S. S. P.; Engler, E. M.; Schumaker, R. R.; Lagier, R.; Lee, V. Y.; Scott, J. C.; Greene, R. L.
Phys. Rev. Lett. **1983**, *50*, 270

Effect of Pressure on the Superconductivity of β-(BEDT-TTF)$_2$I$_3$
Laukhin, V. N.; Kostyuchenko, E. E.; Sushko, Yu. V.; Shchegolev, I. F.; Yagubskii, E. B.
Pis'ma Zh. Eksp. Teor. Fiz. **1985**, *41*, 68; *JETP Lett.* **1985**, *41*, 81

Superconductivity with the Onset at 8 K in the Organic Conductor β-(BEDT-TTF)$_2$I$_3$ Under Pressure
Murata, K.; Tokumoto, M.; Anzai, H.; Bando, H.; Saito, G.; Kajimura, K.; Ishiguro, T.
J. Phys. Soc. Jpn. **1985**, *54*, 1236

Superconductivity with An Onset at 8 K in β-(BEDT-TTF)$_2$I$_3$ Under Pressure
Murata, K.; Tokumoto, M.; Anzai, H.; Bando, H.; Kajimura, K.; Ishiguro, T.; Saito, G.
Synth. Met. **1986**, *13*, 3

Ambient-Pressure Superconductivity at 8 K in the Organic Conductor β-(BEDT-TTF)$_2$I$_3$
Tokumoto, M.; Murata, K.; Bando, H.; Anzai, H.; Saito, G.; Kajimura, K.; Ishiguro, T.
Solid State Commun. **1985**, *54*, 1031

Homogeneous Superconducting State at 8.1 K Under Ambient Pressure in the Organic Conductor β-(BEDT-TTF)$_2$I$_3$
Creuzet, F.; Creuzet, G.; Jérome, D.; Schweitzer, D.; Keller, H. J.
J. Phys. Lett. (Les Ulis, Fr.) **1985**, *46*, L1079

Ambient-Pressure Superconductivity in Organic Metals, BEDT-TTF Trihalides
Tokumoto, M.; Bando, H.; Murata, K.; Anzai, H.; Kinoshita, N.; Kajimura, K.; Ishiguro, T.; Saito, G.
Synth. Met. **1986**, *13*, 9

Metastable State of β-(BEDT-TTF)$_2$I$_3$ with a Superconducting Transition Temperature of 7.5 K
Ginodman, V. B.; Gudenko, A. V.; Zasavitskii, I. I.; Yagubskii, E. B.
Pis'ma Zh. Eksp. Teor. Fiz. **1985**, *42*, 384; *JETP Lett.* **1985**, *42*, 472

Crystalline and Molecular Structure of the Organic Superconductor β-(BEDT-TTF)$_2$I$_3$ Under a Pressure of 9.5 kbar
Molchanov, V. N.; Shibaeva, R. P.; Kachinskii, V. N.; Yagubskii, E. B.; Simonov, V. I.; Vainshtein, B. K.
Dokl. Akad. Sci. USSR **1989**, *286*, 637

Magnetic Susceptibility of α and β Phase of Di[Bis(Ethylenediothiolo)tetrathiafulvalene] Tri-iodide [(BEDT-TTF)$_2$I$_3$] Under Pressure
Rothaemel, B.; Forro, L.; Cooper, J. R.; Schilling, J. S.; Weger, M.; Bele, P.; Brunner, H.; Schweitzer, D.; Keller, H. J.
Phys. Rev. B **1986**, *34*, 704

Shear Induced Superconductivity in β-(BEDT-TTF)$_2$I$_3$
Schirber, J. E.; Kwak, J. F.; Beno, M. A.; Wang, H. H.; Williams, J. M.
Physica **1986**, *143B*, 343

The Structure of β*-(BEDT-TTF)$_2$I$_3$ at 4.5 K and 1.5 kbar
Schultz, A. J.; Wang, H. H.; Williams, J. M.; Filhol. A.
Physica **1986**, *143B*, 354

Pressure Dependence of the Conduction-electron-spin-resonance Linewidth of the α- and β-phases of Di-bis(ethylenediothiolo)tetrathiafulvalene Triiodide
Forro, L.; Sekretarczyk, G.; Krupski, M.; Schweitzer, D.; Keller, H. J.
Phys. Rev. B **1987**, *35*, 2501

Influence of Pressure on the Superconducting Transition Temperature of the β-phase of Bis(ethylenedithiolo)tetrathiafulvalene Iodide of β-(ET)$_2$I$_3$ with $T_c = 7.5$ K
Kononovich, P. A.; Laukhin, V. N.; Sushko, Yu. V.; Shchegolev, I. F.
Sov. Phys. Solid Stat. **1987**, *29*, 534

Superconductivity in (BEDT-TTF)$_3$Cl$_2$2•H$_2$O
Mori, T.; Inokuchi, H.
Solid State Commun. **1987**, *64*, 335

Superconductivity of the Organic Conductor, β-(BEDT-TTF)$_2$I$_3$ Viewed from H$_{c2}$ and the Effect of Impurity
Murata, K.; Tokumoto, M.; Toyota, N.; Nashiyama, I.; Anzai, H.; Saito, G.; Muto, Y.; Kajimura, K.; Ishiguro, T.
Physica **1987**, *148B*, 506

Effect of Pressure on the Superconducting Transition Temperature of κ-(BEDT-TTF)$_2$Cu(NCS)$_2$
Schirber, J. E.; Venturini, E. L.; Kini, A. M.; Wang, H. H.; Whitworth, J. R.; Williams, J. M.
Physica C **1988**, *152*, 157

Superconducting Transition at $T_c = 5.3$ K in the High-pressure Phase of the Organic Metal (BEDT-TTF)$_4$Hg$_3$Cl$_2$
Lyubovskii, R. B.; Lyubovskaya, R. N.; Kapustin, N. V.
Pis'ma Zh. Eksp. Teor. Fiz. **1987**, *93*, 1863; *JETP Lett.* **1987**, *66*, 1063

High-pressure Transport Measurements of α'-BEDT-TTF Salts
Parker, I. D.; Friend, R. H.; Kurmoo, M.; Day, P.
J. Phys. Condens. Matter **1989**, *1*, 5681

Phase Metal-semiconductor Transition with Gigantic Hysteresis in Organic Metal (ET)$_2$IBr$_2$

Avramenko, N. B.; Zvarykina, A. V.; Laukhin, E. E.; Lubovskii, R. B.; Shibaeva, R. P.
Materials Sci., **1988**, *XIV*, 11

The Role of Pressure in the Study of Organic Superconductors
Schirber, J. E.; Williams, J. M.; Wang, H. H.
Mol. Cryst. Liq. Cryst. **1990**, *181*, 285

Some Properties of the Organic Superconductor κ-(BEDT-TTF)$_2$Cu(SCN)$_2$ Under Pressure
Kang, W.; Jérome, D.; Lenoir, C.; Batail, P.
J. Phys.: Condens. Matter **1990**, *2*, 1665

High Pressure Transport Measurements of α'-BEDT-TTF Salts
Parker, I. D.; Friend, R. H.; Kurmoo, M.; Day, P.
Synth. Met. **1988**, *27*, A433

Shear-induced Superconductivity in β-di[bis(ethylenedithio)tetrathiafulvalene]-triiodide [β-(BEDT-TTF)$_2$I$_3$]
Schirber, J. E.; Azevedo, L. J.; Kwak, J. F.; Venturini, E. L.; Leung, P. C. W.; Beno, M. A.; Wang, H. H.; Williams, J. M.
Phys. Rev. B **1986**, *33*, 1987

QUICK REPORT

Ambient Pressure Superconductivity at 5 K in the Organic System: β-(BEDT-TTF)$_2$AuI$_2$
Williams, J. M.; Wang, H. H.; Beno, M. A.; Geiser, U.; Firestone, M. A,.; Webb, K. S.; Nuñez, L.; Crabtree, G. L.; Carlson, K. D.; Azevedo, L. J.; Kwak, J. F.; Schirber, J. E.
Physica **1985**, *135B*, 520

Tetraformyltetraselenafulvalene (TFTSeF): Synthesis and Some Uses as a Precursor of Polyfunctionalized Tetraselenafulvalenes
Salle, M.; Gorgues, A.; Fabre, J.-M.; Bechgaard, K.; Jubault, M.; Texier, F.
J. Chem. Soc., Chem. Commun. **1989**, 1520

RADIATION EFFECTS

The Effect of Disorder on the Metal-Insulator and Superconductor Phase Transitions in the α and β Phases of (BEDT-TTF)$_2$I$_3$
Forro, L.; Bouffard, S.; Schweitzer, D.
Solid State Commun. **1988**, *65*, 1359

RAMAN EFFECT

Resonant Raman Scattering in Organic Conductors α- and β-$(BEDT-TTF)_2X$ ($X = I_3$ and IBr_2)
Sugai, S.; Saito, G.
Solid State Commun. **1986**, *58,* 759; *Synth. Met.* **1987**, *19,* 231

A Stable Superconducting State at 8 K and Ambient Pressure in $α_t$-$(BEDT-TTF)_2I_3$
Schweitzer, D.; Bele, P.; Brunner, H.; Gogu, E.; Haeberlen, U.; Hennig, I.; Klutz, I.; Swietlik, R.; Keller, H. J.
Z. Phys. B - Condens. Matt. **1987**, *67,* 489

Resonance Raman Investigations of the Symmetric Stretching Mode of I_3-Anions in α and β Phases of Di-bis(ethylenedithio)tetrathiafulvalene Tri-iodide
Swietlik, R.; Schweitzer, D.; Keller, H. J.
Phys. Rev. B **1987**, *36,* 6881

Electrochemically Prepared Radical Salts of BEDT-TTF: Molecular Metals and Superconductors
Schweitzer, D.; Gogu, E.; Hennig, I.; Klatz, I.; Keller, H. J.
Ber. Bunsenges. Phys. Chem. **1987**, *91,* 890

BEDT-TTF Radical Salts: Organic Metals and Superconductors
Schweitzer, D.
Proceedings of an International Winter School, Kirchberg, Tirol, *Electronic Properties of Conjugated Polymers*, **1987**, pp. 354–361

Raman Investigations on Single Crystals and Polycrystalline Pressed Samples of Organic Superconductors
Zamboni, R.; Schweitzer, D.; Keller, H. J.
Proceedings of NATO ASI, Spetses, *Lower-Dimensional Systems and Molecular Electronics* **1989**

Transport Properties of Single Crystals and Polycrystalline Pressed Samples of $(BEDT-TTF)_2X$ Salts and Related Coordination Polymers
Schweitzer, D.; Kahlich, S.; Gärtner, S.; Gogu, E.; Grimm, H.; Heinen, I.; Klutz, I.; Zamboni, R.; Keller, H. J.; Renner, G.
Proceedings of NATO ASI, Spetses, *Lower-Dimensional Systems and Molecular Electronics* **1989**

Raman Scattering Investigations of the Organic Crystal $β_d{'}$-$(BEDT-TTF)_2I_3$
Lu, L.; Liu, J.; Ma, B.; Zhang, D.; Wang, X.; Zhu, D.
Synth. Met. **1989**, *31,* 45

Resonance Raman Spectra of I_3^- in TEDA(triethylenediamine)$I_{2.5}$TCNQ and β-$(BEDT-TTF)_2I_3$
Bandrauk, A. D.; Truong, K. D.; Carlone, C.

Organic and Inorganic Low-Dimensional Crystalline Materials, Delhaes, P.; Drillon, M. Eds; Plenum Publishing: **1987,** pp. 329–332

Resonant Raman Scattering on Single Crystals of $(BEDT-TTF)_2Cu(NCS)_2$
Zamboni, R.; Schweitzer, D.; Keller, H. J.
Solid State Commun. **1990**, *73*, 41

Raman Spectra of Radical Ion DBTTF Complexes; Relation Between Raman Frequency and Formal Charge
Tanaka, M.; Shimizu, M.; Saito, Y.; Tanaka, J.
Chem. Phys. Lett. **1986**, *125*, 594

REVIEWS

Organic Amalgams: Substances with Metallic Properties Composed in Part of Non-Metallic Elements
McCoy, H. N.; Moore, W. C.
J. Am. Chem. Soc. **1911**, *33*, 273

The Many Faces of ET
Parkin, S. S. P.; Engler, E. M.; Lee, V. Y.; Schumaker, R. R.
Mol. Cryst. Liq. Cryst. **1985**, *119*, 375

Organic Superconductors: Structural Aspects and Design of New Materials
Williams, J. M.; Beno, M. A.; Wang, H. H.; Leung, P. C. W.; Emge, T. J.; Geiser, U.; Carlson, K. D.
Acc. Chem. Res. **1985**, *18*, 261

Superconductors with Unusual Properties and Possibilities of Increasing the Critical Temperature
Golovashkin, A. I.
Soc. Phys. Usp. **1986**, *29*, 199

Superconductivity in Organic Charge Transfer Salts
Ishiguro, T.
Physica C **1988**, *153*, 1055

Organic Layered Superconductors
Bulaevskii, L. N.
Advances in Physics **1988**, *37*, 443

Introduction to Synthetic Electrical Conductors
Ferraro, J. R.; Williams, J. M.
Academic Press, Orlando, Florida **1987**

New Organic Superconductors
Inokuchi, H.
Angew. Chem. Int. Ed. Engl. **1988**, *27*, 1747

Composite Crystals: What Are They and Why Are They So Common in the Organic Solid State?
Coppens, P.; Maly, K.; Petricek, V.
Mol. Cryst. Liq. Cryst. **1990**, *181*, 81

The Use of Vibrational Spectroscopy in the Characterization of Synthetic Organic Electrical Conductors and Superconductors
Ferraro, J. R.; Williams, J. M.
"Practical Fourier Transform Infrared Spectroscopy," 1990, Academic Press, Eds. Ferraro, J. R.; Krishnan, K., p. 41

Exotic Organic Superconductors Based on BEDT-TTF and the Prospects of Raising T_c's
Williams, J. M.; Beno, M. A.; Wang, H. H.; Geiser, U.; Emge, T. J.; Leung, P. C. W.; Crabtree, G. W.; Carlson, K. D.; Azevedo, L. J.; Venturini, E. L.; Schirber, J. E.; Kwak, J. F.; Whangbo, M.-H.
Physica **1986**, *136B*, 371

Characterization of Synthetic Organic Superconductors By Use of Vibrational Spectroscopy
Ferraro, J. F.; Williams, J. M.
Appl. Spectroscopy **1990**, *44*, 200

Concluding Remarks: Organic Conducting Crystals
Jérome, D.
Synth. Met. **1987**, *19*, 1017

Approaches to the High T_c Superconductivity in β-(BEDT-TTF)$_2X$ Structure
Tokumoto, M.; Anzai, H.; Murata, K.; Bando, H.; Kajimura, K.; Morita, S.; Ishiguro, T.; Saito, G.
Synth. Met. **1987**, *19*, 169

SPECIFIC HEAT

Specific Heat of the Ambient-pressure Organic Superconductor β-di(*bis*-(ethylenedithio)tetrathiafulvalene Triiodide
Stewart, G. R.; O'Rourke, J.; Crabtree, G. W.; Carlson, K. D.; Wang, H. H.; Williams, J. M.; Gross, F.; Andres, K.
Phys. Rev. B **1986**, *33*, 2046

Specific Heat in High Magnetic Field of κ-di[bis(ethylenedithio)tetrathiafulvalene]-di(thiocyanato)cuprate [κ-(ET)$_2$Cu(NCS)$_2$]: Evidence for Strong-coupling Superconductivity
Andraka, B.; Kim, J. S.; Stewart, G. R.; Carlson, K. D.; Wang, H. H.; Williams, J. M.
Phys. Rev. B **1989**, *40*, 11345

Low-temperature Specific Heat of Organic Superconductor κ-(BEDT-TTF)$_2$Cu(NCS)$_2$
Katsumoto, S.; Kobayashi, S.; Urayama, H.; Yamochi, H.; Saito, G.
J. Phys. Soc. Jpn. **1988**, *57*, 3672

Bulk Superconducting Specific-heat Anomaly in β-di[bis(ethylenedithio)tetrathiafulvalene] diiodoaurate [β-(ET)$_2$AuI$_2$]
Stewart, G. R.; Williams, J. M.; Wang, H. H.; Hall, L. N.; Perozzo, M. T.; Carlson, K. D.
Phys. Rev. B **1986**, *34*, 6509

Specific Heat of Superconducting κ-(BEDT-TTF)$_2$Cu(NCS)$_2$ Near T_c [where BEDT-TTF is Bis(ethylenedithio)tetrathiafulvalene]
Graebner, J. E.; Haddon, R. C.; Chichester, S. V.; Glarum, S. H.
Phys. Rev. B **1990**, *41*, 4808

STRUCTURAL PHASE TRANSITIONS

X-Ray Evidence for a Structural Phase Transition in the Organic Two-Dimensional Conductor (BEDT-TTF)$_2$ClO$_4$(C$_2$H$_3$Cl$_3$)$_{0.5}$
Kagoshima, S.; Pouget, J. P.; Saito, G.; Inokuchi, H.
Solid State Commun. **1983**, *45*, 1001

Two-Dimensionality and Suppression of Metal-Semiconductor Transition in a New Organic Metal with Alkylthio Substituted TTF and Perchlorate
Saito, G.; Enoki, T.; Toriumi, K.; Inokuchi, H.
Solid State Commun. **1982**, *42*, 557

A Metal-Insulator Phase Transition Close to Room Temperature: (BEDT-TTF)$_2$SbF$_6$ and (BEDT-TTF)$_2$AsF$_6$
Laversanne, R.; Amiell, J.; Delhaes, P.; Chasseau, D.; Hauw, C.
Solid State Commun. **1984**, *52*, 177

Superconducting Properties of the Orthorhombic Phase of bis-(ethylenedithiolo)-tetrathiafulvalene Triiodide
Yagubskii, E. B.; Shchegolev, I. F.; Pesotskii, S. I.; Laukhin, V. N.; Kononovich, P. A.; Kartsovnik, M. V.; Zvarykina, A. V.
Pis'ma Zh. Eksp. Teor. Fiz. **1984**, *39*, 275; *JETP Lett.* **1984**, *39*, 328

Coexistence of Different Superconducting Phases with Transition Temperatures Between 1.5 and 7 K in the (BEDT-TTF)-I$_3$ Systems

Yagubskii, E. B.; Shchegolev, I. F.; Topnikov, V. N.; Pesotskii, S. I.; Laukhin, V. N.; Kononovich, P. A.; Kartsovnik, M. V.; Zvarykina, A. V.; Dedik, S. G.; Buravov, L. I.
Pis'ma Zh. Eksp. Teor. Fiz. **1985**, *88*, 244; *JETP Lett.* **1985**, *61*, 142

Superconducting Transition in the Dielectric α Phase of Iodine-Doped (BEDT-TTF)$_2$I$_3$ Compound
Yagubskii, E. B.; Shchegolev, I. F.; Laukhin, V. N.; Shibaeva, R. P.; Kostyuchenko, E. E.; Khomenko, A. G.; Sushko, Yu. V.; Zvarykina, A. V.
Pis'ma Zh. Eksp. Teor. Fiz. **1984**, *40*, 387; *JETP Lett.* **1984**, *40*, 1201

(BEDT-TTF)$_2$SbF$_6$ and (BEDT-TTF)$_2$AsF$_6$: A Metal-Insulator Close to Room Temperature
Laversanne, R.; Amiell, J.; Delhaes, P.; Chasseau, D.; Hauw, C.
Mol. Cryst. Liq. Cryst. **1985**, *119*, 405

Superconducting Transitions in β-(BEDT-TTF)$_2$I$_3$
Buravov, L. I.; Kartsovnik, M. V.; Kaminskii, V. F.; Kononovich, P. A.; Kostuchenko, E. E.; Laukhin, V. N.; Makova, M. K.; Pesotskii, S. I.; Shchegolev, I. F.; Topnikov, V. N.; Yagubskii, E. B.
Synth. Met. **1985**, *II*, 207

Competition Between Organic Superconductivity and a Displacive Structural Modulation in Bis(ethylenedithio)tetrathiafulvalene Perrhenate, (BEDT-TTF)$_2$ReO$_4$
Ravy, S.; Moret, R.; Pouget, J. P.; Comes, R.; Parkin, S. S. P.
Phys. Rev. B **1986**, *33*, 2049

Anomalous Increase in the Conductivity of the System α-(BEDT-TTF-d$_8$)$_2$I$_3$ in the Region of the Metal-Insulator Phase Transition
Pokhodnya, K. I.; Sushko, Yu. V.; Tanatar, M. A.; Baram, G. O.; Yurchenko, A. A.
Pis'ma Zh. Eksp. Teor. Fiz. **1986**, *43*, 252; *JETP Lett.* **1986**, *43*, 323

Polymorphic Modifications of Crystals of (BEDT-TTF)$_2$I$_3$ With Transition Metals With Superconductor and Metal-Dielectric Properties
Kaminskii, V. F.; Laukhin, V. N.; Merzhanov, V. A.; Neiland, O. Ya.; Knodorkovskii, V. Yu.; Shibaeva, R. P.; Yagubskii, E. B.
Translated from *Izvestiya Akademii Nauk SSR, Seriya Khimicheskaya, No. 2*, pp. 342–347, Feb., **1986**; Original article submitted 8/20/84.

Transformation of the α-phase (BEDT-TTF)$_2$I$_3$ to the Superconducting β-phase with T_c = 6–7 K
Baram, G. O.; Buravov, L. I.; Degtyarev, L. S.; Kozlov, M. E.; Laukhin, V. N.; Laukhina, E. E.; Onishchenko, V. G.; Pokhodnya, K. I.; Sheinkman, M. K.; Shibaeva, R. P.; Yagubskii, E. B.
Pis'ma Zh. Eksp. Teor. Fiz. **1986**, *44*, 293; *JETP Lett.* **1986**, *44*, 376

X-Ray Evidence For Structural Changes In The Organic Conductors, α-(BEDT-TTF)$_2$I$_3$, α-(BEDT-TTF)$_2$IBr$_2$ and β-(BEDT-TTF)$_2$I$_3$
Nogami, Y.; Kagoshima, S.; Sugano, T.; Saito, G.
Synth. Met. **1986**, *16*, 367

The Role of Two High-Temperature Phase Transitions on Occurrence and Stability of the High-T_c Superconducting State of β-(BEDT-TTF)$_2$I$_3$: Evidence from c*-Resistivity Study
Hamzic, B.; Creuzet, G.; Lenoir, C.
Europhysics Lett. **1987**, *3*, 373

Direct Detection of the β-1.5 β-8 Phase Transition in β-dibis(ethylenedithio)-tetrathiafulvalene triiodide, β-(ET)$_2$I$_3$
Ginodman, V. B.; Gudenko, A. V.; Kononovich, P. A.; Laukhin, V. N.; Shchegolev, I. F.
Pis'ma Zh. Eksp. Teor. Fiz. **1986**, *44*, 523; *JETP Lett.* **1986**, *44*, 674

In Situ Observation of a Phase Transformation in α-(BEDT-TTF)$_2$I$_3$ Using a Contactless Technique
Späth, K.; Gross, F.; Heidmann, C. P.; Andres, K.
Ber. Bunsenges. Phys. Chem. **1987**, *91*, 909

Metal-semiconductor Phase Transition with a Giant Hysteresis in the Organic Metal (ET)$_2$IBr$_2$
Avramenko, N. V.; Zvarykina, A. V.; Laukhin, V. N.; Laukhina, E. E.; Lyubovskii, R. B.; Shibaeva, R. P.
Pis'ma Zh. Eksp. Teor. Fiz. **1988**, *48*, 429; *JETP Lett.* **1988**, *48*, 472

Metal-to-Semiconductor Transition of an Organic Metal, (EDT-TTF)$_2$AuBr$_2$
Mori, T.; Inokuchi, H.
Solid State Commun. **1989**, *70*, 823

Thermal Conversion of α-(BEDT-TTF)$_2$IBr$_2$ to Superconducting β-(BEDT-TTF)$_2$IBr$_2$
Wang, H. H.; Carlson, K. D.; Montgomery, L. K.; Schlueter, J. A.; Cariss, C. S.; Kwok, W. K.; Geiser, U.; Crabtree, G. W.; Williams, J. M.
Solid State Commun. **1988**, *66*, 1113

Electron Spin Resonance, Infrared Spectroscopic, and Molecular Packing Studies of the Thermally Induced Conversiton of Semiconducting α- to Superconducting α$_t$-(BEDT-TTF)$_2$I$_3$
Wang, H. H.; Ferraro, J. R.; Carlson, K. D.; Montgomery, L. K.; Geiser, U.; Williams, J. M.; Whitworth, J. R.; Hill, S.; Whangbo, M.-H.; Evain, M.; Novoa, J. J.
Inorg. Chem. **1989**, *28*, 2267

High T_c Superconducting State in Organic Metal β-(BEDT-TTF)$_2$X

Tokumoto, M.; Murata, K.; Bando, H.; Anzai, H.; Kajimura, K.; Ishiguro, T.
Physica **1986**, *143B*, 338

Contrasted Structural Properties of Organic Superconductors
Ravy, S.; Moret, R.; Pouget, J. P.; Comes, R.
Synth. Met. **1987**, *19*, 237

THERMOELECTRIC STUDIES

Anisotropic Thermopower of the Organic Metal, β-(BEDT-TTF)$_2$I$_3$
Mortensen, K.; Williams, J. M.; Wang, H. H.
Solid State Commun. **1985**, *56*, 105

Electrical Properties of the Organic Conductor (BEDT-TTF)$_3$(ClO$_4$)
Imaeda, K.; Enoki, T.; Saito, G.; Inokuchi, H.
Bull. Chem. Soc., Jpn. **1988**, *61*, 3332

Thermoelectric Power of Organic Superconductors - Calculation on the Basis of the Tight-Binding Theory
Mori, T.; Inokuchi, H.
J. Phys. Soc. Jpn. **1988**, *57*, 3674

THEORY

Two-Dimensional Band Structure of an Organic Metal, Perchlorate Salt of Bis(ethylenedithiolo)tetrathiafulvalene (BEDT-TTF)$_2$ClO$_4$(C$_2$H$_3$Cl$_3$)$_{0.5}$
Mori, T.; Kobayashi, A.; Sasaki, Y.; Kobayashi, H.; Saito, G.; Inokuchi, H.
Chem. Lett. **1982**, 1963

Band Structures of Two Types of (BEDT-TTF)$_2$I$_3$
Mori, T.; Kobayashi, A.; Sasaki, Y.; Kobayashi, H.; Saito, G.; Inokuchi, H.
Chem. Lett. **1984**, 957

The Intermolecular Interaction of Tetrathiafulvalene and Bis(ethylenedithio)-tetrathiafulvalene in Organic Metals. Calculation of Orbital Overlaps and Models of Energy-band Structures
Mori, T.; Kobayashi, A.; Sasaki, Y.; Kobayashi, H.; Saito, G.; Inokuchi, H.
Bull. Chem. Soc. Jpn. **1984**, *57*, 627

One-Dimensional Nature of the Band-Structure Parameters of (BPDT-TTF)$_3$(PF$_6$)$_2$ and (BPDT-TTF)$_2$I$_3$
Mori, T.; Kobayashi, A.; Sasaki, Y.; Kato, R.; Kobayashi, H.
Chem. Lett. **1984**, 1335

Band Structure of β-(BEDT-TTF)$_2$PF$_6$. One-Dimensional Metal Along the Side-by-Side Molecular Array
Mori, T.; Kobayashi, A.; Sasaki, Y.; Kato, R.; Kobayashi, H.
Solid State Commun. **1985**, *53*, 627

Structural Systematics in the Family of (BEDT-TTF)$_2$X Salts
Kistenmacher, T. J.
Solid State Commun. **1985**, *53*, 831

Electronic Structure of the Organic Conductors Based on BMDT-TTF [Bis(methylenedithio)tetrathiafulvalene]
Kato, R.; Kobayashi, H.; Mori, T.; Kobayashi, A.; Sasaki, Y.
Solid State Commun. **1985**, *55*, 387

Superconductivity of Narrow-band Metals and Semiconductors and the Model of Superconducting Glass
Kulik, I. O.
Sov. Phys. Usp. **1985**, *28*, 92

Crystal and Band Electronic Structures of an Organic Salt with the First Three-Dimensional Radical-Cation Donor Network, (BEDT-TTF)Ag$_4$(CN)$_5$
Geiser, U.; Wang, H. H.; Gerdom, L. E.; Firestone, M. A.; Sowa, L. M.; Williams, J. M.; Whangbo, M.-H.
J. Am. Chem. Soc. **1985**, *107*, 8305

Band Electronic Structures of the Ambient Pressure Organic Superconductors β-(ET)$_2$X (X = I$_3^-$, IBr$_2^-$)
Whangbo, M.-H.; Williams, J. M.; Leung, P. C. W.; Beno, M. A.; Emge, T. J.; Wang, H. H.; Carlson, K. D.; Crabtree, G. W.
J. Am. Chem. Soc. **1985**, *107*, 5815

Role of the Intermolecular Interactions in the Two-Dimensional Ambient-Pressure Organic Superconductors β-(ET)$_2$I$_3$ and β-(ET)$_2$IBr$_2$
Whangbo, M.-H.; Williams, J. M.; Leung, P. C. W.; Beno, M. A.; Emge, T. J.; Wang, H. H.
Inorg. Chem. **1985**, *24*, 3502

Crystal and Band Electronic Structures of a New Class of 2:1 Organic Conducting Salts α-(BEDT-TTF)$_2$X, X$^-$ = Ag(CN)$_2^-$, Au(CN)$_2^-$ and AuBr$_2^-$
Beno, M. A.; Firestone, M. A.; Leung, P. C. W.; Sowa, L. M.; Wang, H. H.; Williams, J. M.; Whangbo, M.-H.
Solid State Commun. **1986**, *57*, 735

Crystal and Band Structures Of An Organic Conductor β''-(BEDT-TTF)$_2$AuBr$_2$
Mori, T.; Sakai, F.; Saito, G.; Inokuchi, H.
Chem. Lett. **1986**, 1037

Structural and Electrical Properties of (BEDT-TTF)(TCNQ)
Mori, T.; Inokuchi, H.
Solid State Commun. **1986**, *59*, 355

Structural Trends in (BEDT-TTF)$_2X$ Salts with Linear Anions: A Shear Transformation View
Kistenmacher, T. J.
Solid State Commun. **1986**, *60*, 913

Importance of Intermolecular Hydrogen$\cdots$Hydrogen and Hydrogen$\cdots$Anion Contacts for the Lattice Softness, the Electron-Phonon Coupling, and the Superconducting Transition Temperatures, T_c, of Organic Conducting Salts β-(ET)$_2X$ ($X^- =$ IBr$_2^-$, AuI$_2^-$, I$_3^-$)
Whangbo, M. H.; Williams, J. M.; Schultz, A. J.; Emge, T. J.; Beno, M. A.
J. Am. Chem. Soc. **1987**, *109*, 90

Effects of Randomness on Critical Temperature of Organic Superconductors
Hasegawa, Y.; Fukuyama, H.
J. Phys. Soc. Jpn. **1986**, *55*, 3717

On the Mechanism of Superconductivity in the Organic Conductors Composed of TTF-Analogs
Yamaji, K.
Solid State Commun. **1987**, *61*, 413

Self-Consistent Band Structure For β-(BEDT-TTF)$_2$I$_3$
Kubler, J.; Weger, M.; Sommers, C.
Solid State Commun. **1987**, *62*, 801

Identification of the Superconductivity Type in Organic Superconductors
Burlachkov, L. I.; Gor'kov, L. P.; Lebed, A. G.
Europhys. Lett. **1987**, *4*, 941

Theory of Organic Superconductivity
Mazumdar, S.
Solid State Commun. **1988**, *66*, 427

Explanation of the Phase Diagram of β-(di[bis(ethylenedithio)tetrathiafulvalene]-triiodide Based on the Two Configurations of the Organic Molecule
Ravy, S.; Moret, R.; Pouget, J. P.
Phys. Rev. B **1988**, *38*, 4469

Relation between Geometry and Charge Transfer in Low-dimensional Organic Salts
Umland, T. C.; Allie, S.; Kuhlmann, T.; Coppens, P.
J. Phys. Chem. **1988**, *92*, 6456

Electronic Structure of Counterions in Organic Salts: $Au(CN)_2^-$ and $Au(NC)_2^-$
Gutsev, G. L.
J. Structural Chem. **1988**, *29*, 670

A Specific Cation Stabilization Parameter for the Linear Correlation of Relative Oxidation Potentials
Hinkelmann, K.; Liou, K. K.; Wudl, F.
J. Chem. Soc. Chem. Commun. **1989**, 1744

MNDO Calculations on Tetrathiafulvalenes
Bowadt, S.; Jensen, F.
Synth. Met. **1989**, *32*, 179

Band Electronic Structure Study of the Metal-Insulator Transitions in $(BEDT-TTF)_3Cl_2\cdot 2H_2O$ and $(BEDT-TTF)_4Cl_2\cdot 4H_2O$
Whangbo, M.-H.; Ren, J.; Kang, D. B.; Williams, J. M.
Mol. Cryst. Liq. Cryst. **1990**, *181*, 17

Ab Initio Computational Study of the C-H•••Donor and C-H•••Anion Contact Interactions in Organic Donor Salts
Novoa, J. J.; Whangbo, M.-H.; Williams, J. M.
Mol. Cryst. Liq. Cryst. **1990**, *181*, 25

Electron-phonon Effects in the Infrared Properties of Metals
Allen, P. B.
Phys. Rev. B **1971**, *3* 305

Organic Linear Conductors as Systems for the Study of Electron-Phonon Interactions in the Organic Solid State
Rice, M. J.
Phys. Rev. Lett. **1976**, *37*, 36

Phase Phonons and Intramolecular Electron-Phonon Coupling in the Organic Linear Chain Semiconductor $TEA(TCNQ)_2$
Rice, M. J.; Pietronero, L.; Breüsch, L.
Solid State Commun. **1977**, *21*, 757

Organic Superconductivity: An Experiment Based Theory
Mazumdar, S.; Ramasesha, S.
Synth. Met. **1988**, *27*, A105

Mechanism of Superconductivity in TTF-analog Complexes
Yamaji, K.
Synth. Met. **1988**, *27*, A115

Band Electronic Structures of Unsymmetrical Electron Donor Salts (MET)X, (MPT)$_2$•(THF), and (MET)$_3$(ReO$_4$)$_2$ ($X = $ ClO$_4^-$, PF$_6^-$)
Ren, J.; Evain, M.; Whangbo, M.-H.; Beno, M. A.; Geiser, U.; Kini, A. M.; Wang, H. H.; Williams, J. M.
Solid State Commun. **1989**, *70*, 615

Network of Intermolecular Contacts in Organic Conductors and Superconductors
Helberg, H. W.
Synth. Met. **1987**, *19*, 251

Valence Electronic Structures of Tetrakis(alkylthio)tetrathiafulvalenes
Seki, K.; Tang, T. B.; Mori, T.; Ji, W.-P.; Saito, G.; Inokuchi, H.
J. Chem. Soc., Faraday Trans. **1986**, *82*, 1067

Importance of Weak Hydrogen Bonding C-H•••Donor and C-H•••Anion Interactions in Governing the Structural Properties of Organic Donors BEDT-TTF and BEDO-TTF and Their Charge-Transfer Salts
Whangbo, M.-H.; Jung, D.; Ren, J.; Evain, M.; Novoa, J. J.; Mota, F.; Alvarez, S.; Williams, J. M.; Beno, M. A.; Kini, A. M.; Wang, H. H.; Ferraro, J. R.
The Physics and Chemistry of Organic Superconductors, Saito, G.; Kagoshima, S., Eds., Springer-Verlag, Berlin, **1990**, vol. 51, pp. 262–266

Importance of Intermolecular Hydrogen-Hydrogen and Hydrogen-Anion Contacts for the Lattice Softness and the Superconductivity of θ-(ET)$_2$X ($X = $ I$_3^-$, AuI$_2^-$, IBr$_2^-$,)
Whangbo, M.-H.; Williams, J. M.; Schultz, A. J.; Beno, M. A.
Organic and Inorganic Low-Dimensional Crystalline Materials, Delhaes, P.; Drillon, M., Eds., Plenum Press: New York, 1987, p. 333.

Effects of Anion Size and Disorder on Spin Resonance in the BEDT-TTF Salts
Venturini, E. L.; Schirber, J. E.; Wang, H. H.; Williams, J. M.
Synth. Met. **1988**, *27*, A243

Theory of Superconductivity
Bardeen, J.; Cooper, L. N.; Schrieffer, J. R.
Phys. Rev. **1957**, *108*, 1175

THERMAL EXPANSION

Thermal Expansion and Stepwise Superconducting Transition of β-(BEDT-TTF)$_2$(I$_3$)$_{1-x}$(AuI$_2$)$_x$
Kobayashi, H.; Kato, R.; Kobayashi, A.; Mori, T.; Inokuchi, H.
Solid State Commun. **1986**, *60*, 473

Thermal Expansion of the Organic Conductor β-(BEDT-TTF)$_2$I$_3$
Hamzic, B.; Creuzet, G.; Schweitzer, D.; Keller, H. J.
Solid State Commun. **1986**, *60*, 763

A New Transformable BEDT-TTF Complex, (BEDT-TTF)$_2$(IBr$_2$)$_2$(TCE)$_{0.5}$
Yamochi, H.; Urayama, H.; Saito, G.; Oshima, K.; Kawamoto, A.; Tanaka, A.
Synth. Met. **1988**, *27*, A485

TUNNELING STUDIES

Determination of the Electron Phonon Coupling and the Superconducting Gap in β-(BEDT-TTF)$_2$X
Nowack, A.; Poppe, U.; Weger, M.; Schweitzer, D.; Schwenk, H.
Z. Phys. B. - Condens. Matt. **1987**, *68*, 41

Tunneling Spectroscopic Study on the Superconducting Gap of (BEDT-TTF)$_2$Cu(NCS)$_2$ Crystals
Maruyama, Y.; Inabe, T.; Urayama, H.; Yamochi, H.; Saito, G.
Solid State Commun. **1988**, *67*, 35

Measurement of the Energy Gap in an Organic Superconductor: Evidence for Extremely Strong Coupling
Hawley, M. E.; Gray, K. E.; Terris, B. D.; Wang, H. H.; Carlson, K. D.; Williams, J. M.
Phys. Rev. Lett. **1986**, *57*, 629

X-RAY DIFFRACTION

Crystal Structure of a New Type of Two-Dimensional Organic Metal, (C$_{10}$H$_8$S$_8$)$_2$ClO$_4$(C$_2$H$_3$Cl$_3$)$_{0.5}$
Kobayashi, H.; Kobayashi, A.; Sasaki, Y.; Saito, G.; Enoki, T.; Inokuchi, H.
J. Am. Chem. Soc. **1983**, *105*, 297

Crystal Structure of α-(BEDT-TTF)$_2$PF$_6$
Kobayashi, H.; Kato, R.; Mori, T.; Kobayashi, A.; Sasaki, Y.; Saito, G.; Inokuchi, H.
Chem. Lett. **1983**, 759

The Crystal Structures of (BEDT-TTF)ReO$_4$(THF)$_{0.5}$ and (BEDT-TTF)IO$_4$(THF)$_{0.5}$
Kobayashi, H.; Kobayashi, A.; Sasaki, Y.; Saito, G.; Inokuchi, H.
Chem. Lett. **1984**, 183

Crystal Structure of the Organic Superconductor $(BEDT-TTF)_2I_3$
Kaminskii, V. F.; Prokhorova, T. G.; Shibaeva, R. P.; Yagubskii, E. B.
Pis'ma Zh. Eksp. Teor. Fiz. **1983**, *39*, 15; *JETP Lett.* **1984**, *39*, 17

Two-Dimensional Character of Crystal Structure of Organic Conductor, $(BMDT-TTF)_3PF_6(1,2-Dichloroethane)$
Kato, R.; Kobayashi, A.; Sasaki, Y.; Kobayashi, H.
Chem. Lett. **1984**, 993

Structure of Semiconducting 3,4;3',4'-Bis(ethylenedithio)-2,2',5,5'-tetrathiafulvalene-Hexafluoroarsenate (2:1), $(BEDT-TTF)_2AsF_6$, $(C_{10}H_8S_8)_2AsF_6$
Leung, P. C. W.; Beno, M. A.; Blackman, G. S.; Coughlin, B. R.; Miderski, C. A.; Joss, W.; Crabtree, G. W.; Williams, J. M.
Acta Cryst. **1984**, *C40*, 1331

Structure of 3,4;3',4'-Bis(ethylenedithio)-2,2',5,5'-tetrathiafulvalene-Tetrabromoindate (2:1), $(BEDT-TTF)_2InBr_4$, $(C_{10}H_8S_8)_2InBr_4$
Beno, M. A.; Cox, D. D.; Williams, J. M.; Kwak, J. F.
Acta. Cryst. **1984**, *C40*, 1334

Crystal Structure of $(BMDT-TTF)_3ClO_4(1,2-Dichloroethane)$. Molecular Design for the Multi-Dimensional Molecular Conductor
Kato, R.; Kobayashi, H.; Kobayashi, A.; Sasaki, Y.
Chem. Lett. **1984**, 1693

Novel Structural Modulation in the First Ambient-Pressure Sulfur-Based Organic Superconductor $(BEDT-TTF)_2I_3$
Leung, P. C. W.; Emge, T. J.; Beno, M. A.; Wang, H. H.; Williams, J. M.; Petricek, V.; Coppens, P.
J. Am. Chem. Soc. **1984**, *106*, 7644

The Crystal Structure of the Organic Superconductor Bis(ethylenedithiolo)-tetrathiafulvalene Triiodide $(BEDT-TTF)_2I_3$
Shibaeva, R. P.; Kaminskii, V. F.; Bel'skii, V. K.
Sov. Phys. Crystallogr. **1984**, *29*, 638

Crystal Structure and Properties of the DBTTF Salt with a Novel Chloride-Bridged Trimerized Dimethyltin (IV) Dianion. $[DBTTF]_3[Sn_3(CH_3)_6Cl_8]\cdot C_6H_5CN$
Matsubayashi, G.; Shimizu, R.; Tanaka, T.
Chem. Lett. **1985**, 973

Crystal Structures of Organic Metals and Superconductors of (BEDT-TTF)-I System
Shibaeva, R. P.; Kaminskii, V. F.; Yagubskii, E. B.
Mol. Cryst. Liq. Cryst. **1985**, *119*, 361

Novel Low Temperature Modulated Structure of the Ambient Pressure Superconductor (BEDT-TTF)$_2$I$_3$ and a Design Strategy for New Superconducting Polyhalide Phases
Williams, J. M.; Leung, P. C. W.; Emge, T. J.; Wang, H. H.; Beno, M. A.; Petricek, V.; Coppens, P.
Mol. Cryst. Liq. Cryst. **1985**, *119*, 347

Three-Dimensional Intermolecular Se•••Se Network in (BEDSe-TSeF)PF$_6$
Kato, R.; Kobayashi, H.; Kobayashi, A.; Sasaki, Y.
Chem. Lett. **1985**, 1943

Novel Structural Modulation in the Ambient-Pressure Sulfur-Based Organic Superconductor β-(BEDT-TTF)$_2$I$_3$: Origin and Effects on Its Electrical Conductivity
Leung, P. C. W.; Emge, T. J.; Beno, M. A.; Wang, H. H.; Williams, J. M.; Petricek, P.; Coppens, P.
J. Am. Chem. Soc. **1985**, *107*, 6184

Crystal Structure and Physical Properties of (TTM-TTF)I$_{2.47}$
Wu, P.; Mori, T.; Enoki, T.; Imaeda, K.; Saito, G.; Inokuchi, H.
Chem. Soc. Bull. Jpn. **1986**, *59*, 127

The Crystal and Molecular Structures of Bis(ethylenedithio)tetrathiafulvalene
Kobayashi, H.; Kobayashi, A.; Sasaki, Y.; Saito, G.; Inokuchi, H.
Bull. Chem. Soc. Jpn. **1986**, *59*, 301

Crystal and Band Electronic Structures of a New Class of 2:1 Organic Conducting Salts α-(BEDT-TTF)$_2$X, $X^- $ = Ag(CN)$_2^-$, Au(CN)$_2^-$ and AuBr$_2^-$
Beno, M. A.; Firestone, M. A.; Leung, P. C. W.; Sowa, L. M.; Wang, H. H.; Williams, J. M.; Whangbo, M.-H.
Solid State Commun. **1986**, *57*, 735

The Crystal Structure of β'-(BEDT-TTF)$_2$ICl$_2$. A Modification of the Organic Superconductor, β-(BEDT-TTF)$_2$I$_3$
Kobayashi, H.; Kato, R.; Kobayashi, A.; Saito, G.; Tokumoto, M.; Anzai, H.; Ishiguro, T.
Chem. Lett. **1986**, 89

Crystal Structure of α-(BEDT-TTF)$_2$BrICl
Kobayashi, H.; Kato, R.; Kobayashi, A.; Saito, G.; Tokumoto, M.; Anzai, H.; Ishiguro, T.
Chem. Lett. **1986**, 93

Crystal and Band Structures of an Organic Conductor β''-(BEDT-TTF)$_2$AuBr$_2$
Mori, T.; Sakai, F.; Saito, G.; Inokuchi, H.
Chem. Lett. **1986**, 1037

Crystal and Electronic Structures of (BEDSe-TSeF)$_2$AuBr$_2$
Kato, R.; Kobayashi, H.; Kobayashi, A.
Chem. Lett. **1986**, 785

Crystal Structure of α-(BEDT-TTF)$_3$(ReO$_4$)$_2$
Kanbara, H.; Tajima, H.; Aratani, S.; Yakushi, K.; Kuroda, H.; Saito, G.; Kawamoto, A.; Tanaka, J.
Chem. Lett. **1986**, 437

X-Ray Study of Compressibilities of β-(BEDT-TTF)$_2$I$_3$ Under Hydrostatic Pressure
Tanino, H.; Kato, K.; Tokumoto, M.; Anzai, H.; Saito, G.
J. Phys. Soc. Jpn. **1985**, *54*, 2390

Structure of Crystals of (BEDT-TTF)/$_{3.5}$' the Starting Point for Obtaining the Organic Superconductor (BEDT-TTF)/$_{1.5}$ with T_c at Normal Pressure
Shibaeva, R. P.; Lobkovskaya, R. M.; Yagubskii, E. B.; Kostyuchenko, E. E.
Sov. Phys. Crystallogr. **1986**, *31*, 267

Crystal and Electronic Structures of (BEDT-TTF)AuCl$_2$AuCl$_4$
Mori, T.; Inokuchi, H.
Chem. Lett. **1986**, 2069

Anion Arrangement in a New Molecular Superconductor, θ-(BEDT-TTF)$_2$(I$_3$)$_{1-x}$(AuI$_2$)$_x$ ($x < 0.02$)
Kobayashi, A.; Kato, R.; Kobayashi, H.; Moriyama, S.; Nishio, Y.; Kajita, K.; Sasaki, W.
Chem. Lett. **1986**, 2017

Structure of 3,4; 3',4'-Bis(ethylenedithio)-2,2',5,5'-tetrathiafulvalene-dichlorocuprate (BEDT-TTF)CuCl$_2$
Kawamoto, A.; Tanaka, J.
Acta Cryst. **1987**, *C43*, 205

Crystal and Electronic Structures of a New Molecular Superconductor, κ-(BEDT-TTF)$_2$I$_3$
Kobayashi, A.; Kato, R.; Kobayashi, H.; Moriyama, S.; Nishio, Y.; Kajita, K.; Sasaki, W.
Chem. Lett. **1987**, 459

Crystal Structures of AuCl$_2$ Salts of Bis(ethylenedithio)tetrathiafulvalene (BEDT-TTF). Existence of Divalent Gold, Au(II)
Mori, T.; Inokuchi, H.
Solid State Commun. **1987**, *62*, 525

Crystal Structure of Bis(ethylenedithio)tetrathiafulvalene polyiodide, ζ-(BEDT-TTF)$_2$I$_{10}$
Shibaeva, R. P.; Lobkovskaya, R. M.; Yagubskii, E. B.; Kostyuchenko, E. E.
Sov. Phys. Crystallogr. **1986**, *31*, 657

The Crystal Structure of α-(BEDT-TTF)$_2$IBr$_2$
Shibaeva, R. P.; Lobkovskaya, R. M.; Simonov, M. A.; Yagubskii, E. B.; Ignat'ev, A. A.
Sov. Phys. Crystallogr. **1986**, *31*, 654

Indicators for a Low-temperature Structural Transition at Ambient Pressure in β-(BEDT-TTF)$_2$I$_3$
Kistenmacher, T. J.
Solid State Commun. **1987**, *63*, 977

X-ray Study of the Incommensurate Modulation of the Organic Superconductor β-di[bis(ethylenedithio)tetrathiafulvalene] Triiodide
Ravy, S.; Pouget, J. P.; Moret, R.; Lenoir, C.
Phys. Rev. B **1988**, *37*, 5113

Crystal Structures of the Organic Superconductor, (BEDT-TTF)$_2$Cu(NCS)$_2$, at 298 K and 104 K
Urayama, H.; Yamochi, H.; Saito, G.; Sato, S.; Kawamoto, A.; Tanaka, J.; Mori, T.; Maruyama, Y.; Inokuchi, H.
Chem. Lett. **1988**,463

Synthesis and Structure of Black (BEDT-TTF)$_2$IBr$_2$: A Structural Analogue of β-(BEDT-TTF)$_2$X Organic Superconductors
Terzis, A.; Hountas, A.; Papavassiliou, G. C.
Solid State Commun. **1988**, *66*, 1161

Crystal Structure of 2,5-Bis(dicyanomethylene)-2,5-dihydrothieno [3,2-b]thiophene and Bis(ethylenedithio)tetrathiafulvalene Complex
Aso, Y.; Yui, K.; Ishida, H.; Otsubo, T.; Ogura, F.; Kawamoto, A.; Tanaka, J.
Chem. Lett. **1988**,1069

Crystal Structure and Physical Property of (BEDT-TTF)$_2$(IBr$_2$)$_2$ (1,1,2-Trichloroethane)$_{0.5}$
Yamochi, H.; Urayama, H.; Saito, G.; Oshima, K.; Kawamoto, A.; Tanaka, J.
Chem. Lett. **1988**,1211

Origin of the Nonmetallic Properties of δ-(ET)$_2$AuI$_2$
Whangbo, M.-H.; Evain, M.; Beno, M. A.; Wang, H. H.; Webb, K. S.; Williams, J. M.
Solid State Commun. **1988**, *68*, 421

Crystal Structure of Two-dimensional Organic Metal Based on Bis(ethylenedithio)tetrathiafulvalene (BEDT-TTF)$_2$(Cu$_5$I$_6$) with a Polymer Anion Layer
Shibaeva, R. P.; Lobkovskaya, R. M.
Sov. Phys. Crystallogr. **1988**, *33*, 241

The Crystal Structure of the Radical-cation Salt of Bis(ethylenedithio)tetrathiafulvalene with the Chloromercurate Anion, (BEDT-TTF)$_3$(HgCl$_3$)$_2$
Shibaeva, R. P.; Rozenberg, L. P.; Aldoshina, M. Z.; Lyubsovskaya, R. N.
Sov. Phys. Crystallogr. **1988**, *33*, 71

A Change of the Superstructure and an Associated Rise of the Superconducting Critical Temperature in the Organic Superconductor β-(BEDT-TTF)$_2$I$_3$
Kagoshima, S.; Nogami, Y.; Hasumi, M.; Anzai, H.; Tokumoto, M.; Saito, G.; Mori, N.
Solid State Commun. **1989**, *69*, 1177

Polymorphic Modifications of Crystals of the Cation-radical Salt Bis(propylenedithio)tetrathiafulvalene with ICl$_2$: α(BPDT-TTF)ICl$_2$ and α'(BPDT-TTF)$_2$ICl$_2$
Shibaeva, R. P.; Rozenberg, L. P.; Simonov, M. A.; Kushch, N. D.; Yagubskii, E. B.
Sov. Phys. Crystallogr. **1988**, *33*, 685

The Crystal and Molecular Structures of Bis[1,2-bis(methylthio)vinylenedithio]-tetrathiafulvalene (TMTVT)
Nakano, H.; Nogami, T.; Shirota, Y.; Harada S.; Kasai, N.
Bull. Chem. Soc., Jpn. **1989**, *62*, 2382

Structure of the Conducting Salt of Pyrazinoethylenedithiotetrathiafulvalene (PEDT-TTF): α-(PEDT-TTF)$_2$IBr$_2$, at 98 K
Terzis, A.; Papavassiliou, G.; Kobayashi, H.; Kobayashi, A.
Acta Cryst. **1989**, *C45*, 683

Recent Progress in the Development of Structure-Property Correlations for κ-Phase Organic Superconductors
Williams, J. M.; Wang, H. H.; Kini, A. M.; Carlson, K. D.; Beno, M. A.; Geiser, U.; Whangbo, M.-H.; Jung, D.; Evain, M.; Novoa, J. J.
Mol. Cryst. Liq. Cryst. **1990**, *181*, 59

Salts with Square-Planar Gold(III) Complex Anions: β-(ET)$_2$AuCl$_4$ and (ET)$_2$Au(CN)$_2$Cl$_2$
Geiser, U.; Anderson, B. A.; Murray, A.; Pipan, C. M.; Rohl, C. A.; Vogt, B. A.; Wang, H. H.; Williams, J. M.; Kang, D. B.; Whangbo, M.-H.
Mol. Cryst. Liq. Cryst. **1990**, *181*, 105

The Crystal and Molecular Structure of (BEDT-TTF)Ag$_x$Br$_3$ (x ≈ 2.4)
Geiser, U.; Wang, H. H.; Rust, P. R.; Tonge, L. M.; Williams, J. M.
Mol. Cryst. Liq. Cryst. **1990**, *181*, 117

The Crystal and Molecular Structure of (BEDT-TTF)$_4$(Hg$_2$Br$_6$)(1,1,2-Trichloroethane)
Geiser, U.; Wang, H. H.; Kleinjan, S.; Williams, J. M.
Mol. Cryst. Liq. Cryst. **1990**, *181*, 125

Structures of Two Semiconducting Charge-Transfer Salts Based on 4,5-Methylenedithio-4',5'-propylenedithiotetrathiafulvalene (MPT): (MPT)$_2$ClO$_4$(THF) and (MPT)$_2$PF$_6$(THF)
Beno, M. A.; Kini, A. M.; Wang, H. H.; Tytko, S. F.; Carlson, K. D.; Williams, J. M.
Acta Cryst. **1988**, *C44*, 1223

Crystal Structure of the η Phase of Bis(ethylenedithio)tetrathiafulvalene, η-(BEDT-TTF)I$_3$
Shibaeva, R. P.; Lobkovskaya, R. M.; Yagubskii, E. B.; Laukhina, E. E.
Sov. Phys. Crystallogr. **1987**, *32*, 530

Crystal Structure of the Organic Conductor (BEDT-TTF)I$_3$(C$_2$H$_3$Cl$_3$)$_{0.333}$
Shibaeva, R. P.; Lobkovskaya, R. M.; Kaminskii, V. F.; Lindeman, S. V.; Yagubskii, E. B.
Sov. Phys. Crystallogr. **1986**, *31*, 546

Structure of Electroconducting Complexes Based on Cation-radical VI. Complexes of Dibenzotetrathiafulvalene with Inorganic Anions (SnCl$_6$)$^{2-}$ and (SnBr$_6$)$^{2-}$: (C$_{14}$H$_8$S$_4$)$_8$(SnCl$_6$)$_3$ and (C$_{14}$H$_8$S$_4$)$_3$(SnBr$_6$)
Shibaeva, R. P.; Rozenberg, L. P.; Lobkovskaya, R. M.
Kristallografiya **1980**, *25*, 507; *Sov. Phys. Crystallogr.* **1980**, *25*, 292

The Structure of Conducting Complexes Based on Cation-radicals V. The Crystal and Molecular Structure of 1:1 Ion-radical Salt of Dibenzotetrathiafulvalene with an Inorganic Anion (C$_{14}$H$_8$S$_4$) + [PtBr$_5$S(Ch$_3$)$_2$]$^-$
Shibaeva, R. P.; Rozenberg, L. P.
Sov. Phys. Crystallogr. **1980**, *25*, 156

Preparation and Characterization of Two Structural Phases of (EPT)$_2$ICl$_2$
Schultz, A. J.; Geiser, U.; Kini, A. M.; Wang, H. H.; Schlueter, J.; Cariss, C. S.; Williams, J. M.
Synth. Met. **1988**, *27*, A299

Structure of 3,4-Ethylenedithio-3',4'-propylenedithio-2,2',5,5'-tetrathiafulvalenium Tetraiodoindate (III)
Porter, L. C.; Allen, T. J.; Carlson, K. D.; Chen, M. Y.; Geiser, U.; Kao, H.-C. I.; Kini, A. M.; Schlueter, J. A.; Wang, H. H.; Williams, J. M.
Acta Cryst. **1988**, *C44*, 1712

Structure of 3,4;3',4'-Bis(ethylenedithio)-2,2',5,5'-tetrathiafulvalenium - Hydrogensulfate (3/2). A (BEDT-TTF) Charge-Transfer Salt Containing a Tetrahedral Anion
Porter, L. C.; Wang, H. H.; Miller, M. M.; Williams, J. M.
Acta Cryst. **1987**, *C43*, 2201

Crystal Structure of Cation-radical Salt of Bis(ethylenedithio)tetrathiafulvalene with a Chlorine-mercury Anion (BEDT-TTF)$_4$(Hg$_3$Cl$_8$)

Shibaeva, R. P.; Rozenberg, L. P.

Sov. Phys. Crystallogr. **1988**, *33*, 834

Crystal and Molecular Structure of 2,2'-bis(5,8-dihydro-1,3-dithiolo[4,5-b]-4,9-dithiocyan-2-ylidene) C$_{14}$H$_{12}$S$_8$

Shibaeva, R. P.; Rozenberg, L. P.; Abramov, M. A.; Rubtsova, I. K.; Petrov, M. L.

Sov. Phys. Crystallogr. **1989**, *34*, 68

Molecular Structure of a New Multi-Sulfur π-Donor Molecule, Bis(vinylenedithio)tetrathiafulvalene, VT and Crystal and Electronic Structures of VT$_2$PF$_6$

Kobayashi, H.; Kobayashi, A.; Nakamura, T.; Nogami, T.; Shirota, Y.

Chem. Lett. **1987**, 559

Crystal and Electronic Structures of (BEDSe-TSeF)AuBr$_2$

Kato, R.; Kobayashi, H.; Kobayashi, A.; Mori, T.; Inokuchi, H.

Chem. Lett. **1987**, 277

The Space Group of (TTM-TTF)$^{2+}$(AuCl$_4^-$)$_2$[TTM-TTF = Tetra(methylthio)tetrathiafulvalene

Jones, P. G.

Z. Naturforsch **1989**, *44b*, 243

Quasi Three-dimensional Electrical Conductor Having Alternating Mixed Stacks: Tetrakis(methyltelluro)tetrathiafulvalene (TTeC$_1$ − TTF)•TCNQ Complex

Iwasawa, N.; Shinozaki, F.; Saito, G.; Oshima, K.; Mori, T.; Inokuchi, H.

Chem. Lett. **1988**, 215

The Crystal Structure of a Charge-Transfer Complex of Tetrakis(methylthio)tetrathiafulvalenium Dibromoiodate, TMT-TTF•IBr$_2$

Honda, K.; Goto, M.; Kurahashi, M.; Anzai, H.; Tokumoto, M.; Ishiguro, T.

Bull. Chem. Soc. Jpn. **1988**, *61*, 558

Structure of Di[3,4;3',4'-bis(ethylenedithio)-2,2', 5,5'-tetrathiafulvalenium] Tetraiodogallate(III), (BEDT-TTF)$_2$GaI$_4$

Geiser, U.; Wang, H. H.; Schlueter, J. A.; Hallenbeck, S. L.; Allen, T. J.; Chen, M. Y.; Kao, H.-C.; Carlson, K. D.; Gerdom, L. E.; Williams, J. M.

Acta Cryst. **1988**, *C44*, 1544

Structure of Di[3,4;3',4'-bis(ethylenedithio)-2,2',5,5'-tetrathiafulvalenium]Dichlorocuprate(I), (BEDT-TTF)$_2$CuI$_2$

Geiser, U.; Wang, H. H.; Hammond, C. E.; Firestone, M. A.; Beno, M. A.; Carlson, K. D.; Nuñez, L.; Williams, J. M.

Acta Cryst. **1987**, *C43*, 656

Crystal Structure of the Mixed-Stack Salt of Bis(ethylenedithio)tetrathiafulvalene (BEDT-TTF) and Tetracyanoquinodimethane (TCNQ)
Mori, T.; Inokuchi, H.
Bull. Chem. Soc. Jpn. **1987**, *60*, 402

BEDT-TTF Salts Containing Magnetic Anions $FeCl_4^-$, $FeBr_4^-$ and $CuCl_4^{2-}$
Mallah, T.; Hollis, C.; Bott, S.; Day, P.; Kurmoo, M.
Synth. Met. **1988**, *27*, A381

Synthesis and Structure of Organic Donor-cluster-hybrids
Fuchs, H.; Fuchs, S.; Polborn, K.; Lehnert, Th.; Heidmann, C.-P.; Müller, H.
Synth. Met. **1988**, *27*, A271

Layered Molecular Superconductors θ- and κ-$(BEDT-TTF)_2I_3$
Kobayashi, H.; Kato, R.; Kobayashi, A.; Moriyama, S.; Nishio, Y.
Synth. Met. **1988**, *27*, A283

Molecular Design and Solid State Properties of New Superconductors and Molecular Metals with Ordered Spin Structures
Kobayashi, H.; Kato, R.; Kobayashi, A.; Mori, T.; Inokuchi, H.; Nishio, Y.; Kajita, K.; Sasaki, W.
Synth. Met. **1988**, *27*, A289

Structural Properties of Organic Superconductors $(DMET)_2X$ and $(BEDT-TTF)_2Cu(NCS)_2$
Kagoshima, S.; Nogami, Y.
Synth. Met. **1988**, *27*, A299

The First Polymeric-Anion Derivatives of BEDT-TTF
Geiser, U.; Wang, H. H.; Williams, J. M.; Venturini, E. L.; Kwak, J. F.; Whangbo, M.-H.
Synth. Met. **1987**, *19*, 599

Low Temperature Structure of 3,4;3',4'-Bis-(ethylenedithio)-2,2',5,5'-tetrathiafulvalene diiodoaurate(I) (2:1), β-$(BEDT-TTF)_2AuI_2$
Geiser, U.; Wang, H. H.; Webb, K. S.; Firestone, M. A.; Beno, M. A.; Williams, J. M.
Acta Cryst. **1987**, *C43*, 996

Structure of Tetrathiafulvalenium Bis(5,6-dihydro-1,4-dithiin-2,3-dithiolato)-aurate(III), (BEDT-TTF)[Au(dddt)$_2$]
Geiser, U.; Schultz, A. J.; Wang, H. H.; Beno, M. A.; Williams, J. M.
Acta Cryst. **1988**, *C44*, 259

Structure of the 2/1 Complex Dibenzotetrathiafulvalenium Hexabromodicuprate(II), $2C_{14}H_8S_4^+ \cdot Cu_2Br_6^{2-}$
Honda, M.; Katayama, C.; Tanaka, J.; Tanaka, M.
Acta Cryst. **1985**, *C41*, 688

Structure of the 2/1 Complex Dibenzotetrathiafulvalenium Hexachlorodicuprate(II), $2C_{14}H_8S_4^+ \cdot Cu_2Cl_6$
Honda, M.; Katayama, C.; Tanaka, J.; Tanaka, M.
Acta Cryst. **1985**, *C41*, 197

The Crystal and Molecular Structure of the Complex of Dibenzotetrathiofulvalene with Iodine, $C_{14}H_8S_4 \cdot I_3$
Shibaeva, R. P.; Rozenberg, L. P.; Aldoshina, M. Z.; Lyubovskaya, R. N.; Khidekel, M. L.
Zh. Strukt. Khim. **1979**, *20*, 485

Structures of Two Conducting Salts of Pyrazinoethylenedithiotetrathiafulvalene (PEDTTTF): $(PEDTTTF)_2PF_6$ and $(PEDTTTF)_2BF_4 \cdot (CH_2Cl_2)_{0.5}$
Mentzafos, D.; Psycharis, V.; Terzis, A.
Acta Cryst. **1989**, *C45*, 1333

The Structure of 3,3',4,4'-Tetrakis(ethyltelluro)-2,2',5,5'-tetrathiafulvalene ($TTeC_2$-TTF)
Becker, J. Y.; Bernstein, J.; Bitner, S.; Sarma, J. A. R. P.; Shahal, L.; Shaik, S. S.
Acta Cryst. **1988**, *C44*, 1770

Crystal Structures and Electrical Properties of Tetrakis(methylseleno)-tetrathiafulvalene ($TSeC_1$-TTF) and Two Phases of $(TSeC_1\text{-}TTF)I_3$
Wang, P.; Inabe, T.; Nakano, C.; Maruyama, Y.; Inokuchi, H.; Iwasawa, N.; Saito, G.
Bull. Chem. Soc. Jpn. **1989**, *62*, 2252

A TTF Derivative Acting as a Chelate Ligand: Structure of µ-[Tetra(methylthio)-tetrathiafulvalene]di[diodomercury(II)], $(TTM\text{-}TTF)(HgI_2)_2$ and the Structure of a Second Modification of Free TTM-TTF
Endres, H.
Z. Naturforsch **1986**, *41b*, 1351

Tribromide Salts of the Organic Donor Tetra(methylthio)tetrathiafulvalene, $[TTM\text{-}TTF]^+(Br_3^-)$ and $[TTM\text{-}TTF]^{2+}(Br_3^-)_2$
Endres, H.
Z. Naturforsch **1986**, *41b*, 1437

Crystal Structures of Complexes between Hexacyanobutadiene and Tetramethyltetrathiafulvalene and Tetramethylthiotetrathiafulvalene
Katayama, C.; Honda, M.; Kumagai, H.; Tanaka, J.; Saito, G.; Inokuchi, H.
Bull. Chem. Soc. Jpn. **1985**, *58*, 2272

Two Modifications of Di[tetra(methylthio)tetrathiafulvelenium] Hexachlorodicuprate(II), $[(TTM\text{-}TTF)^+]_2[Cu_2Cl_6]^{2-}$
Endres, H.
Z. Naturforsch **1987**, *42b*, 5

The Crystal Structure of the 1:1 Radical-Cation-Radical-Anion Salt of 2,2'-Bis-1,3-Dithole (TTF) and 7,7,8,8-tetracyanoquidimethane (TCNQ)
Kistenmacher, T. J.; Phillips, T. E.; Cowan, D. O.
Acta Cryst. **1974**, *B30*, 763

XPS (X-RAY PHOTOELECTRON SPECTROSCOPY)

Electronic Structure of (BEDT-TTF)$CuCl_2$ Complex
Tanaka, M.; Kawamoto, A.; Tanaka, J.; Sano, M.; Enoki, T.; Inokuchi, H.
Bull. Chem. Soc. Jpn. **1987**, *60*, 2531

Appendix A

DMIT, Dithiolate Series*

CONDUCTIVITY STUDIES

The First Molecular Superconductor Based on the π-Acceptor Molecule and the Closed Shell Cation, $[(CH_3)_4N][Ni(dmit)_2]_2$
Kobayashi, A.; Kim, H.; Sasaki, Y.; Moriyama, S.; Nishio, Y.; Kajita, K.; Sasaki, W.; Kato, R.; Kobayashi, H.
Synth. Met. **1988**, *27*, B339

EPR/ESR

Single-Crystal EPR Spectra of $(n-Bu_4N)_2[^{63}Cu(dmit)_2]$
Stach, J.; Kirmse, R.; Dietzsch, W.; Olk, R.; Hoyer, E.
Inorg. Chem. **1984**, *23*, 4779

Anisotropic Exchange Interaction in Tetrabutylammonium Bis(Stilbene-1,2-Dithiolato) Nickelate (III), $[NBu_4][Ni(sdt)_2]$
Kuppusamy, P.; Manoharan, P. T.
Chem. Phys. Lett. **1985**, *118*, 159

*Includes donor molecules which form complexes with elemental metals.

MAGNETIC STUDIES

High Field Magnetoresistance of a New Organic Metal: TTF[Ni(dmit)$_2$]$_2$
Ulmet, J. P.; Auban, P.; Khmou, A.; Valade, L.; Cassoux, P.
Phy. Lett. **1985**, *113A*, 217

MULTIPLE DISCIPLINE STUDIES

Molecular Structure and Solid-state Properties of the Two-dimensional Conducting Mixed-valence Complex [NBu$_4$]$_{0.29}$[Ni(dmit)$_2$] and the Neutral [Ni(dmit)$_2$](H$_2$dmit = 4,5-dimercapto-1,3-dithiole-2-thione); Members of an Electron-transfer Series
Valade, L.; Legros, J.-P.; Bousseau, M.; Cassoux, P.; Garbauskas, M.; Interrante, L. V.
J. Chem. Soc. Dalton Trans. **1985**, 783

Electrical and Magnetic Properties of (BEDT-TTF)$_4$Ni(CN)$_4$ Complex
Tanaka, M.; Takeuchi, H.; Sano, M.; Enoki, T.; Suzuki, K.; Imaeda, K.
Bull. Chem. Soc. Jpn. **1989**, *62*, 1432

New Molecular Conductors, α- and β-(EDT-TTF)[Ni(dmit)$_2$] Metal with Anomalous Resistivity Maximum vs. Semiconductor with Mixed Stacks
Kato, R.; Kobayashi, H.; Kobayashi, A.; Naito, T.; Tamura, M.; Tajima, H.; Kuroda, H.
Chem. Lett. **1989**, 1839

The First Molecular Superconductor Based on the π-acceptor Molecule and the Closed Shell Cation, [(CH$_3$)$_4$N][Ni(dmit)$_2$]$_2$
Kobayashi, A.; Kim, H.; Sasaki, Y.; Moriyama, S.; Nishio, Y.; Kajita, K.; Sasaki, W.; Kato, R.; Kobayashi, H.
Synth. Met. **1988**, *27*, B339

(n-Bu$_4$N)x[Pd(dmit)$_2$] ($x = 0.5$ and 0.33): Metal-like Conductivity vs. Stacking of "True" Dimers with Metal-metal Bond: Comparison with δ-TTF[Pd(dmit)$_2$]$_2$
Legros, J.-P.; Valade, L.; Cassoux, P.
Synth. Met. **1988**, *27*, B347

Search for New Molecular Ferromagnets: EPR and Magnetic Studies of Charge Transfer Salts
Laversanne, R.; Chakroborty, A.; Glatzhofer, D. T.; Epstein, A. J.; Miller, J. S.
Synth. Met. **1988**, *27*, B353

Structural and Electronic Properties of a New Two-dimensional Molecular Metal Et$_2$Me$_2$N[Ni(dmit)$_2$]$_2$
Kato, R.; Kobayashi, H.; Kim, H.; Kobayashi, A.; Sasaki, Y.; Mori, T.; Inokuchi, H.
Synth. Met. **1988**, *27*, B359

Condensed Matter Physics (Coordination Chemistry) - A New Type of Molecular Superconductor: TTF[Ni(dmit)$_2$]$_2$
Brossard, L.; Ribault, M.; Bousseau, M.; Valade, L.; Cassoux, P.
C. R. Acad. Sc. Paris **1986**, *302*, 205

New Organic Synthetic Metals Derived from BEDT-TTF, Ni(dsit)$_2$ and BEDO-TTF
Beno, M. A.; Kini, A. M.; Geiser, U.; Wang, H. H.; Carlson, K. D.; Williams, J. M.
The Physics and Chemistry of Organic Superconductors, Saito, G.; Kagoshima, S.; Eds., Springer-Verlag, Berlin, **1990**, vol. 51, pp. 369–372.

PHASE DIAGRAMS

Pressure-temperature Phase Diagram of α'-TTF[Pd(dmit)$_2$]$_2$
Brossard, L.; Hurdequint, H.; Ribault, M.; Valade, L.; Legros, J. P.; Cassoux, P.
Synth. Met. **1988**, *27*, B157

Pressure Induced Superconductivity in Molecular TTF[Pd(dmit)$_2$]$_2$
Brossard, L.; Ribault, M.; Valade, L.; Cassoux, P.
J. Phys. France **1989**, *50*, 1521

POLARIZED REFLECTANCE/ABSORPTION

Reflectance Spectra of [M(dmit)$_2$] Salts
Tajima, H.; Tamura, M.; Naito, T.; Kobayashi, A.; Kuroda, H.; Kato, R.; Kobayashi, H.; Clark, R. A.; Underhill, A. E.
Mol. Cryst. Liq. Cryst. **1990**, *181*, 233

PREPARATION, COMPOSITION, PROPERTIES

A New Two-dimensional Conducting Mixed Valence Compound Derived from a Nickel Bisdithiolato-complex
Valade, L.; Bousseau, M.; Gleizes, A.; Cassoux, P.
J. Chem. Soc. Chem. Commun. **1983**, 110

Trithione- and Isotrithionedithiolate. A New Class of Unsaturated 1,2-Dithiolates. IV. The Crystal Structure of Dipotassium 1,2-Dithiole-3-thion-4,5-Dithiolate, K$_2$C$_3$S$_5$
Sieler, J.; Beyer, F.; Hoyer, E.; Andersen, L.; Lindqvist, O.
Acta Chemica Scandinavica **1985**, *A39*, 153

A New Type of Molecular Superconductor: TTF [Ni(dmit)$_2$]$_2$
Brossard, L.; Ribault, M.; Bousseau, M.; Valade, L.; Cassoux, P.

Condensed Matter Physics (Coordination Chemistry) C. R. Acad. Sci. **1986**, *302*, 205

Highly Conducting Charge-Transfer Compounds of Tetrathiafulvalene and Transition Metal-"dmit" Complexes
Bousseau, M.; Valade, L.; Legros, J.-P.; Cassoux, P.; Garbauskas, M.; Interrante, L.
J. Am. Chem. Soc. **1986**, *108*, 1908

Molecular Designing Analysis of a New Superconducting Metal Dithiolene Complex
Kobayashi, A.; Kim, H.; Sasaki, Y.; Kato, R.; Kobayashi, H.
Solid State Commun. **1987**, *62*, 57

Synthesis and Structure of (ET)[Ni(SC(CN) = C(CN)S)$_2$], a Radical Cation Salt of Bis(ethylenedithio)tetrathiafulvalene
Reith, W.; Polborn, K.; Amberger, E.
Angew. Chem. Int. Ed. Engl. **1988**, *27*, 699

A New Preparation of 5-(Alkylthio)-1,2-dithiole-3-thiones and a Highly Functionalized 1,3-Dithiole-2-thione
Lu, F. L.; Keshavarz, M.; Srdanov, K. G.; Jacobson, R. H.; Wudl, F.
J. Org. Chem. **1989**, *54*, 2165

Uniform, Non-Interacting Antiferromagnetic Chains of Spins in the 1:1 Bisethylenedithiotetrathiafulvalenium Salt of a Monovalent Hexarhenium Chalcohalide Cluster Anion: (BEDT-TTF)$^+$•(Re$_6$Se$_5$Cl$_9$)$^-$•(C$_3$H$_7$ON)$_2$
Penicaud, A.; Lenoir, C.; Batail, P.; Coulon, C.; Perrin, A.
Synth. Met. **1989**, *32*, 25

Synthesis and Structural Characterization of CuI$_2$$^-$
Rath, N. P.; Holt, E. M.
J. Chem. Soc. Chem. Commun. **1986**, 311

New Molecular Conductors Based on Metal Complex Anions
Hawkins, I.; Clark, R. A.; Wainwright, C. E.; Underhill, A. E.
Mol. Cryst. Liq. Cryst. **1990**, *181*, 209

New Aspects of Molecular Conductors of Metal Complexes with Multi-Sulfur π-Donor and π-Acceptor Molecules
Kobayashi, A.; Kato, R.; Kobayashi, H.
Synth. Met. **1987**, *19*, 635

Synthesis of Organometallic Mixed Sulfur-selenium π-acceptor Molecules
Nigrey, P. J.
Synth. Met. **1988**, *27*, B365

New Molecular Conductors, α- and β-(EDT-TTF)[Ni(dmit)$_2$] Metal with Anomalous Resistivity Maximum vs. Semiconductor with Mixed Stacks
Kato, R.; Kobayashi, H.; Kobayashi, A.; Naito, T.; Tamura, M.; Tajima, H.; Kuroda, H.
Chem. Lett. **1989**, 1839

A Novel Macrocycle Containing the 4,5-dithio-1,3-dithiole-2-thione Unit and a Related Macrocycle Incorporating the Tetrathiafulvalene Moiety
Girmay, B.; Kilburn, J. D.; Underhill, A. E.; Varma, K. S.; Hursthouse, M. B.; Harman, M. E.; Becher, J.; Bojesen, G.
J. Chem. Soc., Chem. Commun. **1989,** 1406

Conducting and Superconducting Salts Based on MDTTTF, EDTTTF, VDTTTF, EDTDSDTF, MDSTTF, BMDTTTF, Pd(dmit)$_2$, and Ni(dcit)$_2$
Papavassiliou, G. C.; Mousdis, G.; Kakoussis, V.; Terzis, A.; Hountas, A.; Hilti, B.; Mayer, C. W.; Zambounis, J. S.
Proc. Int. Sym. Org. Superconductors, Springer-Verlag, submitted

PRESSURE STUDY

Transport Properties of [(CH$_3$)$_4$N) (Ni(dmit)$_2$)]$_2$: A New Organic Superconductor
Kajita, K.; Nishio, Y.; Moriyama, S.; Kato, R.; Kobayashi, H.; Sasaki, W.; Kobayashi, A.; Kim, H.; Sasaki, Y.
Solid State Commun. **1988**, *65,* 361

THEORY

Molecular Designing Analysis of a New Superconducting Metal Dithiolene Complex
Kobayashi, A.; Kim, H.; Sasaki, Y.; Kato, R.; Kobayashi, H.
Solid State Commun. **1987**, *62,* 57

Dynamic Localization in Two Low-dimensional Synthetic Semiconductors
Kramer, G. J.; Brom, H. B.
Synth. Met. **1988**, *27,* A133

X-RAY DIFFRACTION

The Crystal Structure and Conductivity of TTF[Ni(dmit)$_2$]$_2$. A New Low-temperature Molecular Metal
Bousseau, M.; Valade, L.; Bruniquel, M.-F.; Cassoux, P.; Garbauskas, M.; Interrante, L.; Kasper, J.
Nouveau Journal De Chimie **1984**, *8,* 3

Crystal Structure of a New Molecular Conductor (DBTTF)[Ni(dmit)$_2$]
Kato, R.; Kobayashi, H.; Kobayashi, A.; Sasaki, Y.
Chem. Lett. **1985**, 131

The Crystal Structure of [BEDT-TTF][Ni(dmit)$_2$]. A New Route to Design of Organic Conductors
Kobayashi, H.; Kato, R.; Kobayashi, A.; Sasaki, Y.
Chem. Lett. **1985**, 191

The First Molecular Superconductor Based on π-Acceptor Molecules and Closed-Shell Cations, [(CH$_3$)$_4$N][Ni(dmit)$_2$]$_2$, Low Temperature X-ray Studies and Superconducting Transition
Kobayashi, A.; Kim, H.; Sasaki, Y.; Kato, R.; Kobayashi, H.; Moriyama, S.; Nishio, Y.; Kajita K.; Sasaki, W.
Chem. Lett. **1987**, 1819

The Crystal and Electronic Structures of a New Molecular Conductor, (HMTTeF)$_2$[Pt(dmit)$_2$]
Kobayashi, A.; Sasaki, Y.; Kato, R.; Kobayashi, H.
Chem. Lett. **1986**, 387

Structural Evidence of Charge Density Waves in the Series of Molecular Conductors and Superconductors: TTF(M(dmit)$_2$)$_2$ (M = Pd, Ni)
Ravy, S.; Pouget, J. P.; Valade, L.; Legros, J. P.
Europhys. Lett. **1989**, *9*, 391

Molecular Design and Solid State Properties of New Superconductors and Molecular Metals with Ordered Spin Structures
Kobayashi, H.; Kato, R.; Kobayashi, A.; Mori, T.; Inokuchi, H.; Nishio, Y.; Kajita, K.; Sasaki, W.
Synth. Met. **1988**, *27*, A289

Appendix A

New and Novel Donors Series*

CONDUCTIVITY STUDIES

A Radical Anion Salt of 2,5-Dimethyl-N,N'-dicyanoquinonediimine with Extremely High Electrical Conductivity
Aumüler, A.; Erk, P.; Klebe, G.; Hünig, S.; Ulrich von Schutz, J.; Werner, H. P.
Angew. Chem. Int. Ed. Engl. **1986**, *25,* 740

On Ambient-pressure Superconductivity in Organic Conductors: Electrical Properties of $(DMET)_2I_3$, $(DMET)_2I_2Br$ and $(DMET)_2IBr_2$
Kikuchi, K.; Murata, K.; Honda, Y.; Namiki, T.; Saito, K.; Ishiguro, T.; Kobayashi, K.; Ikemoto, I.
J. Phys. Soc. Jpn. **1987**, *56,* 3436

Superconductivity in $(DMET)_2AuCl_2$ and $(DMET)_2AuI_2$
Kikuchi, K.; Murata, K.; Honda, Y.; Namiki, T.; Saito, K.; Anzai, H.; Kobayashi, K.; Ishiguro, T.; Ikemoto, I.
J. Phys. Soc. of Jpn. **1987**, *56,* 4241

*Includes unsymmetrical donors.

Superconductivity and the Possibility of Semiconductor-Metal Transition in $(DMET)_2AuBr_2$
Kikuchi, K.; Murata, K.; Honda, Y.; Namiki, T.; Saito, K.; Kobayashi, K.; Ishiguro, T.; Ikemoto, I.
J. Phys. Soc. Jpn. **1987**, *56*, 2627

Superconductivity and Surrounding Phase of Organic Conductor, $(DMET)_2Au(CN)_2$
Kikuchi, K.; Murata, K.; Kikuchi, M.; Honda, Y.; Takahashi, T.; Oyama, T.; Ikemoto, I.; Ishiguro, T.; Kobayashi, K.
Jpn. J. Appl. Phys. **1987**, *26*, 1369

Polymorphism and Electrical Conductivity of the Organic Superconductor $(DMET)_2AuBr_2$
Kikuchi, K.; Honda, Y.; Ishikawa, Y.; Saito, K.; Ikemoto, I.; Murata, K.; Anzai, H.; Ishiguro, T.; Kobayashi, K.
Solid State Commun. **1988**, *66*, 405

EPR/ESR

EPR Studies of Organic Conductors $(DMET)_2X$
Kanoda, K.; Takahashi, T.; Kikuchi, K.; Saito, K.; Ikemoto, I.; Kobayashi, K.
Synth. Met. **1988**, *27*, B385

MAGNETIC STUDIES

Magnetic Properties of $(DIMET)_2SbF_6$: Quantitative Discussion of the Antiferromagnetic Behavior
Laversanne, R.; Coulon, C.; Amiell, J.; Morand, J. P.
Europhys. Lett. **1986**, *2*, 401

Possible Appearance of Spin-Density Waves in the Insulating Phase of the Organic Conductors, $(DMET)_2Au(CN)_2$ and $(DMET)_2AuI_2$
Nogami, Y.; Tanaka, M.; Kagoshima, S.; Kikuchi, K.; Saito, K.; Ikemoto, I.; Kobayashi, K.
J. Phys. Soc. Jpn. **1987**, *56*, 3783

MULTIPLE DISCIPLINE STUDIES

Radical Cation Salts of an Unsymmetrical BEDT-TTF Derivative: Molecular Structure and Physical Properties of $(DIMET)_2ClO_4 \cdot xTHF$
Heid, R.; Endres, H.; Keller, H. J.; Gogu, E.; Heinen, I.; Bender, K.; Schweitzer, D.
J. Naturforsch. **1985**, *40b*, 1703

Superconductivity in $(DMET)_2AuCl_2$ and $(DMET)_2AuI_2$
Kikuchi, K.; Murata, K.; Honda, Y.; Namiki, T.; Saito, K.; Anzai, H.; Kobayashi, K.; Ishiguro, T.; Ikemoto, I.
J. Phys. Soc. Jpn. **1987**, *56*, 4241

NMR and EPR Studies of the Organic Conductor [dimethyl(ethylenedithio)-diselenadithiafulvalene]$_2$Au(CN)$_2$: Evidence of a Spin-density-wave Transition
Kanoda, K.; Takahashi, T.; Tokiwa, T.; Kikuchi, K.; Saito, K.; Ikemoto, I.; Kobayashi, K.
Phys. Rev. B **1988**, *38*, 39

Polymorphism and Electrical Conductivity of the Organic Superconductor $(DMET)_2AuBr_2$
Kikuchi, K.; Honda, Y.; Ishikawa, Y.; Saito, K.; Ikemoto, I.; Murata, K.; Anzai, H.; Ishiguro, T.; Kobayashi, K.
Solid State Commun. **1988**, *66*, 405

Structural and Electronic Properties of κ-Phase Organic Donor Salts: κ-$(DMET)_2AuBr_2$ and κ-$(BEDT-TTF)_4Hg_3I_8$
Whangbo, M.-H.; Jung, D.; Wang, H. H.; Beno, M. A.; Williams, J. M.; Kikuchi, K.
Mol. Cryst. Liq. Cryst. **1990**, *181*, 1

The Organic π-electron Metal System with Interaction Through Mixed-valence Metal Cation: Electronic and Structural Properties of Highly Conducting Anion Radical Salts (2,5-R_1, R_2-DCNQI$_2$Cu (DCNQI) = N, N'-dicyanoquinodiimine; R_1, R_2 = CH_3O, Cl, Br)
Kobayashi, A.; Mori, T.; Inokuchi, H.; Kato, R.; Kobayashi, H.
Synth. Met. **1988**, *27*, B275

$(MDT-TTF)_2AuI_2$: An Ambient Pressure Organic Superconductor (T_c = 4.5 K) Based on an Unsymmetrical Electron Donor
Kini, A. M.; Beno, M. A.; Son, D.; Wang, H. H.; Carlson, K. D.; Porter, L. C.; Welp, U.; Vogt, B. A.; Williams, J. M.; Jung, D.; Evain, M.; Whangbo, M.-H.; Overmyer, D. L.; Schirber, J.E.
Solid State Commun. **1989**, *69*, 503

Electronic Properties of $(DMET)_2X$
Murata, K.; Kikuchi, K.; Takahashi, T.; Kobayashi, K.; Honda, Y.; Saito, K.; Kanoda, K.; Tokiwa, T.; Anzai, H.; Ishiguro, T.; Ikemoto, I.
Proc. Elorma (Tashkent) **1987**, 1

PHASE DIAGRAMS

Phase Diagram of Superconductivity and Spin Density Wave in $(DMET)_2Au(CN)_2$
Honda, Y.; Murata, K.; Kikuchi, K.; Saito, K.; Ikemoto, I.; Kobayashi, K.
Solid State Commun. **1989**, *71*, 1087

Pressure-temperature Phase Diagram of the Organic Conductor (DM-DCNQI)$_2$Cu
Tomic, S.; Jérome, D.; Aumüller, A.; Erk, P.; Hünig, S.; Von Schutz, J. U.
Synth. Met. **1988**, *27*, B281

Pressure-temperature Phase Diagram of (DMDCNQI)$_2$Ag: A Comparative Study with Related Compounds
Henriques, R. T.; Tomic, S.; Kang, W.; Jérome, D.; Brisset, F.; Batail, P.; Erk, P.; Hünig, S.; von Schutz, J. U.
Synth. Met. **1988**, *27*, B333

POLARIZED REFLECTANCE/ABSORPTION

Polarized Reflectance Spectra of DCNQI Salts
Yakushi, K.; Ugawa, A.; Ojima, G.; Ida, T.; Tajima, H.; Kuroda, H.; Kobayashi, A.; Kato, R.; Kobayashi, H.
Mol. Cryst. Liq. Cryst. **1990**, *181*, 217

Reflectance Spectra of [M(dmit)$_2$] Salts
Tajima, M.; Tamura, M.; Naito, T.; Kobayashi, A.; Kuroda, H.; Kato, R.; Kobayashi, H.; Clark, R.A.; Tamura, M.; Underhill, A.E.
Mol. Cryst. Liq. Cryst. **1990**, *181*, 233

PREPARATION, COMPOSITION, PROPERTIES

A Radical Anion Salt of 2,5-Dimethyl-N,N'-dicyanoquinonediimine with Extremely High Electrical Conductivity
Aumüler, A.; Erk, P.; Klebe, G.; Hünig, S.; von Schutz, J.U.; Werner, H. P.
Angew. Chem. Int. Ed. Engl. **1986**, *25*, 740

Unsymmetrically Substituted Dithiadiselenafulvalene Donors For Organic Metals. Synthesis and Conductivity of the Charge-Transfer Salts
Kikuchi, K.; Namiki, T.; Ikemoto, I.; Kobayashi, K.
J. Chem. Soc., Chem. Commun. **1986**, 1472

Novel Quinone-Type Acceptors Fused With Sulphur Heterocycles and Their Highly Conductive Complexes With Electron Donors
Yamashita, Y.; Suzuki, T.; Saito, G.; Mukai, T.
J. Chem. Soc., Chem. Commun. **1986**, 1489

New Organic Superconductor, (DMET)$_2$Au(CN)$_2$
Kikuchi, K.; Kikuchi, M.; Namiki, T.; Saito, K.; Ikemoto, I.; Murata, K.; Ishiguro, T.; Kobayashi, K.
Chem. Lett. **1987**, 931

3-Oxo-4-thioxo-1,2,5,6-tetrathiapentalene (OTTP): A Novel Thiocarbon with an Unusual Chalcogen Network in its Solid State Structure
Closs, F.; Srdanov, G.; Wudl, F.
J. Chem. Soc. Chem. Commun. **1989**, 1716

Organic Superconductors, (DMET)$_2X$
Ikemoto, I.; Kikuchi, K.; Saito, K.; Kanoda, K.; Takahashi, T.; Murata, K.; Kobayashi, K.
Mol. Cryst. Liq. Cryst. **1990**, *181*, 185

New π-Electron Donors
Bechgaard, K.; Lerstrup, K.; Jorgensen, M.; Johannsen, I.; Christensen, J.; Larsen, J.
Mol. Cryst. Liq. Cryst. **1990**, *181*, 161

Preparation and Structure of Highly Conductive Anion Radical Salts, M(2,5 − R$_1$, R$_2$-DCNQI)$_2$ (DCNQI = N, N'-dicyanoquinonediimine; R$_1$, R$_2$ = Me, MeO, Halogen; M = Ag, Li, Na, K,NH$_4$)
Kato, R.; Kobayashi, H.; Kobayashi, A.; Mori, T.; Inokuchi, H.
Synth. Met. **1988**, *27*, B263

Structural and Electrical Properties of Some Cation Radical Salts Based on EDTT-TF, PEDTTTF, and Similar π-Donors
Terzis, A.; Hountas, A.; Underhill, A. E.; Clark, A.; Kaye, B.; Hilti, B.; Mayer, C.; Pfeiffer, J.; Yiannopoulos, S. Y.; Mousdis, G.; Papavassiliou, G. C.
Synth. Met. **1988**, *27*, B97

Organic Metal Based on a New Unsymmetrical S, Se-containing Donor — Bis-(ethylenedithiodimethyldithiadiselenafulvalene)Triiodide
Aldoshina, M. Z.; Atovmyan, L. O.; Gol'denberg, L. M.; Krasochka, O. N.; Lyubovskaya, R. N.; Lyubovskii, R. B.; Khidekel, M. L.
Doklady Akademii Nauk SSSR **1986**, *289*, 1140

Synthesis and Structure of Black (EDTTTF)$_2$IBr$_2$: A Structural Analogue of β-(BEDT-TTF)$_2X$ Organic Superconductors
Terzis, A.; Hountas, A.; Papavassiliou, G. C.
Solid State Commun. **1988**, *11*, 1161

Structures of the Conducting Salts of Pyrazinoethylenedithiotetrathiafulvalene (PEDT-TTF) and [4,5-*b*]Pyridinoethylenedithiotetrathiafulvalene ([4,5-*b*]PEDT-TTF): α-(PEDT-TTF)$_2$IBr$_2$ and α-([4,5-*b*]PEDT-TTF)$_2$IBr$_2$
Terzis, A.; Psycharis, V.; Hountas, A.; Papavassiliou, G.
Acta Cryst. **1988**, *C44*, 128

4,5-Methylenedithio-4',5'-ethylenedithiotetrathiafulvalene: A New, Unsymmetrical Electron Donor
Kini, A. M.; Beno, M. A.; Williams, J. M.
J. Chem. Soc. Chem. Commun. **1987**, 335

Physical Properties of Radical Cation Salts of the Dimethylethylenedithiotetrathiafulvalene
Laversanne, R.; Coulon, C.; Amiell, J.; Dupart, E.; Delhaes, P.; Morand, J. P.; Manigand, C.
Solid State Commun. **1986**, *58,* 765

Conducting Solids Based on Some New Unsymmetrical Tetraheterofulvalenes
Papavassiliou, G. C.; Mousdis, G. A.; Yiannopoulos, S. Y.; Kakoussis, V. C.; Zambounis, J. S.
Synth. Met. **1988**, *27,* B373

Low Temperature Measurements of the Electrical Conductivities of Some Charge Transfer Salts with the Asymmetric Donors MDT-TTF, EDT-TTF, and EDT-DSDTF. (MDT-TTF)$_2$AuI$_2$, A New Superconductor (T_c = 3.5 K at Ambient Pressure)
Papavassiliou, G. C.; Mousdis, G. A.; Zambounis, J. S.; Terzis, A.; Hountas, A.; Hilti, B.; Mayer, C. W.; Pfeiffer, J.
Synth. Met. **1988**, *27,* B379

A New Series of Radical-cation Salts Based on Asymmetrical Ethylenedithiodimethyltetrathiafulvalene (EDTDM-TTF). Properties and Crystal Structure of (EDTDM-TTF)$_2$PF$_6$
Aldoshina, M. Z.; Atovmyan, L. O.; Goldenberg, L. M.; Krasochka, O. N.; Lyubovskaya, R. N.; Lyubovskii, R. B.; Merzhanov, V. A.; Khidekel, M. L.
J. Chem. Soc. Chem. Commun. **1985**, 1658

A New Series of Radical-Cation Salts Based on an Asymmetrical Donor DMET-DSDTF
Kikuchi, K.; Ikemoto, I.; Kobayashi, K.
Synth. Met. **1987**, *19,* 551

Synthesis of a New Series of Unsymmetrical Dithiadiselenafulvalenes
Kobayashi, K.; Kikuchi, K.; Namiki, T.; Ikemoto, I.
Synth. Met. **1987**, *19,* 555

New Series of Organic Metal. Radical Salts of Unsymmetrical Donors Containing Pyrazino Group
Kikuchi, K.; Kamio, H.; Saito, K.; Yiannopoulos, S.; Papavassiliou, G. C.; Kobayashi, K.; Ikemoto, I.
Bull. Chem. Soc. Jpn. **1988**, *61,* 741

Electronic Properties of (DMET)$_2X$
Murata, K.; Kikuchi, K.; Takahashi, T.; Kobayashi, K.; Honda, Y.; Saito, K.; Kanoda, K.; Tokiwa, T.; Anzai, H.; Ishiguro, T.; Ikemoto, I.
J. Mol. Electronics **1988**, *4,* 173

Physical Properties of (DMET)$_2X$
Kikuchi, K.; Saito, K.; Ikemoto, I.; Murata, K.; Ishiguro, T.; Kobayashi, K.
Synth. Met. **1989**, *27,* B269

Conducting and Superconducting Salts Based on MDTTTF, EDTTTF, VDTTTF, EDTDSDTF, MDSTTF, BMDTTTF, Pd(dmit)$_2$, and Ni(dcit)$_2$
Papavassiliou, G. C.; Mousdis, G.; Kakoussis, V.; Terzis, A.; Hountas, A.; Hilti, B.; Mayer, C. W.; Zambounis, J. S.
The Physics and Chemistry of Organic Superconductors, Saito, G,; Kagoshina, S., Eds, Springer-Verlag, Berlin, **1990**, 247

New Conducting Solids Based on Some Symmetrical and Unsymmetrical π-Donors
Mousdis, G. C.; Kakoussis, V. C.; Papavassiliou, G. C.
Proc. NATO-ASIOM Lower Dsgn. Systems and Mol. Electronics, Spetses, Greece, June, 1989, Metzger, R. M.; Day, P.; Papavassiliou, G., Eds., Plenum Press, 1990

Design and Synthesis of Polyheterotetraheterafulvalenes, Metal, 1,2-diheterolenes, and Their Low-Dimensional Conducting and Superconducting Salts
Papavassiliou, G. C.
Proc. NATO-ASIOM Lower Dsgn. Systems and Mol. Electronics, Spetses, Greece, June, 1989, Metzger, R. M.; Day, P.; Papavassiliou, G., Eds., Plenum Press, 1990

New Tetraheterafulvalenes, Metal 1,2-diheterolenes and Some of Their Products
Papavassiliou, G. C.
Pure and Appl Chem., submitted

4,5-Ethylenedioxy-4',5'-ethylenedithiotetrathiafulvalene (EOET): A New Unsymmetrical Electron Donor
Kini, A. M.; Mori, T., Geiser, U.; Budz, S. M.; Williams, J. M.
J. Chem. Soc. Chem. Commun. **1990**, 647

Methylenediselenotetrathiafulvalene and Similar Unsymmetrical π-Donors
Papavassiliou, G. C.; Kakoussis, V. C.; Zambounis, J. S.; Mousdis, G. A.
Chemica Scripta **1989,** to be published

4,5-Bis(methylthio)-4',5'-bis(alkylthio)tetrathiafulvalenes
Papavassiliou, G. C.; Kakoussis, V. C.; Mousdis, G. A.; Zambounis, J. S.; Mayer, C. W.
Chemica Scripta **1989**, *29,* 71

PRESSURE STUDIES

Superconductivity and Surrounding Phase of Organic Conductor, $(DMET)_2Au(CN)_2$
Kikuchi, K.; Murata, K.; Kikuchi, M.; Honda, Y.; Takahashi, T.; Oyama, T.; Ikemoto, I.; Ishiguro, T.; Kobayashi, K.
Jpn. J. Appl. Phys. **1987**, *26*, Supplement 26-3, 1369

Pressure-Induced One-dimensional Instability in $(DMDCNQI)_2Cu$
Mori, T.; Imaeda, K.; Kato, R.; Kobayashi, A.; Inokuchi, H.
J. Phys. Soc. Jpn. **1987**, *56*, 3429

High-Pressure Optical Study of Partially Oxidized Metallophthalocyanines and Metallotetrabenzo-Prophyrins
Ida, T.; Yamakado, H.; Masuda, H.; Yakushi, K.; Kanazawa, D.; Tajima, H.; Kuroda, H.
Mol. Cryst. Liq. Cryst. **1990**, *181*, 243

REVIEW

Composite Crystals: What are They and Why are They so Common in the Organic Solid State?
Coppens, P.; Maly, K.; Petricek, V.
Mol. Cryst. Liq. Cryst. **1990**, *181*, 81

New Organic Superconductors, $(DMET)_2X$
Ikemoto, I.
JJAP Series 1, Superconducting Materials **1988**, 170

THEORY

Electronic Properties of $(DMET)_2X$
Murata, K.; Kikuchi, K.; Takahashi, T.; Kobayashi, K.; Honda, Y.; Saito, K.; Kanoda, K.; Tokiwa, T.; Anzai, H.; Ishiguro, T.; Ikemoto, I.
Proc. Elorma (Tashkent) **1987**

Organic Conductors and Superconductors: A Comparative Survey
Pouget, J. P.; Ravy S.; Moret, R.
Phase Transitions **1989**, *14*, 261

X-RAY DIFFRACTION

Possible Appearance of Spin-Density Waves in the Insulating Phase of the Organic Conductors, $(DMET)_2Au(CN)_2$ and $(DMET)_2AuI_2$
Nogami, H.; Tanaka, M.; Kagoshima, S.; Kikuchi, K.; Saito, K.; Ikemoto, I.; Kobayashi, K.
J. Phys. Soc. Jpn. **1987**, *56*, 3783

Crystal Structure of 2,5-Bis(dicyanomethylene)-2,5-dihydrothieno [3,2-b]thiophene and Bis(ethylenedithio)tetrathiafulvalene Complex

Aso, Y.; Yui, K.; Ishida, H.; Otsubo, T.; Ogura, F.; Kawamoto, A.; Tanaka, J.
Chem. Lett. **1988**, 1069

Structure of the Organic Superconductor $(DMET)_2AuI_2$

Ishikawa, Y.; Kikuchi, K.; Saito, K.; Ikemoto, I.; Kobayashi, K.
Acta Cryst. **1989**, *C45,* 572

The Synthesis of 4,5-Ethylenedithio-4′,5′-vinylenedithio-tetrathiafulvalene (EVT) and Its Methyl and Dimethyl Derivatives (EMVT, EDMVT), and the Molecular and Crystal Structures of EVT

Nakano, H.; Miyawaki, K.; Nogami, T.; Shirota, Y.; Harada, S.; Kasai, N.
Bull. Chem. Soc. Jpn., in press

First Unsymmetrical Donor Which Gives Organic Superconductors: DMET

Saito, K.; Ishikawa, Y.; Kikuchi, K.; Ikemoto, I.; Kobayashi, K.
Acta Cryst. **1989**, *C45,* 1403

Structure of 4,5-Dimethyl-9,10-ethylenedithio-1,3-diselena-6,8-dithiafulvalene Dicyanoaurate(I) (2:1), $(DMET)_2Au(CN)_2$

Kikuchi, K.; Ishikawa, Y.; Saito, K.; Ikemoto, I.
Acta Cryst. **1988**, *C44,* 466

Crystal Structures of $(DMET)_2X$

Kikuchi, K.; Ishikawa, Y.; Saito, K.; Ikemoto, I.; Kobayashi, K.
Synth. Met. **1988**, *27,* B391

Structure of Dimethylethylenedithiodiselenadithiafulvalene Perchlorate $(DMEDT\text{-}DSDTF)_2ClO_4$

Takhirov, T. G.; Krasochka, O. N.; D'yachenko, O. A.; Atovmyan, L. O.; Petrov, M. L.; Rubtsova, I. K.; Lyubovskaya, R. N.
Zh. Strukt. Khlmii **1989**, *30,* 114; *J. Struct. Chem.* **1989**, *30,* 455

Structures of the Conducting Salts of Pyrazinoethylenedithiotetrathiafulvalene (PEDTTTF) and Dimethylpyrazinoethylenedithiotetrathiafulvalene (DMPEDTT-TF): β-$(PEDTTTF)_3I_3$ and β-$(DMPEDTTTF)_3I_3$

Psycharis, V.; Hountas, A.; Terzis, A.; Papavassiliou, G.
Acta Cryst. **1988**, *C44,* 125

Structure of the 2/1 Complex Dibenzotetrathiafulvalenium Hexabromodicuprate-(II), $2C_{14}H_8S_4^+ \cdot Cu_2Br_6^{2-}$

Honda, M.; Katayama, K.; Tanaka, J.; Tanaka, M.
Acta Cryst. **1985**, *C41,* 688

The Double-Stack Structure of Di(3,4-ethylenedithio-3′,4′-dimethyl-2,2′,5,5′-tetrathiafulvalenium)Perchlorate, $(DIMET)_2ClO_4$
Endres, H.; Heid, R.; Keller, H. J.; Heinen, I.; Schweitzer, D.
Acta Cryst. **1987**, *C43*, 115

Structures of Dimethyltetramethylenetetrathiafulvalene Perchlorate, $(DMCTTF)_2ClO_4$, and Tetrafluoroborate, $(DMCTTF)_2BF_4$, Salts
Granier, T.; Gallois, B.; Fabre, J. M.
Acta Cryst. **1989**, *C45*, 1376

Structural Instabilities in Molecular Conductors: Silver and Copper Salts of dicyanoquinonediimine, $(DCNQI)_2X$ (X = Ag or Cu)
Moret, R.
Synth. Met. **1988**, *27*, B301

$(DMEDT-TTF)_2X$: Structures of Two Radical Cation Salts with Centre-symmetrical Anions $X = AsF_6^-$
Gallois, B.; Gaultier, J.; Bechtel, F.; Chasseau, D.; Hauw, C.; Ducasse, L.
Synth. Met. **1987**, *19*, 419

Possible Appearance of Spin-Density Waves in the Insulating Phase of the Organic Conductors, $(DMET)_2Au(CN)_2$ and $(DMET)_2AuI_2$
Nogami, Y.; Tanaka, M.; Kagoshima, S.; Kikuchi, K.; Saito, K.; Ikemoto, I.; Kogayashi, K.
J. Phys. Soc. Jpn. **1987**, *56*, 3783

Structure of 4,5-Dimethyl-9,10-ethylenedithio-1,3-diselena-6,8-dithiafulvalene Dicyanoaurate(I)(2:1), $(DMET)_2Au(CN)_2$
Kikuchi, K.; Ishikawa, Y.; Saito, K.; Ikemoto, I.; Kobayashi, K.
Acta Cryst. **1988**, *C44*, 466

Appendix A

BO Series*

MULTIPLE DISCIPLINE STUDIES

Charge-Transfer Salts Derived from the New Electron Donor Molecule BEDO-TTF: ESR, Superconductivity and Electrical Properties, and Crystal and Band Electronic Structure
Beno, M. A.; Wang, H. H.; Carlson, K. D.; Kini, A. M.; Frankenbach, G. M.; Ferraro, J. R.; Larson, N.; McCabe, G. D.; Thompson, J.; Purnama, C.; Vashon, M.; Williams, J. M.; Jung, D.; Whangbo, M.-H.
Mol. Cryst. Liq. Cryst. **1990**, *181*, 145

$(BEDO)_{2.4}I_3$: The First Robust Organic Metal of BEDO-TTF
Wudl, F.; Yamochi, H.; Suzuki, T.; Isotalo, H.; Fite, C.; Kasmai, H.; Liou, K.; Srdanov, G.; Coppens, P.; Maly, K.; Frost-Jensen, A.
J. Am. Chem. Soc. **1990**, *112*, 2461

The First Ambient Pressure Organic Superconductor Containing Oxygen in the Donor Molecule, $\beta_m - (BEDO-TTF)_3Cu_2(NCS)_3$, $T_c = 1.06$ K
Beno, M. A.; Wang, H. H,; Kini, A. M.; Carlson, K. D.; Geiser, U.; Kwok, W. K.; Thompson, J. E.; Williams, J. M.

*Includes oxygen analogues of ET, and symmetrical oxygen related compounds.

Inorg. Chem. **1990**, *29*, 1599

New Organic Synthetic Metals Derived from BEDT-TTF, Ni(dsit)$_2$ and BEDO-TTF

Beno, M. A.; Kini, A. M.; Geiser, U.; Wang, H. H.; Carlson, K. D.; Williams, J. M.

The Physics and Chemistry of Organic Superconductors, Saito, G.; Kagoshima, S.; Eds., Springer-Verlag, Berlin, **1990**, vol. 51, pp. 369–372

PREPARATION, COMPOSITION, PROPERTIES

Bis(ethylenedioxy)tetrathiafulvalene: The First Oxygen-Substituted Tetrathiafulvalene

Suzuki, T.; Yamochi, H.; Srdanov, G.; Hinkelmann, K.; Wudl, F.

J. Am. Chem. Soc. **1989**, *111*, 3108

Synthesis and Crystal Structure of Bis(oxapropylenedithio)Tetrathiafulvalene Hexafluorophosphate

Medne, R. S.; Khodorkovskii, V. Yu.; Neiland, O. Ya.; Aldoshina, M. Z.; Gol'denberg, L. M.; Lyubovskaya, R. N.; Takhirov, T. G.; D'yachenko, O. A.; Atovmyan, L. O.

Izvestiya Akademii Nauk SSSR, Seriya Khimicheskaya **1989**, *1*, 174

Appendix B

List of Researchers in the Field of Organic Superconductivity

The following list is an attempt to include the many researchers in the field of organic superconductivity, to the extent as it is covered in this book. Given the limited amount of time we had to compile this list, it is unlikely that we have named every principal investigator, but we would like to emphasize that no omissions are intentional. In fact, we would appreciate receiving additions and corrections in order to include them in possible future editions. We have tried to restrict the list to principal investigators (Ph.D. or equivalent), but in many cases it was impossible from the author list of a publication to discern which names were students. In such instances we have included all names in the list. Some names, mainly Russian, have been transcribed into the Roman alphabet in more than one way. The spelling most commonly found has been adopted here. The list is alphabetically grouped by country, and within each country, institutions are listed alphabetically.

Université de Constantine
Laboratoire des Matériaux
   Moléculaires et de
   Crystallochimie
Institut de Chimie
Algeria
	Mustapha Bencharif

University of British Columbia Department of Physics Vancouver, BC V6T 2A6 Canada	Doug Bonn, John Eldridge, T. Pfiz, D. Ll. Williams
Academia Sinica Institute of Chemistry Beijing China (People's Republic)	Min-Xie Qian, Meixiang Wan, Ping Wang (also at Inst. Mol. Sci. Japan), Xiao-Hong Wang, You-Xin Yao, Zhaolou Yu, Daoben Zhu, Naijue Zhu, Yu-Lan Zhu
Academia Sinica Institute of Physics Beijing China (People's Republic)	Hong-Min Duan, Shu-Yuan Lin, Jing-Qing Liu, Li Lu, Bei-Hai Ma, Peiji Wu, Dian-Lin Zhang
Yianbian University Department of Chemistry Yianji Jilin China (People's Republic)	Zhu Shi Li
Czechoslovak Academy of Sciences Institute for Macromolecular Chemistry 16206 Prague 6 Czechoslovakia	Vladimír Kubánek, Stanislav Nespurek, Olen Ryba, Miloslav Sorm
Odense University Department of Chemistry Odense M Denmark	J. Becher, G. Bojesen
Technical University of Denmark Chemistry Department B 2800 Lyngby Denmark	Grethe Rindorf, Niels Thorup
Technical University of Denmark Physics Laboratory 3 2800 Lyngby Denmark	Claus S. Jacobsen, Kell Mortensen
University of Copenhagen Department of General and Organic Chemistry Universitætsparken 5 2100 Copenhagen Denmark	Klaus Bechgaard, T. Bjørnholm, Ib Johannsen, M. Jørgensen, Knud A. Lerstrup

Centre de Recherche Paul Pascal CNRS Domaine Universitaire de Bordeaux I 33405 Talence Cédex France	J. Amiell, Claude Coulon, Pierre Delhaès, E. Dupart, Chantal Garrigou-Lagrange, R. Laversanne, G. Maceno
École Nationale Supérieure de C.P.B. 351 Cours de la Libération 33405 Talence Cédex France	C. Manigand, J. P. Morand
École Nationale Supérieure de Chimie de Paris CNRS UA 403 11 Rue Pierre et Marie Curie 75231 Paris Cédex 05 France	M. Lequan, R. M. Lequan
GDPC-USTL Place E. Bataillon 34060 Montpellier Cédex France	Patrick Bernier
Université de Bordeaux I Laboratoire de Cristallographie 150 Cours de la Libération 33405 Talence Cédex France	F. Bechtel, Daniel Chasseau, Bernard Gallois, J. Gaultier, C. Hauw
Université de Bordeaux I Laboratoire de Physico Chimie Théorique 33405 Talence Cédex France	L. Ducasse
Université de Paris VI-Jussieu Laboratoire de Physicochimie Inorganique 75230 Paris France	Anne Davidson (see also Orsay), Gilbert Hervé
Université de Paris-Sud Laboratoire de Chimie Théoretique (CNRS) 91405 Orsay Cédex France	Enric Canadell

Université de Paris-Sud Laboratoire de Physique des Solides (CNRS) 91405 Orsay Cédex France	Pascale Auban, Patrick Batail, Kamal Boubekeur, C. Bourbonnais, Robert Comes, F. Creuzet, G. Creuzet, Anne Davidson (see also U. Paris VI), Patrick Davidson, H. Hurdequint, Denis Jérome, W. Kang, Christine Lenoir, Anne-Marie Levelut, R. Moret, Alain Pénicaud, Jean Paul Pouget, S. Ravy
Université de Rennes I Laboratoire de Chimie Minérale B UA CNRS 254 35042 Rennes France	André Perrin, Christiane Perrin
Université de Rennes I Laboratoire du Solide Inorganique Moléculaire UA CNRS 254 35042 Rennes France	Daniel Grandjean, Lahcène Ouahab, Jean Padiou
Université Paul Sabatier Laboratoire de Chimie de Coordination du CNRS 205 Route de Narbonne 31077 Toulouse France	Michèle Bousseau, Patrick Cassoux, Jean-Pierre Legros, Lydie Valade
Demokritos National Research Center for Physical Sciences Institute of Materials Science Aghia Paraskevi Attikis 15310 Athens Greece	Athanasios Hountas, V. Psycharis, Aristides Terzis
National Hellenic Research Foundation Theoretical and Physical Chemistry Institute 48, Vassileos Constantinou Ave. Athens 116/35 Greece	V. C. Kakoussis, G. A. Mousdis, George C. Papavassiliou, S. Y. Yiannopoulos, J. S. Zambounis (see also Ciba-Geigy, Switzerland)
Ben-Gurion University of the Negev Department of Chemistry Beer Sheva 84105 Israel	Eliezer Aharon-Shalom, J. Y. Becker, J. Bernstein, S. Bittner, J. A. R. P. Sarma, L. Shahal, S. S. Shaik

Hebrew University of Jerusalem Racah Institute of Physics 91904 Jerusalem Israel	M. Weger
Electrotechnical Laboratory 1-1-4 Umezono Tsukuba Ibaraki 305 Japan	Hiroshi Bando, Yoshiaki Honda (see also Tokyo Metro. U.), Koji Kajimura, Nobumori Kinoshita, Keizo Murata, Kazuhiro Takahashi, Hiroshi Tanino, Madoka Tokumoto, Kunihiko Yamaji
Gakushuin University Department of Physics 1-5-1 Mejiro, Toshima-ku Tokyo 171 Japan	Kazushi Kanoda, Yutaka Maniwa, Toshihiro Takahashi, Toru Tokiwa
Himeji Institute of Technology Basic Research Laboratory Himeji 671-22 Japan	Hiroyuki Anzai, Yoshiki Higuchi, Noritake Yasuoka
Hiroshima University Department of Applied Chemistry Faculty of Engineering Saijo, Higashi Hiroshima 724 Japan	Yoshio Aso, Hideki Ishida, Fumio Ogura, Tetsuo Otsubo, Koji Yui
Institute for Molecular Science Myodaiji, Okazaki 444 Japan	Kenichi Imaeda, Tamotsu Inabe, Hiroo Inokuchi, Yusei Maruyama, Takehiko Mori, Kazuhiko Seki, Tong Bor Tang, K. Toriumi, Kyuya Yakushi, Ping Wang, Peiji Wu
International Superconductivity Technology-Center Mutsuno, Atsuta-ku Nagoya 456 Japan	Hatsumi Mori (née Urayama)
Japan Carlit Co. Ltd. Central Research Laboratory Shibukawa Gunma 377 Japan	Masahi Oshima

Kumamoto University Department of Chemistry Faculty of Science Kurokami Kumamoto 860 Japan	Susumu Matsuzaki, Mizuka Sano
Kyoto University Department of Chemistry Sakyo-ku Kyoto 606 Japan	T. Nakamura, Gunzi Saito
Kyoto University Department of Physics Faculty of Science Kitashirakawa Oiwake-cho Sakyoku, Kyoto 606 Japan	Takehiko Ishiguro, Yoshio Nogami
Nagoya University College of General Education Chikusa-ku Nagoya 464 Japan	Mitsuru Sano, Hideo Takeuchi, Masashi Tanaka
Nagoya University Department of Chemistry Faculty of Science Chikusa-ku Nagoya 464 Japan	Shamim Akhtar, Masako Honda, Chuji Katayama, Atsushi Kawamoto, Hiroaki Kumagai, Yasuyoshi Saito, Masaaki Shimizu, Chizuko Tanaka, Jiro Tanaka
National Chemical Laboratory for Industry Tsukuba Ibaraki 305 Japan	Midori Goto, Kazumasa Honda, Masayasu Kurahashi
National Institute for Research in Inorganic Materials Sakura-mura, Niihari-gun Ibaraki 305 Japan	Katsuo Kato
Okayama University Department of Physics Okayama 700 Japan	Kokichi Oshima

Osaka University Department of Applied Chemistry Faculty of Engineering Yamadaoka, Suita Osaka 565 Japan	Shigeharu Harada, Nobutami Kasai, Gen-etsu Matsubayashi, Tadashi Nakamura, Hideyuki Nakano, Takashi Nogami, Ryuichi Shimizu, Yashuhiko Shirota, Toshio Tanaka
Osaka University Department of Chemistry Faculty of Science Toyonaka Osaka 560 Japan	Ichiro Murata, Kazuhiro Nakasuji
Osaka University Department of Physics Faculty of Science Machikaneyama Toyonaka 560 Japan	S. Sugai
Osaka University Institute for Protein Research Suita 565 Japan	Koji Inaka
Perkin-Elmer Japan Kitasaiwai, Nishi-ku Yokohama 220 Japan	Koichi Nishikida
Sumitomo Electric Industries R & D Group 1-1-3 Shimaya Konohanaku Osaka 554 Japan	M. Kaji, H. Kusuhara, Y. Sakata, K. Tada, Y. Ueba
Toho University Department of Chemistry Faculty of Science Funabashi Chiba 274 Japan	Reizo Kato, Hayao Kobayashi, Miyuki Takahashi

Toho University Department of Physics Faculty of Science Funabashi Chiba 274 Japan	Koji Kajita, Shinji Moriyama, Yutaka Nishio, Waturu Sasaki, T. Takahashi
Tohoku University Department of Chemistry Faculty of Science Aramaki Sendai 980 Japan	T. Miyashi, T. Suzuki
Tohoku University Institute for Materials Research Katahira 2-1-1 Sendai 980 Japan	Yoshio Muto, Takahiko Sasaki, Naoki Toyota
Tokyo Institute of Technology Department of Chemistry Faculty of Science Megro-ku Tokyo 152 Japan	Toshiaki Enoki, Kazuya Suzuki
Tokyo Metropolitan University Department of Chemistry Faculty of Science Fukazawa, Setagaya-ku Tokyo 158 Japan	Yoshiaki Honda (see also El.Techn. Lab.), Isao Ikemoto, Yoshimitsu Ishikawa, Hiroyuki Kamio, Koichi Kikuchi, Mikio Kikuchi, Takahisa Namiki, Kazuya Saito
University of Tokyo Department of Chemistry College of Arts and Sciences Komaba, Meguro-ku Tokyo 153 Japan	Keiji Kobayashi
University of Tokyo Department of Chemistry Faculty of Science Hongo, Bunkyo-ku Tokyo 113 Japan	Koji Kajita (see also Toho U.), Reizo Kato, Akiko Kobayashi, Haruo Kuroda, Toshio Naito, Yuji Okawa, Yukiyoshi Sasaki, Hiroyuki Tajima, Masafumi Tamura, Akito Ugawa, Michiko Yoshitake

University of Tokyo Department of Pure and Applied Science Komaba 3-8-1, Meguro-ku Tokyo 153 Japan	M. Hasumi, Seiichi Kagoshima, A. Kawasumi, T. Osada, Masanori Tanaka, R. Yagi
University of Tokyo Institute of Solid State Physics Ropongi, Minato-ku Tokyo 106 Japan	Hakuro Hayashi, Y. Iye, Naoko Iwasawa, Minoru Kinoshita, Shinji Mino, N. Miura, Nobuo Mori, Kiyokazu Nozawa, Naoko Okada, Fumiko Sakai, Shoichi Sato, Fumihiko Shinozaki, Tadashi Sugano, Hiroshi Takenouchi, Kenji Terui, T. Tsuji, Hideki Yamochi
University of Sonora CIPM AP 130, Hermosillo, Sonora Mexico	Michiko B. Inoue, Motomichi Inoue
Polish Academy of Sciences Institute of Molecular Physics Smoluchowskiego 17/19 60-179 Poznan Poland	A. Graja, L. Firlej, R. Swietlik, J. Wolak
Ciba-Geigy AG Zentrale Forschung 4002 Basel Switzerland	M. Bürkle, Bruno Hilti, Carl W. Meyer, Jürgen Pfeiffer, J. Zambounis
Bristol University School of Chemistry Cantock's Close Bristol BS8 1TS United Kingdom	Judith A. K. Howard, Andrew M. Stringer
Cambridge University Cavendish Laboratory Madingley Road Cambridge CB3 0HE United Kingdom	Margaret Allan, Richard H. Friend, H. P. Hughes, David S. Obertelli, I. D. Parker
Oxford University Chemical Crystallography 9 Parks Road Oxford OX1 3PD United Kingdom	David Watkin

Oxford University Inorganic Chemistry Laboratory South Parks Road Oxford OX1 3QR United Kingdom	Simon Bott, Peter Day (currently at Inst. Laue-Langevin, Grenoble, France), Claire Hollis, Mohamedally Kurmoo, Talal Mallah, K. L. Pritchard, Matthew J. Rosseinsky
Queen Mary College Department of Chemistry Mile End Road London E1 4NS United Kingdom	M. E. Harman, M. B. Hursthouse
University College of North Wales, Bangor Department of Chemistry Gwynedd LL57 2UW, Wales United Kingdom	A. Clark, B. Girmay, B. Kaye, J. D. Kilburn, Allan E. Underhill, K. S. Varma
Argonne National Laboratory Chemistry and Materials Science Divisions 9700 S. Cass Ave. Argonne, IL 60439 USA	Mark A. Beno, K. Douglas Carlson, John R. Ferraro, Urs Geiser, Aravinda M. Kini, Arthur J. Schultz, Robert J. Thorn, Hau H. Wang, Jack M. Williams
Argonne National Laboratory Materials Science Division 9700 S. Cass Ave. Argonne, IL 60439 USA	George W. Crabtree, Kenneth E. Gray, W. K. Kwok
AT&T Bell Laboratories Murray Hill, NJ 07974 USA	S. V. Chichester-Hicks, Sherman D. Cox, S. H. Glarum, J. E. Graebner, R. C. Haddon, D. R. Harshman, R. N. Kleiman, M. L. Kaplan, L. W. Rupp, Jr.
Boston University Physics Department Boston, MA 02215 USA	J. S. Brooks
Columbia University Physics Department 538 West 120th Street New York, NY 10027 USA	Y. T. Uemura
General Electric Corp. R & D Center P. O. Box 8 Schenectady, NY 12301 USA	Margaret L. Blohm, Mary Garbauskas, (Leonard V. Interrante, now at Rensselaer Poly.), Oscar Le Blanc

GTE Laboratories, Inc. 40 Sylvan Road Waltham, MA 02254 USA	D. T. Sandman
IBM-Almaden Research Center 650 Harry Road San Jose, CA 95120-6099 USA	E. M. Engler, V. Y. Lee, S. S. P. Parkin
Indiana University Department of Chemistry Bloomington, IN 47405 USA	Lawrence K. Montgomery
Johns Hopkins University Department of Chemistry Baltimore, MD 21218 USA	Dwaine O. Cowan
Johns Hopkins University M. S. Eisenhower Research Center Applied Physics Laboratory Laurel, MD 20707-6099 USA	Thomas J. Kistenmacher, Theodore E. Poehler
North Carolina State University Department of Chemistry Raleigh, NC 27695-8204 USA	Michel Evain, Juan J. Novoa, Jinquin Ren, Myung-Hwan Whangbo
Northwestern University Department of Chemistry and Materials Research Center Evanston, IL 60208 USA	William E. Broderick, Martin R. Godfrey, Brian M. Hoffman, James A. Ibers, Tobin J. Marks, Ellen M. McGhee
Princeton University Department of Physics Princeton, NJ 08554 USA	Paul M. Chaikin
San Jose State University San Jose, CA 95192 USA	J. V. Acrivos
Sandia National Laboratories Albuquerque, NM 87185 USA	Richard J. Baughman, James F. Kwak, Bruno Morosin, D. L. Overmyer, James F. Schirber, Eugene L. Venturini
Stanford University Department of Applied Physics Stanford, CA 94305 USA	William A. Little, D. B. Mitzi

State University of New York at Buffalo Department of Chemistry Buffalo, NY 14214 USA	Philip Coppens
University of Alabama Department of Chemistry Tuscaloosa, AL 35487 USA	Patrick J. Carroll, M.V. Lakshimikantham, Michael P. Cava, Robert M. Metzger
University of Arizona Department of Chemistry Tucson, AZ 85721 USA	Quintus Fernando, Kenneth W. Nebesny
University of Arizona Department of Physics Tucson, AZ 85721 USA	S. Mazdumar
University of California at Santa Barbara Institute for Polymers and Organic Solids Departments of Physics and Chemistry Santa Barbara, CA 93106 USA	A. Heeger, K. Hinkelmann, G. Srdanov, T. Suzuki, F. Wudl
University of California, Los Angeles Department of Physics and Solid State Science Centers Los Angeles, CA 90024 USA	G. Grüner, K. Holczer, O. Klein, D. Quinlivan
University of Florida Department of Chemistry Gainesville, FL 32611 USA	Daniel R. Talham
University of Florida Department of Physics Gainesville, FL 32611 USA	Bohdan Andraka, J. S. Kim, Gregory R. Stewart, David B. Tanner
University of Houston Houston, TX 77004 USA	Thomas Jones
University of Illinois Department of Physics 1110 West Green Street Urbana, IL 61801 USA	Charles P. Slichter
University of Maryland College Park, MD USA	R. L. Greene

List of Researchers in the Field of Organic Superconductivity

University of New Mexico Department of Chemistry Albuquerque, NM 87131 USA	E. Duesler
Academy of Sciences of the Ukrainian SSR Institute of Semiconductors Kiev USSR	G. O. Baram, L. S. Degtyarev, M. É. Kozlov, V. G. Onishchenko, K. I. Pokhodnya, Yu. V. Sushko, M. A. Tanatar, A. A. Yurchenko, M. K. Sheinkman
USSR Academy of Sciences Institute of Chemical Physics 142432 Chernogolovka USSR	M. A. Abramov, M. Z. Aldoshina, Lev Oganovich Atovmyan, N. V. Avramenko, V. K. Bel'skii (Karpov Phys. Chem. Inst.?), Yu. G. Borodko, L. I. Buravov, Oleg Anatolyevich D'yachenko, O. N. Eremenko, L. (D.?) N. Fedutin, A. A. Galimzyanov, L. M. Goldenberg, A. A. Ignatiev, Vladimir Fedorovich Kaminskii, M. G. Kaplunov, M. V. Kartsovnik, M. L. Khidekel', A. G. Khomenko, P. A. Kononovich, V. E. Korotkov, E. E. Kostyuchenko, A. I. Kotov, N. D. Kushch, Oleg Nikolayevich Krasochka (deceased), V. N. Laukhin, E. É. Laukhina, S. V. Lindeman, R. M. Lobkovskaya, R. N. Lyubovskaya, R. B. Lyubovskii, M. K. Makova, V. A. Merzhanov, V. N. Molchanov, S. I. Pesostskii, M. L. Petrov, A. A. Romanyukha, L. P. Rozenberg, I. K. Rubtsova, I. F. Shchegolev, Rimma Pavlovna Shibaeva, M. A. Simonov (Moscow State U.?), E. M. Sogomonyan, T. G. Takhirov, L. S. Veretennikova, É. B. Yagubskii, E. I. Zhilyaeva, A. V. Zvarykina
USSR Academy of Sciences Institute of Crystallography Leninsky pr. 59 Moscow 117234 USSR	Vitol'd Nikolayevich Kachinskii, Vladimir Nikolayevich Molchanov, Valentin Ivanovich Simonov, Boris Konstantinovich Vainshtein
USSR Academy of Sciences L. D. Landau Institute of Theoretical Physics 117334 Moscow V-334 USSR	L. P. Gor'kov
USSR Academy of Sciences P. N. Lebedev Physical Institute Leninsky Prospect 53 Moscow USSR	L. N. Bulaevskii, V. B. Ginodman, A. V. Gudenko, I. I. Zasavitskii
USSR Academy of Sciences Ural Scientific Research Center Institute of Metal Physics Kovalevskaya 18 Sverdlovsk, GSP 170 USSR	Alexander A. Romanyukha, Yuri N. Shvachko, Alexander V. Skripov, V. V. Ustinov

V. I. Ulyanov-Lenin Institute of Electrical Engineering Leningrad USSR	I. B. Vendik, A. N. Ermolenko, V. V. Esipov, V. M. Pchelkin, M. F. Sitnikova
Max-Planck-Institut für Festkörperforschung Heisenbergstrasse 1 7000 Stuttgart 80 Germany	M. Gehrke, R. K. Kremer
Max-Planck-Institut für Kohlenforschung Kaiser-Wilhelm-Platz 1 4330 Mülheim (Ruhr) 1 Germany	K. Angermund, C. Krüger
Max-Planck-Institut für Medizinische Forschung Abteilung für Molekulare Physik Jahnstrasse 29 6900 Heidelberg 1 Germany	P. Bele (see also U. Heidelberg), K. Bender, H. Brunner (see also U. Heidelberg), S. Gärtner, Emil Gogu, Hans Grimm, U. Häberlen, I. Heinen, I. Hennig, Siegfried Kahlich, T. Klutz, K. Polychroniadis, R. Zamboni
Technische Universität München Anorganische Chemie 8046 Garching Germany	H. P. Fritz, H. Müller
Universität Bochum Experimentalphysik IV 4630 Bochum 1 Germany	B. Rothaemel
Universität Heidelberg Anorganisch-Chemisches Institut Im Neuenheimer Feld 270 6900 Heidelberg Germany	P. Bele (see also MPI Med. F.), H. Brunner (see also MPI Med. F.), Karlheinz Dietz, H. Endres, R. Heid, M. Hiller, Heimo J. Keller, B. Nuber, A. Weber
Universität Karlsruhe Institut für angewandte Physik Kaiserstrasse 12 7500 Karlsruhe Germany	H. P. Geserich

Universität München Anorganisch-Chemisches Institut Meiserstrasse 1 8000 München 2 Germany	Eberhard Amberger, Helmut Fuchs, S. Fuchs, Th. Lehnert, Kurt Polborn, Walter Reith
Universität München Sektion Physik Schellingstrasse 4 8000 München 40 Germany	S. Klotz, J. S. Schilling, R. Sieburger
Universität Stuttgart Physikalisches Institut Pfaffenwaldring 57 7000 Stuttgart 80 Germany	Dieter Schweitzer, J. U. von Schütz, H.-P. Werner, C. C. Wolf
Universität Würzburg Institut für Organische Chemie Am Hubland 8700 Würzburg Germany	A. Aumüller, P. Erk, S. Hünig
Walther-Meissner-Institut für Tieftemperaturforschung 8048 Garching Germany	Klaus Andres, H. Fuchs, Frieder Gross, Claus-Peter Heidmann, E. Hess, S. Klotz, A. Lerf, H. Müller, J. S. Schilling, Helmut Schwenk, H. Veith
University of Zagreb Ruter Boskovic Institute of Physics P.O. Box 304 41001 Zagreb Yugoslavia	J. R. Cooper, L. Forró, B. Korin-Hamzic

Appendix C

Superconducting Fullerenes (C_{60})

Since the completion of the main text of this book, there have occurred remarkable new developments in molecular superconductors of such importance that we are compelled to append a survey of this topic for the sake of completeness. This development concerns superconductivity in alkali metal salts of C_{60}, buckminsterfullerene, discovered first in K_3C_{60} ($T_c \sim 18$ K) by Hebard and associates[1] at AT&T Bell Laboratories. Since the discovery[2] of the soccer ball-like, 60 carbon-atom molecule C_{60}, and similar spheroidal carbon atom species, commonly called fullerenes, and the development of methods to produce macroscopic quantities of these materials,[3,4] scientists have been intrigued by the properties and the potential applications of C_{60} based systems. The discovery of superconducting derivatives illustrates such potentialities and reinforces one's expectation that additional new superconducting materials are easily within reach. Whether or not C_{60} is an organic molecule, sometimes defined as a molecule containing both C and H atoms, or an inorganic homonuclear carbon molecule, the superconducting alkali metal fullerides are similar to the conventional organic superconductors to the extent that they are charge-transfer salts and are molecular-based synthetic metals in which the conduction band is derived from the C_{60} electron-acceptor molecule. In contrast, however, the superconducting fullerides have cubic (isotropic) structures and therefore are three-dimensional conductors. In addition, the T_c's of the fullerides are substantially higher than those presently known for the ET-based organic superconductors (maximum $T_c = 12.8$ K).

The superconducting alkali metal fullerides are face-centered cubic compounds of the type A_3C_{60}, where A represents an alkali metal atom. The anchor members of this class are the well-characterized salts K_3C_{60},[1] with T_c in the range 18.0 – 19.6K,[5–8] and Rb_3C_{60},[9] with T_c in the range 28.0 – 29.8K.[5, 7–11] The variability in T_c represents a problem of phase purity; generally, the higher T_c is characteristic of the purer phase. In addition, this class includes an uncharacterized salt, Cs_xC_{60},[12] a variety of mixed alkali metal compositions with K, Rb, and Cs of the type $A_{3-x}A'_xC_{60}$ (x = 0 – 3),[7, 11, 13] which extend the range of T_c up to 33 K, and a thallium-doped Rb-C_{60} salt[14] (possibly $RbTl_2C_{60}$) with $T_c = 45$ K. These salts and their T_c's (generally the temperature for the onset of the superconducting transition) are summarized in Table C.1. One observes in this table a rich variety of superconducting fullerides with an impressive range of remarkably high T_c's extending from 18 K to a present maximum of 45 K.

TABLE C.1 Superconducting Alkali Metal Fullerides

Salt	T_c (K)	Reference
K_3C_{60}	18.0–19.6	1, 5–8
K_2RbC_{60}	21.8–22.5	7, 11
Rb_2KC_{60}	24.4–28.0	7, 11
$Rb_{1.5}K_{1.5}C_{60}$	25.1	7
Rb_3C_{60}	28.0–29.8	5, 7–11
Rb_2CsC_{60}	31.3	7, 13
Cs_2RbC_{60}	33	13
Cs_xC_{60}[a]	29.5	12
$Rb_xTl_yC_{60}$[b]	45	14

[a] x not characterized.
[b] Possibly x = 1, y = 2.

The alkali metal fullerides can be synthesized by a variety of methods involving the reaction of the metal with solid or dissolved C_{60}. These methods require the use of a dry box with an inert atmosphere or the use of high-vacuum techniques because the products, as well as the alkali metal reagents, are reactive to the ambient atmosphere. Macroscopic quantities of C_{60} can be easily prepared from soot generated in the contact-arc[15] or plasma ignition of graphite rods.[16] The soot is dissolved in an organic solvent such as toluene, and the C_{60} is isolated by the use of column chromatography and purified of solvent.[17] Actually, the first of the conducting alkali metal fullerides to be reported were "doped" thin films of C_{60} produced from vaporized C_{60} deposited on glass substrates and reacted with alkali metal vapors.[18] The impetus to these syntheses was derived from both structural and electronic considerations[18, 19] of the nature of solid C_{60}, which is normally a semiconductor (gap = 1.5 eV). Solid C_{60} (fullerite) has a face-centered cubic (fcc) structure (a = 14.1 Å)[20, 21] with two vacant tetrahedral sites and one vacant octahedral site, per molecule, suitable for the accommodation of small electron-donor species such as K^+. Furthermore, molecular orbital calculations[22] for C_{60} yield a low-lying, triply-degenerate, unoccupied level that should be easily reducible by electron transfer of up to a maximum of 6 electrons, thus leading to a partially occupied (conduction) band in the solid for a small amount of electron transfer or to an insulating state for complete charge transfer. Indeed, the alkali metal-doped films of C_{60} at suitably low doping levels, were found[18] to have reasonably good conductivities at room temperature of 4-500 $(\Omega \text{ cm})^{-1}$, which are values comparable to those of the organic synthetic metals. Large concentrations of the alkali metal produced insulating films, presumably due to the full reduction of C_{60} to C_{60}^{6-} and thus to the production of a completely filled band. Further studies[1] of the temperature dependence of the resistivity of thin-film K_xC_{60} (x unknown) showed the onset of a superconducting transition near 16 K and a broad transition to zero resistance at temperatures below 5 K.

Bulk quantities of the superconducting fullerides have been synthesized typically by either a vapor-solid phase technique,[1, 5, 9] in which alkali metal vapors are reacted with purified C_{60} powder in appropriate stoichiometric ratios, or by a solution phase route,[6, 10] in which alkali metal chips are added to a warm toluene solution of purified C_{60} and brought to refluxing temperatures for short periods of time to complete the reduction and precipitation of the product. The products are small-particle, polycrystalline materials, which are compacted and sintered for periods of hours at elevated temperatures, $\sim 200 - 400°C$. The compacting and sintering of samples have been found to be very important in controlling the stoichiometry and phase purity. No reports have yet appeared on the synthesis of single-crystal specimens—a great future challenge! A novel procedure has been reported[8] for the control of stoichiometry that involves the synthesis of the fully reduced C_{60} salt, A_6C_{60}, by the use of the vapor transport technique and the use of a temperature gradient to eliminate excess metal. This step is followed by the mixing and sintering of A_6C_{60} and C_{60} in the correct stoichiometric ratios. Two syntheses, one for the preparation of Cs_xC_{60} with $T_c \sim 30$ K[12] and the other for the preparation of the Tl-doped Rb-C_{60} salt with $T_c = 45$ K,[14] have employed binary alloys of the type A/M, where M is Hg, Tl, or Bi, for the source of the metal atoms. Attempts by others to synthesize Cs_3C_{60} from pure Cs have failed to detect evidence of superconductivity, so that superconducting Cs_xC_{60} may have a different stoichiometry from the other fullerides or possibly it may incorporate the alloying constituent.[7]

Superconductivity in the bulk alkali metal fullerides has been determined by ac susceptibility and dc magnetization measurements. Figure C.1 illustrates the dc magnetization of an unsintered, polycrystalline sample of Rb_3C_{60} prepared by the solution phase method.[10] This figure shows the mass susceptibility χ, expressed in units of cm^3/g, determined with an applied magnetic field of 5 Oe (G) as a function of temperature for both zero field (zfc) and field cooling (fc) of the sample. One observes an onset T_c of 28.6 K and a very broad transition that reaches completion only near 5 K. At this temperature, the zfc-determined χ, which gives the diamagnetic shielding susceptibility, indicates a superconducting fraction of about 5% of a perfect superconductor (volume susceptibility $= -\frac{1}{4}\pi$) on the basis of a density of ~ 2 g/cm^3, and the fc-determined χ indicates a Meissner fraction of about 14% of the superconducting phase. Low values of T_c, very broad superconducting transitions, and low superconducting fractions are typical signatures of samples with

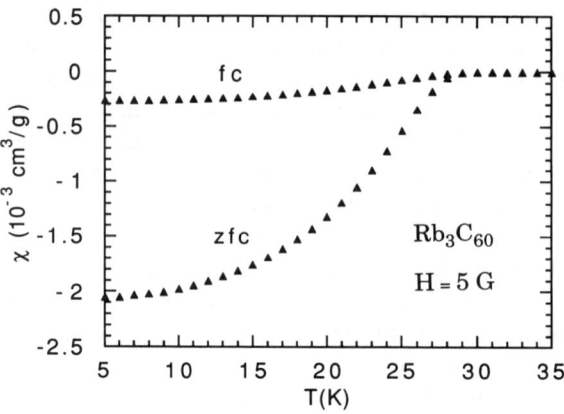

Figure C.1 Temperature dependence of the mass dc susceptibility χ in a magnetic field of 5 Oe (G) for a field cooled (fc) and zero field cooled (zfc) sample of unsintered, polycrystalline Rb_3C_{60} (data taken from the authors' work, ref. 10).

considerable phase impurities. The monitoring of the diamagnetic shielding has proved to be very useful in the development of procedures for the grinding, compacting, and sintering of samples to achieve materials of high phase purity,[23, 24] with volume superconducting fractions reportedly approaching 100%. In some cases, low diamagnetic shielding fractions may be the result of the small particle size, comparable to the dimensions of the London penetration depth ($\lambda_L = 0.24$ μm),[5, 25] and to the granular nature of the materials, leading to weak electrical links between the superconducting grains.[23]

X-ray powder diffraction techniques applied to materials of high phase purity with the use of conventional and synchrotron light sources have elucidated the crystallographic structures of superconducting K_3C_{60},[26] Rb_3C_{60},[8, 11] and several of the salts of mixed alkali metal compositions.[7] In agreement with earlier speculations,[18] these salts indeed have fcc structures derived from the lattice of solid C_{60} with the alkali metal ions intercalated into all of the tetrahedral and octahedral sites, as illustrated in Figure C.2(a). This intercalated structure indicates the possibility of occupational disorder in the mixed alkali metal compositions, a possibility not yet explored in experiments, but it has been suggested[7] that the larger alkali metal ions should prefer the larger octahedral sites. In addition to the superconducting fullerides, the structures have been determined for the fully reduced C_{60} salt, A_6C_{60} with A = K and Cs,[27] and a new salt of the type A_4C_{60} with A = K, Rb, and Cs.[28] The A_6C_{60} and A_4C_{60} salts, which are illustrated in Figures C.2(b) and C.2(c), are nonconducting fullerides having body-centered cubic (bcc) and body-centered tetragonal (bct) structures, respectively, with alkali metal atoms occupying distorted tetrahedral sites. These salts, and possibly others, can occur as impurities in the superconducting phase.

The structural studies of the superconducting fullerides, A_3C_{60}, indicate that these salts comprise an isostructural series of compounds having a systematic correlation between changes in structural parameters and superconducting properties.[7] With the incorporation of the metal ions into the host lattice of C_{60}, there is an

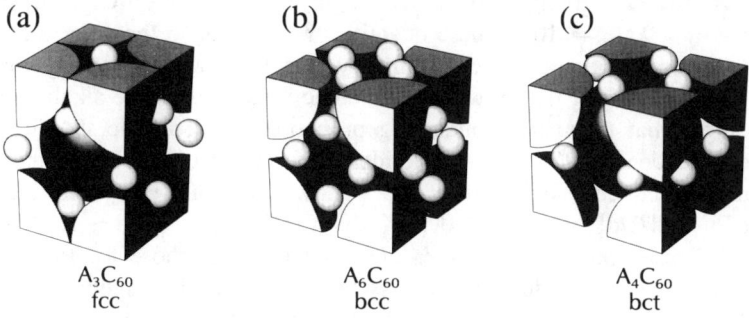

Figure C.2 A representation of the cubic structures of the alkali metal superconducting fullerides (a) A_3C_{60} (face-centered cubic) and the nonconducting fullerides (b) A_6C_{60} (body-centered cubic) and (c) A_4C_{60} (body-centered tetragonal), where A represents the alkali metal cation. This figure was kindly supplied by D. W. Murphy, AT&T Bell Laboratories.

expansion of the unit cell which increases with the increasing size of the cation. Furthermore, accompanying this increase in unit cell size there occurs an increase in T_c. This correlation is very well illustrated in Figure C.3 by a nearly linear relationship between T_c and the cubic lattice parameter a_o,[7] and it is quite similar to the structure-property correlation one finds for replacements of the anion acceptor species X in the β-(ET)$_2$X superconducting salts.

Magnetization measurements[25] of the critical magnetic fields of granular K_3C_{60} show that this material is an extreme type II superconductor. The extrapolated critical magnetic fields at zero temperature are reported[5, 25] to be H_{c2} = 49 T

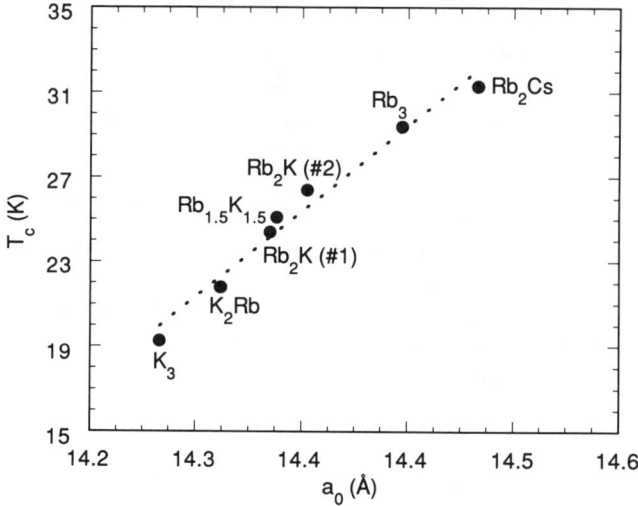

Figure C.3 Change in the superconducting transition temperature (T_c) with change in lattice parameter a_o for the superconducting alkali metal fullerides, A_3C_{60} (data taken from ref. 7).

and $H_{c1} = 132$ Oe, giving a coherence length $\xi = 26$Å and a Ginzburg-Landau parameter $\kappa = \lambda_L/\xi = 100$. These quantities are similar to those for the organic superconductors for properties determined within the plane of the organic donor molecules (see Chapter 4). However, the cubic structures of the alkali metal fullerides imply that their superconducting properties are isotropic. Studies of the pressure dependence of T_c for the fullerides, reported first by Schirber and associates[29] and subsequently by Sparn et al.,[23] show a decrease in T_c with increasing pressure P: $- dT_c/dP = 0.63 - 0.78$ K/kbar for K_3C_{60}[23, 29] and ~ 1.0 K/kbar for Rb_3C_{60}.[23] These pressure derivatives are comparable to those for the ET-based β-phase organic superconductors but considerably smaller in magnitude than those for the ET-based κ-phase organic superconductors with $T_c \geq 10$ K.[30] As exemplified by the organic superconductors for increasing T_c with increasing size of the anion, the negative pressure dependence of T_c for the fullerides suggests that T_c should increase with increasing size of the alkali metal cation (or unit cell volume), and indeed this is experimentally confirmed. Various arguments based on weak-coupling BCS theory suggest that the change in T_c with change in cation size (or lattice parameter) or with change in the applied pressure is related to changes in the density of states at the Fermi level.[7, 23] In contrast to this, the change in T_c of the organic superconductors, at least those of the β-phase salts, can be attributed to changes in the electron-phonon coupling strength because the density of states is essentially constant.[31, 32]

With the discovery of the superconducting alkali metal fullerides, one can now identify *three different classes* of carbon-based molecular superconductors: the one-dimensional salts of the radical cation donor molecule TMTSF; the two-dimensional salts of the radical cation donor molecule BEDT-TTF (ET); and the three-dimensional salts of the radical anion acceptor-molecule C_{60}. The T_c's generally increase with the dimensionality, and one wonders whether the three-dimensional class offers the best opportunity for materials with even higher T_c's. It has been pointed out that the three-dimensional materials have the best chance of achieving the superconducting ground state because they are least likely to possess the common structural instabilities of charge-density- or spin-density-wave states.[19] On the other hand, the highest-T_c materials presently known, the ceramic copper oxide superconductors, are two dimensional systems. In any case, the search for new carbon-based superconductors likely will be successful in yielding new surprising results in the near future. The discovery of superconducting materials with increasing T_c has been remarkably fast-paced in recent years, as exemplified by Figure C.4. Whether or not the rapid increase in T_c will be maintained, it seems likely that discoveries of additional superconducting materials of some importance will continue at a steady pace.

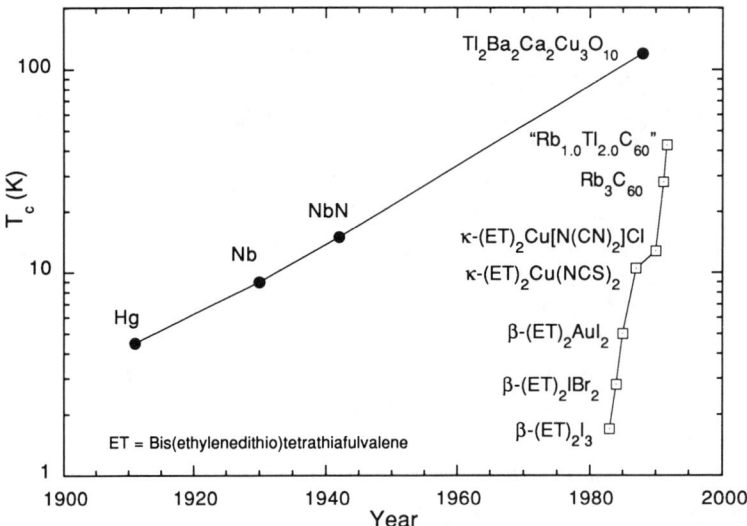

Figure C.4 Superconducting transition temperature T_c vs. year of discovery for representative examples of conventional, high-T_c oxide, and C_{60} carbon-based superconductors.

REFERENCES

1. Hebard, A. F.; Rosseinsky, M. J.; Haddon, R. C.; Murphy, D. W.; Glarum, S. H.; Palstra, T. T. M.; Ramirez, A. P.; Kortan, A. R. *Nature* **1991**, *350*, 600.
2. Kroto, H. W.; Heath, J. R.; O'Brien, S. C.; Curl, R. F.; Smalley, R. E. *Nature* **1985**, *318*, 162.
3. Kratschmer, W.; Lamb, L. D.; Fostiropoulos, K.; Huffman, D. R. *Nature* **1990**, *347*, 354.
4. Meijer, G.; Bethune, D. S. *J. Chem. Phys.* **1990**, *93*, 7800.
5. Holczer, K.; Klein, O.: Huang, S.-M.; Kaner, R. B.; Fu, K.-J.; Whetten, R. L.; Diederich, F. *Science* **1991**, *252*, 1154.
6. Wang, H. H.; Kini, A. M.; Savall, B. M.; Carlson, K. D.; Williams, J. M.; Lykke, K. R.; Wurz, P.; Parker, D. H.; Pellin, M. J.; Gruen, D. M.; Welp, U.; Kwok, W.-K.; Fleshler, S.; Crabtree, G. W. *Inorg. Chem.* **1991**, *30*, 2838.
7. Fleming, R. M.; Ramirez, A. P.; Rosseinsky, M. J.; Murphy, D. W.; Haddon, R. C.; Zahurak, S. M.; Makhija, A. V. *Nature* **1991**, *352*, 787.
8. McCauley, Jr., J. P.; Zhu, Q.; Coustel, N.; Zhou, O.; Vaughan, G.; Idziak, J.; Fischer, J. E.; Tozer, S. W.; Groski, D. M.; Bykovetz, N.; Lin, C. L.; McGhie, A. R.; Allen, B. H.; Romanow, W. J.; Denenstein, A. M., Smith, III, A. B. *J. Am. Chem. Soc.* **1991**, *113*, 8537.
9. Rosseinsky, M. J.; Ramirez, A. P.; Glarum, S. H.; Murphy, D. W.; Haddon, R. C.; Hebard, A. F.; Palstra, T. T. M.; Kortan, A. R.; Zahurak, S. M.; Makhija, A. V. *Phys., Rev. Lett.* **1991**, *60*, 2830.
10. Wang, H. H.; Kini, A. M.; Savall, B. M.; Carlson, K. D.; Williams, J. M.; Lathrop, M. W.; Lykke, K. R.; Parker, D. H.; Wurz, P.; Pellin, M. J.; Gruen, D. M.; Welp, U.; Kwok, W.-K., Fleshler, S.; Crabtree, G. W. *Inorg. Chem.* **1991**, *30*, 2962.

11. Holczer, K.; Chalmers, G. R.; Wiley, J. B.; Huang, S.-M.; Kaner, R. B.; Strouse, C. E.; Diederich, F.; Whetten, R. L. *Science* **1991**, submitted for publication.
12. Kelty, S. P.; Chen, C.-C.; Lieber, C. M. *Nature* **1991**, *352*, 223.
13. Tanigaki, K.; Ebbesen, T. W.; Saito, S.; Mizurki, J.; Tsai, J. S.; Kubo, Y.; Kuroshima, S. *Nature* **1991**, *352*, 222.
14. Iqbal, Z.; Baughman, R. H.; Ramakrishna, B. L.; Khare, S.; Murthy, N. S.; Bornemann, H. J.; Morris, D. E. *Science* **1991**, 254, 826.
15. Haufler, R. E.; Conceicao, J.; Chibante, L. P. F.; Chai, Y.; Byrne, N. E.; Flanagan, S.; Haley, M. M.; O'Bien, S. C.; Panb, C.; Xiao, Z.; Billups, W. E.; Ciufolini, M. A.; Hauge, R. H.; Margrave, J. L.; Wilson, L. J.; Curl, R. F.; Smalley, R. E. *J. Phys. Chem.* **1990**, *94*, 8634.
16. Parker, D. H.; Wurz, P.; Chatterjee, K.; Lykke, K. R.; Hunt, J. E.; Pellin, M. J.; Hemminger, J. C.; Gruen, D. M.; Stock, L. M. *J. Am. Chem. Soc.* **1991**, *113*, 7499.
17. Hare, J. P.; Kroto, H. W.; Tayler, R. *Chem. Phys. Lett.* **1991**, *177*, 394.
18. Haddon, R. C.; Hebard, A. F.; Rosseinsky, M. J.; Murphy, D. W.; Duclos, S. J.; Lyons, K. B.; Miller, B.; Rosamillia, J. M.; Fleming, R. M.; Kortan, A. R.; Glarum, S. H.; Makhija, A. V.; Muller, A. J.; Eick, R. H.; Zahurak, S. M.; Tycko, R.; Dabbagh, G.; Thiel, F. A. *Nature* **1991**, *350*, 320.
19. Murphy, D. W.; Rosseinsky, M. J.; Haddon, R. C.; Ramirez, A. P.; Hebard, A. F.; Tycko, R.; Fleming, R. M.; Dabbagh, G., submitted to *Physica C*.
20. Fleming, R. M.; Siegrest, T.; Marsh, P. M.; Hessen, B.; Kortan, A. R.; Murphy, D. W.; Haddon, R. C.; Tycko, R.; Dabbagh, G.; Mujsce, A. M.; Kaplan, M. L.; Zahurak, S. M. *Proc. Mat. Res. Soc. Symp. G, Boston* **1991**, *206*, 691.
21. Fischer, J. E.; Heiney, P. A.; McGhie, A. R.; Romanow, W. J.; Denenstein, A. M.; McCauley, Jr., J. P.; Smith III, A. B. *Science* **1991**, *252*, 1288.
22. Haddon, R. C.; Brus, L. E.; Raghavachari, K. *Chem. Phys. Lett.* **1986**, *125*, 459.
23. Sparn, G.; Thompson, J. D.; Huang, S.-M.; Kaner, R. B.; Diederich, F.; Whetten, R. L.; Grüner, G.; Holczer, K. *Science* **1991**, *252*, 1829.
24. Rosseinsky, M. J.; Ramirez; A. P.; Glarum, S. H.; Murphy, D. W.; Haddon, R. C.; Hebard, A. F.; Palstra, T. T. M.; Kortan, A. R.; Zahurak, S. M.; Makhiga, A. V. *Phys. Rev. Lett.* **1991**, *66*, 2830.
25. Holczer, K.; Klein, O.; Grüner, G.; Thompson, J. D.; Diederich, F.; Whetten, R. L. *Phys. Rev. Lett.* **1991**, *67*, 271.
26. Stephens, P.; Mihaly, L.; Lee, P. L.; Whetten, R. L.; Huang, S.-M.; Kaner, R.; Diederich, F.; Holczer, K. *Nature* **1991**, *351*, 632.
27. Zhou, O.; Fischer, J. E.; Coustel, N.; Kycia, S.; Zhu, Q.; McGhie, A. R.; Romanow, W. J.; McCauley, Jr., J. P.; Smith III, A. B.; Cox, D. E. *Nature* **1991**, *351*, 462.
28. Fleming, R. M.; Rosseinsky, M. J.; Ramirez, A. P.; Murphy, D. W.; Tully, J. C.; Haddon, R. C.; Siegrist, T.; Tycko, R.; Glarum, S. H.; Marsh, P.; Dabbagh, G.; Zahurak, S. M.; Makhija, A. V.; Hampton, C. *Nature* **1991**, *352*, 701.
29. Schirber, J. E.; Overmyer, D. L.; Wang, H. H.; Williams, J. M.; Carlson, K. D.; Kini, A. M.; Welp, U.; Kwok, W.-K. *Physica C* **1991**, *178*, 137.
30. Schirber, J. E.; Wang, H. H.; Williams, J. M. in *Organic Superconductivity*, ed. by Kresin, V. Z.; Little, W. A., Plenum Press, New York, **1990**, 117.

31. Whangbo, M.-H.; Williams, J. M.; Schultz, A. J.; Emge, T. J.; Beno, M. A. *J. Am. Chem. Soc.* **1987**, *109*, 90.
32. Andraka, B.; Stewart, G. R.; Carlson, K. D.; Wang, H. H.; Vashon, M. D.; Williams, J. M. *Phys. Rev. B: Condens. Matter* **1990**, *42*, 9963.

Index

A

Activation energy:
    definition, 122
    table, for several ET salts, 124
α-phase salts, 74, 80, 81, 130–31, 181, 184, 186, 188–89, 194, 203

B

BADT-TTF [bis(*cis*-acetylene-dimethyldithio)tetrathiafulvalene], 68, 86, 87
Band theory 119-122, 251–54
    elementary aspects, 119–22
    tight-binding method, 251–54
Bardeen, Cooper, and Shrieffer (BCS) theory, 161–65, 258, 266
BEDO-TTF [bis(ethylenedioxo)-tetrathiafulvalene], 6, 11, 12, 69, 70, 84, 87, 273–75
    compounds, 69, 70, 84, 100, 101
    (BEDO-TTF)$_2$AuBr$_2$, 84, 100, 272-273
    (BEDO-TTF)$_2$ClO$_4$, 84, 100
    β_m-(BEDO-TTF)$_3$Cu$_2$(NCS)$_3$, 6, 69, 84, 100, 101
    molecular structure, 69, 70, 87
BEDSe-TseF [bis(ethylenediseleno)tetraselenafulvalene]:
    compounds, 83
    molecular structure, 69, 86
BEDSe-TTF [bis(ethylenediseleno)tetrathiafulvalene], 29, 69, 86
    compounds, 83
    (BEDSe-TTF)$_2$AuI$_2$, 69, 83
    molecular structure, 69, 86
BEDT-TSeF [bis(ethylenedithio)tetraselenafulvalene], 69
BEDT-TTF [bis(ethylenedithio)tetrathiafulvalene], 3, 11, 12, 28
    See ET

BEDT-TTF salts, structural factors
affecting superconductivity,
271–75
α-(BEDT-TTF)$_2$I$_3$, 184, 278
α'-(BEDT-TTF)$_2$AuBr$_2$, 278–80
α$_t$-(BEDT-TTF)$_2$I$_3$, 278
β-(BEDT-TTF)$_2$AuI$_2$, 259–63, 265
β-(BEDT-TTF)$_2$I$_3$, 259–65
β*-(BEDT-TTF)$_2$I$_3$, 259–60, 262–65, 277–78
β-(BEDT-TTF)$_2$IBr$_2$, 259–63, 265
δ-(BEDT-TTF)$_2$AuBr$_2$, 278–79
κ-(BEDT-TTF)$_2$Cu[N(CN)$_2$]Br, 3, 265, 268, 270
κ-(BEDT-TTF)$_2$κ-Cu[N(CN)$_2$]Cl, 3
κ-(BEDT-TTF)$_2$Cu(NCS)$_2$, 3, 265–66, 268–70
κ-(BEDT-TTF)$_4$Hg$_3$Cl$_8$, 265, 268–70
κ-(BEDT-TTF)$_2$I$_3$, 265–68, 270
β-phase salts, 74, 80, 81, 88–90, 127–30, 141–48, 259, 265
Binaphtho[1, 6-d, e]-1.3-diselenin-2-ylidene, 38
Binaphtho[1, 6-d, e]-1.3-dithiin-2-ylidene, 38
Bi-TTF, 37
Bloch orbital, 251–53
κ-(BMDT-TTF)$_2$Au(CN)$_2$, 265–66, 268–70
BOPDT-TTF [bis(2-oxapropylenedithio)-tetrathiafulvalene], 70, 84, 87
BPDT-TTF. *See* PT
o-Bromanil, 33
BVDT-TTF. *See* VT

C

Chandrasekhar-Clogston limit, 145
Charge carriers, 120–21
Charge density wave, 116, 256–58
Charge transfer, 1, 2, 6, 119, 247
Charge-transfer salts:
  micro FT-IR reflectance techniques, 217–24
  experimental, 224–25
  κ-(ET)$_2$Cu(NCS)$_2$, 220–22
  (MDT-TTF)$_2$AuI$_2$, 222–24
  molecular structure:
    κ-(ET)$_2$Cu(NCS)$_2$, 221–22
    κ(MDT-TTF)$_2$AuI$_2$, 223
  plasma frequencies, 216–17
  polarized reflectance spectra, 214–17
    experimental, 224–25
  Raman studies, 213–14, 224
  vibrational studies, 213–14
C-H···anion interactions, 259, 263, 265–68, 270
C-H···donor interactions, 259, 263, 266–68, 270
o-Chloranil, 33
Chronology of organic superconductors, 8
Coherence length (ξ), 145–47, 152, 156
Complex dielectric function, 230
Complex reflectance, 230
Compositional effects, 166–68
Conducting organic polymers, 118, 120
Conductivity, electrical:
  anisotropy, 116, 122
  band theory relation to, 119–22
  definition, 120, 122
  measurement of, 122
  nature of, in organic salts, 118–23, 181
  temperature dependence, 121
  variations in, 123
  See also Resistivity
Conductivity, experimental values:
  DMET salts, 138–39
  ET salts:
    of α-phase structure, 130–31
    of β-phase structure, 127–30
    of κ-phase structure, 134–38
    of modified β-phase structure, 131
    with polyhedral anions, 123–27
    with polyiodide anions, 132–34
  TTF-Metal (dmit)$_2$ salts, 139
Cooper pairs, 160, 162, 258
Corey-Winter reaction, 40
Coulomb repulsion energy (U), 38
Critical current density (J$_c$), 141
Critical magnetic field:
  anisotropy, 141, 145–46
  β-phase superconductors, 144–48
  β*-(ET)$_2$I$_3$, 151–52

H_{c1} (lower critical magnetic field), 141, 144–45
H_{c2} (upper critical magnetic field), 141, 163–65
κ-phase superconductors, 156–59
Crystal packing patterns, 270
Crystal structures, 65-114

D

Damping constant, 231
DBTTF (dibenzotetrathiafulvalene), 19, 26, 73, 84, 85, 87
Density of carriers, 231
Density of states, 254
  at Fermi level, 260, 268
d_4-Tetraselenafulvalene, 21, 22
d_4-Tetrathiafulvalene, 21, 22
Diamagnetic shielding effect, 142
Dichalcogenoacenes, 35
Difulvathiane, miscellaneous donors, 39, 41
DIMET (3,4-dimethyl-3',4'-ethylenedithio-tetrathiafulvalene), 70, 71, 83, 87
(DIMET)$_2$ClO$_4$, 71, 83
Dirty superconductor, 152, 164
Disorder, 81, 88, 196
1,6-Dioxapyrene, 39, 42
1,6-Dithiapyrene, 39, 42
1,7-Dithiaperylene, 39, 42
DMET (3,4-dimethyl-3',4'-ethylenedithio-2,5-diselena-2',5'-dithiafulvalene) 6, 11, 12, 31
  compounds, 83, 99, 100, 138, 139
  (DMET)$_2$AuCl$_2$, 83, 99
  (DMET)$_2$Au(CN)$_2$, 83, 99
  (DMET)$_2$AuI$_2$, 83, 99
  (DMET)$_2$ClO$_4$, 71, 83, 99
  (DMET)$_2$I$_3$, 83, 99, 100
  (DMET)$_2$IBr$_2$, 83, 100
  (DMET)$_2$I$_2$Br, 83, 100
  (DMET)$_2$SCN, 83, 100
  κ-(DMET)$_2$AuBr$_2$, 83, 92, 99, 100, 265, 268-270
  p-(DMET)$_2$AuBr$_2$, 83, 100
  molecular structure, 6, 11, 12, 31, 70, 71, 87
DMTCNQ, 3

Donors:
  infrared spectroscopy:
    assignments, 211–12
    (BEDT-TTF), 211–12
    (BEDT-TTF-d_8), 211–12
  Raman spectroscopy:
    (BEDT-TTF), 211–14
    (BEDT-TTF-d_8), 211–12
    experimental, 224
    I_3^- vibrational analysis, 214
Drude model, 230–31
d_{12}-Tetramethyltetrathiafulvalene, 25

E

EDOEDT-TTF (ethylenedioxo-ethylenedithio-tetrathiafulvalene), 70, 84
EDTPDT-TTF. See EPT
EDT-TTF (ethylenedithio-tetrathiafulvalene), 70, 84, 87
Effective number of carriers, 232
Electrical transport, 120
Electrocrystallization, 43-45
Electronic instability, 250, 255, 257
Electronic specific-heat coefficient (γ), 148, 159
Electron-phonon coupling, 160, 162, 235–36, 247, 258, 259–61, 263, 266–67
Electron-scattering mean-free path, 152, 164
Energy bands, 250, 253–54
Energy gap:
  BCS, 148, 162
  insulator, 121–22
  semiconductor, 232–33
  superconductor, 148, 160, 162
EPT (ethylenedithio-propylenedithio-tetrathiafulvalene):
  compounds, 82
  molecular structure, 67, 86
ESR lineshape of ET salts:
  Dysonian lineshape, 181, 184, 196
  Lorentzian lineshape, 180, 181, 194
ESR linewidths of ET salts, 183, 185–93
ESR of organic metals, 199–201

ESR of organic semiconductors, 201–3
  Bonner Fisher behavior, 186–88, 202
  Curie-Weiss behavior, 186–88, 201
  spin Peierls transition, 186, 203
ESR of organic superconductors, 194–99
ET [bis(ethylenedithio)tetrathiafulvalene]:
  compounds, 73–81, 88–89
    α'-$(ET)_2AuBr_2$, 73, 75, 184, 186, 278–290
    α-$(ET)_2I_3$, 74, 80, 81, 93, 97, 98, 184, 186
    α_t-$(ET)_2I_3$, 236, 239, 240, 278
    β-$(ET)_2AuI_2$, 74, 80, 88, 90, 184, 186, 259–63, 265
    β-$(ET)_2IBr_2$, 74, 80, 88, 90, 184, 186, 242, 259–63, 265
    β-$(ET)_2I_2Br$, 74, 80, 81, 88, 186
    β-$(ET)_2I_3$, 74, 80, 81, 88–90, 184, 186
    β^*-$(ET)_2I_3$, 88–90, 186
    β-$(ET)_2X$, 74, 80, 81, 88–90, 103
    $(ET)_3Br_2 \cdot 2H_2O$, 77, 95–97, 184, 188
    $(ET)_3Cl_2 \cdot 2H_2O$, 77, 95–97, 188
    $(ET)_4Cl_2 \cdot 4H_2O$, 77, 95, 97
    $(ET)_4Hg_{3-x}Br_8$, 78, 92, 94, 95, 184, 188
    $(ET)_5(HgBr_4)_2(HgBr_3)$, 73, 77, 265, 268–270
    $(ET)_4Hg_{3-x}Cl_8$, 78, 92, 94, 95, 246–47, 265, 268–270
    $(ET)_2KHg(SCN)_4$, 78, 97, 98
    $(ET)_2(NH_4)Hg(SCN)_4$, 78, 97, 98, 169, 170, 181–84, 188, 193–196
    $(ET)_2(TlI_4)(I_3)$, 73, 76, 94
    γ-$(ET)_3Br_2$, 77, 97, 184, 188
    γ-$(ET)_3(I_3)_{2.5}$, 74, 94
    δ-$(ET)_2AuBr_2$ 278–79
    κ-$(ET)_2Ag(CN)_2 \cdot H_2O$, 78, 92, 93, 95, 188
    κ-$(ET)_2Cu(NCS)_2$, 3, 78, 90–92, 98, 100–102, 184, 188, 221–22, 236, 265, 266, 268–270
    κ-$(ET)_2Cu[N(CN)_2]Br$, 3, 78, 101–3, 265–66, 268, 270
    κ-$(ET)_2Cu[N(CN)_2]Cl$, 3, 78, 101, 103, 188
    κ-$(ET)_2I_3$, 74, 90, 92, 95, 98, 265–68, 270
    θ-$(ET)_2(I_3)_{1-x}(AuI_2)_x$, 74, 92–94, 98, 186, 236, 239–241
    θ-$(ET)_2(I_3)_{1-x}(IBr)_x$, 74, 93
  crystallographic data, 74–79
  molecular structure, 66, 67, 70
Ethylene group conformation, 66, 67, 88
Extinction coefficient, 230

F

Fermi level, 120, 121, 250–51, 255
Fermi surface, 120, 121, 251, 255–57, 260–61, 266–67
  nesting vector, 256–57
Fermi surface nesting, 251, 255, 257–58
First Brillouin zone, 253–56
First primitive zone, 253–54

G

Ginsburg-Landau (GL) analysis, 144–46

H

HMTSF (hexamethylenetetraselenafulvalene), 19, 71, 73, 86, 87
HMTTeF (hexamethylenetetratellurafulvalene), 71, 86, 87
HOMO, 254

I

Incommensurate structure, 81, 88, 89, 94
Index of refraction, 230
Insulator:
  energy gap, 121–22
  Mott-Hubbard, 119
Isotope effect, 162, 266

K

κ-(ET)$_2$Cu(NCS)$_2$, 3, 4, 11, 78, 90–92, 98, 100–102, 184, 188, 221–22, 236, 265–66, 268–270
κ-(ET)$_2$Cu[N(CN)$_2$]Br, 3, 4, 11, 265–66, 268, 270
κ-(ET)$_2$Cu[N(CN)$_2$]Cl, 3, 265–66, 268, 270
κ parameter (GL κ parameter), 147–48
κ-phase salts, 74, 78, 82, 83, 85, 90–92, 94, 95, 98, 103, 134-38, 153–61, 184, 188, 196
Kramers-Kronig analysis, 218, 220, 225
Kramers-Kronig transformation, 230

L

Lattice softness, 261, 265
Lorentz oscillator, 230, 232
Low temperature, 81, 88–90, 102, 193
LUMO, 254
  κ-(MDT-TTF)$_2$AuI$_2$, 265, 268–70

M

M(dmit)$_2$$^{n-}$, 6
MDTEDT-TTF. *See* MET
MDTPDT-TTF. *See* MPT
MDT-TTF (methylenedithiotetrathiafulvalene), 6, 11, 12, 31, 70, 87
  compounds, 85, 98, 99
  κ-(MDT-TTF)$_2$AuI$_2$, 70, 85, 90, 92, 98–100, 222, 265, 286–70
  molecular structure, 70, 87
Mechanism for formation of tetrachalcogenafulvalenes:
  electrochemical coupling, 15
  photochemical coupling, 14
  with trivalent phosphorus reagents, 13
Meissner effect, 2, 142
MET (methylenedithio-ethylenedithiotetrathiafulvalene):
  compounds, 82
  molecular structure, 67, 86
Metal-to-semiconductor transition, 115–16, 120, 126–27, 199, 200
MPT (methylenedithio-propylenedithiotetrathiafulvalene):
  compounds, 82
  (MPT)$_2$ClO$_4$(THF), 68, 82
  molecular structure, 67, 86
M,S,L-scheme, 73–79
MT [bis(methylenedithio)tetrathiafulvalene]:
  compounds, 82
  (MT)$_2$Au(CN)$_2$, 82, 90, 92, 265–66, 268–70
  molecular structure, 67, 70, 86

N

n-type conductivity, 121
Number density (of charge carriers), 120–21

O

One-dimensional (1D) conductor, 116
1,3-dithiole-2-ylidene compounds miscellaneous donor, 38–40
Optical absorption, 229
Optical conductivity:
  α-(ET)$_2$I$_3$, 237, 239–40
  α-(ET)$_3$(ReO$_4$)$_2$, 241, 244
  β-(BMDT-TTF)AsF$_6$, 242, 245
  β-(BMDT-TTF)SbF$_6$, 242, 245
  β-(ET)$_2$IBr$_2$, 240, 242
  β-(ET)$_2$I$_3$, 237, 238–39
  β-(ET)$_2$PF$_6$, 246
  definitions, 229–35
  (ET)$_3$(ClO$_4$)$_2$, 242, 244
  (ET)$_2$ClO$_4$(C$_2$H$_5$Cl$_3$)$_{0.5}$, 245
  (ET)$_2$ClO$_4$(TCE)$_{0.5}$, 242
  (ET)$_4$Hg$_3$Cl$_8$, 246–47
  (ET)$_2$I$_3$, 237
  semiconductor, 232–33
  θ-(ET)$_2$I$_3$, 239–41
Optical constants, 230

Order-disorder effects on superconductivity:
  β-phase salts, 128–29
  ET salts, 117, 168–69
  κ-phase salts, 136
  $(TMTSF)_2ClO_4$, 116
Organic superconductors (compilation), 7
Orientational studies of ET salts, 181–84

P

Pauli paramagnetism, 186, 188, 195, 197
Peierls transition, 116, 119
Phase identification, 185–93
Phase transition, 81, 88
Plasma frequency, 216–17, 232
  α-$(ET)_2I_3$, 217, 236
  κ-$(ET)_2Cu(SCN)_2$, 236
  θ-$(ET)_2I_3$, 236
Polarizability, 230–31
Polaron, 235, 248
Pressure, 81, 88–9, 94–5, 99, 103
Principal g-values of ET salts, 181–84
PT [bis(propylenedithio)tetrathiafulvalene]:
  compounds, 82–3
  molecular structure, 67, 86
p-type conductivity, 121

R

Reduced upper critical magnetic field, 163–65
Reflection spectra:
  α-$(ET)_2I_3$, 239
  α-$(ET)_2(ReO_4)$, 241
  β-$(ET)_2IBr_2$, 240
  β-$(ET)_2I_3$, 239
  $(ET)_4Hg_3Cl_8$, 246
  κ-$(ET)_2Cu(SCN)_2$, 240, 243
  θ-$(ET)_2I_3$, 239
Relaxation, 231, 234–35, 239–40
Residual linewidth, 195–96
Resistivity:
  anisotopy, 122
  definition, 122
  ratio, 141
  See also Conductivity

rf penetration depth measurement, 142–43, 154–55, 170

S

S⋯S contacts, 65, 81, 91, 99, 101, 102
Semiconductor, 121–22, 201–203
Semimetal, 120
Specific heat discontinuity, 148, 159, 160, 162, 170
Spin density wave (SDW), 116, 256–58
Spin susceptibilities of ET salts, 186–88
Strong-coupling superconductor, 160, 161, 165
Superconducting state, 257
Superconducting transition temperature (T_c):
  α-$(ET)_2(NH_4)Hg(SCN)_4$, 169–70
  β_m-$(BEDO-TTF)_3Cu_2(NCS)_3$, 169–70
  β-phase superconductors, 141–44
  β-phase under pressure, 149–52
  definition, 140–41
  inductive measurement, 140
  κ-phase superconductors, 153–56
  κ-phase under pressure, 160–61
  pressure dependence, 139, 149, 160
  resistive measurement, 140
  salts of ET/I system, 166–69
  transition width, 140, 142
Superconductivity, 69, 70, 80, 88, 90, 92, 94, 95, 97–101, 103

T

T_c. *See* Superconducting transition temperature (T_c)
TCNE (tetracyanoethylene), 33
TCNQ (tetracyanoquinodimethane), 2, 3, 33, 71, 75, 85–87
Tetrachalcogenafulvalenes:
  cross-coupling of 1,3-dithiole-2-phosphonates:
    with 1,3-dithiolium ions, 31
    with 1,3-dithiolium-2-iminium ions, 31, 32
  deprotonation of 1,3-dithiolium salts, 16
  derivatization of TTF and TSeF, 26–30

electrochemical coupling, 15
lithiation of TTF and TSeF, 27–30
mechanism for formation of:
  electrochemical coupling, 15
  photochemical coupling, 14
  trivalent phosphorus reagents, use of, 13
oxidative coupling of 2-methylene-1,3-diselenole, 17
photochemical coupling, 14
reaction of acetylenes:
  with CS_2 and CSe_2, 20–23
  with CS_2, CSe_2, and tributylphoshine, 22
  with CS_2, CSe_2 under high pressure, 21
  with η^2-CS_2 complexes of Fe and Ni, 23
reaction of 1,2-dichalcogenolates with tetrachloroethylene, 19
reaction of reactive cycloalkynes with CS_2, 25, 26
reaction of 2-alkylthio-1,3-dithiolim salts with zinc and bromine, 17
synthetic methods, 12–33
thermolysis:
  of hexathioorthooxalates, 18, 33
  of 2-alkoxy-1,3-dithioles, 17
  of 2-tosylhydrazono-1,3-dithioles, 16, 17
thiapendione, use of, 23–25
transition metal carbonyls, use of, 14, 15
trivalent phosphorus reagents, use of, 12, 13
Tetrachalcogenoacenes:
  reaction of alkali metal dichalcogenides with tetrachloroacenes, 34
  reaction of Se with dichlorotetracene, 34
  synthetic methods, 33–35
1,4,5,8-Tetrathiaanthracene, 39, 42
Tetratelluratetralin, 20
Tetrathiatetralin, 20
Thermal conversion of ET salts, 168–69, 203–5
θ-phase salts, 74, 92, 93

Thiapendione, 23
3,4,9,10-Tetrathiaperylene, 39, 42
3,9-Dithiaperylene, 39, 42
3,10-Dithiaperylene, 39, 42
TMTSF (tetramethyltetraselenafulvalene), 3, 11, 12, 68–70
(TMTSF)(DMTCNQ), 3
(TMTSF)(TCNQ), 3
$(TMTSF)_2ClO_4$, 1, 2
$(TMTSF)_2X$: 1, 3
  X = PF_6^-, AsF_6^-, TaF_6^-, NbF_6^-, SbF_6^-, ReO_4^-, ClO_4^-, BF_4^-, BrO_4^-, IO_4^-, NO_3^-, FSO_3^-, $CFSO_3^-$,
$(TMTSF)_2X$ superconductors, 116, 123
Triselenaphenanthrene, 39, 42
Trithiaphenanthrene, 39, 42
TSeF:
  tetrakis(trimethylsilyl), 21, 22
  tetralithio, 29
  tetraselenafulvalene, 17, 18, 26
TSeT (tetraselenotatracene), 34
$TTC_n TTF$ [tetrakis (alkylthio)-tetrathiafulvalene], 71
TTeT (tetratellurotetracene), 34
TTF, 2, 3, 6
  BODM-TTF, bis(oxymethylene), 30
  monolithio, 27
  tetrabromo, 30
  terachloro, 30
  tetrakis(trimethylsilyl), 21, 22
  tetralithio, 28, 30
  tetrathiafulvalene, 26, 27
  tetrathiotetracene, 33
$TTF[Ni(dmit)_2]$, 254
$TTF[Pd(dmit)_2]$, 254
TTF-type, miscellaneous donor, 35–38
TTF-TCNQ, 2
TTM-TTF [tetrakis(methylthio)tetrathiafulvalene]:
  compounds, 85
    TTM-TTF $(HgI_2)_2$, 71, 85
  molecular structure, 71
Two-dimensional (2D) conductor, 116–17
2,7-Bis(methylthio)-1,6-dithiapyrene, 39, 42
2,3:7,8-Bis(ethylenedithio)-1,6-dithiapyrene, 39, 42

TXF·TCNQ, 275–77
Type II superconductor, 141

V

Van der Waals radius, 81, 90
VT [bis(vinyldithio)tetrathiafulvalene]:
    compounds, 86
    molecular structure, 67, 72, 87

W

Weak-coupling superconductor, 148, 159, 160, 164